New Frontiers in Human Genetics

New Frontiers in Human Genetics

Editor: Lesley Easton

STATES
ACADEMIC PRESS
www.statesacademicpress.com

States Academic Press,
109 South 5th Street,
Brooklyn, NY 11249, USA

Visit us on the World Wide Web at:
www.statesacademicpress.com

ISBN: 978-1-63989-374-4 (Hardback)

Cataloging-in-Publication Data

New frontiers in human genetics / edited by Lesley Easton.
p. cm.
Includes bibliographical references and index.
ISBN 978-1-63989-374-4
1. Human genetics. 2. Genetics. 3. Human biology. I. Easton, Lesley.
QH431 .N49 2022
599.935--dc23

Table of Contents

Preface

Human genetics is concerned with the study of the inheritance of characteristics from parents to children. This inheritance in humans depends upon discrete units called factors or genes. Human genes refer to a set of nucleic acid sequences that are encoded as DNA in the twenty-three chromosome pairs. The field of human genetics attempts to understand the genetics of human life and the development of diseases and its effective treatment. The discipline overlaps with other fields such as cytogenetics, biochemical genetics, population genetics, genetic counseling, developmental genetics, genomics, etc. There has been rapid progress in this field and its applications are finding their way across multiple industries such as medicine and biotechnology. The book studies, analyzes and upholds the pillars of human genetics and its utmost significance in modern times. It contains some path-breaking studies in the field of human genetics. In this book, using case studies and examples, constant effort has been made to make the understanding of the difficult concepts of human genetics as easy and informative as possible, for the readers.

This book has been the outcome of endless efforts put in by authors and researchers on various issues and topics within the field. The book is a comprehensive collection of significant researches that are addressed in a variety of chapters. It will surely enhance the knowledge of the field among readers across the globe.

It gives us an immense pleasure to thank our researchers and authors for their efforts to submit their piece of writing before the deadlines. Finally in the end, I would like to thank my family and colleagues who have been a great source of inspiration and support.

Editor

1

Structure of precursor microRNA's terminal loop regulates human Dicer's dicing activity by switching DExH/D domain

Zhongmin Liu[1,2], Jia Wang[2], Gang Li[1✉], Hong-Wei Wang[2✉]

[1] Department of Biochemistry and Molecular Biology, School of Basic Medical Sciences, Peking University Health Science Center, Beijing 100191, China
[2] Tsinghua-Peking Joint Center for Life Sciences, School of Life Sciences, Tsinghua University, Beijing 100084, China
✉ Correspondence: ligang55@bjmu.edu.cn (G. Li), hongweiwang@tsinghua.edu.cn (H.-W. Wang)

ABSTRACT

Almost all pre-miRNAs in eukaryotic cytoplasm are recognized and processed into double-stranded microRNAs by the endonuclease Dicer protein comprising of multiple domains. As a key player in the small RNA induced gene silencing pathway, the major domains of Dicer are conserved among different species with the exception of the N-terminal components. Human Dicer's N-terminal domain has been shown to play an auto-inhibitory function of the protein's dicing activity. Such an auto-inhibition can be released when the human Dicer protein dimerizes with its partner protein, such as TRBP, PACT through the N-terminal DExH/D (ATPase-helicase) domain. The typical feature of a pre-miRNA contains a terminal loop and a stem duplex, which bind to human Dicer's DExH/D (ATPase-helicase) domain and PAZ domain respectively during the dicing reaction. Here, we show that pre-miRNA's terminal loop can regulate human Dicer's enzymatic activity by interacting with the DExH/D (ATPase-helicase) domain. We found that various editing products of pre-miR-151 by the ADAR1P110 protein, an A-to-I editing enzyme that modifies pre-miRNAs sequence, have different terminal loop structures and different activity regulatory effects on human Dicer. Single particle electron microscopy reconstruction revealed that pre-miRNAs with different terminal loop structures induce human Dicer's DExH/D (ATPase-helicase) domain into different conformational states, in correlation with their activity regulatory effects.

KEYWORDS human Dicer, DExH/D (ATPase-helicase) domain, pre-miRNA-151, A-to-I editing, single particle electron microscopy

INTRODUCTION

MicroRNAs are emerging as an important regulator of almost all biological processes ranging from cell differentiation to senescence (He and Hannon, 2004; Schraml and Grillari, 2012). Thousands of precursor microRNAs are precisely processed into microRNAs with a typical length of 20–25 nt by RNase III enzyme Dicer (Griffiths-Jones et al., 2008; Schraml and Grillari, 2012). The multi-domain Dicer is conserved in all eukaryotic species with the RNase domain, PAZ domain and dsRBD domain but differ in their N-terminal compositions (Carthew and Sontheimer, 2009). The human Dicer (hDicer) with 220 kDa contains an N-terminal DExH/D (ATPase-helicase) domain, a DUF283 domain, a PAZ domain, two RNase-III domain and a double-stranded RNA-binding domain (dsRBD) (Carthew and Sontheimer, 2009). The two RNase-III domains form an intra-molecular dimer to function as an endonuclease processing center for double-stranded RNA substrates (Zhang et al., 2004).

The typical secondary structure of a pre-miRNA contains a terminal loop and a stem with the 5′ phosphate and 2-nucleotide (nt) 3′ overhang. In a pre-miRNA maturation process, Dicer's PAZ domain recognizes and binds with the 5′ phosphate and 2-nucleotide (nt) 3′ overhang of a pre-miRNA (Park et al., 2011), and the N-terminal DExH/D (ATPase-helicase) domain interacts with the terminal loop of pre-miRNA

(Tsutsumi et al., 2011) to align the pre-miRNA to the RNase III site for precise cleavage. The terminal loop sequence and structure of pre-miRNA could directly impact on the binding affinity of RNA to Dicer protein and even on the cleavage of Dicer (Zhang and Zeng, 2010; Castilla-Llorente et al., 2013). Furthermore, the terminal loop also plays a critical role for accuracy of Dicer processing (Gu et al., 2012).

Recently, non-canonical pre-miRNAs with 5′-cap termini (Xie et al., 2013), 5′ overhanging termini (Ando et al., 2011), and uridylated termini (Heo et al., 2008; Heo et al., 2012) were reported to be diced into mature miRNAs. In order to accurately processing thousands of pre-miRNAs with different features into mature miRNAs, the only single isoform of Dicer in human cells (Griffiths-Jones et al., 2008) needs precise activity regulation upon different miRNA precursors in various cellular environment. The N-terminal DExH/D (ATPase-helicase) domain has been shown to auto-inhibit the dicing activity of hDicer (Ma et al., 2008). Human Dicer's regulatory proteins such as TRBP (Chendrimada et al., 2005), PACT (Yoontae Lee et al., 2006), and ADAR1 (Ota et al., 2013) interact with hDicer via its N-terminal DExH/D (ATPase-helicase) domain, and increasing dicing activity has been observed within these heterodimers. The regulatory mechanism of the DExH/D (ATPase-helicase) domain on hDicer's activity, however, remains elusive.

Post-transcriptional RNA editing introduces the nucleotide changes in specific RNA sequence in tRNA, rRNA and miRNA and also plays an important role in gene expression (Axel Brennicke et al., 1999; A. A. H. Su and L.Randau, 2011). ADAR family of enzymes mediate adenosine to inosine changes in double-stranded RNAs, contributing the major form of RNA editing events in human cells (Nishikura, 2010). ADAR-mediated RNA editing of the double-stranded portion of primary and precursor microRNAs were shown to result in the redirection of the silencing target of microRNA (Blow et al., 2006; Das and Carmichael, 2007; Kawahara et al., 2007b). How and why the sequence modification of miRNA precursors affect their silencing property are yet to be fully understood.

One of the ADAR family protein, ADAR1P110, was shown to induce the RNA editing of pre-miR-151 to generate three different products, pre-miR-151A1I, pre-miR-151A3I and pre-miR-151A13I by both *in vivo* and *in vitro* assays (Kawahara et al., 2007a). These products were reported to bind with hDicer-TRBP complex but cannot be further diced into mature miR-151. Until now, there is little explanation for this observation. In this work, we discovered that edited and unedited pre-miR-151 can directly bind with hDicer. *In vitro* dicing assay shows that while pre-miR-151A1I and pre-miR-151A13I were processed with greatly reduced dicing rate compared to pre-miR-151, the dicing of pre-miR-151A3I was not much affected. Single particle electron microscopy reconstruction showed that hDicer-pre-miR-151 and hDicer-pre-miR-151A3I complex has a similar 3D structure with the DExH/D (ATPase-helicase) domain in an open state. In comparison, the DExH/D (ATPase-helicase) domain in the hDicer-pre-miR-151A1I and hDicer-pre-miR-151A13I complexes appeared in a close state. The structural variation in good correlation with our dicing activity assay results indicated that the pre-miRNA's terminal loop structure has a strong regulatory effect of hDicer's activity via induced conformational change of the protein's DExH/D (ATPase-helicase) domain.

RESULTS

Electron microscopy of human Dicer with a well-defined N-terminal DExH/D domain

We expressed and purified hDicer using the insect cell system to high purity (Fig. 1A and 1B). The purified hDicer was verified to have endonuclease activity on its classical pre-miRNA substrate, pre-let-7 (Fig. 1C). We examined the purified hDicer using negative staining transmission electronic microscopy (TEM) and found that the protein were mono-dispersed and homogenous in shape and dimension (Fig. 1D). We thus performed single particle reconstruction of hDicer from 50,403 particle images to verify the protein's structural features with previously reconstructed models (Wang et al., 2009; Lau et al., 2012; Taylor et al., 2013). This practice also set a consistent protocol for the EM study of the different RNA-hDicer complexes as discussed below. Different from the previous hDicer reconstructions, we used the most recent version of RELION image processing package (Scheres, 2012) to perform maximum-likelihood 3D classification of the particle images in order to overcome the conformational heterogeneity and negative stain artifact problem of hDicer as revealed before (Taylor et al., 2013). We found the latter to be the major issue in our structural determination. After classification of all the particles into four classes, we found that all the four classes appeared to share similar L-shape as the previous reported 3D reconstructions (Lau et al., 2012; Taylor et al., 2013), but the third class had the best defined features and the highest resolution and was composed of the majority of particles (Figs. 1E and S1). The structure revealed a clear shape of the N-terminal DExH/D domain that forms a V-shaped arrangement with hDicer's RNase domains from the bottom view of the molecule. We found that the change of this structural arrangement of DExH/D (ATPase-Helicase) domain reflects the interaction of different pre-miRNA substrates with hDicer as described below.

RNA editing causes different pre-miRNA terminal loop structures

ADAR enzymes deaminate the adenosine into inosine (A-to-I) within dsRNAs (Nishikura, 2010). It is known that ADAR deaminase can modulate the processing and expression of microRNA and redirect the silencing targets (Blow et al., 2006; Kawahara et al., 2007b). Human ADAR1P110 edits pre-miR-151 to generate three major products based on the A-to-I editing positions relative to the dicing site on the RNA stem, namely, pre-miR-151A1I at −1 site, pre-miR-151A3I at +3 site, and pre-miR-151A13I at −1 and +3 sites, respectively (Fig. 2A) (Kawahara et al., 2007a). We exploited secondary structure prediction

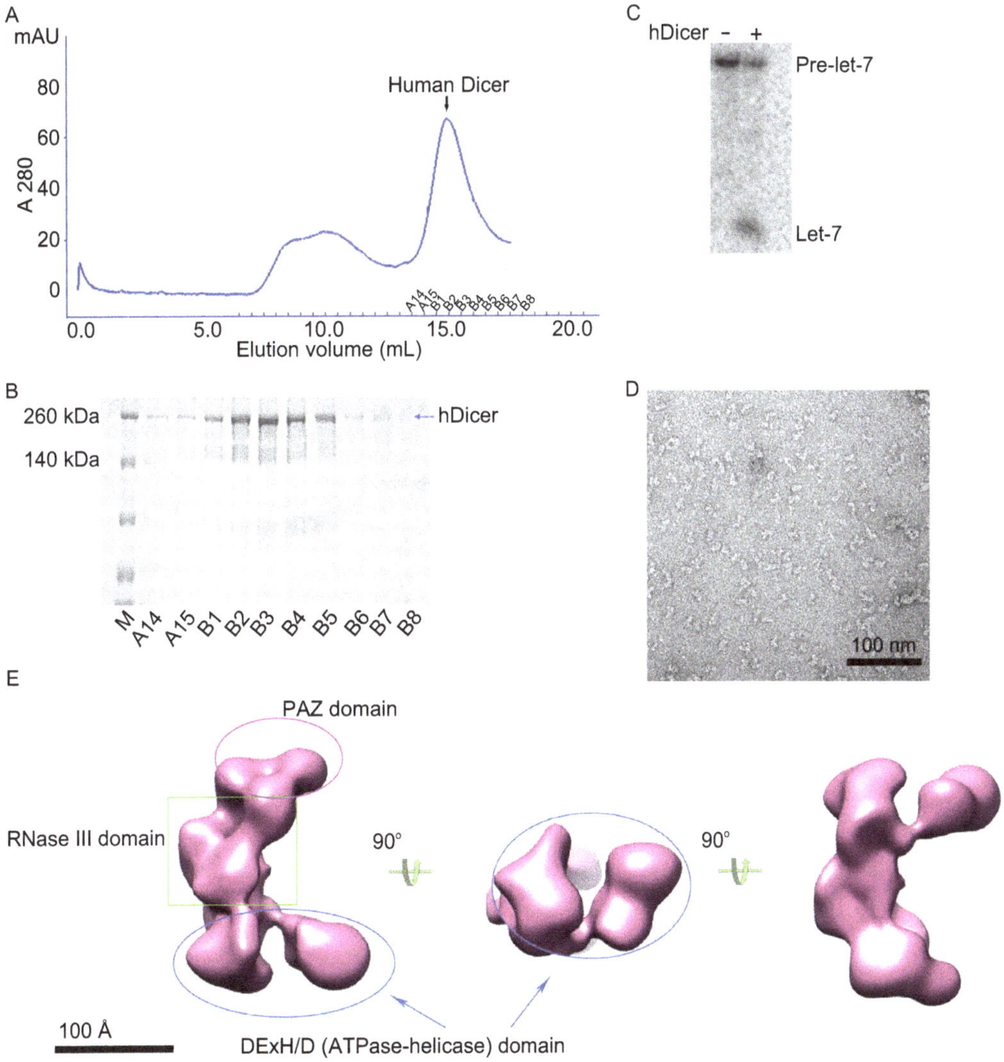

Figure 1. Purification and 3D reconstruction of human Dicer protein. (A and B) hDicer protein fractions eluted from size exclusion column (Superose 6 10/300 GL) and 8% SDS-PAGE gel of the fractions stained by Coomassie brilliant blue. (C) hDicer's dicing assay on 60 nmol/L radioactive labeled pre-let-7. (D) Raw images of apo-hDicer was recorded with CCD at a nominal magnification of 49,000×, defocus value of −1 to −3 μm. (E) 3D reconstruction of hDicer protein. The location of domains are labeled in the model.

of the four types of RNA molecules including the unedited and the three edited pre-miR-151 products using Mfold RNA structural folding server (Zuker, 2003). The structural prediction showed that A-to-I changes at position +3 have little effect on the secondary structure of pre-miR-151. In contrast, the A-to-I editing at position −1 clearly destabilizes the stem end close to the loop therefore generates different terminal loop structures from that of the unedited pre-miR-151. As a result, the edited product pre-miR-151A3I adopts a similar terminal loop structure with pre-miR-151, whereas pre-miR-151A1I and pre-miR-151A13I may have either a big loop or a hammerhead-shaped terminal loop (Fig. 2B). In order to study these RNAs' transactions with human Dicer, we synthesized the pre-miR-151 and its A-to-I editing products which all appeared as single folding species on native gel electrophoresis (Fig. 2C).

Human Dicer has different dicing activity on edited pre-miR-151

In order to reveal the effect of A-to-I editing on pre-miRNA's processing activity by hDicer, we performed *in vitro* dicing activity assay of hDicer over pre-miR-151 and its related A-to-I editing products. Using 5′ radioactive ^{32}P labeled pre-miR-151, pre-miR-151A1I, pre-miR-151A3I and pre-miR-151A13I, we found that the RNA editing on position +3 adenosine had little effect on hDicer's dicing activity of miR-151 compared to that of unedited substrate, but editing on position −1 or both −1 and +3 adenosine greatly reduced the cleavage activity of hDicer (Fig. 3A). It is possible that the different dicing activity may be due to the various binding affinity of hDicer with the RNA substrates. We therefore

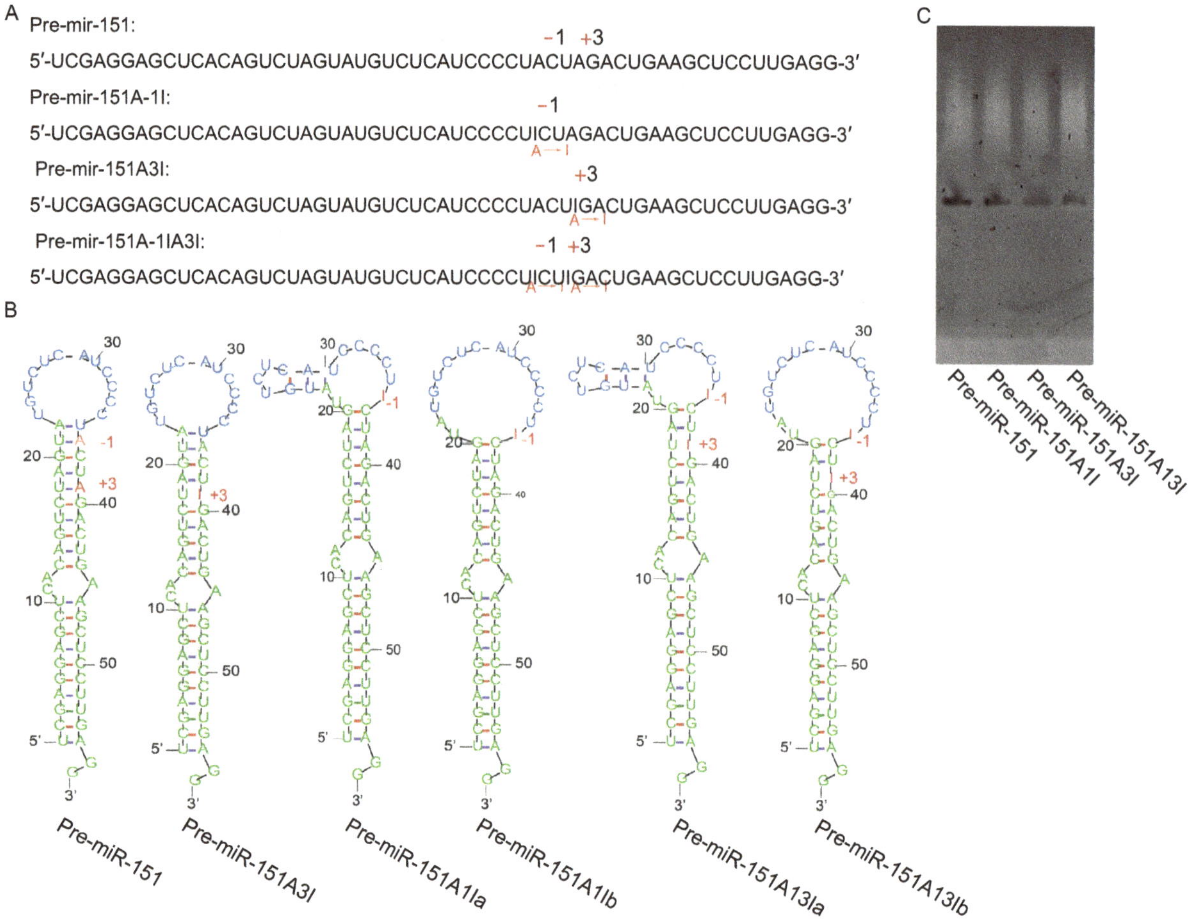

Figure 2. The RNA editing sites on pre-miR-151 and the secondary structure of edited and unedited pre-miR-151. (A) Pre-miR-151 and its A-to-I edited partners. Green boxes represent the sequences in the diced product. −1 Stands for the first nucleotide outside the dicing site; +3 represents the third nucleotide inside the dicing site from 5′ end. −1 And +3 labelled adenosine can be converted into inosine by ADAR1P110. (B) Secondary structural prediction of the unedited and edited pre-miR-151 RNAs using Mfold web server. Note the two possible structures of terminal loop of pre-miR-151A1I and pre-miR-151A13I. (C) Native gel electrophoresis of synthesized pre-miR-151 and its edited partners.

performed electrophoresis mobility shift assay (EMSA) to examine the interaction of hDicer with different RNA substrates. Our results suggested that, within the error range of the assays, the edited pre-miR-151A1I, pre-miR-151A3I and pre-miR-151A13I have approximately the same *K*d value as the unedited pre-miR-151 for binding to hDicer (Fig. 3B). This also indicates that the different dicing activities of various RNA species by hDicer are likely due to the effect on hDicer directly by the secondary structural difference of terminal loops of the RNAs.

RNA editing results in conformational change of hDicer

We have shown previously that pre-let7 and dsRNA precursors induce hDicer into distinct conformations and thus cause hDicer's different activity on these substrates (Taylor et al., 2013). In order to investigate the mechanism of RNA editing's influence on hDicer's activity, we performed single particle analysis of hDicer incubated with edited and unedited RNA substrates at 4°C and in the presence of 2 mmol/L EDTA, a condition under which hDicer protein has no endonuclease activity (Zhang et al., 2002) but can still form stable complex with RNA. Using similar single particle analysis approaches as for hDicer only, we obtained the 3D reconstructions of hDicer in complex with pre-miR-151, pre-miR-151A1I, pre-miR-151A3I and pre-miR-151A13I, respectively, using the same analysis procedure as for apo-hDicer, i.e. classifying the particles into four classes and choosing the best model for further analysis (Fig. 4). Overall shapes of all the structures are similar as that of hDicer only but close examination revealed interesting conformational difference, especially in the N-terminal DExH/D (ATPase-helicase) domain.

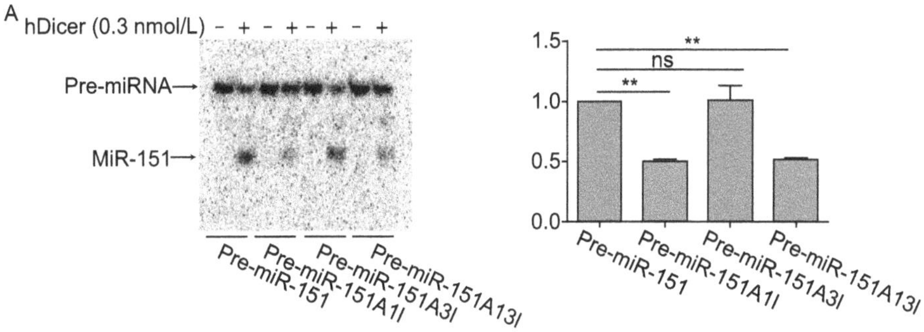

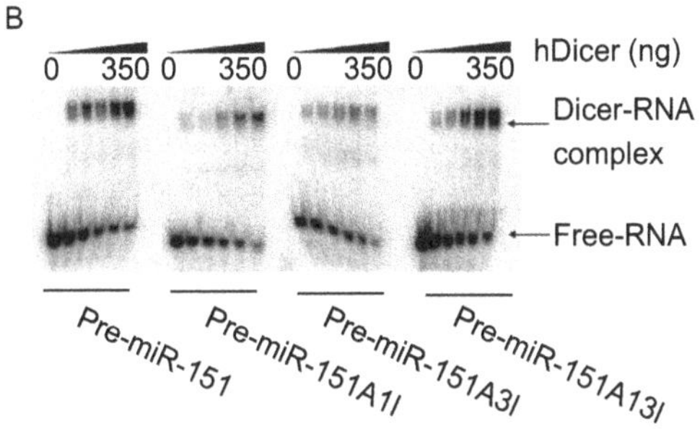

hDicer-RNA complex	K_d (mean ± SD)
hDicer-pre-miR-151	91.66 ± 53
hDicer-pre-miR-151A1I	172.22 ± 11
hDicer-pre-miR-151A3I	109.33 ± 10
hDicer-pre-miR-151A13I	79.66 ± 32.7

Figure 3. *In vitro* cleavage of hDicer and RNA binding analysis. (A) Dicing assay of pre-miR-151 and its A-to-I edited partners by human Dicer. The ratio of diced product microRNA-151 against precursor substrate pre-miR-151 was normalized to "1". Data present mean ± SD (n = 3). (**) were highly significant when compared with the substrate pre-miR-151, in both cases $P < 0.002$. (B) EMSA assay of pre-miR-151 and its A-to-I edited partners binding to human Dicer. Varying amounts of purified hDicer (0, 7, 14, 35, 70 and 350 ng) binding to edited and unedited pre-miR-151 were examined by electrophoresis mobility shift assay using a native 6% polyacrylamide gel. Data present mean ± SD (n = 3).

Compared with the hDicer only structure, pre-miR-151-bound-hDicer adopts a dramatic conformational change of the N-terminal domain which appears to rotate into a perpendicular arrangement with the RNase domain (Fig. 4D). Such a conformational change causes the elongation of hDicer's base-branch from ~115 Å in the apo-hDicer to ~130 Å in the RNA-bound hDicer. We termed this conformation as the open state of hDicer. A feature worth noting of the open-state is a density filling the space between the DExH/D and RNase domain (pointed by the red arrowheads in Fig. 4D and 4E). Another conformational variant happens in the complex of hDicer and pre-miR-151A1I, in which the N-terminal domain rotates into a more parallel arrangement with the RNase domain and makes the space between the DExH/D and RNase domain very narrow (Fig. 4B). We termed this conformation the close state of hDicer.

Interestingly, the RNA substrate induced conformational changes are correlated with the specific structural feature of the RNA terminal loops. It was obvious that the pre-miR-151 and pre-miR-151A3I with the same predicted terminal loop structure both induced hDicer into the open state (Fig. 4D and 4E). In contrast, the pre-miR-151A1I and pre-miR-151-A13I with a different predicted terminal loop structure from that of pre-miR-151 both induced hDicer into the close state (Fig. 4B and 4C). Such a conformational change variation also correlated to the dicing activities of these substrates by hDicer as shown earlier. Therefore, the terminal loop structural change caused by A-to-I edition of pre-miR-151 induces specific conformational change of hDicer that represents the enzyme's different activity states.

DISSCUSSION

ADAR1P110 is a constitutive expressed isoform of ADAR1 protein which are proved to be a homo-dimer when functions as an RNA editing enzyme (Nishikura, 2010). ADAR1 mediated A-to-I deamination reaction on dsRNA is the major RNA editing form in human cells (Blow et al., 2006; Kawahara et al., 2007b; Nishikura, 2010). Pri-miRNA and pre-miRNA share a feature structure of dsRNA which is susceptible to be substrates of ADAR1 family (Nishikura, 2010). Many pri-miRNA and pre-miRNA are predicted to be targets of ADAR1 proteins and some pri-miRNA and pre-miRNA, for example pre-miR-142 (Yang et al., 2006), pri-miR-151 (Kawahara et al., 2007a) and pri-miR-376 (Kawahara et al., 2007b), have been proved to be targets of ADAR1 by biochemical assays. The adenosine converted into inosine happening near the terminal loop of pre-miR-151 could make terminal structure rearranged following the nucleotide changed in terminal loop (Fig. 2B). Our structural prediction data confirmed the structural changes of terminal loop within edited pre-miR-151. The nucleotide sequence and structure of the terminal loop have been shown to play a critical role in the cleavage of pre-microRNA (Zhang and Zeng, 2010; Gu et al., 2012; Castilla-Llorente et al., 2013). Agreeing with the

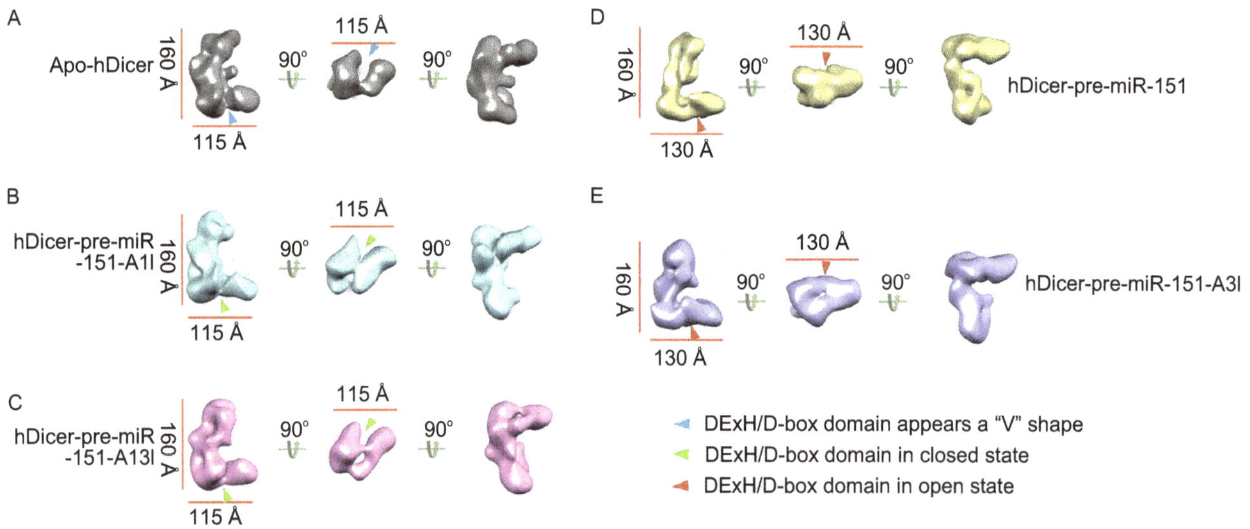

Figure 4. 3D reconstruction of hDicer-RNA complexes. (A) 3D structures of human apo-Dicer appear as an L shape with the DExH/D (ATPase-helicase) domain like a "V" shape, and the distance between two branches of DExH/D (ATPase-helicase) domain is about 115 Å. (B and C) hDicer-pre-miR-151A1I complex and hDicer-pre-miR-151A13I complex appear as an L shape with DExH/D (ATPase-helicase) domain in a close state. The distance between two branches of DExH/D (ATPase-helicase) domain is less than 115 Å. (D and E) DExH/D (ATPase-helicase) domain in hDicer-pre-miR-151 complex and hDicer-pre-miR-151A1I complex appears an open state. The distance between two branches of DExH/D (ATPase-helicase) domain is about 130 Å.

terminal loop structure's importance, our *in vitro* cleavage assay showed that the pre-miR-151A1I and pre-miR-151A13I with changed terminal structure reduced the endonuclease activity of hDicer (Fig. 3B), while the A-to-I editing at +3 site only with little terminal loop structural change does not affect hDicer's activity.

In human Dicer, the N-terminal DExH/D (ATPase-helicase) domain plays an important auto-inhibition regulatory function on the enzyme's dicing activity (Ma et al., 2008). Our previous work has shown that pre-miRNA and dsRNA substrates interacting differently with the N-terminal domain induce hDicer into dramatic different conformations either in an inhibitory state or an active state (Taylor et al., 2013). We also showed how hDicer's interaction cofactor proteins such as TRBP and PACT may enhance the protein's activity by changing its N-terminal domain conformation. In the current work, for the first time, we dissected the regulatory mechanism of hDicer's activity by structural elements of the pre-miRNA. Our results combining the structural prediction, *in vitro* dicing activity assays, and single particle EM reconstruction revealed a strong correlation of pre-miRNA's terminal loop structure with its effect on hDicer's conformation and activity (Fig. 5). Although the resolution of our 3D reconstruction at current stage is not high enough for us to see the RNA bound to hDicer, the dramatic conformational change induced by pre-miR-151 and its edited partners with different terminal loop structures may well reflect the terminal loop's effect on hDicer's N-terminal domain via direct interaction. The pre-miRNA terminal loop on one end induces hDicer's N-terminal domain structural change, on the other end, helps pre-miRNA's stem region align against hDicer's RNase domain. Structural changes in terminal loop thus probably have an allosteric effect of the RNA substrate's precise recognition by hDicer's RNase site. A seemingly small change such as the A-to-I mutation at −1 site of pre-miR-151 could cause a dramatic structural variation of the stem terminal loop, therefore resulting the drop of hDicer's processing activity. In contrast, the A-to-I mutation at +3 site of pre-miR-151, which does not affect the terminal loop's structure, showed almost no effect on hDicer's processing activity. This further underlines the importance of pre-miRNA's terminal loop structure in regulating hDicer activity. Our discovery here that the structural changes of terminal loop within pre-miRNA correlating the switches of DExH/D (ATPase-helicase) domain of hDicer regulate hDicer's activity provides insight into our understanding of how human Dicer processing pre-microRNA and opens new questions about miRNA terminal loops role in gene silencing regulation.

MATERIALS AND METHODS

Protein expression and purification

Human dicer was expressed and purified as described (MacRae et al., 2008). Briefly, Bac-to-Bac baculovirus expression system (Life technologies) was used to express the N-terminally His_6-tagged human Dicer proteins. SF-21 cells were infected for 60 h with hDicer recombinant baculovirus before collecting them for sonication. Supernatant was applied to Ni^{2+}-NTA resin after centrifugation at 3500 rpm for 1 h. Eluted hDicer protein was dialyzed with TEV

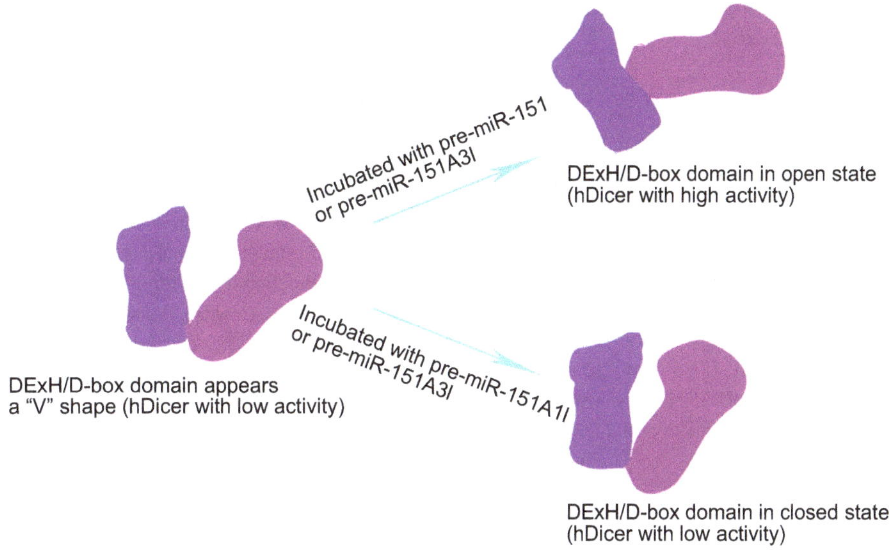

Figure 5. Model of DExH/D (ATPase-helicase) domain regulating the dicing activity of human Dicer protein. From the bottom view of human Dicer as in Fig. 4, the relative position between the branch (pink) relative to the main body (purple) of human Dicer switches depending on the RNA substrate's different loop structures.

enzyme against dialysis buffer overnight. The cleaved product was then applied to a 5 mL His Trap HP column, from which the flow-through was collected as the fraction containing His6-tag deleted hDcier. The protein was then concentrated to 1 mL and applied to a Sepharose 6 10/300 column for a final gel filtration step. The eluted protein solution in aliquots was snap-frozen in liquid nitrogen for storage at −80 centigrade.

Synthetic RNA substrates

All of our RNA substrates were synthesized by TAKARA BIOTECHNOLOGY. Human pre-let-7: 5′-UGAGGUAGUAGGUUGUAUAGUUUUAGGGUCACACCCACCACUGGGAGAUAACUAUACAAUCUACUGUCUUACC-3′, human pre-miR-151: 5′-CGAGGAGCUCACAGUCUAGUAUGUCUCAUCCCCUACUAGACUGAAGCUCCUUGAGG-3′, human pre-miR-151A1I: 5′-UCGAGGAGCUCACAGUCUAGUAUGUCUCAUCCCCUICUAGACUGAAGCUCCUUGAGG-3′, human pre-miR-151A3I: 5′-UCGAGGAGCUCACAGUCUAGUAUGUCUCAUCCCCUACUIGACUGAAGCUCCUUGAGG-3′, human pre-miR-151A13I: 5′-CGAGGAGCUCACAGUCUAGUAUGUCUCAUCCCCUICUIGACUGAAGCUCCUUGAGG-3′. The RNA substrates were annealed by fast-cooling to fold them into the right structure. They were verified for purity and the correct folding by Gelsafe staining (Yuanpinghao Bio) after native gel electrophoresis and 5′-end-labeled with γ-^{32}P-ATP (PerkinElmer) using T4 polynucleotide kinase (from TAKARA BIOTECHNOLOGY) before all related assays described in this work.

In vitro cleavage of RNA substrates

The *in vitro* cleavage assay of different RNA substrates by hDicer were performed at 37°C for 1 h in 15 μL of reaction volume containing 60 nmol/L Dicer, 60 nmol/L γ-^{32}p-labeled RNA substrates, 30 mmol/L Tris-HCl (pH 6.8), 50 mmol/L NaCl, 3 mmol/L $MgCl_2$, 0.1% Triton X-100, 15% glycerol, 1 mmol/L DTT. Reactions were stopped with RNA loading buffer containing 8 mol/L Urea, 1× TBE, 0.05% Bromphenol blue and 0.05% Xylene cyanol, boiled for 4 min, then chilled on ice. RNA products were separated on an 18% polyacrylamide, 8 mol/L urea denaturing gel, visualized on a phosphor screen (Amersham Biosciences) with a Typhoon Trio Imager (Amersham Biosciences).

Electrophoretic mobility shift assay (EMSA)

For EMSA, RNA substrates were incubated with hDicer on ice for 40 min in a 20 μL volume containing 30 mmol/L Tris-HCl (Ph 6.8), 25 mmol/L NaCl, 2 mmol/L $MgCl_2$, 1 mmol/L DTT and 2 mmol/L EDTA. Samples were resolved through 6% native acrylamide gels in 0.5× Tris-glycine buffer under an electric field of 15 V/cm for 70 min visualized on a phosphor screen (Amersham Biosciences) with a Typhoon Trio Imager (Amersham Biosciences).

Negative staining electron microscopy

Human Dicer-RNA complexes were assembled by pre-incubating 150 nmol/L hDicer with a 500% excess of unedited and edited pre-miR-151 in a specific volume containing 20 mmol/L Tris-HCl, pH 6.8, 25 mmol/L NaCl, 2 mmol/L $MgCl_2$, 1 mmol/L DTT and 2 mmol/L EDTA and 5% glycerol. All the samples were negatively stained on holey carbon grids covered by a thin layer of continuous carbon over holes with 2% (*w*/*v*) uranyl acetate solution. All micrographs were collected on a Tecnai F20 Twin transmission electron microscope (FEI) running at 200 kV or an Tecnai-12 Biotwin electron microscope (FEI) operated at 120 kV, using a nominal magnification of 50,000× or 49,000×, respectively. The micrographs were collected on Ultrascan 4000 CCD camera (Gatan Inc.) at specimen-level pixel sizes of

0.223 nm or 0.29 nm, respectively, using a dose of ~30 e^{-} $Å^{-2}$ and a nominal defocus range of −1 to −3 μm.

Image processing and model reconstructions

For 3D classification of apo-hDicer and hDicer-RNA complexes, about 70,000 particles of each sample were picked together from each set of micrographs. EMAN2 (Tang et al., 2007) was used to do the particle picking. Two-dimensional classification and alignment of particle images were performed using IMAGIC-4D (van Heel et al., 1996). Three dimension classification and refinement was performed with Relion1.2 (Scheres, 2012). The starting 3D model was generated from previous structure (EMD-5601) (Taylor et al., 2013) low-pass filtered to 60 Å with SPIDER (Shaikh et al., 2008).

Statistical analysis

All the biochemical assays were repeated more than three times and student's *t*-test was used for statistical analysis. Statistical significance was defined by a two tailed *P*-value of 0.05.

ACKNOWLEDGEMENTS

We thank Dr. Enbo Ma and Dr. Jennifer Doudna at UC-Berkeley for the human Dicer recombinant pFastBac vector. We thank Yanji Xu, Xiaomei Li for the maintenance of electron microscopy facility. We acknowledge the China National Center for Protein Sciences Beijing and "Explorer 100" cluster system of Tsinghua National Laboratory for Information Science and Technology for providing the facility support. This work was supported by the National Natural Science Foundation of China (Grant No. 31270765) and the National Basic Research Program (973 Program) (No. 2010CB912401).

ABBREVIATIONS

dsRBD, double-stranded RNA-binding domain; EMSA, electrophoresis mobility shift assay; hDicer, human Dicer.

COMPLIANCE WITH ETHICS GUIDELINES

ZhongMin Liu, Jia Wang, Gang Li and Hong-Wei Wang declare that they have no conflict of interest.

This article does not contain any studies with human or animal subjects performed by the any of the authors.

REFERENCES

Ando Y, Maida Y, Morinaga A, Burroughs AM, Kimura R, Chiba J, Suzuki H, Masutomi K, Hayashizaki Y (2011) Two-step cleavage of hairpin RNA with 5' overhangs by human DICER. BMC Mol Biol 12:6

Blow MJ, Grocock RJ, van Dongen S, Enright AJ, Dicks E, Futreal PA, Wooster R, Stratton MR (2006) RNA editing of human microRNAs. Genome Biol 7:R27

Brennicke A, Marchfelder A, Binder S (1999) RNA editing. FEMS Microbiol Rev 23:297–316

Carthew RW, Sontheimer EJ (2009) Origins and mechanisms of miRNAs and siRNAs. Cell 136:642–655

Castilla-Llorente V, Nicastro G, Ramos A (2013) Terminal loop-mediated regulation of miRNA biogenesis: selectivity and mechanisms. Biochem Soc Trans 41:861–865

Chendrimada TP, Gregory RI, Kumaraswamy E, Norman J, Cooch N, Nishikura K, Shiekhattar R (2005) TRBP recruits the Dicer complex to Ago2 for microRNA processing and gene silencing. Nature 436:740–744

Das AK, Carmichael GG (2007) ADAR editing wobbles the microRNA world. ACS Chem Biol 2:217–220

Griffiths-Jones S, Saini HK, van Dongen S, Enright AJ (2008) miRBase: tools for microRNA genomics. Nucleic Acids Res 36: D154–D158

Gu S, Jin L, Zhang Y, Huang Y, Zhang F, Valdmanis PN, Kay MA (2012) The loop position of shRNAs and pre-miRNAs is critical for the accuracy of dicer processing in vivo. Cell 151:900–911

He L, Hannon GJ (2004) MicroRNAs: small RNAs with a big role in gene regulation. Nat Rev Genet 5:522–531

Heo I, Joo C, Cho J, Ha M, Han J, Kim VN (2008) Lin28 mediates the terminal uridylation of let-7 precursor MicroRNA. Mol Cell 32:276–284

Heo I, Ha M, Lim J, Yoon MJ, Park JE, Kwon SC, Chang H, Kim VN (2012) Mono-uridylation of pre-microRNA as a key step in the biogenesis of group II let-7 microRNAs. Cell 151:521–532

Kawahara Y, Zinshteyn B, Chendrimada TP, Shiekhattar R, Nishikura K (2007a) RNA editing of the microRNA-151 precursor blocks cleavage by the Dicer–TRBP complex. EMBO Rep 8:763–769

Kawahara Y, Zinshteyn B, Sethupathy P, Iizasa H, Hatzigeorgiou AG, Nishikura K (2007b) Redirection of silencing targets by adenosine-to-inosine editing of miRNAs. Science 315:1137–1140

Lau P-W, Guiley KZ, De N, Potter CS, Carragher B, MacRae IJ (2012) The molecular architecture of human Dicer. Nat Struct Mol Biol 19:436–440

Lee Y, Hur I, Park S-Y, Kim Y-K, Suh MR, Kim AVN (2006) The role of PACT in the RNA silencing pathway. EMBO J 25:522–532

Ma E, MacRae IJ, Kirsch JF, Doudna JA (2008) Autoinhibition of human dicer by its internal helicase domain. J Mol Biol 380:237–243

MacRae IJ, Ma E, Zhou M, Robinson CV, Doudna JA (2008) In vitro reconstitution of the human RISC-loading complex. Proc Natl Acad Sci USA 105(2):512–517

Nishikura K (2010) Functions and regulation of RNA editing by ADAR deaminases. Annu Rev Biochem 79:321–349

Ota H, Sakurai M, Gupta R, Valente L, Wulff BE, Ariyoshi K, Iizasa H, Davuluri RV, Nishikura K (2013) ADAR1 forms a complex with Dicer to promote microRNA processing and RNA-induced gene silencing. Cell 153:575–589

Park JE, Heo I, Tian Y, Simanshu DK, Chang H, Jee D, Patel DJ, Kim VN (2011) Dicer recognizes the 5′ end of RNA for efficient and accurate processing. Nature 475:201–205

Scheres SHW (2012) RELION: implementation of a Bayesian approach to cryo-EM structure determination. J Struct Biol 180:519–530

Schraml E, Grillari J (2012) From cellular senescence to age associated diseases: the miRNA connection. Longev Heal 1:1–15

Shaikh TR, Gao H, Baxter WT, Asturias FJ, Boisset N, Leith A, Frank J (2008) SPIDER image processing for single-particle reconstruction of biological macromolecules from electron micrographs. Nat Protoc 3:1941–1974

Su AAH, Randau L (2011) A-to-I and C-to-U editing within transfer RNAs. Biochemistry 76:932–937

Tang G, Peng L, Baldwin PR, Mann DS, Jiang W, Rees I, Ludtke SJ (2007) EMAN2: an extensible image processing suite for electron microscopy. J Struct Biol 157:38–46

Taylor DW, Ma E, Shigematsu H, Cianfrocco MA, Noland CL, Nagayama K, Nogales E, Doudna JA, Wang H-W (2013) Substrate-specific structural rearrangements of human Dicer. Nat Struct Mol Biol 20:662–670

Tsutsumi A, Kawamata T, Izumi N, Seitz H, Tomari Y (2011) Recognition of the pre-miRNA structure by Drosophila Dicer-1. Nat Struct Mol Biol 18:1153–1158

van Heel M, Harauz G, Orlova EV, Schmidt R, Schatz M (1996) A new generation of the IMAGIC image processing system. J Struct Biol 116:17–24

Wang HW, Noland C, Siridechadilok B, Taylor DW, Ma E, Felderer K, Doudna JA, Nogales E (2009) Structural insights into RNA processing by the human RISC-loading complex. Nat Struct Mol Biol 16:1148–1153

Xie M, Li M, Vilborg A, Lee N, Shu MD, Yartseva V, Sestan N, Steitz JA (2013) Mammalian 5′-capped microRNA precursors that generate a single microRNA. Cell 155:1568–1580

Yang W, Chendrimada TP, Wang Q, Higuchi M, Seeburg PH, Shiekhattar R, Nishikura K (2006) Modulation of microRNA processing and expression through RNA editing by ADAR deaminases. Nat Struct Mol Biol 13:13–21

Zhang X, Zeng Y (2010) The terminal loop region controls microRNA processing by Drosha and Dicer. Nucleic Acids Res 38:7689–7697

Zhang H, Kolb FA, Vincent Brondani EBA, Filipowicz W (2002) Human Dicer preferentially cleaves dsRNAs at their termini without ATP. EMBO J 21:5875–5885

Zhang H, Kolb FA, Jaskiewicz L, Westhof E, Filipowicz W (2004) Single processing center models for human Dicer and bacterial RNase III. Cell 118:57–68

Zuker M (2003) Mfold web server for nucleic acid folding and hybridization prediction. Nucleic Acids Res 31:3406–3415

2

Essential role of the iron-sulfur cluster binding domain of the primase regulatory subunit Pri2 in DNA replication initiation

Lili Liu[1,2], Mingxia Huang[2]✉

[1] Department of Pharmacology and Chemical Biology, University of Pittsburgh School of Medicine, Pittsburgh, PA 15213, USA
[2] Department of Biochemistry and Molecular Genetics, University of Colorado School of Medicine, Aurora, CO 80045, USA
✉ Correspondence: mingxia.huang@ucdenver.edu (M. Huang)

ABSTRACT

DNA primase catalyzes *de novo* synthesis of a short RNA primer that is further extended by replicative DNA polymerases during initiation of DNA replication. The eukaryotic primase is a heterodimeric enzyme comprising a catalytic subunit Pri1 and a regulatory subunit Pri2. Pri2 is responsible for facilitating optimal RNA primer synthesis by Pri1 and mediating interaction between Pri1 and DNA polymerase α for transition from RNA synthesis to DNA elongation. All eukaryotic Pri2 proteins contain a conserved C-terminal iron-sulfur (Fe-S) cluster-binding domain that is critical for primase catalytic activity *in vitro*. Here we show that mutations at conserved cysteine ligands for the Pri2 Fe-S cluster markedly decrease the protein stability, thereby causing S phase arrest at the restrictive temperature. Furthermore, Pri2 cysteine mutants are defective in loading of the entire DNA pol α-primase complex onto early replication origins resulting in defective initiation. Importantly, assembly of the Fe-S cluster in Pri2 is impaired not only by mutations at the conserved cysteine ligands but also by increased oxidative stress in the *sod1Δ* mutant lacking the Cu/Zn superoxide dismutase. Together these findings highlight the critical role of Pri2's Fe-S cluster domain in replication initiation *in vivo* and suggest a molecular basis for how DNA replication can be influenced by changes in cellular redox state.

KEYWORDS primase, Pri2, iron-sulfur cluster, replication initiation

INTRODUCTION

A unique challenge in DNA replication initiation is the inability of DNA polymerases to synthesize new DNA strands *de novo* (Hubscher et al., 2002; Wang, 1996). The problem is overcome by DNA primase, a specialized DNA-dependent RNA polymerase that synthesizes a short RNA primer for DNA polymerase to extend new strand synthesis (Arezi et al., 1999; Arezi and Kuchta, 2000; Frick and Richardson, 2001; Zerbe and Kuchta, 2002). The eukaryotic DNA primase is part of the DNA polymerase α-primase (pol-prim) complex that is composed of four conserved subunits, referred to as Pol1, Pol12, Pri1 and Pri2 in the budding yeast *Saccharomyces cerevisiae* (Burgers, 1998; Foiani et al., 1997; Hubscher et al., 2002; Kawasaki and Sugino, 2001; Wang, 1996). The structural and catalytic properties of all four pol-prim subunits are highly conserved among eukaryotes; all four subunits are essential for viability in yeast cells (Budd and Campbell, 1987; Foiani et al., 1994; Foiani et al., 1989; Francesconi et al., 1991; Lucchini et al., 1987). The catalytic primase subunit Pri1 and its regulatory subunit Pri2 form the heterodimeric DNA primase. Pol1 is the catalytic subunit of DNA polymerase α that is tightly bound to its regulatory subunit Pol12, also known as the B subunit.

Replication of a eukaryotic genome is initiated from multiple specific sites along the chromosomes known as replication origins during the S phase of the mitotic cell cycle. The pol-prim complex associates with replication forks in S phase to initiate nucleic acid synthesis of the leading strand as well as each Okazaki fragment of the lagging strand

(Muzi-Falconi et al., 2003). Primase synthesizes a 7–12 nt RNA primer (Arezi and Kuchta, 2000; Frick and Richardson, 2001) that is transferred to DNA Pol α for limited extension to ~30 nt. The resulting RNA-DNA hybrid primer is subsequently handed over to Pol ε and Pol δ on the leading and lagging strand, respectively, for processive DNA replication (Arezi et al., 1999; Hubscher et al., 2002; Zerbe and Kuchta, 2002).

The transition from primase-catalyzed RNA synthesis to Pol α-catalyzed DNA synthesis is mediated by Pri2 that bridges Pri1 and Pol α within the pol-prim complex (Arezi et al., 1999). Electron microscopy studies and computational modeling of the pol-prim complex reveal a dumbbell-shaped architecture with flexible association between the DNA polymerase and primase lobes (Nunez-Ramirez et al., 2011). Degradation of a Pri2 temperature-sensitive mutant protein at the restrictive temperature abolishes co-immunoprecipitation of Pol1-Pri1, whereas Pol1-Pri2 interaction remains intact when Pri1 is degraded (Longhese et al., 1993). Thus, the association between primase and Pol α appears to be directly mediated by Pri2, in keeping with co-immunoprecipitation studies of human pol-prim subunits that were ectopically expressed in insect cells (Copeland and Wang, 1993).

Pri2 also plays essential roles in stabilization of Pri1 as well as in the initiation, elongation and length counting of the RNA primer. The N-terminal half of Pri2 binds directly to Pri1 via a conserved protein-protein interface as revealed by a crystallographic study of the archeaon core primase (Lao-Sirieix et al., 2005). The Pri2 CTD (residues 307–528 of the *S. cerevisiae* protein), which is absent in the core Pri1-Pri2 crystal structure, contains a [4Fe-4S] cluster that is coordinated by four conserved cysteine ligands (Klinge et al., 2007; Sauguet et al., 2010). Biochemical studies of recombinant Pri2 proteins show that individual cysteine substitutions cause only moderate reduction of Fe binding and iron-sulfur (Fe-S) cluster content. In contrast, mutating two of the four cysteines nearly abolishes iron binding and significantly reduces initiation of RNA synthesis in an *in vitro* primase activity assay (Klinge et al., 2007). These studies were performed with proteins that are heterologously expressed in bacteria with a Fe-S assembly machinery equivalent to that of the mitochondrial iron-sulfur cluster (ISC) biogenesis pathway in eukaryotic cells. However, bacteria lack the eukaryotic cytosolic Fe-S assembly pathway and in some cases are unable to express eukaryotic proteins with native Fe-S cluster assembly (Urzica et al., 2009). Thus, the *in vivo* function of Pri2's putative Fe-S cluster-binding domain and the relative contribution of the cysteine ligands in DNA replication initiation remain unclear.

Fe-S cluster-binding domains have been found in an increasing number of nuclear proteins that are involved in DNA replication and repair processes, including the helicases XPD/Rad3 and FancJ (Liu et al., 2008; Rudolf et al., 2006), the nucleases AddAB and Dna2 (Messick et al., 2002; Yeeles et al., 2009), and all four of the eukaryotic class B DNA polymerases: α, δ, ε and ζ (Baranovskiy et al., 2012; Netz et al., 2012). Biochemical and structural studies suggest that the Fe-S cluster binding domains in these nuclear enzymes serve a structural rather than a redox-active role, likely by stabilizing local domain conformation that may mediate protein-protein or protein-nucleic acid interactions. Nevertheless, the Fe-S clusters in some proteins such as yeast aconitases Aco1 and Lys4 are known to be targets of oxidative damage (Brazzolotto et al., 1999; Lu and Cortopassi, 2007; Missirlis et al., 2003; Wallace et al., 2004). Therefore, we were interested in investigating whether the Fe-S clusters in the DNA replication enzymes may also be susceptible to oxidative damage.

In this study, we characterize the *in vivo* role of the Fe-S cluster binding domain in Pri2 function by systematic site-directed mutagenesis of the putative cysteine ligands. We followed Fe-S assembly directly in yeast cells by ^{55}Fe radiolabeling and determined the consequence of the cysteine mutations on S phase progression and pol-prim loading at early replication origins by chromatin immunoprecipitation (ChIP). We find that Fe-S cluster assembly in Pri2 is critical for maintaining its protein stability *in vivo* and is dependent on the function of Tah18, a diflavin NADPH reductase that is an essential component of the cytosolic iron-sulfur assembly (CIA) (Nunez-Ramirez et al.) machinery (Netz et al., 2010). Cysteine mutants of Pri2 exhibit temperature-sensitive growth defects. At the permissive temperature, the mutants are defective in ^{55}Fe incorporation into nascent Pri2 polypeptide and exhibit reduced loading of multiple pol-prim subunits onto early replication origins. The Pri2 mutant proteins are further destabilized at the restrictive temperature, leading to permanent cell cycle arrest in the S phase with a G_1 DNA content. We also find that Pri2 protein is unstable under iron-deficient growth conditions and in *sod1Δ* mutant cells lacking the cytosolic Cu/Zn superoxide dismutase. These findings demonstrate the importance of the conserved cysteine residues for Fe-S assembly into Pri2 and their essential role in maintaining the active pol-prim complex at replication origins. Our data also suggest that the initial step of DNA replication may be modulated by oxidative stress.

RESULTS

Mutations in conserved cysteine residues of the Pri2 Fe-S cluster binding domain cause growth defect at restrictive temperatures

The four conserved cysteine residues in Pri2's C-terminal Fe-S cluster binding domain are located at positions 336, 417, 434 and 474 (Fig. 1A). To determine the physiological importance of these residues, we first made single cysteine-to-alanine substitutions at each position and integrated both the wild-type and the mutant alleles into the chromosomal *PRI2* locus with an inframe triple HA epitope at the C-terminus. Three of the four single mutants had no apparent effect on cell fitness at growth temperatures ranging from 23°C to 37°C (Fig. 1B, top panel). In contrast, the *pri2*

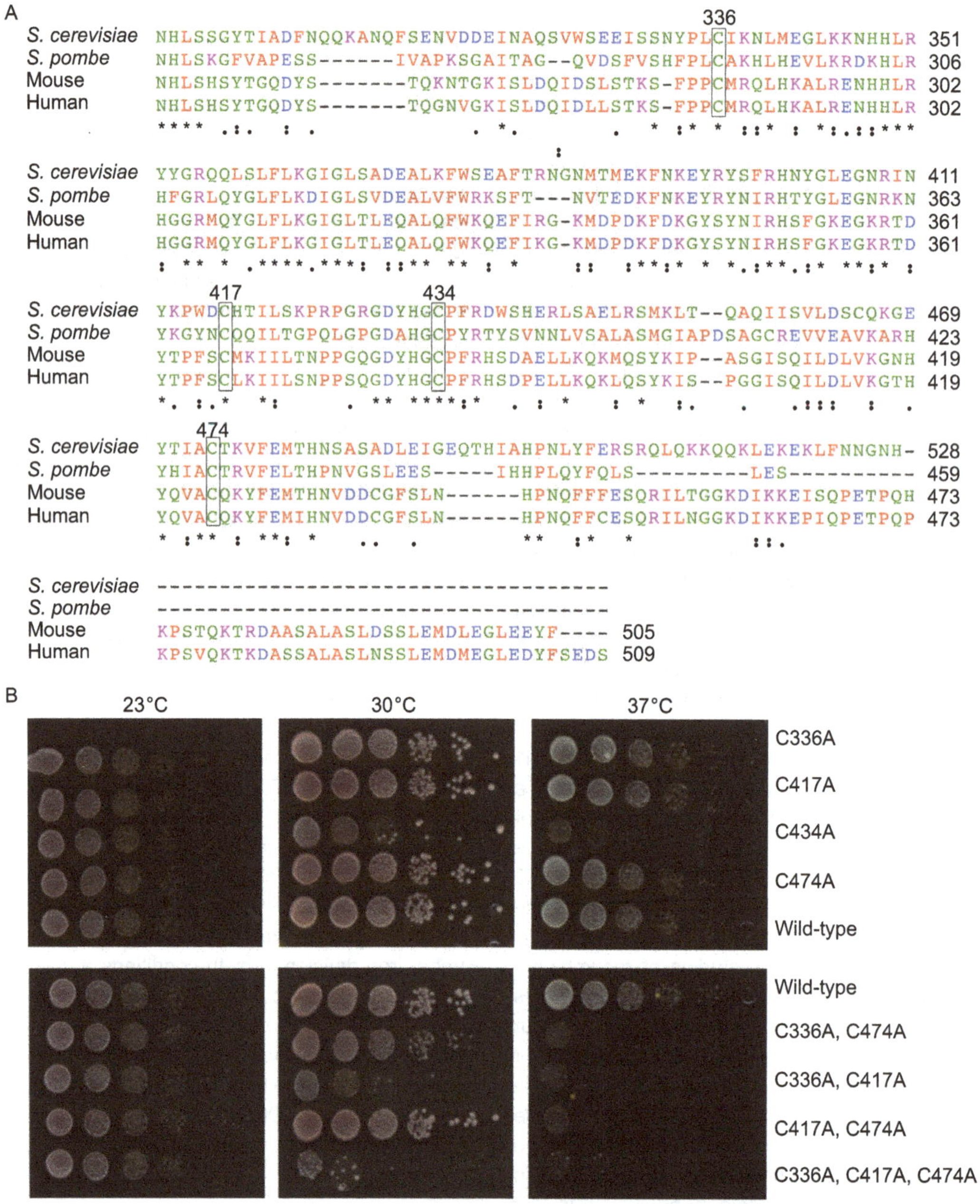

Figure 1. Importance of conserved cysteine residues of Pri2's C-terminal Fe-S cluster binding domain for mitotic viability. (A) Multiple sequence alignment of primase large subunit C-terminal domains. Sequences are shown for *S. cerevisiae*, *S. pombe*, *H. sapiens* (Human), *M. musculus* (Mouse). The conserved cysteines were marked in boxes. The residue numbers are those of *S. cerevisiae* Pri2 protein. (B) Comparison of growth between congenic wild-type and *pri2* mutants with various Cys-to-Ala substitutions at different temperatures. Cells were grown in liquid YPD media to log phase (OD_{600} ~1) and harvested. Cell pellets were washed with water. Ten-fold serial dilutions starting at 10^5 cells were dot-plated on YPD plates and incubated at 23°C, 30°C and 37°C for two days before being photographed.

(*C434A*) single mutant exhibited severe growth defects at 30°C and 37°C. Interestingly, a mutation at the same residue, *C434Y*, was previously shown to cause lethality at 37°C arresting cells with large-budded morphology (Francesconi et al., 1991), a characteristic phenotype of mutants defective in S phase progression. Cys-434 thus appears to play a more critical role in stabilizing the Fe-S cluster and/or local protein conformation relative to the other three cysteines in Pri2's CTD. We then made double and triple mutants between cysteines 336, 417 and 474. The triple mutant *pri2*(*C336A*, *C417A*, *C474A*) showed severe growth defects at 30°C and 37°C. All three double mutants are temperature-sensitive (ts) at 37°C, with *pri2*(*C336A, C417A*) showing a growth defect comparable to those of *pri2*(*C434A*) and *pri2*(*C336A, C417A, C474A*) even at 30°C (Fig. 1B, bottom panel). Together, these results indicate important but varying contributions of the four Fe-S ligand cysteines to Pri2 function and mitotic viability.

Cysteine mutations reduce Pri2 protein stability and arrest cells in the S phase at the restrictive temperature

A couple of double cysteine substitutions in the Pri2 CTD have previously been shown to cause destabilization of the recombinant protein expressed in bacteria (Klinge et al., 2007). To assess the effect of the cysteine substitutions on Pri2 protein stability in yeast cells, we first compared Pri2 levels in log-phase cells between the wild-type and mutants at the permissive temperature (23°C). The *C434A* single, *C336A, C417A* double, and *C336A, C417A, C474A* triple mutants displayed significant decrease in Pri2 protein levels at 23°C (Fig. 2A). The three mutants also showed gradual destabilization of Pri2 protein when being shifted from 23°C to 37°C relative to the wild-type control (Fig. 2B), consistent with their growth defect at the restrictive temperature.

Mutants defective in DNA primase function are usually associated with DNA synthesis impairment and S phase delay (Longhese et al., 1993). To determine the contribution of the Pri2 Fe-S cluster binding domain to DNA replication, we compared the *pri2*(*C336A, C417A, C474A)* triple mutant and wild-type cells in progression through the first cell cycle after being released from an α factor-mediated G_1-arrest by using flow cytometry. Both the wild-type and the mutant moved through the first S phase with similar pace after being released from G_1 at 23°C (Fig. 2C). Upon being released at 37°C, the wild-type cells moved through the first S phase at a faster pace compared to 23°C (Fig. 2D). In contrast, the *pri2 (C336A, C417A, C474A)* mutant permanently arrested with a G_1 DNA content after being released from G_1 at 37°C, indicative of a severe defect in initiation of DNA replication. The arrested unsynchronized mutant cells were mostly dumbbell shaped (data not shown), consistent with an S phase-arrested morphology. Taken together, these results suggest that disruption of Pri2's Fe-S domain causes destabilization of the Pri2 protein, which impairs the initial step of DNA synthesis resulting in cell cycle arrest in S phase with a G_1 DNA content.

The pri2 cysteine mutants impair loading of the DNA Pol α-primase complex onto early replication origins

The growth defects and cell cycle arrest of the *pri2* cysteine mutants at the restrictive temperature suggested a critical role of its Fe-S cluster binding domain in replication initiation. A possible explanation is that lower stability and disturbance of local conformation of the *pri2* mutant proteins compromise loading of the pol-prim complex onto replication origins. To test this notion, we monitored association of individual pol-prim subunits with two well characterized early replication origins ARS305 and ARS607 by using chromatin immunoprecipitation (ChIP) and real-time PCR in *pri2*(*C336A, C417A, C474A*) and *pri2*(*C434A*), the two mutants with the most severe ts defect. The strains harbor *PRI2-HA* and *PRI1-Myc* or *POL1-Myc* that were integrated at their respective chromosomal loci. Cells from log-phase cultures were synchronized in G_1 phase with α factor and then released into the cell cycle by washing out the α factor at 23°C. The relative enrichment of ARS DNA in the ChIP over input samples was used to evaluate change in association between various pol-prim subunits and the replication origins at different time points after release from G_1.

In the wild-type cells, association of Pri1 and Pri2 at the two early origins was low in G_1 and peaked at 30–45 min after G_1 release (Fig. 3B, dark blue columns), which was before the bulk of DNA replication (based on FACS analysis, Fig. 3A). Dissociation of Pri1 and Pri2 from the two early origins coincided with a further increase of DNA bulk synthesis at the 60 min and 75 min time points. The temporal changes of the two primase subunits at ARS305 and ARS607 are consistent with previously observed kinetics of pol-prim complex loading at the two origins (Aparicio et al., 1999; Ricke and Bielinsky, 2004). In contrast, loading of Pri1 and Pri2 at the two early origins remained low in the *pri2*(*C336A, C417A, C474A*) mutant cells throughout the time course after G_1 release, lacking any significant peak (Fig. 3B).

We next asked whether loading of DNA Pol α to the early replication origins was also affected in the *pri2* mutant. Since the pol-prim complex contains two heterodimers with distinct biochemical activities, it is unknown whether origin association of Pol1-Pol12 is independent of that of Pri1-Pri2. We found that there is a significant reduction of Pol1 association with ARS305 and ARS607 in *pri2*(*C336A, C417A, C474A*) mutant cells during the first 45 min after G_1 release (Fig. 4). It is not clear the extent to which the relatively higher level of Pol1 binding at the origins at time zero reflects initial loading of Pol1, which may later be destabilized because of the failure of an increased loading of Pri1-Pri2. Similar to *pri2* (*C336A, C417A, C474A*), loading of Pri2 and Pol1 at the early origins were both defective in the *pri2*(*C434A*) mutant (Fig. 5). Since we have not observed any difference in levels of the two proteins between the wild-type and *pri2* mutants at 23°C (Fig. S1), the decreased loading of Pri1 and Pol1 at the two origins is unlikely due to variation in the endogenous protein levels. The results support the notion that decreased protein levels and origin loading of Pri2 can lead to destabilization of the whole pol-prim complex at the origins.

The pri2 cysteine mutants are defective in Fe-S cluster assembly *in vivo*

Previous biochemical studies have shown that some substitutions of the four conserved cysteines in Pri2 CTD

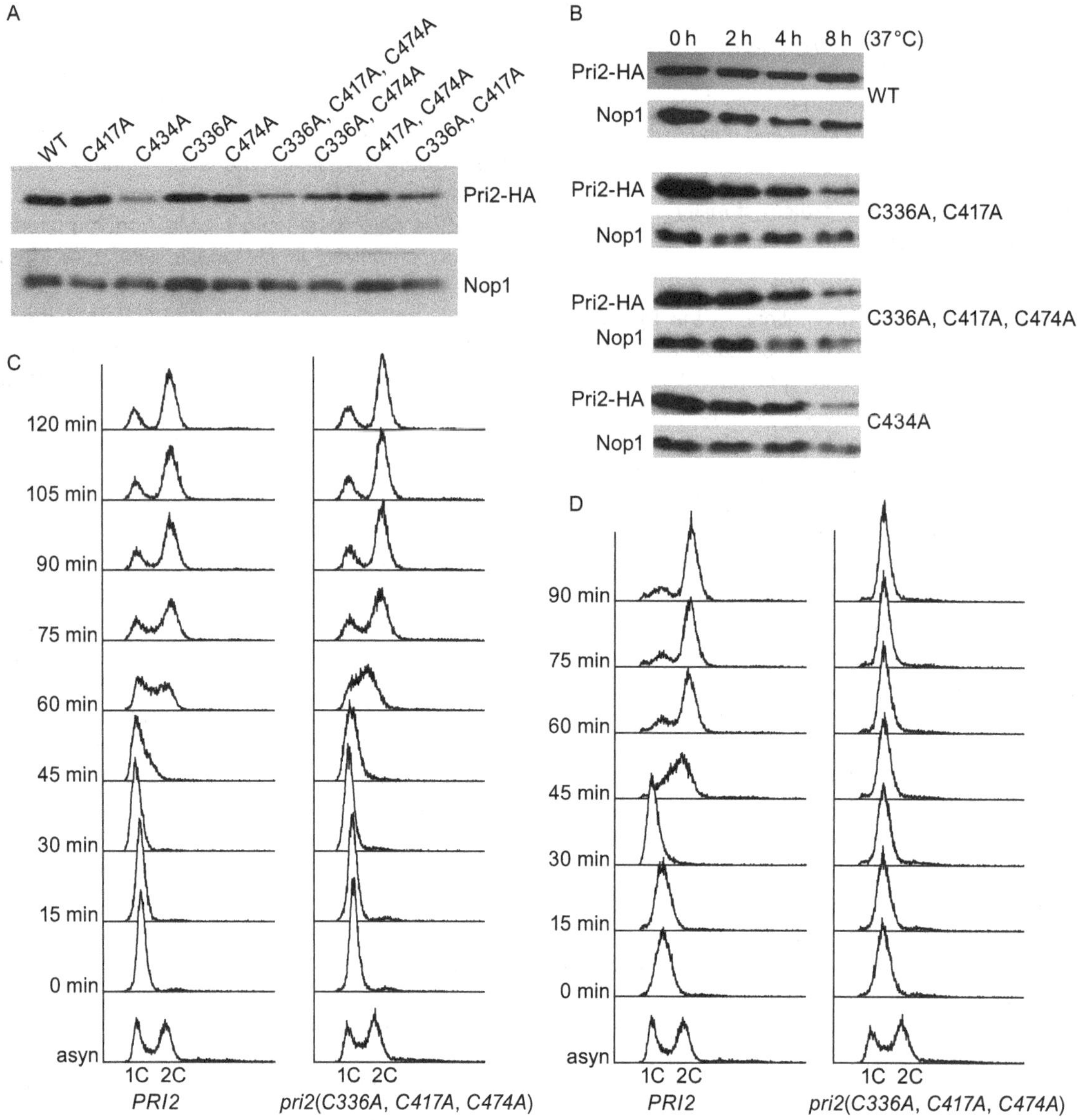

Figure 2. Mutations of the cysteine residues of Pri2's Fe-S cluster domain destabilize the Pri2 protein and impede DNA replication. (A) Each *pri2* Cys-to-Ala mutant contained a C-terminal triple HA epitope and was integrated at the chromosomal *PRI2* locus. Congenic wild-type and *pri2* mutants were grown at 23°C to log phase. Protein extracts from harvested cells were resolved by SDS-PAGE and probed with anti-HA and anti-Nop1 (as a loading control). (B) Congenic wild-type and *pri2* mutants were grown at 23°C to log phase and then shifted to 37°C for an additional 2, 4 and 8 h. Protein extracts from harvested cells were subjected to immunoblotting for Pri2 and Nop1 proteins. Four-fold cell equivalents were loaded for each mutant relative to wild-type. (C and D) Comparison of cell cycle progression between wild-type (*PRI2*) and *pri2(C336A,C417A, C474A)* mutant cells at 23°C and 37°C. Cells were grown at 23°C to early log phase (OD_{600} ~0.3) and synchronized in G_1 phase by using α factor. G_1-arrested cells were split into two halves, released into cell cycle by washing out α factor at 23°C (C) and 37°C (D), and collected at the indicated time points for flow cytometry analysis.

disrupted Fe-S assembly in the recombinant protein and hindered Pri2 purification because of protein destabilization (Klinge et al., 2007). One caveat is the recombinant protein may lack the native Fe-S cluster (Urzica et al., 2009). We sought to follow Pri2 Fe-S cluster assembly in yeast cells. The level and activity of some iron-containing proteins are known to be altered under iron-limiting growth conditions (Kaplan et al., 2006). Indeed, we show that protein levels of Pri2 and Pol1, both containing a Fe-S cluster binding domain, were markedly reduced in cells grown in iron-deficient medium in comparison to iron-replete medium (Fig. 6A).

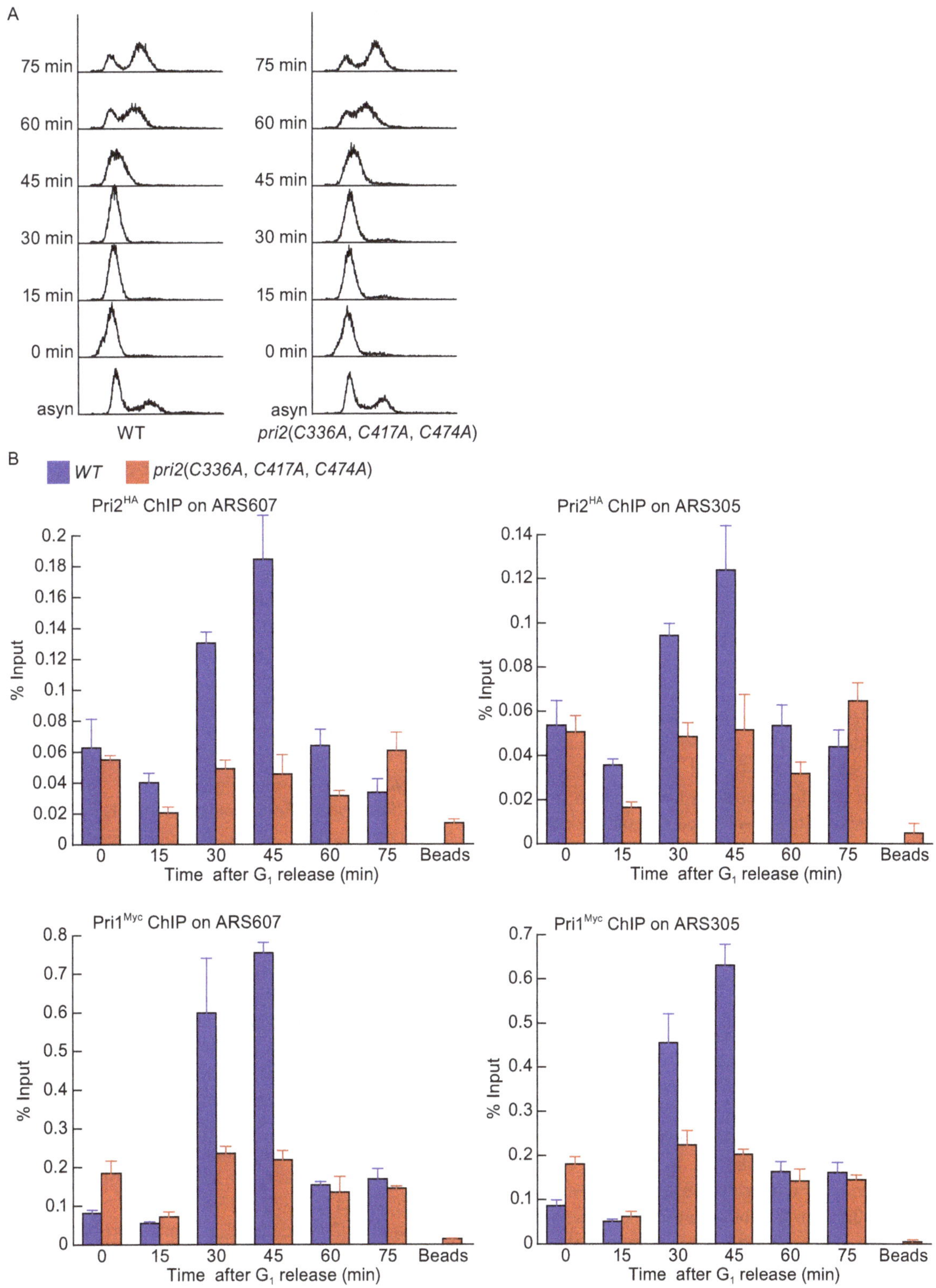

We then examined whether deficiencies in the CIA machinery would have an effect on Pri2 protein levels by using a ts mutant of *TAH18*. Tah18 is a diflavin NADPH reductase that forms a complex with the Fe-S cluster protein Dre2 and together constitutes an essential and early step in cytosolic Fe-S protein biogenesis. Depletion of Dre2-Tah18 results in reduction of Fe-S assembly into many CIA substrates and subsequent destabilization of some of the apo-proteins (Netz et al., 2010). While Pri2 protein levels remained relatively unchanged between 23°C and 37°C in

Figure 3. Impaired loading of both DNA primase subunits onto early replication origins in the *pri2(C336A, C417A, C474A)* mutant. Congenic wild-type (*PRI1-Myc, PRI2-HA*) and mutant (*PRI1-Myc, pri2(C336A, C417A, C474A)-HA*) cells were synchronized in G_1 with α factor and released at 23°C. Cells were collected at the time points indicated. (A) FACS analysis indicates the DNA content of cells throughout the time course. (B) Chromatin-containing extracts were prepared from formaldehyde cross-linked cells collected at the indicated time points. Pri2-HA and Pri1-Myc were immunoprecipitated with anti-HA and anti-Myc monoclonal antibodies, respectively. The recovery efficiency of two early chromosomal replication origins, ARS305 and ARS607, in the immunoprecipitated material relative to the input material was determined by real-time PCR. Background was determined by calculating the amount of target DNA in the mock-IP sample relative to the Input sample (Beads). The results are an average of three independent experiments with standard deviations.

wild-type cells, we observed significant reduction in Pri2 levels when the *tah18-ts* mutant was shifted from 23°C to 37°C (Fig. 6B). These results suggest that Pri2 is a substrate of CIA and that the apo form of Pri2 is unstable *in vivo*.

We also compared growth of wild-type and different *pri2* cysteine mutants on synthetic medium containing 1 mmol/L of iron chelator ferrozine, which has maximum chelating capacity of 333 µmol/L ferrous iron (Stearman et al., Ö 1996), and varied concentrations of supplemental ferrous ammonium sulfate at 30°C. Of the eight mutants, *pri2 (C434A)*, *pri2(C336A, C417A)* and *pri2(C336A, C417A, C474A)* exhibited growth defects on iron-deprived medium (0, 100 µmol/L) relative to the wild-type and other single and double cysteine mutants (Fig. 6C). These were the same three mutants that exhibited the most severe ts growth defects on rich medium (Fig. 1B). The defect of the three mutants was not rescued by growth on an iron-rich (350 µmol/L) medium (Fig. 6C), suggesting that Fe-S cluster assembly in these mutants is defective, which cannot be remedied by an increase in ferrous iron supply. To directly test this hypothesis, we used ^{55}Fe radiolabeling and immunoprecipitation assay to measure and compare iron incorporation into newly synthesized Pri2 proteins. Both the wild-type and two mutant Pri2(*C434A* and *C417A, C474A*) were N-terminally tagged with a triple Myc epitope and expressed from a high copy number plasmid to improve signal to background ratios of ^{55}Fe labeling. Cells were grown in iron-free medium at 30°C overnight before being pulse-labeled with $^{55}FeCl_3$ for 4 h. Pri2 were immunoprecipitated from whole cell extracts and the associated radioactivity was quantified by scintillation counting. Significantly more ^{55}Fe were bound to the wild-type Pri2 protein relative to the two cysteine mutants. After corrected for differences in protein stability, the *C434A* and *C417A, C474A* mutant proteins had a 60% and 90% reduction in ^{55}Fe binding, respectively (Fig. 6D). Together, these results demonstrate that Pri2 contains a *bona fide* Fe-S cluster *in vivo* that are coordinated by the conserved cysteine residues in its CTD. Our data also strongly suggest that the apo-form of Pri2 is unstable *in vivo*, which underlies the observed pol-prim loading defect at the permissive temperature and the replication initiation defect at the restrictive temperature.

Fe-S cluster assembly into Pri2 is sensitive to oxidative stress

The redox chemistry of Fe-S clusters means they are inherently vulnerable to oxidation and degradation. Superoxide has been shown to oxidize exposed [4Fe-4S] clusters in some proteins such as aconitase (Aco1) and isopropylmalate isomerase (Leu1), causing loss of iron from the cluster and inactivation of the enzyme (Djaman et al., 2004). Since both Pri2 and replicative DNA polymerases contain a [4Fe-4S] cluster, we were interested in determining whether the cluster assembly and stability of these proteins are affected by increased oxidative stress in the *sod1Δ* mutant lacking the Cu/Zn superoxide dismutase. We observed a significant decrease of Pri2 protein levels in *sod1Δ* relative to the wild-type cells. In contrast, there was no difference in protein levels of Pri1 between wild-type and *sod1Δ* (Fig. 7A). The proteins levels of two other [4Fe-4S] containing replication enzymes Pol1 and Pol2 were not affected by *sod1Δ* either. It thus seems that Pri2 is more vulnerable to oxidative damage relative to other Fe-S containing proteins involved in DNA replication. Indeed, we find that Pri2 is further destabilized when *sod1* cells were treated with oxidants paraquat and hydrogen peroxide (Fig. S2).

To determine if Fe-S cluster assembly in Pri2 is sensitive to oxidative stress, we carried out ^{55}Fe labeling and Pri2 immunoprecipitation assays in wild-type and *sod1Δ* mutant cells expressing transformed Myc-tagged Pri2. ^{55}Fe incorporation into newly translated Pri2 is 5-fold lower in *sod1Δ* mutant relative to the wild-type control after being corrected for difference in Pri2 protein levels (Fig. 7B). The *sod1Δ* mutant has previously been shown to have a prolonged S phase (Carter et al., 2005). Our results suggest that impaired Fe-S cluster assembly into nascent Pri2 polypeptide and the resulted destabilization of the apo-Pri2 may contribute to the S phase defect in *sod1Δ* mutant. In keeping with this notion, we found that the *pri2(C4343A)* mutant allele exhibited synthetic growth defect with *sod1Δ*; the double mutant is lethal at 30°C while each single mutant showed only mild growth defect (Fig. 7C). The synthetic phenotype between *sod1* and *pri2-ts* points to a connection between oxidative stress and DNA replication initiation.

DISCUSSION

Precise temporal and spatial control of the initiation events is critical for optimal replication of the genome and maintenance of genomic stability (Casper et al., 2008; Foiani et al., 1997; Lemoine et al., 2008). The four-subunit pol-prim

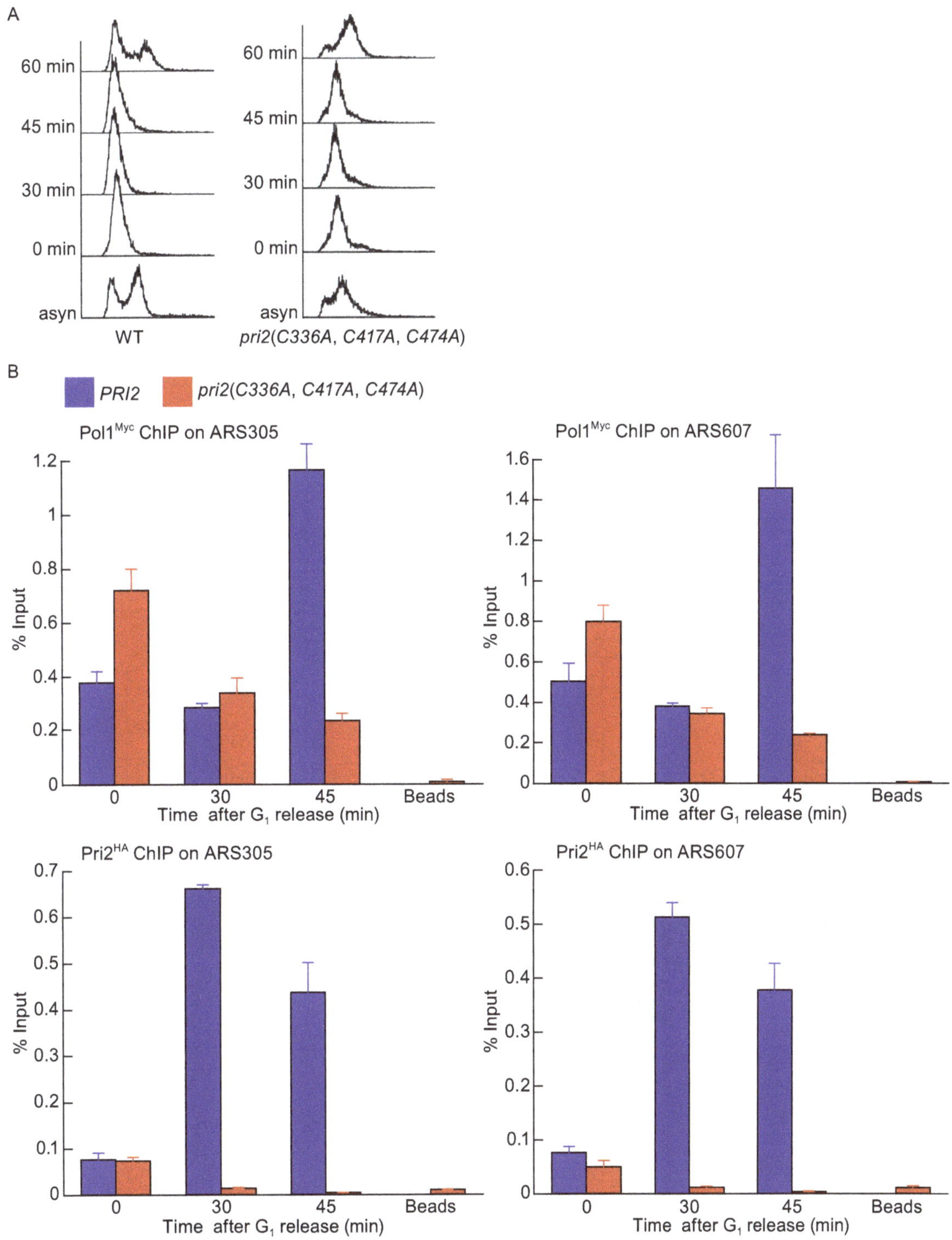

Figure 4. Destabilized association between DNA polymerase α and early replication origins in the *pri2(C336A, C417A, C474A)* mutant. Congenic wild-type (*POL1-Myc*, *PRI2-HA*) and mutant (*POL1-Myc*, *pri2(C336A, C417A, C474A)-HA*) cells were synchronized in G_1 with α factor and released at 23°C. Cells were collected at the indicated time points for FACS (A) and ChIP (B) analyses as described in Fig. 3. Pol1- and Pri2-associated chromatin was immunoprecipitated using anti-Myc and anti-HA monoclonal antibodies, respectively.

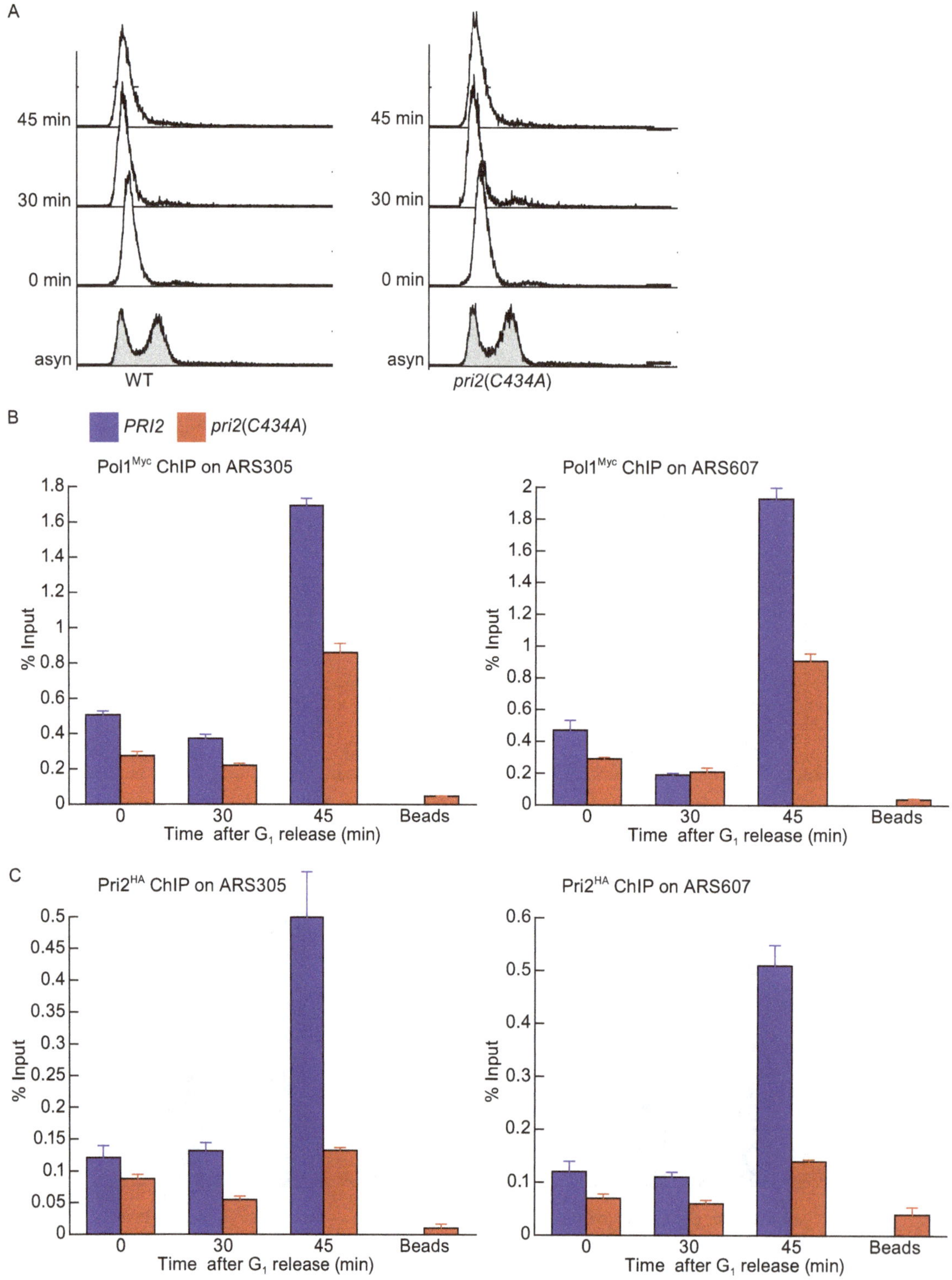

Figure 5. Defective loading of Pri2 and Pol α onto early replication origins in the *pri2(C434A)* mutant. Congenic wild-type (*POL1-Myc*, *PRI2-HA*) and mutant (*POL1-Myc*, *pri2(C434A)-HA*) cells were synchronized in G_1 with α factor and released at 23°C. Cells were collected at the indicated time points for FACS (A) and ChIP (B) analyses as described in Fig. 3. Pol1- and Pri2-associated chromatin was immunoprecipitated using anti-Myc and anti-HA monoclonal antibodies, respectively.

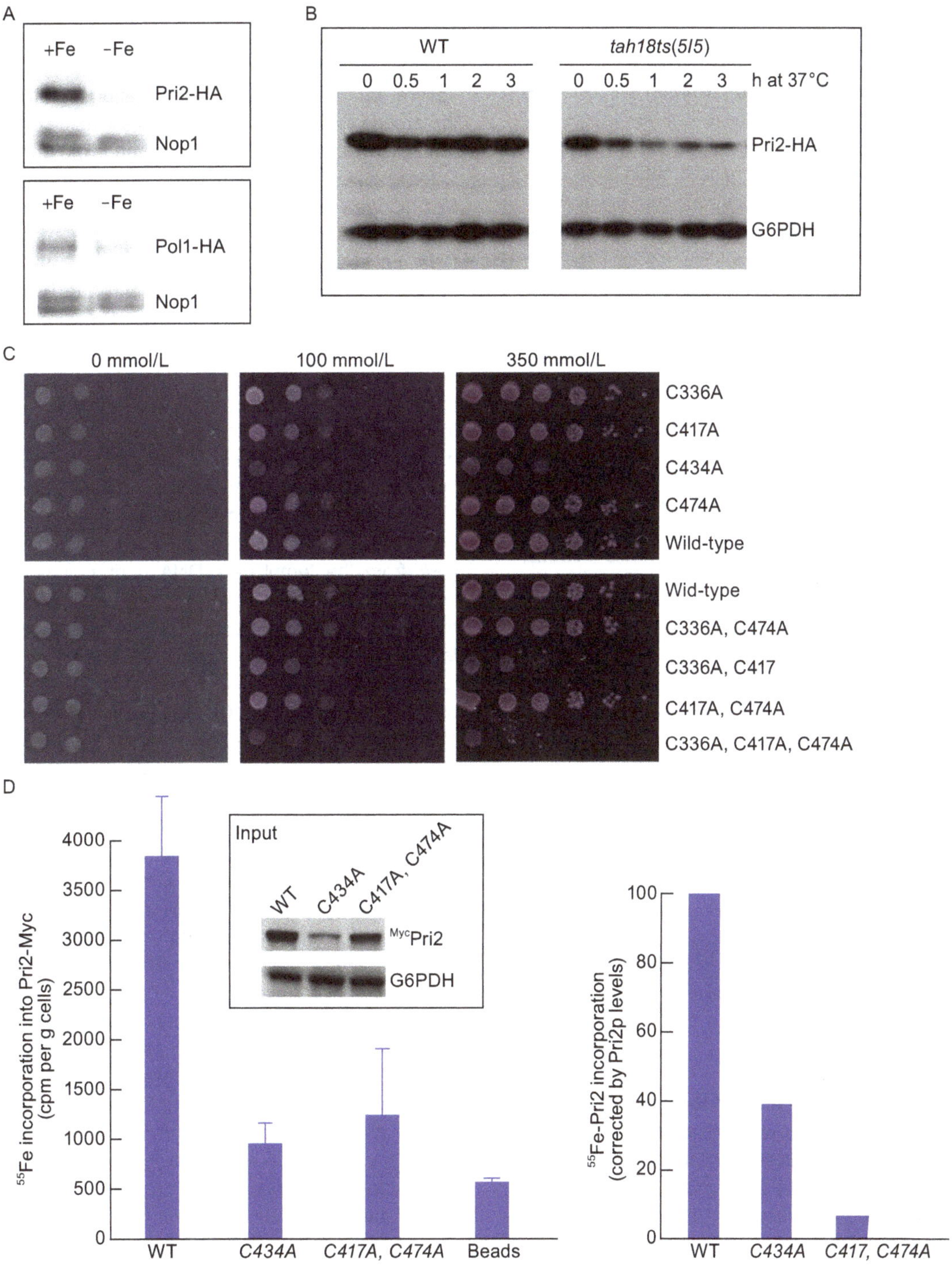

complex performs a unique function in the initial step of DNA synthesis. How the complex is assembled onto and maintained at the replication origins is not fully understood. In this report, we provide genetic and molecular evidence that the Pri2 Fe-S cluster domain is critical for stable association of the pol-prim complex with the origins during replication initiation. Mutations at the conserved cysteine residues that disrupt Fe-S cluster assembly in Pri2 destabilize the protein and reduce association of the entire pol-prim complex with the origins. We also show that Fe-S cluster assembly into newly synthesized Pri2 is sensitive to increased oxidative stress thus linking replication initiation to cellular redox state.

Our mutagenesis studies clearly show that the four conserved cysteine ligands of the Pri2 Fe-S cluster exhibit different roles in maintaining the Fe-S cluster and local protein conformation. Of the single cysteine substitutions, only *C434A* causes a severe growth defect and replication impairment at the restrictive temperature. Crystallographic analysis of the yeast Pri2 CTD reveals a larger N-terminal (residues 321–433) and a smaller C-terminal (residues 434–

◀ **Figure 6. Importance of Pri2 C-terminal conserved cysteine residues for *de novo* iron loading into the nascent Pri2 protein.** (A) Decreased Pri2 and Pol1 protein levels in cells grown under iron-deficient conditions. Wild-type cells containing integrated *PRI2-HA* or *POL2-HA* at the respective chromosomal loci were grown in regular Fe-supplemented SC media (+Fe) or in iron-depleted SC medium to log-phase before being harvested for protein extraction. Proteins were resolved by SDS-PAGE and probed with anti-HA for Pri2-HA and anti-Nop1 as a loading control. (B) Decreased Pri2 protein levels in the *tah18-5I5* mutant at the restrictive temperature. Both the wild-type and *tah18-5I5 ts* mutant contained C-terminal triple HA epitope integrated at the chromosomal *PRI2* locus. Cells were grown to log phase at 23°C before being shifted to 37°C at time zero. Protein extracts from cells collected at the indicated time points at 37°C were resolved by SDS-PAGE and probed with anti-HA for Pri2-HA and anti-G6PDH as a loading control. (C) The growth defects of *pri2* cysteine mutants are independent of Fe levels in media. Congenic wild-type and various *pri2* Cys-to-Ala mutants were dot-plated in 10-fold serial dilutions starting at 10^5 cells onto selective media containing 1 mmol/L ferrozine (maximal ferrous chelating capacity at 333 µmol/L) and supplemental ferrous ammonium sulfate at 0 µmol/L, 100 µmol/L (iron-poor) and 350 µmol/L (iron-rich). The plates were incubated at 30°C for 2 days before being photographed. (D) The Pri2 cysteine mutants are defective in iron incorporation into newly translated Pri2 polypeptide. Wild-type cells that harbor a high-copy number vector (pRS426-P_{TDH3}) expressing the Myc-tagged wild-type (pLL98) or *C434A* (pLL108) and *C417A*, *C474A* (pLL112) mutant Pri2 proteins were grown in synthetic iron-poor medium supplemented with 24 µmol/L BPS for 16 h at 30°C. Radiolabeling with ^{55}Fe was conducted for 4 h at 30°C in a BPS-free medium and Myc-Pri2 proteins were immunoprecipitated from cell extracts using an anti-Myc monoclonal antibody. The radioactivity associated with Pri2 was quantified by scintillation counting (left panel) and corrected for the differences in protein levels (right panel), which were determined by immunoblotting and quantitative densitometry (insert).

512) subdomains, which are bridged together with the Fe-S cluster situated at the seam with two cysteines from each subdomain acting as iron ligands (Sauguet et al., 2010). Cysteines 336, 417 and 474 are located within α-helices, which are unlikely to be disrupted by a Cys-to-Ala substitution. This structural organization is consistent with the normal growth phenotype of the three single mutants (Fig. 1B) and previously observed moderate loss of iron binding *in vitro* (Klinge et al., 2007). It is possible that a [3Fe-4S] cluster, likely resulted from a single cysteine substitution in Pri2, is sufficient to maintain local conformation and stability in the three single mutants. On the contrary, the unique position of Cys-434 at the linker region between the two subdomains suggests a more important role in maintaining the Fe-S cluster in place. In addition to destabilizing the Fe-S cluster, substitutions at Cys-434 may weaken shielding of the Fe-S cluster within a hydrophobic environment by exposing it to the aqueous environment, leading to Pri2 protein instability.

The Pri2 CTD has been implicated in association with several proteins involved in replication initiation. Although the archea Pri1 and Pri2 NTD form a stable complex *in vitro* (Lao-Sirieix et al., 2005), in the human Pri2 both the N- and C-terminal regions are involved in physical association with Pri1 (Copeland, 1997). Within the pol-prim complex, Pri2 binds Pol α providing a link between the primase and the DNA polymerase (Copeland and Wang, 1993; Longhese et al., 1993; Mizuno et al., 1999; Tan and Wang, 2000). A recent cryo-EM study suggests that Pri2's CTD is in direct contact with Pol α (Nunez-Ramirez et al., 2011). The atomic structure of the human Pri2 CTD (p58C) reveals a strikingly basic surface suggesting DNA or RNA binding activity (Vaithiyalingam et al., 2010). In fact, the human p58C does bind nucleic acids *in vitro*, with a preference for ssRNA/dsDNA junction substrates (Vaithiyalingam et al., 2010). The human p58C also interacts with the ssDNA binding protein RPA, which may help recruiting primase to the origin and releasing RPA from the template ssDNA during priming step. It is conceivable that these interactions are compromised by mutations that disrupt the Fe-S cluster assembly and proper folding of the Pri2 CTD.

How the pol-prim complex is brought onto replication origins remains to be elucidated. Previous studies show that Mcm10 associates with and stabilizes Pol α regardless of the stage of the cell cycle (Ricke and Bielinsky, 2004; Ricke and Bielinsky, 2006). Thus, individual subunits of the pol-prim complex may exist as monomers independent of the other subunits at some stages of the cell cycle. The loading of Pol α to replication origins requires Mcm10 and Ctf4 (Zhu et al., 2007) while Pri1-Pri2 loading may involve RPA (Tanaka and Nasmyth, 1998). It is unclear whether the Pol1-Pol12 and Pri1-Pri2 heterodimers are recruited to the origins separately or as one concerted step. Nevertheless, our finding that Pri2 cysteine mutants disrupting origin association of Pri1-Pri2 also severely destabilize origin binding of Pol α strongly suggest that the interaction between the pol and prim heterodimers, likely mediated by Pri2, is critical in maintaining/stabilizing the entire pol-prim complex at the origins.

We find that some *pri2* cysteine mutants exhibit normal S phase progression pace at the permissive temperature even though the Pri2 protein level is much lower than the wild-type control. It would thus appear that the level of pol-prim subunit is not usually rate-limiting during S phase entry. Consistent with the notion, these mutants become permanently arrested with a G_1 DNA content with further diminishing Pri2 protein levels when shifted to the restrictive temperature. Previous studies show that the Pol1 protein levels remain relatively unchanged throughout the cell cycle despite of periodic fluctuation in *POL1* mRNA levels (Muzi Falconi et al., 1993). The assembly of the pol-prim complex is not restricted to the S phase either (Ferrari et al., 1996). Nevertheless, lower pol-prim levels or weakened origin association that approaches below a minimal threshold may alter the temporal and spatial

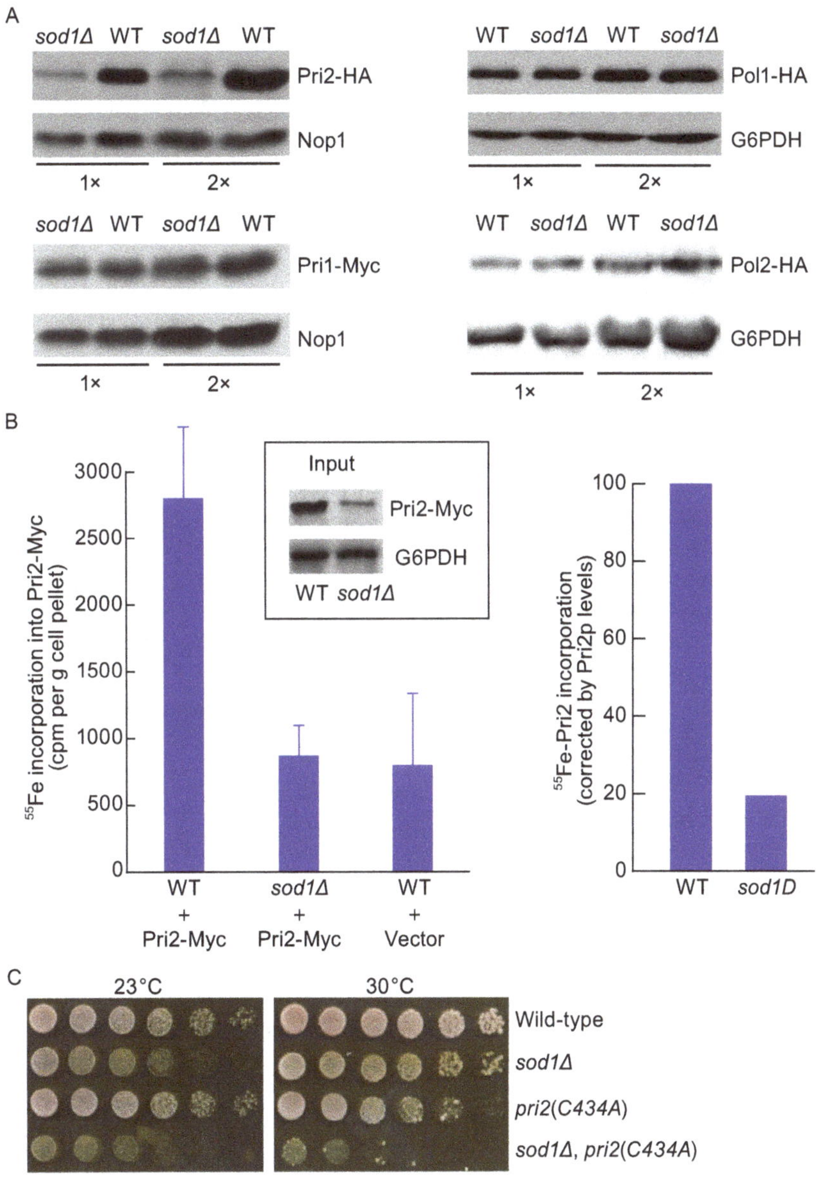

Figure 7. Impaired Pri2 protein stability and iron binding in *sod1Δ* mutant cells. (A) Comparison of protein levels of Pri2, Pri1 and Pol1 between log-phase congenic wild-type and *sod1*Δ mutant cells. Pri2-HA, Pri1-Myc, Pol1-HA and Pol2-HA were expressed from their respective chromosomal loci under the endogenous promoters. Protein extracts were subjected to immunoblotting using monoclonal anti-HA and anti-Myc antibodies. Nop1 and G6PDH were probed as loading controls. (B) Fe-S cluster assembly on Pri2 protein is compromised in *sod1*Δ mutant. Congenic wild-type and *sod1*Δ cells harboring pRS426-P_{TDH3}-Myc-Pri2 were grown and treated as described in the legend of Fig. 5B except for that radiolabeling with ^{55}Fe was conducted for 2 h. The amount of ^{55}Fe bound to immunoprecipitated Pri2 was quantified by scintillation counting (left panel) and corrected for the differences in Pri2 protein levels between wild-type and *sod1*Δ cells (right panel), as determined by immunoblotting and quantitative densitometry (insert). (C) Synthetic lethality between *sod1*Δ and *pri2(C434A)* mutant alleles. Congenic wild-type (LLY263), *pri2(C434A)* (LLY260), *sod1*Δ (LLY338) and *pri2(C434A)*, *sod1*Δ (LLY427) mutant cells were harvested from log phase cultures and dot-plated on YPD medium in 10-fold serial dilutions starting at 10^5 cells. The plates were incubated at 23°C and 30°C for 2 days before being photographed.

pattern of origin firing in S phase, as was shown for the key origin firing factors Cdc45 and Sld3 (Mantiero et al., 2011; Tanaka et al., 2011).

The recent discoveries of a Fe-S cluster binding domain in Pri2 and three major replicative DNA polymerases raise the possibility that DNA replication may be subjected to redox regulation. Crystal structure of the human Pri2 CTD shows that the Fe-S cluster is buried within the hydrophobic core, with very little exposure to the surface (Vaithiyalingam et al., 2010). Nevertheless, the Fe-S cluster may still be accessible to small molecules that may cause damage. Superoxide radical is known to be able to inactivate many [4Fe-4S] cluster-containing enzymes by oxidation of the cluster, leading to loss of the labile iron (Fridovich, 1995; Gardner and Fridovich, 1991; Longo et al., 1999). Our finding of decreased Fe-S assembly and protein stability of Pri2 in the *sod1Δ* mutant suggests that Pri2 may be susceptible to oxidative damage. An alternative explanation is that a diminished cytosolic pool of bioavailable iron in *sod1Δ* mutant (De Freitas et al., 2000; Srinivasan et al., 2000) may impair Fe-S assembly in Pri2. The protein levels of Pol1 and Pol2 (i.e. Pol α and Pol ε), both containing a Fe-S cluster, remain unchanged in *sod1Δ* mutant relative to the wild-type. Some apo proteins retain stability but lose activity when deprived of their Fe-S clusters (Kispal et al., 1999). Therefore, we cannot rule out the possibility of oxidative damage to the Fe-S clusters in Pol1 and Pol2 in *sod1Δ* mutant. The Fe-S cluster in Pri2 is likely more vulnerable to oxidative damage relative to Pol1. We propose that Pri2 is a key component of the replisome that acts to link regulation of replication initiation to changes in cellular redox state.

MATERIALS AND METHODS

Yeast strains, cell growth and plasmids

Yeast strains and plasmids used in this study were listed in Tables 1 and 2, respectively. All yeast strains were derived from a W303 parental strain Y300 except for LLY277 (*8C2, PRI2-HA*) and LLY278 (*tah18-5I5, PRI2-HA*), which were generated by integrating *PRI2-3HA* into the chromosomal *PRI2* locus in strains 8C2 and 5I5 (Vernis et al., 2009). Epitope-tagging, promoter replacement and gene deletion were constructed by PCR-based homologous recombination method as described earlier (Longtine et al., 1998). Substitutions of Pri2's Fe-S domain conserved cysteine residues were introduced by using the QuikChange site-directed mutagenesis kit (Stratagene) according to manufacturer's instruction. DNA sequencing was used to verify the desired sequences of all mutant *PRI2* alleles.

Rich medium (YPD) contained 1% yeast extract, 2% peptone and 2% glucose. Synthetic complete (SC) medium contained 0.17% yeast nitrogen base without amino acids and $(NH_4)_2SO_4$ (MP Biomedicals), 0.5% $(NH_4)_2SO_4$, all twenty amino acids (Sigma) at concentrations as described (Burke et al., 2000) and 2% of glucose. Selective (i.e., drop-out) media were SC omitting one or multiple amino acid as indicated. For solid media, 2% Bacto Agar was added before autoclaving. Iron-buffered medium consisted of modified synthetic complete medium to which MES (50 mmol/L 2-(N-morpholino)ethanesulfonic acid, pH 6.1) buffer and iron chelator (1 mmol/L ferrozine) were added. Different amounts of ferrous ammonium sulfate (0, 100, 350 µmol/L) were added back to establish windows of available ferrous iron as described earlier (Stearman et al., 1996).

Protein extraction and immunoblotting

Yeast cells were harvested from early- to mid-log phase cultures (OD_{600} 0.5~1.0). Protein extracts were prepared by using the trichloroacetic acid method as described earlier (Wu et al., 2011). Proteins were resolved by 8%–12% SDS-PAGE, transferred to nitrocellulose membranes, and probed with primary and secondary antibodies. Blots were developed with an enhanced chemiluminescence substrate (Perkin-Elmer). Primary antibodies were used at the following dilutions: monoclonal anti-HA 12CA5 (Roche) at 1:10,000, monoclonal anti-Myc 9E10 (Roche) at 1:2,000, monoclonal anti-Nop1 (EnCor Biotechnology) at 1:2,500 and anti-G6PDH (Sigma) at 1:200,000.

Cell cycle synchronization and flow cytometry

For α factor block and release experiments, yeast cells (all *bar1Δ* mutants) were harvested from a log-phase culture, re-suspended in fresh YPD containing 300 nmol/L of α factor at a final OD_{600} of 0.3, and incubated at for 3–4 h at the indicated temperatures until >95% of the cells were arrested in G_1. Cells were released from G_1 arrest by washing out the α factor with fresh YPD containing 0.1 mg/mL pronase E (Sigma) at the indicated temperatures and harvested at different time points. DNA content was measured as described earlier (Wu et al., 2011) with a Beckman Coulter Epics XL-MCL flow cytometer and DeltaGraph software.

Chromatin immunoprecipitation (ChIP) and Real-Time PCR analyses

ChIP was performed as described earlier (Aparicio et al., 2004). Fifty mL of cells (2.0×10^7 cells/mL) were treated with formaldehyde to cross-link proteins to DNA. Cells were lysed by glass beads disruption in 500 µL FA-lysis buffer (50 mmol/L HEPES-KOH at pH 7.5, 140 mmol/L NaC1, 1 mmol/L EDTA, 1% Triton X-100, 0.1% sodium deoxycholate) supplemented with 1 mmol/L PMSF and protease inhibitor cocktail (Roche). The suspension was sonicated with a Bioruptor (Diagenode) for 40 min (30 sec on/off) each, resulting in an average fragment size of 0.3~0.5 kb). DNA bound to HA and Myc-tagged Pri2, Pri1 and Pol1 was immunoprecipitated with anti-HA (12CA5) and anti-Myc (9E10), respectively, at 1:100 *v/v* ratio from 2 mg of lysates at 4°C overnight. One-tenth of the lysate was set aside as input. Precipitates were successively washed for 15 min each with FA-lysis buffer, FA-lysis buffer with 500 mmol/L NaCl, LiCl buffer (10 mmol/L Tris-HCl at pH 8.0, 0.25 mol/L LiC1, 0.5% NP-40, 0.5% sodium deoxycholate, 1 mmol/L EDTA) and TE (20 mmol/L Tris-HC1 at pH 8.0, 1 mmol/L EDTA) and eluted by incubation in TE with 1% SDS at 65°C for 15 min. The input and immunoprecipitated samples were then processed for DNA purification as described (Aparicio et al., 2004). Finally, the DNA was purified using a QiaQuick PCR purification kit (Qiagen).

The DNA was diluted (1:3 for immunoprecipitated DNA and 1:30 for input DNA) and used in quantitative real-time PCR using the Bio-Rad CFX96 detection system based on SYBR Green fluorescence according to the manufacturer's protocol. PCR reactions were performed in triplicate on 96-well plates, and an average cycle threshold values was calculated for each reaction. The amount of target DNA in the ChIP sample relative to the Input sample was calculated. Background was determined by calculating the amount of target DNA in the mock-IP sample relative to the Input sample. The primer pair for amplification of the region surrounding ARS305 was 5′-TTTGGAGCTCAAGTGGATTG-3′ and 5′-CACACCGGACAG

Table 1. Yeast strains used in this study

Strain	Genotype	Parental strain
Y300	*MATa, can1-100, ade2-1, his3-11,15, leu2-3,112, trp1-1, ura3-1*	
LLY263	*MATa, pri2::PRI2-3HA-TRP1, bar::LEU2*	Y300
LLY271	*MATa, pri2::PRI2-C417A-3HA-TRP1, ba1r::LEU2*	Y300
LLY260	*MATa, pri2::PRI2-C434A-3HA-TRP1, bar1::LEU2*	Y300
LLY378	*MATa, pri2::PRI2-C336A-3HA-TRP1, bar1::LEU2*	Y300
LLY366	*MATa, pri2::PRI2-C474A-3HA-TRP1, bar1::LEU2*	Y300
LLY368	*MATa, pri2::PRI2-C336A,C417A,C474A-3HA-TRP1, ba1r::LEU2*	Y300
LLY360	*MATa, pri2::PRI2-C336A,C474A-3HA-TRP1, bar1::LEU2*	Y300
LLY288	*MATa, pri2::PRI2-C417A,C474A-3HA-TRP1, bar1::LEU2*	Y300
LLY292	*MATa, pri2::PRI2- C336A,C417A-3HA-TRP1, bar1::LEU2*	Y300
LLY343	*MATa, pri2::PRI2-3HA-TRP1, pri1:: PRI1-13Myc-HIS5, bar1::LEU2*	Y300
LLY372	*MATa, pri2::PRI2-C336A,C417A,C474A-3HA-TRP1, pri1::PRI1-13Myc-HIS5, bar1::LEU2*	Y300
LLY382	*MATa, pri2::PRI2-3HA-TRP1, pol1::POL1-13Myc-HIS5, bar1::LEU2*	Y300
LLY387	*MATa, pri2::PRI2-C336A,C417A,C474A-3HA-TRP1, pol1::POL1-13Myc-HIS5, bar1::LEU2*	Y300
LLY328	*MATa, pri2::PRI2-3HA-TRP1, pri1::PRI1-13Myc-HIS5, bar1::LEU2, sod1::KAN*	Y300
LLY330	*MATa, pri2::PRI2-3HA-TRP1, pri1::PRI1-13Myc-HIS5, bar1::LEU2*	Y300
LLY398	*MATa, pol1::POL1-3HA-TRP1, bar1::LEU2*	Y300
LLY399	*MATa, pol1::POL1-3HA-TRP1, bar1::LEU2, sod1::KAN*	Y300
LLY333	*MATa, sod1::KAN*	Y300
LLY338	*MATa, bar1::LEU2, sod1::KAN*	Y300
LLY426	*MATa, bar1::LEU2*	Y300
LLY427	*MATa, bar1::LEU2, sod1::KAN, pri2::PRI2-C434A-3HA-TRP1*	Y300
LLY428	*MATa, bar1::LEU2, sod1::KAN, pri2::PRI2-C336A,C417A,C474A-3HA-TRP1*	Y300
LLY277	*MATα, ura,3 leu2, trp,1 lys2,* $cyh2^R$*, PRI2-HA*	8C2 (Vernis et al., 2009)
LLY278	*MATα, ura3 leu2, trp1, lys,2* $cyh2^R$*, PRI2-HA, tah18-5I5*	5I5 (Vernis et al., 2009)

Table 2. Plasmids used in this study

Plasmid	Description	Reference
pLL112	pRS426-P_{TDH3}-3Myc-PRI2-C417,474A	This study
pLL108	pRS426-P_{TDH3}-3Myc-PRI2-C434A	This study
pLL98	pRS426-P_{TDH3}-3Myc-PRI2	This study
PRI2 (WT)	pRS404-Pri2-CTD (residues 201–528)-3HA	Ricke and Bielinsky, 2006
PRI2 (C336A)	pRS404-Pri2-CTD-C336A-3HA	This study
PRI2 (C417A)	pRS404-Pri2-CTD-C417A-3HA	This study
PRI2 (C434A)	pRS404-Pri2-CTD-C434A-3HA	This study
PRI2 (C474A)	pRS404-Pri2-CTD-C474A-3HA	This study
PRI2 (C336A/C474A)	pRS404-Pri2-CTD-C336A,C474A-3HA	This study
PRI2 (C336A/C417A)	pRS404-Pri2-CTD-C336A,C417A-3HA	This study
PRI2 (C474A/C417A)	pRS404-Pri2-CTD-C474A,C417A-3HA	This study
PRI2 (C336A/C417A/C474)	pRS404-Pri2-CTD-C336A,C417A,C474-3HA	This study

TACATGAAA-3′. The primer pair for amplification of the region surrounding ARS607 was 5′-CATTTACGCACTCTAACTGGC-3′ and 5′-AAACCAATAGCAGGATCGACC-3′.

^{55}Fe labeling and immunoprecipitation of Pri2

Radiolabeling of yeast cells with $^{55}FeCl_3$ (Perkin-Elmer) and measurement of ^{55}Fe incorporation into proteins by immunoprecipitation were performed as described earlier (Zhang et al., 2011). The standard iron-depleted medium was supplemented with 24 μmol/L bathophenanthrolinedisulfonate (BPS) to chelate any extra ferrous iron in order to make iron-free medium. Wild-type cells harboring a 2-micron plasmid (pRS426-P_{TDH3}) that overexpresses N-terminally Myc-tagged Pri2 (pLL98), Pri2-C434A (pLL108), or Pri2-C417A, C474A (pLL112) mutant proteins were grown in the iron-free medium to log-phase and harvested. For each sample, 1–2 × 10^9 cells (wet weight 0.1–0.2 g) were labeled with 20 μCi of $^{55}FeCl_3$ in for 4 h at 30°C. Wild-type and mutant Myc-Pri2 proteins were immunoprecipitated from 2 mg of whole cell lysates using monoclonal anti-Myc (9E10) at 1:100 *v*/*v* ratio. Pri2-bound ^{55}Fe in the immunoprecipitates were measured by scintillation counting.

ACKNOWLEDGEMENTS

This research was supported by National Institutes of Health grant R01CA125574 to M.H. We thank Dr. Anja-Katrin Belinsky of University of Minnesota for providing us with the Pri2-CTD-3HA plasmid and Dr. Laurence Vernis of Université Paris-Sud for the *tah18-5l5* strain.

ABBREVIATIONS

ChIP, chromatin immunoprecipitation; CIA, cytosolic iron-sulfur cluster protein assembly; Fe-S, iron-sulfur; pol-prim, DNA polymerase α-primase.

COMPLIANCE WITH ETHICS GUIDELINES

The authors declare that they have no conflict of interest. This article does not contain any studies with human or animal subjects performed by the any of the authors.

REFERENCES

Aparicio OM, Stout AM, Bell SP (1999) Differential assembly of Cdc45p and DNA polymerases at early and late origins of DNA replication. Proc Natl Acad Sci USA 96:9130–9135

Aparicio O, Geisberg JV, Struhl K (2004) Chromatin immunoprecipitation for determining the association of proteins with specific genomic sequences in vivo. Current protocols in cell biology, Chapter 17: Unit 17 7

Arezi B, Kuchta RD (2000) Eukaryotic DNA primase. Trends Biochem Sci 25:572–576

Arezi B, Kirk BW, Copeland WC, Kuchta RD (1999) Interactions of DNA with human DNA primase monitored with photoactivatable cross-linking agents: implications for the role of the p58 subunit. Biochemistry 38:12899–12907

Baranovskiy AG, Lada AG, Siebler H, Zhang Y, Pavlov YI et al (2012) DNA polymerases delta and zeta switching by sharing the accessory subunits of DNA polymerase delta. J Biol Chem 287 (21):17281–17287

Brazzolotto X, Gaillard J, Pantopoulos K, Hentze MW, Moulis JM (1999) Human cytoplasmic aconitase (Iron regulatory protein 1) is converted into its [3Fe-4S] form by hydrogen peroxide in vitro but is not activated for iron-responsive element binding. J Biol Chem 274:21625–21630

Budd M, Campbell JL (1987) Temperature-sensitive mutations in the yeast DNA polymerase I gene. Proc Natl Acad Sci USA 84:2838–2842

Burgers PM (1998) Eukaryotic DNA polymerases in DNA replication and DNA repair. Chromosoma 107:218–227

Burke D, Dawson D, Stearns T (2000) Methods in yeast genetics: a cold spring harbor laboratory course manual. Cold Spring Harbor Laboratory Press, Cold Spring Harbor

Carter CD, Kitchen LE, Au WC, Babic CM, Basrai MA (2005) Loss of SOD1 and LYS7 sensitizes *Saccharomyces cerevisiae* to hydroxyurea and DNA damage agents and downregulates MEC1 pathway effectors. Mol Cell Biol 25:10273–10285

Casper AM, Mieczkowski PA, Gawel M, Petes TD (2008) Low levels of DNA polymerase alpha induce mitotic and meiotic instability in the ribosomal DNA gene cluster of *Saccharomyces cerevisiae*. PLoS Genet 4:e1000105

Copeland WC (1997) Expression, purification, and characterization of the two human primase subunits and truncated complexes from *Escherichia coli*. Protein Expr Purif 9:1–9

Copeland WC, Wang TS (1993) Enzymatic characterization of the individual mammalian primase subunits reveals a biphasic mechanism for initiation of DNA replication. J Biol Chem 268:26179–26189

De Freitas JM, Liba A, Meneghini R, Valentine JS, Gralla EB (2000) Yeast lacking Cu-Zn superoxide dismutase show altered iron homeostasis. Role of oxidative stress in iron metabolism. J Biol Chem 275:11645–11649

Djaman O, Outten FW, Imlay JA (2004) Repair of oxidized iron-sulfur clusters in Escherichia coli. J Biol Chem 279:44590–44599

Ferrari M, Lucchini G, Plevani P, Foiani M (1996) Phosphorylation of the DNA polymerase alpha-primase B subunit is dependent on its association with the p180 polypeptide. J Biol Chem 271:8661–8666

Foiani M, Santocanale C, Plevani P, Lucchini G (1989) A single essential gene, PRI2, encodes the large subunit of DNA primase in *Saccharomyces cerevisiae*. Mol Cell Biol 9:3081–3087

Foiani M, Marini F, Gamba D, Lucchini G, Plevani P (1994) The B subunit of the DNA polymerase alpha-primase complex in *Saccharomyces cerevisiae* executes an essential function at the initial stage of DNA replication. Mol Cell Biol 14:923–933

Foiani M, Lucchini G, Plevani P (1997) The DNA polymerase alpha-primase complex couples DNA replication, cell-cycle progression and DNA-damage response. Trends Biochem Sci 22:424–427

Francesconi S, Longhese MP, Piseri A, Santocanale C, Lucchini G et al (1991) Mutations in conserved yeast DNA primase domains impair DNA replication in vivo. Proc Natl Acad Sci USA 88:3877–3881

Frick DN, Richardson CC (2001) DNA primases. Annu Rev Biochem 70:39–80

Fridovich I (1995) Superoxide radical and superoxide dismutases. Annu Rev Biochem 64:97–112

Gardner PR, Fridovich I (1991) Superoxide sensitivity of the *Escherichia coli* aconitase. J Biol Chem 266:19328–19333

Hubscher U, Maga G, Spadari S (2002) Eukaryotic DNA polymerases. Annu Rev Biochem 71:133–163

Kaplan J, McVey Ward D, Crisp RJ, Philpott CC (2006) Iron-dependent metabolic remodeling in *S. cerevisiae*. Biochim Biophys Acta 1763:646–651

Kawasaki Y, Sugino A (2001) Yeast replicative DNA polymerases and their role at the replication fork. Mol Cells 12:277–285

Kispal G, Csere P, Prohl C, Lill R (1999) The mitochondrial proteins Atm1p and Nfs1p are essential for biogenesis of cytosolic Fe-S proteins. EMBO J 18:3981–3989

Klinge S, Hirst J, Maman JD, Krude T, Pellegrini L (2007) An iron-sulfur domain of the eukaryotic primase is essential for RNA primer synthesis. Nat Struct Mol Biol 14:875–877

Lao-Sirieix SH, Nookala RK, Roversi P, Bell SD, Pellegrini L (2005) Structure of the heterodimeric core primase. Nat Struct Mol Biol 12:1137–1144

Lemoine FJ, Degtyareva NP, Kokoska RJ, Petes TD (2008) Reduced levels of DNA polymerase delta induce chromosome fragile site instability in yeast. Mol Cell Biol 28:5359–5368

Liu H, Rudolf J, Johnson KA, McMahon SA, Oke M et al (2008) Structure of the DNA repair helicase XPD. Cell 133:801–812

Longhese MP, Jovine L, Plevani P, Lucchini G (1993) Conditional mutations in the yeast DNA primase genes affect different aspects of DNA metabolism and interactions in the DNA polymerase alpha-primase complex. Genetics 133:183–191

Longo VD, Liou LL, Valentine JS, Gralla EB (1999) Mitochondrial superoxide decreases yeast survival in stationary phase. Arch Biochem Biophys 365:131–142

Longtine MS, McKenzie A 3rd, Demarini DJ, Shah NG, Wach A et al (1998) Additional modules for versatile and economical PCR-based gene deletion and modification in *Saccharomyces cerevisiae*. Yeast 14:953–961

Lu C, Cortopassi G (2007) Frataxin knockdown causes loss of cytoplasmic iron-sulfur cluster functions, redox alterations and induction of heme transcripts. Arch Biochem Biophys 457:111–122

Lucchini G, Francesconi S, Foiani M, Badaracco G, Plevani P (1987) Yeast DNA polymerase–DNA primase complex; cloning of PRI 1, a single essential gene related to DNA primase activity. EMBO J 6:737–742

Mantiero D, Mackenzie A, Donaldson A, Zegerman P (2011) Limiting replication initiation factors execute the temporal programme of origin firing in budding yeast. EMBO J 30:4805–4814

Messick TE, Chmiel NH, Golinelli MP, Langer MR, Joshua-Tor L et al (2002) Noncysteinyl coordination to the [4Fe-4S]2+ cluster of the DNA repair adenine glycosylase MutY introduced via site-directed mutagenesis. Structural characterization of an unusual histidinyl-coordinated cluster. Biochemistry 41:3931–3942

Missirlis F, Hu J, Kirby K, Hilliker AJ, Rouault TA et al (2003) Compartment-specific protection of iron-sulfur proteins by superoxide dismutase. J Biol Chem 278:47365–47369

Mizuno T, Yamagishi K, Miyazawa H, Hanaoka F (1999) Molecular architecture of the mouse DNA polymerase alpha-primase complex. Mol Cell Biol 19:7886–7896

Muzi Falconi M, Piseri A, Ferrari M, Lucchini G, Plevani P et al (1993) De novo synthesis of budding yeast DNA polymerase alpha and POL1 transcription at the G1/S boundary are not required for entrance into S phase. Proc Natl Acad Sci USA 90:10519–10523

Muzi-Falconi M, Giannattasio M, Foiani M, Plevani P (2003) The DNA polymerase alpha-primase complex: multiple functions and interactions. Sci World J 3:21–33

Netz DJ, Stumpfig M, Dore C, Muhlenhoff U, Pierik AJ et al (2010) Tah18 transfers electrons to Dre2 in cytosolic iron-sulfur protein biogenesis. Nat Chem Biol 6:758–765

Netz DJ, Stith CM, Stumpfig M, Kopf G, Vogel D et al (2012) Eukaryotic DNA polymerases require an iron-sulfur cluster for the formation of active complexes. Nat Chem Biol 8:125–132

Nunez-Ramirez R, Klinge S, Sauguet L, Melero R, Recuero-Checa MA et al (2011) Flexible tethering of primase and DNA Pol alpha in the eukaryotic primosome. Nucleic Acids Res 39:8187–8199

Ricke RM, Bielinsky AK (2004) Mcm10 regulates the stability and chromatin association of DNA polymerase-alpha. Mol Cell 16:173–185

Ricke RM, Bielinsky AK (2006) A conserved Hsp10-like domain in Mcm10 is required to stabilize the catalytic subunit of DNA polymerase-alpha in budding yeast. J Biol Chem 281:18414–18425

Rudolf J, Makrantoni V, Ingledew WJ, Stark MJ, White MF (2006) The DNA repair helicases XPD and FancJ have essential iron-sulfur domains. Mol Cell 23:801–808

Sauguet L, Klinge S, Perera RL, Maman JD, Pellegrini L (2010) Shared active site architecture between the large subunit of eukaryotic primase and DNA photolyase. PLoS One 5:e10083

Srinivasan C, Liba A, Imlay JA, Valentine JS, Gralla EB (2000) Yeast lacking superoxide dismutase(s) show elevated levels of "free iron" as measured by whole cell electron paramagnetic resonance. J Biol Chem 275:29187–29192

Stearman R, Yuan DS, Yamaguchi-Iwai Y, Klausner RD, Dancis A (1996) A permease-oxidase complex involved in high-affinity iron uptake in yeast. Science 271:1552–1557

Tan S, Wang TS (2000) Analysis of fission yeast primase defines the checkpoint responses to aberrant S phase initiation. Mol Cell Biol 20:7853–7866

Tanaka T, Nasmyth K (1998) Association of RPA with chromosomal replication origins requires an Mcm protein, and is regulated by Rad53, and cyclin- and Dbf4-dependent kinases. EMBO J 17:5182–5191

Tanaka S, Nakato R, Katou Y, Shirahige K, Araki H (2011) Origin association of Sld3, Sld7, and Cdc45 proteins is a key step for determination of origin-firing timing. Curr Biol 21:2055–2063

Urzica E, Pierik AJ, Muhlenhoff U, Lill R (2009) Crucial role of conserved cysteine residues in the assembly of two iron-sulfur clusters on the CIA protein Nar1. Biochemistry 48:4946–4958

Vaithiyalingam S, Warren EM, Eichman BF, Chazin WJ (2010) Insights into eukaryotic DNA priming from the structure and functional interactions of the 4Fe-4S cluster domain of human DNA primase. Proc Natl Acad Sci USA 107:13684–13689

Vernis L, Facca C, Delagoutte E, Soler N, Chanet R et al (2009) A newly identified essential complex, Dre2-Tah18, controls mitochondria integrity and cell death after oxidative stress in yeast. PLoS One 4:e4376

Wallace MA, Liou LL, Martins J, Clement MH, Bailey S et al (2004) Superoxide inhibits 4Fe-4S cluster enzymes involved in amino acid biosynthesis. Cross-compartment protection by CuZn-superoxide dismutase. J Biol Chem 279: 32055–32062

Wang TSF (1996) Cellular DNA polymerases. In: DePamphilis ML (ed) DNA replication in eukaryotic cells. Cold Spring Harbor Laboratory Press, Cold Spring Harbor, NY, pp 461–493

Wu X, Liu L, Huang M (2011) Analysis of changes in protein level and subcellular localization during cell cycle progression using the budding yeast *Saccharomyces cerevisiae*. Methods Mol Biol 782:47–57

Yeeles JT, Cammack R, Dillingham MS (2009) An iron-sulfur cluster is essential for the binding of broken DNA by AddAB-type helicase-nucleases. J Biol Chem 284:7746–7755

Zerbe LK, Kuchta RD (2002) The p58 subunit of human DNA primase is important for primer initiation, elongation, and counting. Biochemistry 41:4891–4900

Zhang Y, Liu L, Wu X, An X, Stubbe J et al (2011) Investigation of in vivo diferric tyrosyl radical formation in *Saccharomyces cerevisiae* Rnr2 protein: requirement of Rnr4 and contribution of Grx3/4 AND Dre2 proteins. J Biol Chem 286:41499–41509

Zhu W, Ukomadu C, Jha S, Senga T, Dhar SK et al (2007) Mcm10 and And-1/CTF4 recruit DNA polymerase alpha to chromatin for initiation of DNA replication. Genes Dev 21:2288–2299

3

Structural dynamics of the yeast Shwachman-Diamond syndrome protein (Sdo1) on the ribosome and its implication in the 60S subunit maturation

Chengying Ma[1], Kaige Yan[1], Dan Tan[2,3], Ningning Li[1], Yixiao Zhang[1], Yi Yuan[1], Zhifei Li[1], Meng-Qiu Dong[2,3], Jianlin Lei[1], Ning Gao[1✉]

[1] School of Life Sciences, Tsinghua University, Beijing 100084, China
[2] National Institute of Biological Sciences, Beijing 102206, China
[3] Graduate Program in Chinese Academy of Medical Sciences and Peking Union Medical College, Beijing 100730, China
✉ Correspondence: ninggao@tsinghua.edu.cn (N. Gao)

ABSTRACT

The human Shwachman-Diamond syndrome (SDS) is an autosomal recessive disease caused by mutations in a highly conserved ribosome assembly factor SBDS. The functional role of SBDS is to cooperate with another assembly factor, elongation factor 1-like (Efl1), to promote the release of eukaryotic initiation factor 6 (eIF6) from the late-stage cytoplasmic 60S precursors. In the present work, we characterized, both biochemically and structurally, the interaction between the 60S subunit and SBDS protein (Sdo1p) from yeast. Our data show that Sdo1p interacts tightly with the mature 60S subunit in vitro through its domain I and II, and is capable of bridging two 60S subunits to form a stable 2:2 dimer. Structural analysis indicates that Sdo1p bind to the ribosomal P-site, in the proximity of uL16 and uL5, and with direct contact to H69 and H38. The dynamic nature of Sdo1p on the 60S subunit, together with its strategic binding position, suggests a surveillance role of Sdo1p in monitoring the conformational maturation of the ribosomal P-site. Altogether, our data support a conformational signal-relay cascade during late-stage 60S maturation, involving uL16, Sdo1p, and Efl1p, which interrogates the functional P-site to control the departure of the anti-association factor eIF6.

KEYWORDS ribosome biogenesis, SBDS, SDS, Sdo1, cryo-electron microscopy (cryo-EM)

INTRODUCTION

Protein biosynthesis is catalyzed by the ribosome in all living organisms. Thousands of ribosomes must be synthesized per minute to maintain and update the tremendous protein inventory in rapidly growing cells. Ribosomal subunit assembly *in vivo* is a highly complex process, which composes of a large number of intertwined ribosomal protein (RP) binding and rRNA maturation (rRNA folding, processing, modification and assembly) events. In eukaryotes, additional complexity is added, as the assembly starts in the nucleolus and involves the import of RPs and associated factors, as well as the export of premature ribosomal particles (66S and 43S pre-ribosomes) across nuclear membrane. Final maturation of the pre-ribosomes takes place in cytoplasm, and is coupled to the regulation of translation initiation (Karbstein, 2013; Lebaron et al., 2012; Miluzio et al., 2009; Soudet et al., 2010; Strunk et al., 2012). In *Saccharomyces cerevisiae*, the complete maturation of translationally active 60S and 40S subunits from extremely long rRNA transcripts and 79 RPs requires more than 70 small nucleolar RNAs (snoRNAs) and 200 assembly factors (AFs) (Panse and Johnson, 2010; Woolford and Baserga, 2013). Many of these factors function primarily on the quality control of subunit production, acting at various maturation checkpoints during the assembly process. These checkpoints often involve regulated binding and release of a series of factors (Karbstein, 2013; Lo et al., 2010; Matsuo et al., 2014).

Cellular defects in ribosome biogenesis caused by assembly factor insufficiency or mutations result in arrests of assembly at different checkpoints during cell cycle progression (Bernstein et al., 2007; Dez and Tollervey, 2004; Jorgensen et al., 2002). More importantly, disorders in ribosome biogenesis, which induce a nucleolar stress (Boulon et al., 2010) that is monitored by the Mdm2/Hdm2-p53 pathway (Chakraborty et al., 2011; Deisenroth and Zhang, 2010), were shown to be associated with increased cancer susceptibility in animal cells (Montanaro et al., 2008; Ruggero and Pandolfi, 2003). In human, a diverse collection of genetic diseases, named as ribosomopathies, have been linked to mutations in ribosomal proteins or AFs (Chakraborty et al., 2011; Freed et al., 2010; Narla and Ebert, 2010; Teng et al., 2013). Besides their specific clinical phenotypes, patients with ribosomopathies have a predisposition to a variety of cancers. Among these diseases, Shwachman-Diamond syndrome (SDS) is an autosomal recessive disease with multi-system disorders caused by mutations in the highly conserved Shwachman-Bodian-Diamond Syndrome gene (SBDS) (Boocock et al., 2003). Clinical characteristics associated with SDS are pancreatic insufficiency, skeletal abnormalities and bone marrow failure with neutropenia, ineffective hematopoiesis, and increased risk of leukemia (Narla and Ebert, 2010). Most of SDS patients (~90%) are associated with mutations of *SBDS* gene that result in premature truncation of SBDS protein (Austin et al., 2005; Boocock et al., 2003).

SBDS is a highly conserved protein in archaea and eukaryotes (Boocock et al., 2006; Shammas et al., 2005). Converging cell biology data on several SBDS homologues, including yeast (Sdo1p) (Lo et al., 2010; Luz et al., 2009; Menne et al., 2007; Moore et al., 2010; Savchenko et al., 2005), mouse (Finch et al., 2011), *Dictyostelium discoideum* (Wong et al., 2011) and human SDS patient cells (Burwick et al., 2012; Ganapathi et al., 2007; Wong et al., 2011) have implicated a functional role of SBDS in the maturation of the 60S ribosomal subunit. Specifically, SBDS was proposed to coordinate with elongation factor-like 1 (Efl1p) to release eIF6 (Tif6p in yeast), an important 60S shuttling factor, from late cytoplasmic pre-60S particles. Failure in the timely release and recycling of Tif6p impairs the subunit joining and subsequent translation initiation (Karbstein, 2013; Miluzio et al., 2009). The structures of SBDS from several species have been resolved (de Oliveira et al., 2010; Finch et al., 2011; Ng et al., 2009; Shammas et al., 2005), which contain three structural domains, I to III (numbered from the N-terminus). The N-terminal domain of SBDS was shown to be involved in RNA binding (de Oliveira et al., 2010) and domains II-III were found to interact with an insertion domain of Efl1p (Asano et al., 2014). Also, recent data revealed a functional link between Sdo1p and uL16 (RPL10), which is a late-binding protein during 60S assembly (Gamalinda et al., 2014). It was shown that the loop of uL16 residing in the ribosomal P-site is important for the activation of Efl1p to induce the release of Tif6p (Bussiere et al., 2012). Furthermore, uL16 was shown to be involved in the recruitment of Sdo1p (Sulima et al., 2014a), and a role of Sdo1p/SBDS as a nucleotide exchange factor to stabilize the binding of GTP to Efl1p was proposed (Gijsbers et al., 2013).

Despite the functional framework that defines a pathway for SBDS protein family as described above, its biochemical property that contributes to its involvement in the cytoplasmic recycling of Tif6p and ribosomal P-site maturation remains unclear. In this report, using cryo-electron microscopy (cryo-EM) and several complementary approaches, we performed structural and biochemical characterization of the interaction between the yeast SBDS homologue, Sdo1p and the 60S subunit. Our data reveal that Sdo1p binds to the ribosomal P-site and directly contacts H69 and H38 of the 25S rRNA. Moreover, owing to its structural flexibility, Sdo1p displays a dynamic behavior on the 60S subunit, with wobbling terminal domains (II and III). Very interestingly, Sdo1p is able to induce dimerization of the 60S subunits in a very specific manner. Together with published data, our results suggest that Sdo1p is an essential surveillance factor for the 60S maturation, which monitors the conformational status of the ribosomal P-site, including surrounding uL16, H69 and H38, and couples, through Efl1p, the maturation of the ribosomal P-site to the release of Tif6p.

RESULTS

Sdo1p binds to the mature 60S subunit *in vitro*

Previous data showed that Sdo1p co-fractionated with the pre-60S fractions (Menne et al., 2007) and bound to all forms of rRNA *in vitro* (Luz et al., 2009). It is not clear whether Sdo1p could bind to the mature form of the 60S subunit. This prompted us to examine the interaction between the mature 60S subunit and Sdo1p by gel filtration analysis. As shown in Fig. 1A and 1B, Sdo1p co-elutes with the 60S subunit, and forms a stable complex with the mature 60S subunit.

Sdo1p induces dimerization of 60S subunits *in vitro*

Next, we applied cryo-EM to analyze the 60S complex bound with Sdo1p, aiming at determining the Sdo1p binding site in a minimal system. The resulting images of the 60S-Sdo1p complex show a relatively even distribution, without undesired aggregation or precipitation (Fig. 1C). However, unexpectedly, defined oligomers of 60S subunits were also detected. In addition to well separated mono-disperse 60S monomers, many 60S subunits were found to exist in a dimeric or trimeric state (Fig. 1C). The gel filtration analysis does not have enough resolution at its high molecular weight boundary, therefore, we employed sucrose density gradient centrifugation (SDGC) to confirm this observation. However, in contrast to the cryo-EM results, there were no apparent changes in the sedimentation profiles of the 60S subunit in

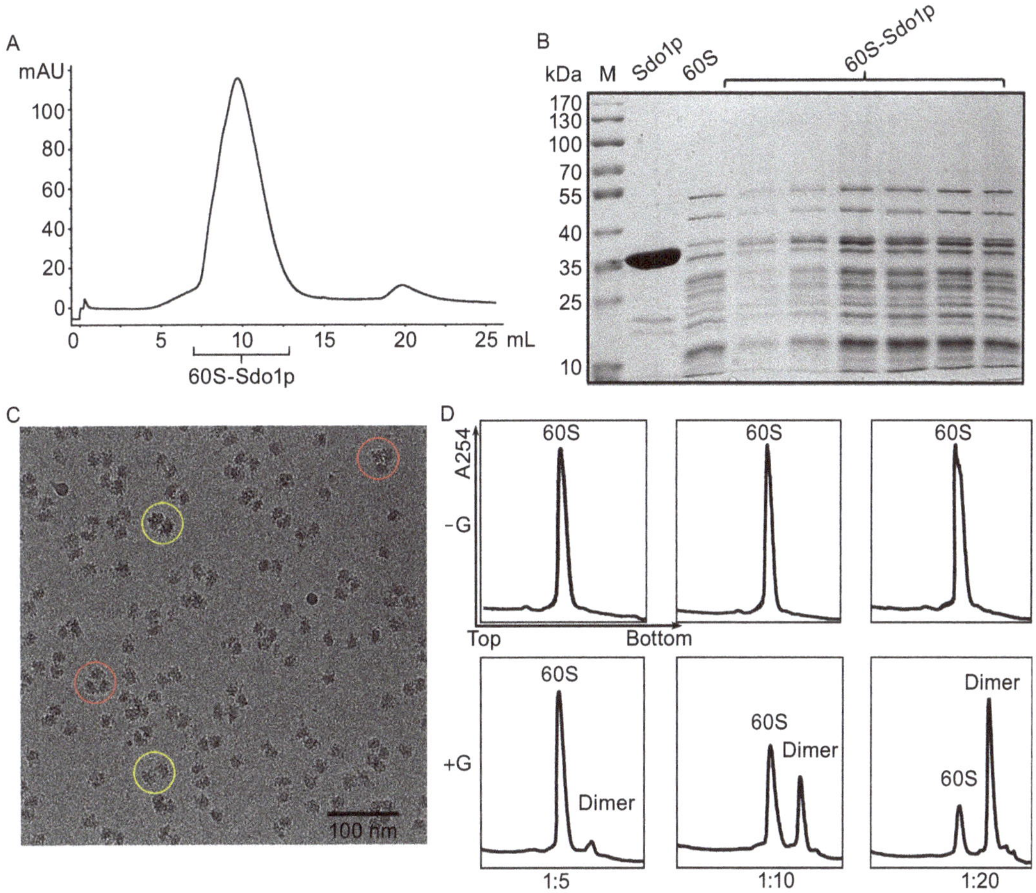

Figure 1. Sdo1p binds to the mature 60S subunit *in vitro*. (A) Gel filtration trace of the 60S-Sdo1p complex monitored by UV absorption at 280 nm. (B) SDS-PAGE examination of fractions labelled in (A). M, maker lane; Sdo1p, loading input containing Sdo1p only; 60S, loading input containing 60S only. (C) A representative cryo-EM micrograph of the 60S-Sdo1p complex (without crosslinking). Particles of 60S dimers and trimers are highlighted by yellow and red circles, respectively. (D) Examination of the formation of 60S dimers by Sdo1p using sucrose density gradient centrifugation. 5-, 10- or 20-fold excess of Sdo1p was incubated with the 60S subunit in the absence (−G) or presence (+G) of 0.1% glutaraldehyde.

the presence of 5-, 10- and 20-fold excess of Sdo1p (Fig. 1D, upper panel). We reason that the seemingly discrepancy between the two approaches might be explained by the weak association between the two 60S subunits which might not sustain a long-time centrifugation. Indeed, upon addition of a chemical crosslinker (0.1% glutaraldehyde) in the reaction system, an additional peak, corresponding to the 60S dimer, appears on the gradient profile. Dimerization of 60S subunits by Sdo1p is clearly concentration dependent: in 5-fold excess of Sdo1p, most of the 60S subunits are still in monomeric state; in 10-fold excess, around 40% of 60S subunits are dimers; in 20-fold excess, the majority of the 60S subunits are in dimer, and tiny peaks corresponding to higher order oligomers start to emerge (Fig. 1D, lower panel). And the empty 60S subunits could not form dimer with 0.1% glutaraldehyde (Fig. S1).

Sdo1p does not induce oligomerization of 40S or 80S ribosomes

Following above observations, we tested whether Sdo1p had any effects on the sedimentation profiles of a mixture of 40S and 60S subunits (2.5 mmol/L Mg^{2+}) or 80S ribosomes (12 mmol/L Mg^{2+}) using SDGC. As shown in Fig. 2, addition of Sdo1 has no effect on the 40S or the 80S ribosome, except that it again preferentially dimerizes the 60S subunits in the reaction mixtures. And the effect is more apparent in the presence of glutaraldehyde and when the free 60S subunit is abundant (Fig. 2C, lower panel). Due to the equilibrium of 80S association and dissociation, there would be a certain amount free 60S subunit in the 80S mixture even in the associating condition (12 mmol/L Mg^{2+}), and as expected, Sdo1p only specifically dimerizes those 60S

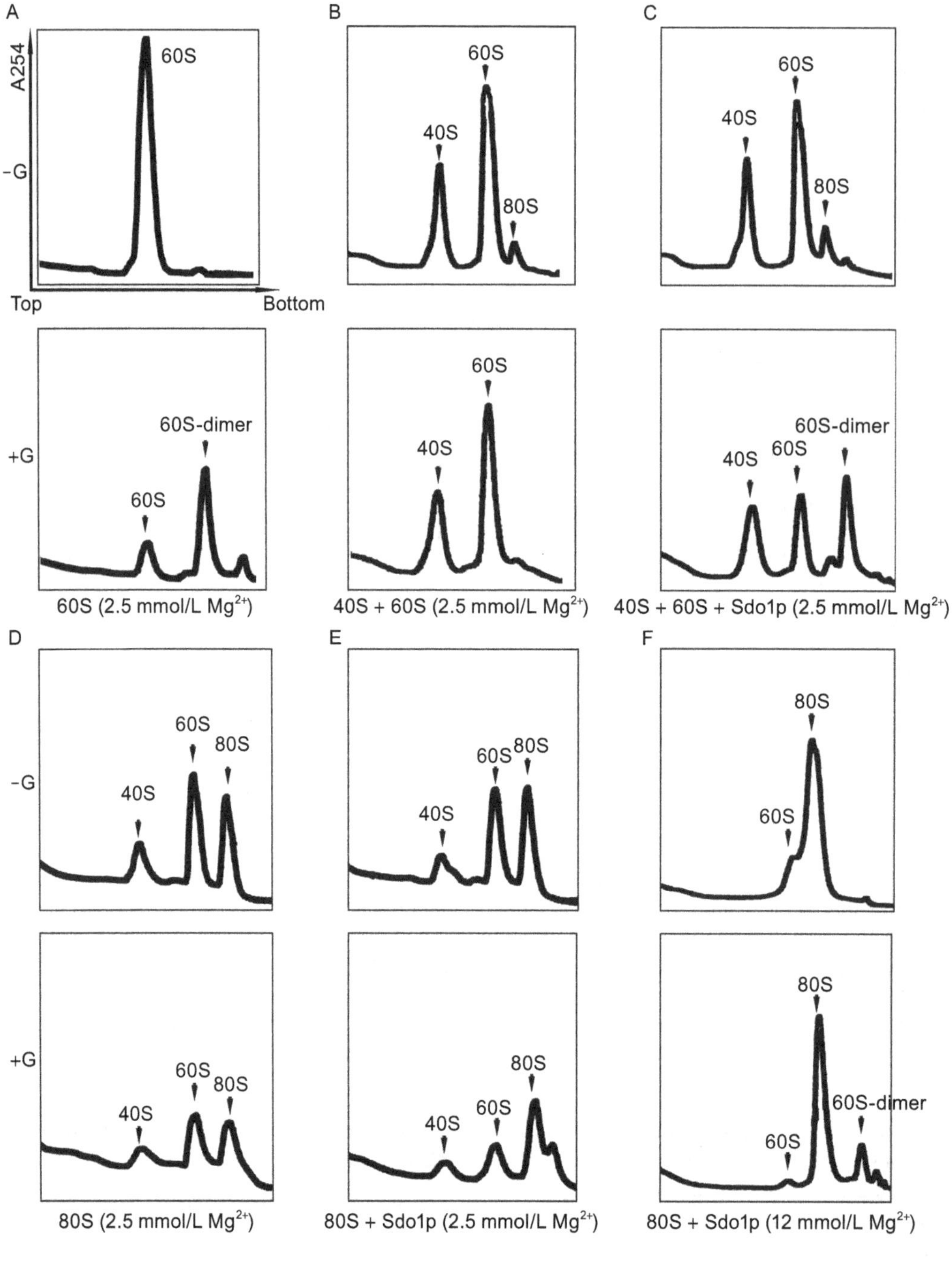

Figure 2. Sdo1p only specifically induces 60S dimerization but not 40S or 80S ribosome. (A) Sdo1p induces the formation of dimers of 60S subunits. (B) A mixture of the 40S and 60S subunits in dissociating condition (2.5 mmol/L Mg^{2+}), in the absence (−G, top panel) or presence (+G, bottom panel) of 0.1% glutaraldehyde, were subjected to SDGC, indicating that glutaraldehyde alone has no effect on dimer formation. (C) Same as (B), but with 20-fold excess of Sdo1p supplied. (D) 80S ribosomes in dissociating condition, showing a partial dissociation of 80S into 40S and 60S subunits. (E) Same as (D), but with 20-fold excess of Sdo1p supplied indicated that Sdo1p do not induce 40S and 80S ribosome dimerization but only 60S. (F) 80S ribosomes in associating condition (12 mmol/L Mg^{2+}), treated with Sdo1p in the absence or presence of glutaraldehyde also showed Sdo1p do not induce 80S ribosome dimerization.

subunits without apparent effect on the 80S ribosomes (Fig. 2F). The same experiment of Sdo1p on the 80S ribosomes was repeated in low Mg^{2+} condition (2.5 mmol/L). 80S ribosomes largely dissociate into 60S and 40S subunits in this dissociating condition (Fig. 2D), and the dissociated 60S subunits could be cross-linked by Sdo1p (Fig. 2E).

However, in both of the high and low Mg^{2+} conditions, Sdo1p did not change the peak heights of the 80S ribosome (Fig. 2E and 2F), suggesting that the binding site of Sdo1p on the 60S subunit is highly likely at the inter-subunit face of the 60S subunit.

Based on the above observations, we conclude that Sdo1p preferentially binds to the 60S subunit, and is capable of promoting weak dimerization of mature 60S subunits. This reminds us of the recent observation that 80S ribosomes are capable of forming 80S-80S and 80S-60S dimers *in vivo* under amino acid depletion (Krokowski et al., 2011). Therefore, the 60S-60S dimer might also be a form of resting translation machinery induced by certain stresses, which appears to be in general consistent with previous reports showing that loss of SBDS in HEK293 or HeLa cells leads to increased sensitivity to various stresses (Ball et al., 2009; Watanabe et al., 2009).

Domains I and II of Sdo1p are responsible for the 60S binding and dimerization

To map the domains of Sdo1p responsible for the binding and dimerization of the 60S subunit, we further characterized the binding of individual domains of Sdo1p to the 60S subunit. In human, SDS is an autosomal recessive disorder and mostly associated with two predominant mutations, 183-184TA to CT that results in an in-frame stop codon (K62X) and 258 + 2T to C that produces premature truncation (84Cfs3) (Boocock et al., 2003), both of which would render truncated forms of SBDS lack of C-terminal domains. Therefore, we constructed a series of Sdo1p truncations to test the ability of mutant proteins in the 60S binding and dimerizing abilities, including three individual domain constructs (I, II and III), and two truncations of terminal domains (I-II and II-III) (Fig. 3A).

Firstly, we tested the ability of these mutants in the 60S subunit dimerization by SDGC in the presence of chemical crosslinker. As shown in Fig. 3B, only the construct of domain I-II is capable of inducing the formation of 60S-60S dimer, and its dimerizing activity is also concentration-dependent (Fig. 3C). At a higher excess of domains I-II mutant (20-fold), a significant peak corresponding to trimers appears on the gradient profile. Next, we examined the binding of Sdo1p mutants to the 60S subunit by co-sedimentation assay. The reaction mixtures were subjected to a 33% sucrose cushion-based centrifugation and the binding was detected by Tricine-SDS-PAGE (Fig. 3D). Our results showed that similar to the full-length Sdo1p, constructs of domain I, II, and I-II all display clear binding to the 60S subunit. Very weak binding of construct II-III could also be detected. In contrast, no binding of domain III construct was detected. Lastly, in order to quantitatively compare the binding of these mutants to the 60S subunit, we measured the affinities of these mutants to the 60S subunit using Bio-layer interferometry (BLI) (Figs. 3E and S2). Highly consistent with the co-sedimentation data, construct of the domain I-II display the highest affinity to the 60S subunit (K_D = 15 nmol/L), comparable to the full-length protein (K_D = 18 nmol/L) (Fig. S2A and S2E). The truncated domain I, II and II-III display modest level of affinities (K_D ranging from 25–40 nmol/L) (Fig. S2B, S2C and S2F). In sharp contrast, domain III had no detectable binding (Fig. S2D).

From these results, a clear conclusion could be drawn: domain I of Sdo1p is the major RNA binding domain, while domain III has no RNA binding activity.

2D image analysis of the 60S-Sdo1p complexes

Knowing that different oligomeric states of the 60S-Sdo1p complexes could form, we processed cryo-EM images according to the particle sizes. Particles representing dimers were selected and subjected to reference-free alignment and classification by RELION (See MATERIALS AND METHODS). As shown in Fig. 4A, many populated class-average images show well resolved structural features for both 60S subunits, indicating that inter-subunit connection could be rather rigid (Fig. 4A). Interestingly, these 2D average images display an apparent 2-fold symmetry. This observation demonstrates that the dimer induced by Sdo1p is a specific structural entity, and not from random aggregation of 60S subunits or Sdo1p. Nevertheless, many average images have defined structural details only for one 60S subunit, with a smeared density blob for the other, suggesting that a certain extent of flexibility for the inter-subunit connection. A simple explanation for different orientations of two subunits within dimers is that Sdo1p and the 60S subunit could exist in different stoichiometry. In rigid dimers, the 60S subunit and Sdo1p are present in a 2:2 ratio, as two copies of Sdo1p would further "staple" two subunits to fix their relative orientation. In contrast, in loose dimers, one copy of Sdo1p is capable of bridging the two subunits, but the high inter-domain flexibility of Sdo1p (de Oliveira et al., 2010) would allow a wobbling between two subunits.

3D structure of the 60S-60S dimer

From 2D image analysis, the 60S subunit and Sdo1p could form stable 60S-60S dimers, which should enable the 3D structural determination of dimers by single particle analysis. Towards this end, different 3D classification strategies were tested on particles of dimers, and both the symmetry-free and 2-fold symmetric reconstructions were explored (Fig. S3). The 3D classification was done using RELION in a multi-round way. After the first round (C2-imposed), particles with 3D class structures showing relatively rigid dimers were kept for further analysis. Following steps were carried out without imposing two-fold symmetry. As a result, only 11% of particles displayed structural details for both 60S subunits, and structural refinement from these particles rendered a final density map at 14-Å resolution (Fig. S3B). This is expected, as 2D image analysis indicates that many dimers

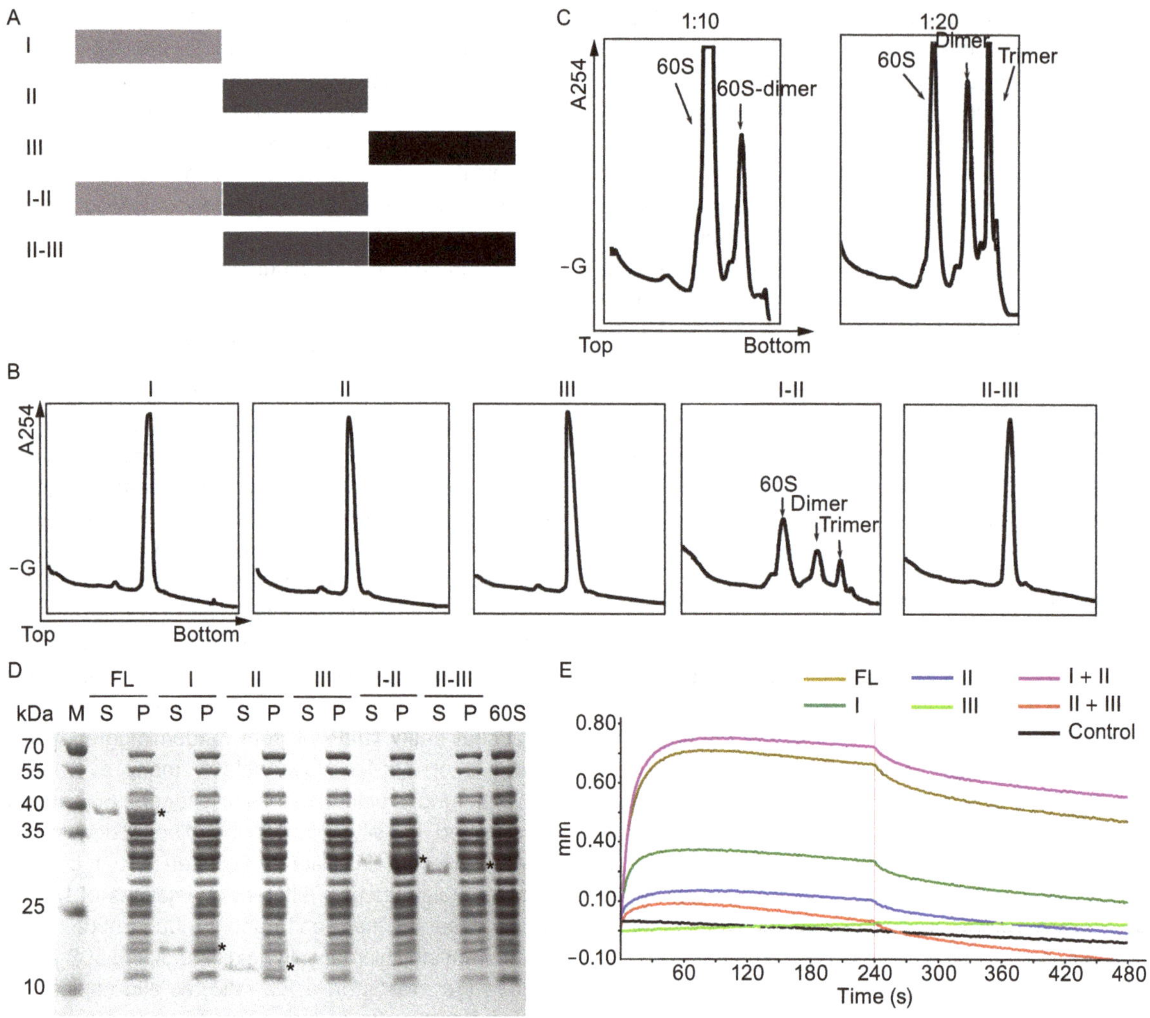

Figure 3. Sdo1p induced 60S dimerization through domain I-II. (A) Schematic overview of the Sdo1p variants. (B) Formation of 60S dimers by Sdo1p variants (20-fold excess) examined by SDGC. (C) The formation of 60S dimers by Sdo1p domains I-II is concentration dependent. (D) The binding of Sdo1p variants to the mature 60S subunit examined by co-sedimentation assay. FL, full-length; S, supernatant; P, pellet. (E) The binding of Sdo1p variants to the mature 60S subunit examined by Bio-layer interferometry. Both the association and dissociation reactions were carried out for 240 s.

are not rigid dimmers (Fig. 4A). Using the crystal structure of the 60S subunit as template (Ben-Shem et al., 2011), additional densities between the two 60S subunits can be segmented (Fig. 4B–D). This mass of densities is sandwiched between H69 of one 60S subunit and H38 of the other 60S subunit.

The homology model of yeast Sdo1p was predicted by I-TASSER (Roy et al., 2010) and fitted into the cryo-EM density map (Fig. 4D). The additional densities between two 60S subunit allowed the docking of two copies of Sdo1p. Although the resolution of our map (14 Å) could not provide unambiguous assignment of individual domains into the density map, the relative orientation of individual domains on the 60S subunit could be fixed (Fig. 3) by integrating the very recent cryo-EM data of a chimerical 60S-SBDS complex (Weis et al., 2015), which revealed that domain I lies in the P-site. Very interestingly, the recent data also showed that domain II-III could exist in very different conformations, owing to the flexible linker between domain I and domain II (Weis et al., 2015). As a result, we built a model for the 60S-Sdo1p dimer (Fig. 4D), which requires a large reorientation of domain II-III (Fig. 4E), compared with the homology model of Sdo1p. The model could well explain the structural basis for the observed dimerization of 60S subunits by Sdo1p: domain I interacts with the ribosomal P-site of one subunit, while domains II-III (mostly II) interact with H38 of the other subunit.

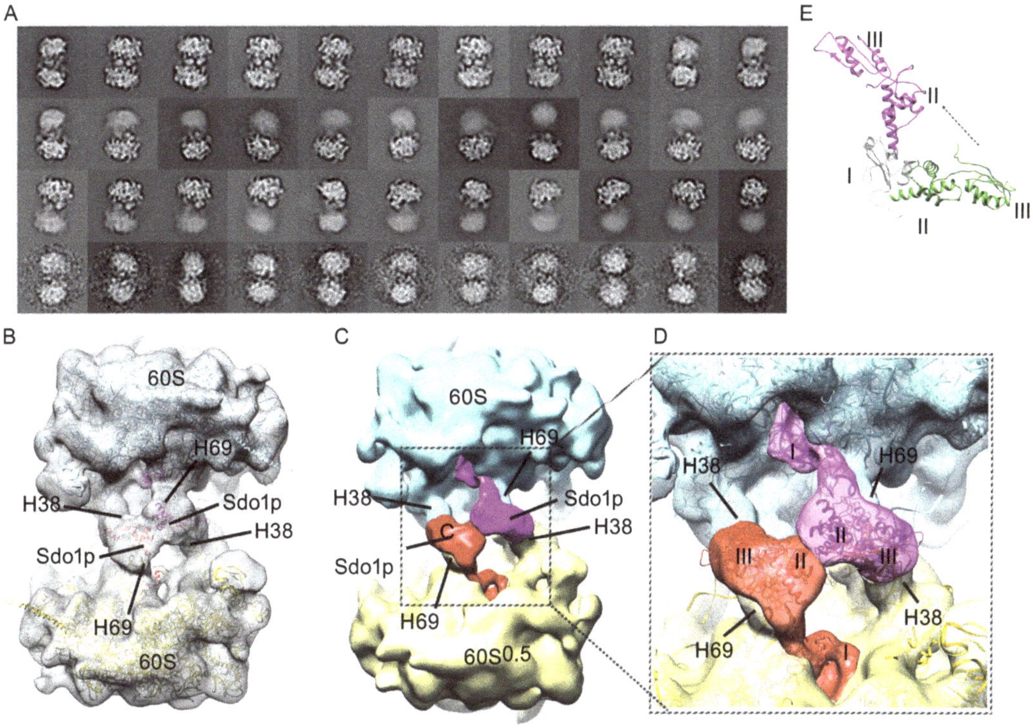

Figure 4. The cryo-EM structure of dimeric 60S-Sdo1p complex. (A) Reference-free 2D classification of selected particles of the 60S dimers. (B) Overview of cryo-EM density map of the dimeric 60S-Sdo1p complex, superimposed with fitted models. (C) Same as (B), but with two 60S subunits and two copies of Sdo1p separately colored. (D) Zoom-in view of boxed region in (C). (E) Comparison of Sdo1p in its 60S-bound conformation with that derived from homology modelling (see MATERIALS AND METHODS).

Interaction of Sdo1p with the 60S subunit is highly dynamic and might involve uL16

The relatively low population of the stable 60S dimers in cryo-EM images hampered the resolution of the dimer density map. We, therefore, sought to improve the structure by only processing the monomeric 60S particles. As a result, we prepared another batch of sample using 10-fold excess of Sdo1p in the 60S binding and only picked well-separated 60S particles from cryo-EM images. Although with a much larger dataset (305,370 particles), exhaustive 3D classification did not reveal high-resolution densities for Sdo1p. When more particles were included for final refinement, as expected, we could only obtain a high-resolution structure for the 60S subunit. In contrast, when a small homogenous fraction of particles were used, an additional piece of density could be seen in the ribosomal P-site. (Fig. S4). Based on our domain assignment in the 60S dimer, it is likely domain I of Sdo1p.

To further confirm our structural and biochemical data that Sdo1p binds to the ribosomal P-site, we employed cross-linking based mass spectrometry (CX-MS) (Singh et al., 2010), which is particularly useful for structural analysis of flexible and transient interactions. Therefore, we treated the 60S-Sdo1p complex with cross-linking reagents disuccinimidyl suberate (DSS) and bis(sulfosuccinimidyl) suberate (BS^3), and subjected the linked sample to mass spectrometry. The reactive sites of the cross-linkers interact with the primary amino group of lysine and N-terminal tail of the protein. Considering the 11.4-Å spacer arm between reactive sites, lysine side-chains within 30 Å could be potentially cross-linked.

We performed chemical cross-linking of the 60S complexes prepared with the full-length, domain I or domain I-II of Sdo1p. The CX-MS results show that a predominant cross-linking is between Sdo1p and uL16 through Lys62 (Sdo1p)-Lys184 (uL16) (Figs. 5A, 5B and S5), which is highly consistent with previous data that uL16 is involved in Sdo1p recruitment in the late-stage maturation of the 60S subunit (Sulima et al., 2014a). Another crosslinking was seen between Lys68 of Sdo1p and Lys of uL5 (Fig. S5).

Altogether, these results demonstrate that the interaction of Sdo1p with the 60S subunit is highly dynamic, which might be essential for its function in probing the conformational status of the ribosomal P-site during the late stage of cytoplasmic maturation of the 60S subunit.

DISCUSSION

According to the established framework of Sdo1p in the 60S maturation, the function of Sdo1p is, by interacting with

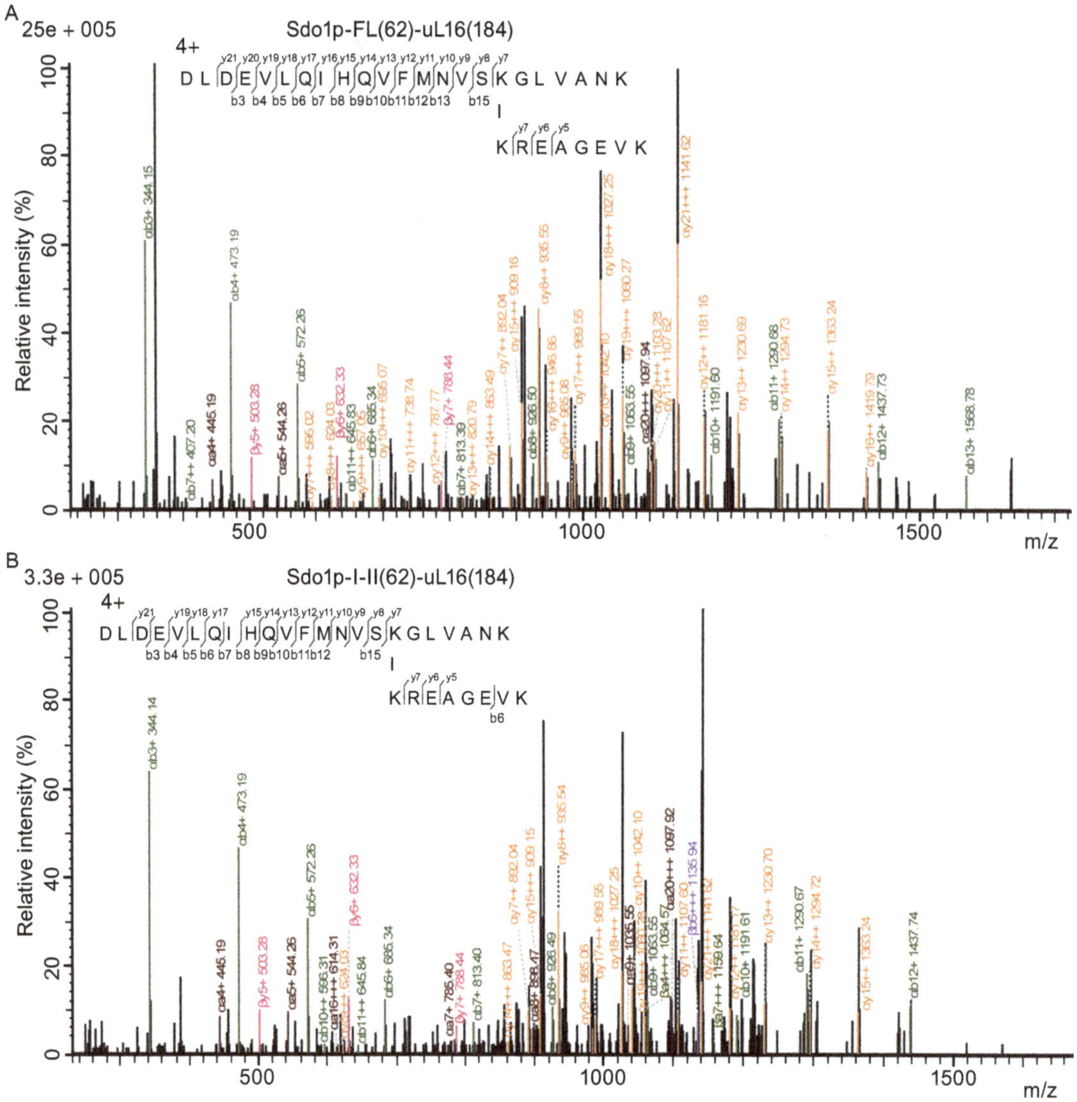

Figure 5. Crosslinking of Sdo1p with uL16 identified in CX-MS. (A) Representative MS spectra of the cross-linked peptides between Sdo1p and uL16 detected in the samples of the 60S-Sdo1p-FL (full-length) complex (A), the 60S-Sdo1p-I-II (Domain I-II) complex (B). Primary sequences of linked peptides are shown, with sites of cleavages labelled in different colors in both the sequences and spectra. 4+ or 3+ indicates the charge of the cross-linked peptides.

Efl1p, to promote the release of nucleolar shuttling factor Tif6p (Menne et al., 2007). Domains II-III of SBDS were reported to directly interact with Efl1p (Asano et al., 2014), and it was proposed that Sdo1p/SBDS might act as nucleotide exchange factor to stabilize the binding of GTP to Efl1p (Gijsbers et al., 2013). Because the binding site of Tif6p (Gartmann et al., 2010; Klinge et al., 2011) is rather distant from the binding site of Sdo1p on the 60S subunit, these data have put Sdo1p in a central position of this maturation pathway. Our data show that Sdo1p interacts with the ribosomal P-site (H69) through its domain I, with terminal domain II-III capable of moving around (Fig. 4E), suggesting that the essential role of Sdo1p is to probe the conformational status of the ribosomal P-site (Fig. 6G). Once the correct/native conformation of the ribosomal P-site is established, which in turn stabilizes domain II-III of Sdo1p to induce Tif6p-releasing activity of Efl1p. This hypothesis is further supported by the essential role of uL16 in the maturation of the 60S subunit. The uL16 was also found to cooperate with Sdo1p and Efl1p to release Tif6p, as well as Nmd3 (Bussiere et al., 2012; Hedges et al., 2005; Menne et al., 2007; Sulima et al., 2014a; Sulima et al., 2014b; West et al., 2005). More importantly, uL16 (R98S) mutant in yeast causes a defect in late-stage 60S subunit maturation and targets mutant ribosomes for degradation (Sulima et al., 2014b), suggesting a checkpoint role of uL16 in the quality

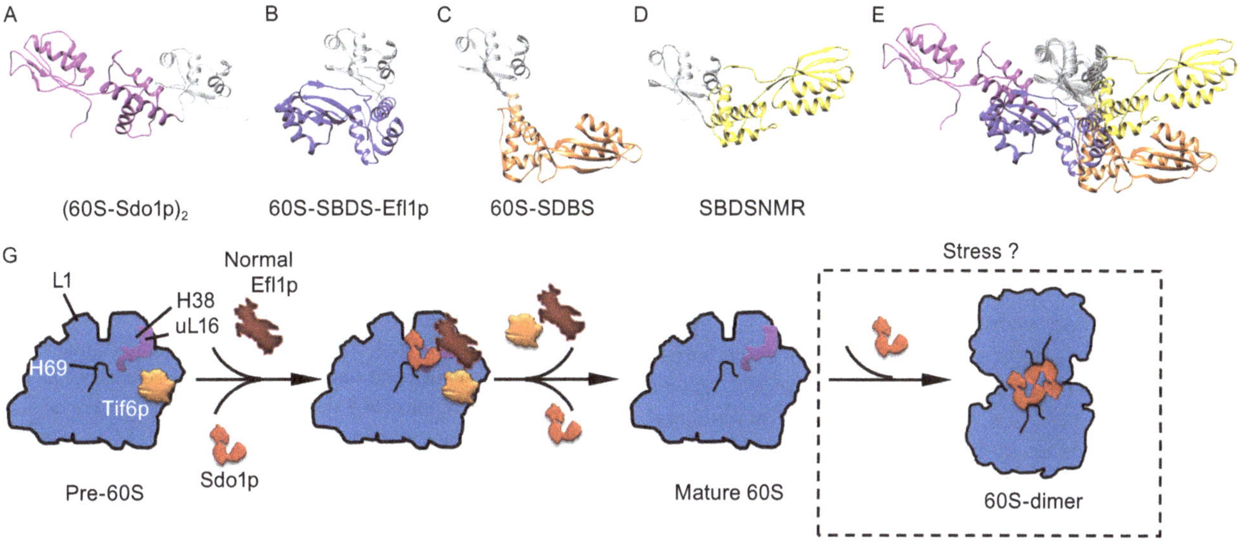

Figure 6. Proposed model of the action of Sdo1p in the late-stage maturation of the 60S subunit. (A–E) Conformational states of Sdo1p/SBDS. Coordinates of SBDS in heterologous 60S-SBDS-Efl1p (B) and 60S-SBDS (C) complexes were taken from a very recent cryo-EM work (Weis et al., 2015). One snapshot of NMR structures of human SBDS (D) was taken from a previous study (de Oliveira et al., 2010). All the structures were aligned using the domain I of Sdo1p/SBDS as reference. (G) In normal condition, Sdo1p, as a member of a conformational signal relay system, probes the maturation status of the ribosomal P-site, and passes the signal to Efl1p to release Tif6p from the nearly mature 60S subunit. Under a certain stress, Sdo1p might induce the formation of inactive 60S dimers to limit the cellular translation capacity.

control of the 60S maturation (Sulima et al., 2014a). Our CX-MS data indicate that Sdo1p (domain I) is in close contact with uL16. Taking into consideration of the critical role of the P-site loop of uL16 in Sdo1p binding (Sulima et al., 2014a) and Tif6p release (Bussiere et al., 2012), it is plausible that Sdo1p senses the incorporation and conformational maturation of uL16 on the 60S subunit and pass the signal to Efl1p. Also, our structural data reveal that H69 and H38 are two binding partners of Sdo1p, both of which are highly conserved (structurally and functionally) in eukaryotes and prokaryotes. And these two helices are known to be highly dynamic and adopt very different conformations in the pre-60S (yeast) (Bradatsch et al., 2012; Greber et al., 2012; Leidig et al., 2014) or pre-50S (bacterial) (Jomaa et al., 2014; Li et al., 2013; Zhang et al., 2014a) particles. Taken together, our data support a conformational signal relay system for these factors (uL16-Sdo1p-Efl1p-Tif6p), in which Sdol1p and Efl1p couple two spatially distant events during the late-stage maturation of the 60S subunit, the ribosomal P-site maturation and the release of anti-association factor Tif6p (Gartmann et al., 2010). The dynamic nature of Sdo1p on the 60S subunit ensures its participation as a probing factor to sense the conformational maturation of its surrounding rRNA and RP.

Our structural data suggest that Sdo1p remains very dynamic on the 60S subunit. Very interestingly, in comparison with the very recent data on chimerical 60S-SBDS and 60S-SBDS-Efl1p (Weis et al., 2015), domains II-III of Sdo1p/SBDS could adopt very different conformations on the 60S subunit (Fig. 6A–E). These observations are consistent with the NMR spectroscopic analysis of SBDS in solution (de Oliveira et al., 2010). Altogether, these results support our hypothesis that Sdo1p is a dynamic probe of the maturation status of the 60S subunit, and the inter-domain flexibility of Sdo1p/SBDS is essential for its function.

Our CX-MS data also indicate a proximity for uL5 and Sdo1p. This observation might be of physiological significance to the disease of SDS in human. It was generally believed that partial loss of SBDS function in SDS might induce a nucleolar stress from defective ribosome biogenesis and activate the RP (ribosomal protein)-p53-HDM2 pathway, and uL5 (RPL11) is one of RPs involved in stabilizing p53 following nucleolar stress (Deisenroth and Zhang, 2010; Holmberg Olausson et al., 2012; Nakhoul et al., 2014).

In the present work, we found that Sdo1p could induce the formation of 60S-60S dimers *in vitro* through crosslinking H69 and H38 from the two participating 60S subunits. In bacterial and mammalian cells, it was shown that higher-order organization of ribosomes is employed by the cells as a means to regulate the translation capacity by limiting the number of active ribosomes. In *E. coli*, upon transition to stationary phase, cells start to accumulate 100S particles (Yoshida and Wada, 2014), composed of 70S dimers induced by protein factors called ribosome modulation factor

(RMF) (Wada et al., 1995) and hibernation promoting factor (HPF) (Ueta et al., 2005). When mammalian (rat) cells were challenged by amino acid starvation, a form of 110S particles, including 80S dimers and 80S-60S heterodimers, were also detected (Krokowski et al., 2011). These previous data indicate a conserved mechanism of translation modulation by organizing ribosomes into resting higher-order assemblies. Previous work also suggested SBDS has a second role in multiple cellular stress response pathways, independent of its primary role in ribosome biogenesis (Ball et al., 2009). HEK293 cells with the deletion of SBDS are hypersensitive to DNA damage and endoplasmic reticulum stress (Ball et al., 2009). HeLa cells with SBDS-knockdown, as well as SDS patient cells are hypersensitive to Fas-mediated apoptosis (Watanabe et al., 2009; Watanabe and Dror, 2005). In yeast, Sdo1p physically interacts with Btn1p (CLN3 in human) and was suggested to involved in cellular responses to pH and nutrient changes (Vitiello et al., 2010). Therefore, it is possible that the 60S dimers induced by Sdo1p, if not an *in vitro* artifact, might represent a form of inactive storage of ribosomal subunits upon a certain stress (Fig. 6G). This attractive hypothesis merits further investigation.

MATERIALS AND METHODS

Gene cloning and protein purification

The genes encoding full-length and truncated Sdo1p were amplified using standard Polymerase Chain Reaction (PCR) from *S. cerevisiae*. Primers were designed according to respective domains of yeast Sdo1p to clone the full-length (1–250), domain I (1–95), domain II (96–169), domain III (170–250), domain I-II (1–169) and domain II-III (96–250) constructs. The fragments were cloned into pET21b vector to yield pET21b-Sdo1p constructs with C-terminal 6× -His-tag for purification. The plasmids were transformed into *E. coli* BL21 (DE3) cells for overexpression.

For protein purification, cells were cultured in LB media containing 50 μg/mL of ampicillin at 30°C to an OD_{600} of 1.0 and induced by 0.5 mmol/L IPTG for 4 h. Cells were pelleted and suspended in buffer A (50 mmol/L Tris-HCl [pH 7.5], 500 mmol/L NaCl, 1 mmol/L PMSF) containing 30 mmol/L imidazole and disrupted by an ultrasonic processor. Cell lysates were clarified by centrifugation at 13,000 rpm (Avanti J-26 XP, JA25.50 rotor, Beckman Coulter) for 1 h. The clarified lysates were load onto a Ni-NTA column (GE Healthcare) and eluted with buffer B (same as buffer A but containing 500 mmol/L imidazole). The target proteins were pooled, concentrated with Millipore Amicon Ultra Centrifugal filters and subjected to gel filtration chromatography (Superdex75 10/300 GL, GE Healthcare) with buffer C (25 mmol/L Tris-HCl [pH7.5], 100 mmol/L KCl, 2.5 mmol/L $MgCl_2$). Purified proteins were split into aliquots and stored at −80°C.

Ribosomal subunit purification

Yeast cells (*S. cerevisiae* S288C) were grown in 3 L of YEPD (1% yeast extract, 2% Bacto peptone and 2% glucose) medium to an OD_{600} of 1.0. Ribosome purification was carried out as previously described (Zhang et al., 2014b). Cells were harvested and washed twice with ice-cold Lysis buffer (50 mmol/L Tris-acetate [pH7.0], 50 mmol/L NH_4Cl, 12 mmol/L $MgCl_2$, 1 mmol/L DTT, 1 pill/50-mL complete protease inhibitor cocktail [Roche]). Resuspended cells were disrupted with a high-pressure homogenizer (5000 p.s.i.) for three times. Cell lysates were centrifuged at 13,000 rpm for 1 h at 4°C in a JA 25.50 motor (Beckman Coulter). Supernatants were layered over a 10%–50% linear sucrose gradient (50 mmol/L Tris-acetate [pH 7.0], 100 mmol/L NH_4Cl, 12 mmol/L $MgCl_2$, 1 mmol/L DTT), and centrifuged at 30,000 rpm for 5 h in an SW32 rotor (Beckman Coulter) at 4°C. The gradient profile was monitored at 254-nm wavelength and fractionated in a gradient collector (Teledyne Isco). Fractions of 80S ribosomes were collected, concentrated with Amicon Ultra Centrifugal filters (Millipore) and washed with separation buffer (50 mmol/L Tris-HCl [pH 7.4], 500 mmol/L KCl, 2.5 mmol/L $MgCl_2$, 1 mmol/L DTT). After the buffer was changed to separation buffer, the mixture was incubated at 37°C for 10 min. Reaction mixtures were layered over a 10%–40% linear sucrose gradient (50 mmol/L Tris-HCl [pH 7.4], 500 mmol/L KCl, 2.5 mmol/L $MgCl_2$, 1 mmol/L DTT) and centrifuged at 30,000 rpm in an SW32 rotor (Beckman Coulter) at 4°C for 7 h. The gradient was similar analyzed and fractions of 40S and 60S subunits were separately collected. After concentrating with Millipore Amicon Ultra Centrifugal filters, samples were split into aliquots and stored at −80°C for further use.

In vitro binding with gel filtration

The 60S subunit (150 pmol) and full-length Sdo1p (20-fold excess) were mixed in 500-μL buffer C (25 mmol/L Tris-HCl [pH 7.5], 100 mmol/L KCl, 2.5 mmol/L $MgCl_2$), and incubated at 30°C for 15 min. The mixture was loaded onto a gel filtration column (Superose6 10/300 GL, GE Healthcare). The peak fractions were subjected to SDS-PAGE.

Co-sedimentation assay

60S ribosomal subunits 50 pmol) were incubated with full-length or truncated Sdo1p proteins in a ratio of 1:20 for 15 min at 30°C in 90 μL buffer. The mixtures were then loaded onto a sucrose cushion (25 mmol/L Tris-HCl [pH 7.5], 100 mmol/L KCl, 2.5 mmol/L $MgCl_2$, 33% sucrose) and centrifuged at 95,000 rpm for 2 h in a TLA-100 rotor (Beckman Coulter) at 4°C. The supernatants (1/18) and the pellets (1/3) were collected separately and resolved by Tricine-SDS-PAGE.

Sucrose density gradient centrifugation (SDGC)

Samples of ribosomes and Sdo1p variants were changed to a Hepes-KOH buffer system (25 mmol/L Hepes-KOH [pH7.5], 100 mmol/L KCl, 2.5 mmol/L $MgCl_2$). 37.5 pmol ribosomes (40S, 60S or 80S) were incubated with full-length or truncated Sdo1p proteins in a ratio of 1:20 for 15 min at 30°C in 200 μL buffer. The mixtures were incubated with or without 0.1% glutaraldehyde for another 15 min, layered onto a 10%–50% linear sucrose gradient (25 mmol/L Hepes-KOH [pH 7.5], 100 mmol/L KCl, 2.5 mmol/L $MgCl_2$), and centrifuged at 39,000 rpm for 4 h in an SW41 rotor

(Beckman Coulter) at 4°C. The gradient was analyzed and fractionated in an ISCO gradient collector.

In vitro binding with Bio-layer Interferometry

The BLI-based experiments were performed with Octet RED96 System (ForteBio, Pall Corp.) according to the manufacturer's instructions. Sdo1p variants (wild type and mutants) in Hepes-KOH buffer system (25 mmol/L Hepes-KOH [pH 7.5], 100 mmol/L KCl, 2.5 mmol/L $MgCl_2$) were immobilized to the optic biosensors of Anti-Penta-HIS (HIS1 K, ForteBio). Before each experiment, biosensors were pre-equilibrated with binding buffer, followed by equilibrium binding with Sdo1p variants (2 μg/mL). Pre-experiments were carried to determine the optimal concentration range of the 60S subunit suitable for the measurements. For the single concentration experiments, a final concentration of 48 nmol/L 60S subunit was used. For the concentration gradient experiments, 2-fold dilution series of the 60S subunit were used (48 nmol/L, 24 nmol/L, 12 nmol/L, 6 nmol/L, 3 nmol/L, and 1.75 nmol/L). Both the association and dissociation of each measurement were carried out for 240 s. The data was analyzed and dissociation constants were calculated by the software of the Data analysis 7.0 provided by ForteBio.

Cryo-sample preparation and data collection

The 60S subunit (100 nmol/L) was incubated with 20-fold or 10-fold excess of full-length Sdo1p at 30°C for 15 min. The mixture was diluted to a final concentration of 60 nmol/L for the 60S complex in binding buffer. 4-μL aliquots of samples were applied to 300-mesh 2/2 glow-discharged Quantifoil grids (Quantifoil Micro Tools) which were pre-coated with a thin layer of carbon. The grids were blotted and plunged into liquid ethane with an FEI Mark IV Vitrobot operated at 4°C. For the sample of dimeric 60S-Sdo1p complexes (20-fold excess of Sdo1p), data collection was performed with an FEI F20 at 80,000× magnification with a Gatan UltraScan 4000 CCD camera. For the sample of monomeric 60S-Sdo1p complexes, data collection was done with an FEI Titan Krios equipped with an FEI Eagle 4 K × 4 K CCD camera at 75,000× magnification. All the images were recorded under low-dose conditions (~20 e-/Å^2) with Auto-EMation package (Lei and Frank, 2005).

Image processing and structural analysis

Micrograph screening, estimation of contrast transfer function parameters, and initial particle picking were performed with SPIDER package (Shaikh et al., 2008). An artificial 60S dimer was used as the template for automatic particle picking (Rath and Frank, 2004). For 60S-Sdo1p dimers, 116,271 particles (from 3,294 micrographs) (2.76 Å/pixel with a binning factor of two) were picked and subjected reference-free 2D classification using XMIPP (Scheres et al., 2008), EMAN2 (Tang et al., 2007) and RELION (Scheres, 2012) packages, which rendered essentially similar results with many of the dimers displaying considerable flexibility on the orientation of the two 60S subunits. For the 3D classification, all the particles were classified into 6 classes without or with C2 symmetry imposed using RELION package. For C2-imposed classification, two classes display reliable details for both two 60S subunits were combined (30,452 particles) and subject to another round of 3D classification. The second round was done without imposing the C2-symmetry, resulting into 4 classes. Two of them were combined (13,219 in total) and subjected to structural refinement by RELION. The final resolution of the refined map is 14 Å based on the 0.143 cutoff criteria of the gold-standard Fourier shell correlation (FSC).

For monomeric 60S-Sdo1p complexes, 305,371 particles (2.33 Å/pixel with a binning factor of two) (from 7,994 micrographs) were classified into 10 classes using RELION. One class (40,186 particles) displayed substantial additional density at the ribosomal P-site. After structural refinement, the final reported resolution is 9 Å based on the 0.143 cutoff criteria of the gold-standard FSC.

The yeast homology model of Sdo1p was modelled using I-TASSER (Roy et al., 2010). The fitting of two copies of Sdo1p was done manually using UCSF Chimera (Pettersen et al., 2004), and optimized by the "fit in map" module of Chimera. Coordinates of the yeast ribosome were from a previous crystallography study (Ben-Shem et al., 2011).

Cross-linking mass spectrometry analysis (CX-MS)

50 pmol 60S ribosome was incubated with full-length or truncated Sdo1p at 1:20 molar ratio for 15 min at 30°C (25 mmol/L Hepes-KOH [pH 7.5], 100 mmol/L KCl, 2.5 mmol/L $MgCl_2$). The mixture was cross-linked with DSS or BS3 (Thermo Fisher Scientific) at 1:1 mass ratio at room temperature for 1 h. The reaction was then quenched with 20 mmol/L ammonium bicarbonate. Cross-linking products were analyzed by SDS-PAGE to evaluate the cross-linking efficiency. For MS analysis, proteins were precipitated with acetone, resuspended in 8 mol/L urea, 100 mmol/L Tris, pH 8.5, and digested with trypsin. LC-MS/MS analyses were carried on an EASY-nLC 1000 system (Thermo Fisher Scientific) interfaced with a Q-Exactive mass spectrometer (Thermo Fisher Scientific). Peptides were separated on an analytical capillary column (75 μm × 10 cm, 1.8 μm C18) using a 60 min linear gradient at a flow rate of 200 nL/min. The mass spectrometer was operated in data-dependent mode with one full MS scan followed by ten HCD MS/MS scans with a dynamic exclusion time of 30 s. Precursors of the +1, +2, or unassigned charge states were rejected. pLink (Yang et al., 2012) was used for identification of cross-linked peptides by requiring FDR <5%.

ACCESSION CODE

The density map of the 60S-Sdo1p dimer has been deposited in the EMdataBank under accession codes of EMD-3280.

ACKNOWLEDGMENTS

We acknowledge the China National Center for Protein Sciences (Beijing) and the "Explorer 100" cluster system of Tsinghua National Laboratory for Information Science and Technology for providing computation resource. This work was supported by grants from the

National Natural Science Foundation of China (Grant Nos. 31422016 and 31470722 to N.G.), the National Basic Research Program (973 Program) (No. 2013CB910404 to N.G.), Beijing Higher Education Young Elite Teacher Project (YETP0131 to N.G.), and Tsinghua University (20131089278 to N.G.).

COMPLIANCE WITH ETHICS GUIDELINES

Chengying Ma, Kaige Yan, Dan Tan, Ningning Li, Yixiao Zhang, Yi Yuan, Zhifei Li, Meng-Qiu Dong, Jianlin Lei, and Ning Gao declare no conflict of interest.

This article does not contain any studies with human or animal subjects performed by the any of the authors.

AUTHOR CONTRIBUTION

NG., J.L. and C.M. conceived and designed the project. C.M., K.Y., N.L., Y.Z., Y.Y., and Z.L. performed experiments. D.T. and M-Q.D. performed CXMS. C.M. and N.G. wrote the paper.

REFERENCES

Asano N, Atsuumi H, Nakamura A, Tanaka Y, Tanaka I, Yao M (2014) Direct interaction between EFL1 and SBDS is mediated by an intrinsically disordered insertion domain. Biochem Biophys Res Commun 443:1251–1256

Austin KM, Leary RJ, Shimamura A (2005) The shwachman-diamond SBDS protein localizes to the nucleolus. Blood 106:1253–1258

Ball HL, Zhang B, Riches JJ, Gandhi R, Li J, Rommens JM, Myers JS (2009) Shwachman-Bodian diamond syndrome is a multi-functional protein implicated in cellular stress responses. Hum Mol Genet 18:3684–3695

Ben-Shem A, Garreau de Loubresse N, Melnikov S, Jenner L, Yusupova G, Yusupov M (2011) The structure of the eukaryotic ribosome at 3.0 a resolution. Science 334:1524–1529

Bernstein KA, Bleichert F, Bean JM, Cross FR, Baserga SJ (2007) Ribosome biogenesis is sensed at the start cell cycle checkpoint. Mol Biol Cell 18:953–964

Boocock GR, Morrison JA, Popovic M, Richards N, Ellis L, Durie PR, Rommens JM (2003) Mutations in SBDS are associated with Shwachman-Diamond syndrome. Nat Genet 33:97–101

Boocock GR, Marit MR, Rommens JM (2006) Phylogeny, sequence conservation, and functional complementation of the SBDS protein family. Genomics 87:758–771

Boulon S, Westman BJ, Hutten S, Boisvert FM, Lamond AI (2010) The nucleolus under stress. Mol Cell 40:216–227

Bradatsch B, Leidig C, Granneman S, Gnadig M, Tollervey D, Bottcher B, Beckmann R, Hurt E (2012) Structure of the pre-60S ribosomal subunit with nuclear export factor Arx1 bound at the exit tunnel. Nat Struct Mol Biol 19:1234–1241

Burwick N, Coats SA, Nakamura T, Shimamura A (2012) Impaired ribosomal subunit association in Shwachman-Diamond syndrome. Blood 120:5143–5152

Bussiere C, Hashem Y, Arora S, Frank J, Johnson AW (2012) Integrity of the P-site is probed during maturation of the 60S ribosomal subunit. J Cell Biol 197:747–759

Chakraborty A, Uechi T, Kenmochi N (2011) Guarding the 'translation apparatus': defective ribosome biogenesis and the p53 signaling pathway. Wiley interdiscip Rev RNA 2:507–522

de Oliveira JF, Sforca ML, Blumenschein TM, Goldfeder MB, Guimaraes BG, Oliveira CC, Zanchin NI, Zeri AC (2010) Structure, dynamics, and RNA interaction analysis of the human SBDS protein. J Mol Biol 396:1053–1069

Deisenroth C, Zhang Y (2010) Ribosome biogenesis surveillance: probing the ribosomal protein-Mdm2-p53 pathway. Oncogene 29:4253–4260

Dez C, Tollervey D (2004) Ribosome synthesis meets the cell cycle. Curr Opin Microbiol 7:631–637

Finch AJ, Hilcenko C, Basse N, Drynan LF, Goyenechea B, Menne TF, Gonzalez Fernandez A, Simpson P, D'Santos CS, Arends MJ et al (2011) Uncoupling of GTP hydrolysis from eIF6 release on the ribosome causes Shwachman-Diamond syndrome. Genes Dev 25:917–929

Freed EF, Bleichert F, Dutca LM, Baserga SJ (2010) When ribosomes go bad: diseases of ribosome biogenesis. Mol BioSyst 6:481–493

Gamalinda M, Ohmayer U, Jakovljevic J, Kumcuoglu B, Woolford J, Mbom B, Lin L, Woolford JL Jr (2014) A hierarchical model for assembly of eukaryotic 60S ribosomal subunit domains. Genes Dev 28:198–210

Ganapathi KA, Austin KM, Lee CS, Dias A, Malsch MM, Reed R, Shimamura A (2007) The human Shwachman-Diamond syndrome protein, SBDS, associates with ribosomal RNA. Blood 110:1458–1465

Gartmann M, Blau M, Armache JP, Mielke T, Topf M, Beckmann R (2010) Mechanism of eIF6-mediated inhibition of ribosomal subunit joining. J Biol Chem 285:14848–14851

Gijsbers A, Garcia-Marquez A, Luviano A, Sanchez-Puig N (2013) Guanine nucleotide exchange in the ribosomal GTPase EFL1 is modulated by the protein mutated in the Shwachman-Diamond syndrome. Biochem Biophys Res Commun 437:349–354

Greber BJ, Boehringer D, Montellese C, Ban N (2012) Cryo-EM structures of Arx1 and maturation factors Rei1 and Jjj1 bound to the 60S ribosomal subunit. Nat Struct Mol Biol 19:1228–1233

Hedges J, West M, Johnson AW (2005) Release of the export adapter, Nmd3p, from the 60S ribosomal subunit requires Rpl10p and the cytoplasmic GTPase Lsg1p. Embo J 24:567–579

Holmberg Olausson K, Nister M, Lindstrom MS (2012) p53 -dependent and -independent nucleolar stress responses. Cells 1:774–798

Jomaa A, Jain N, Davis JH, Williamson JR, Britton RA, Ortega J (2014) Functional domains of the 50S subunit mature late in the assembly process. Nucl Acids Res 42:3419–3435

Jorgensen P, Nishikawa JL, Breitkreutz BJ, Tyers M (2002) Systematic identification of pathways that couple cell growth and division in yeast. Science 297:395–400

Karbstein K (2013) Quality control mechanisms during ribosome maturation. Trends Cell Biol 23:242–250

Klinge S, Voigts-Hoffmann F, Leibundgut M, Arpagaus S, Ban N (2011) Crystal structure of the eukaryotic 60S ribosomal subunit in complex with initiation factor 6. Science 334:941–948

Krokowski D, Gaccioli F, Majumder M, Mullins MR, Yuan CL, Papadopoulou B, Merrick WC, Komar AA, Taylor D, Hatzoglou M (2011) Characterization of hibernating ribosomes in mammalian cells. Cell Cycle 10:2691–2702

Lebaron S, Schneider C, van Nues RW, Swiatkowska A, Walsh D, Bottcher B, Granneman S, Watkins NJ, Tollervey D (2012) Proofreading of pre-40S ribosome maturation by a translation initiation factor and 60S subunits. Nat Struct Mol Biol 19:744–753

Lei J, Frank J (2005) Automated acquisition of cryo-electron micrographs for single particle reconstruction on an FEI Tecnai electron microscope. J Struct Biol 150:69–80

Leidig C, Thoms M, Holdermann I, Bradatsch B, Berninghausen O, Bange G, Sinning I, Hurt E, Beckmann R (2014) 60S ribosome biogenesis requires rotation of the 5S ribonucleoprotein particle. Nat Commun 5:3491

Li N, Chen Y, Guo Q, Zhang Y, Yuan Y, Ma C, Deng H, Lei J, Gao N (2013) Cryo-EM structures of the late-stage assembly intermediates of the bacterial 50S ribosomal subunit. Nucl Acids Res 41:7073–7083

Lo KY, Li Z, Bussiere C, Bresson S, Marcotte EM, Johnson AW (2010) Defining the pathway of cytoplasmic maturation of the 60S ribosomal subunit. Mol Cell 39:196–208

Luz JS, Georg RC, Gomes CH, Machado-Santelli GM, Oliveira CC (2009) Sdo1p, the yeast orthologue of Shwachman-Bodian-Diamond syndrome protein, binds RNA and interacts with nuclear rRNA-processing factors. Yeast 26:287–298

Matsuo Y, Granneman S, Thoms M, Manikas RG, Tollervey D, Hurt E (2014) Coupled GTPase and remodelling ATPase activities form a checkpoint for ribosome export. Nature 505:112–116

Menne TF, Goyenechea B, Sanchez-Puig N, Wong CC, Tonkin LM, Ancliff PJ, Brost RL, Costanzo M, Boone C, Warren AJ (2007) The Shwachman-Bodian-Diamond syndrome protein mediates translational activation of ribosomes in yeast. Nat Genet 39:486–495

Miluzio A, Beugnet A, Volta V, Biffo S (2009) Eukaryotic initiation factor 6 mediates a continuum between 60S ribosome biogenesis and translation. EMBO Rep 10:459–465

Montanaro L, Trere D, Derenzini M (2008) Nucleolus, ribosomes, and cancer. Am J Pathol 173:301–310

Moore JBT, Farrar JE, Arceci RJ, Liu JM, Ellis SR (2010) Distinct ribosome maturation defects in yeast models of Diamond-Blackfan anemia and Shwachman-Diamond syndrome. Haematologica 95:57–64

Nakhoul H, Ke J, Zhou X, Liao W, Zeng SX, Lu H (2014) Ribosomopathies: mechanisms of disease. Clin Med Insights Blood Disorders 7:7–16

Narla A, Ebert BL (2010) Ribosomopathies: human disorders of ribosome dysfunction. Blood 115:3196–3205

Ng CL, Waterman DG, Koonin EV, Walters AD, Chong JP, Isupov MN, Lebedev AA, Bunka DH, Stockley PG, Ortiz-Lombardia M et al (2009) Conformational flexibility and molecular interactions of an archaeal homologue of the Shwachman-Bodian-Diamond syndrome protein. BMC Struct Biol 9:32

Panse VG, Johnson AW (2010) Maturation of eukaryotic ribosomes: acquisition of functionality. Trends Biochem Sci 35:260–266

Pettersen EF, Goddard TD, Huang CC, Couch GS, Greenblatt DM, Meng EC, Ferrin TE (2004) UCSF Chimera–a visualization system for exploratory research and analysis. J Comput Chem 25:1605–1612

Rath BK, Frank J (2004) Fast automatic particle picking from cryo-electron micrographs using a locally normalized cross-correlation function: a case study. J Struct Biol 145:84–90

Roy A, Kucukural A, Zhang Y (2010) I-TASSER: a unified platform for automated protein structure and function prediction. Nat Protoc 5:725–738

Ruggero D, Pandolfi PP (2003) Does the ribosome translate cancer? Nat Rev Cancer 3:179–192

Savchenko A, Krogan N, Cort JR, Evdokimova E, Lew JM, Yee AA, Sanchez-Pulido L, Andrade MA, Bochkarev A, Watson JD et al (2005) The Shwachman-Bodian-Diamond syndrome protein family is involved in RNA metabolism. J Biol Chem 280:19213–19220

Scheres SHW (2012) RELION: implementation of a Bayesian approach to cryo-EM structure determination. J Struct Biol 180:519–530

Scheres SHW, Nunez-Ramirez R, Sorzano COS, Carazo JM, Marabini R (2008) Image processing for electron microscopy single-particle analysis using XMIPP. Nat Protoc 3:977–990

Shaikh TR, Gao H, Baxter WT, Asturias FJ, Boisset N, Leith A, Frank J (2008) SPIDER image processing for single-particle reconstruction of biological macromolecules from electron micrographs. Nat Protoc 3:1941–1974

Shammas C, Menne TF, Hilcenko C, Michell SR, Goyenechea B, Boocock GR, Durie PR, Rommens JM, Warren AJ (2005) Structural and mutational analysis of the SBDS protein family. Insight into the leukemia-associated Shwachman-Diamond Syndrome. J Biol Chem 280:19221–19229

Singh P, Panchaud A, Goodlett DR (2010) Chemical cross-linking and mass spectrometry as a low-resolution protein structure determination technique. Anal Chem 82:2636–2642

Soudet J, Gelugne JP, Belhabich-Baumas K, Caizergues-Ferrer M, Mougin A (2010) Immature small ribosomal subunits can engage in translation initiation in Saccharomyces cerevisiae. EMBO J 29:80–92

Strunk BS, Novak MN, Young CL, Karbstein K (2012) A translation-like cycle is a quality control checkpoint for maturing 40S ribosome subunits. Cell 150:111–121

Sulima SO, Gulay SP, Anjos M, Patchett S, Meskauskas A, Johnson AW, Dinman JD (2014a) Eukaryotic rpL10 drives ribosomal rotation. Nucl Acids Res 42:2049–2063

Sulima SO, Patchett S, Advani VM, De Keersmaecker K, Johnson AW, Dinman JD (2014b) Bypass of the pre-60S ribosomal quality control as a pathway to oncogenesis. Proc Natl Acad Sci USA 111:5640–5645

Tang G, Peng L, Baldwin PR, Mann DS, Jiang W, Rees I, Ludtke SJ (2007) EMAN2: An extensible image processing suite for electron microscopy. J Struct Biol 157:38–46

Teng T, Thomas G, Mercer CA (2013) Growth control and ribosomopathies. Curr Opin Genet Dev 23:63–71

Ueta M, Yoshida H, Wada C, Baba T, Mori H, Wada A (2005) Ribosome binding proteins YhbH and YfiA have opposite functions during 100S formation in the stationary phase of *Escherichia coli*. Genes Cells 10:1103–1112

Vitiello SP, Benedict JW, Padilla-Lopez S, Pearce DA (2010) Interaction between Sdo1p and Btn1p in the *Saccharomyces cerevisiae* model for Batten disease. Hum Mol Genet 19:931–942

Wada A, Igarashi K, Yoshimura S, Aimoto S, Ishihama A (1995) Ribosome modulation factor: stationary growth phase-specific inhibitor of ribosome functions from Escherichia coli. Biochem Biophys Res Commun 214:410–417

Watanabe KI, Dror Y (2005) Characterization of siRNA-mediated SBDS-knockdown cells: Specific hypersensitivity to Fas stimulation. Blood 106:116a

Watanabe K, Ambekar C, Wang H, Ciccolini A, Schimmer AD, Dror Y (2009) SBDS-deficiency results in specific hypersensitivity to Fas stimulation and accumulation of Fas at the plasma membrane. Apoptosis 14:77–89

Weis F, Giudice E, Churcher M, Jin L, Hilcenko C, Wong CC, Traynor D, Kay RR, Warren AJ (2015) Mechanism of eIF6 release from the nascent 60S ribosomal subunit. Nat Struct Mol Biol 22:914–919

West M, Hedges JB, Chen A, Johnson AW (2005) Defining the order in which Nmd3p and Rpl10p load onto nascent 60S ribosomal subunits. Mol Cell Biol 25:3802–3813

Wong CC, Traynor D, Basse N, Kay RR, Warren AJ (2011) Defective ribosome assembly in Shwachman-Diamond syndrome. Blood 118:4305–4312

Woolford JL Jr, Baserga SJ (2013) Ribosome biogenesis in the yeast Saccharomyces cerevisiae. Genetics 195:643–681

Yang B, Wu YJ, Zhu M, Fan SB, Lin J, Zhang K, Li S, Chi H, Li YX, Chen HF et al (2012) Identification of cross-linked peptides from complex samples. Nat Methods 9:904–906

Yoshida H, Wada A (2014) The 100S ribosome: ribosomal hibernation induced by stress. Wiley Interdiscip Rev RNA 5:723–732

Zhang X, Yan K, Zhang Y, Li N, Ma C, Li Z, Zhang Y, Feng B, Liu J, Sun Y et al (2014a) Structural insights into the function of a unique tandem GTPase EngA in bacterial ribosome assembly. Nucl Acids Res 42:13430–13439

Zhang Y, Ma C, Yuan Y, Zhu J, Li N, Chen C, Wu S, Yu L, Lei J, Gao N (2014b) Structural basis for interaction of a cotranslational chaperone with the eukaryotic ribosome. Nat struct Mol Biol 21:1042–1046

4

Reprogramming cell fates by small molecules

Xiaojie Ma, Linghao Kong, Saiyong Zhu✉

Life Sciences Institute, Zhejiang University, Hangzhou 310058, China
✉ Correspondence: saiyong@zju.edu.cn (S. Zhu)

ABSTRACT

Reprogramming cell fates towards pluripotent stem cells and other cell types has revolutionized our understanding of cellular plasticity. During the last decade, transcription factors and microRNAs have become powerful reprogramming factors for modulating cell fates. Recently, many efforts are focused on reprogramming cell fates by non-viral and non-integrating chemical approaches. Small molecules not only are useful in generating desired cell types *in vitro* for various applications, such as disease modeling and cell-based transplantation, but also hold great promise to be further developed as drugs to stimulate patients' endogenous cells to repair and regenerate *in vivo*. Here we will focus on chemical approaches for generating induced pluripotent stem cells, neurons, cardiomyocytes, hepatocytes and pancreatic β cells. Significantly, the rapid and exciting advances in cellular reprogramming by small molecules will help us to achieve the long-term goal of curing devastating diseases, injuries, cancers and aging.

KEYWORDS reprogramming, small molecules, stem cells, cell fates

INTRODUCTION

Studies of developmental biology have demonstrated that the zygote is totipotent and can gradually differentiate into all tissues and organs of the human body, that is composed by more than 200 types of specialized cells. However, the demonstration of interchanging cell fates and converting one cell type into another by different strategies, such as nuclear transfer or cell fusion, has challenged the traditional cognition of development and regeneration (Gurdon, 1962; Tada et al., 1997). In 2006, Yamanaka and Takahashi reported the development of the induced pluripotent stem cell (iPSC) technology, which can convert somatic cells to iPSCs using four transcription factors (TFs) (Takahashi and Yamanaka, 2006). During the last 10 years, iPSC technology has provided many fundamental insights into the molecular mechanisms of cell fate transition, and already demonstrated promising results in various applications, including disease modeling, drug screening, and cell-based therapy (Takahashi and Yamanaka, 2016). Recently, different combinations of lineage-specific TFs have been screened and applied to generate various cell types, including neurons, cardiomyocytes, hepatocytes, and pancreatic β cells (Huang et al., 2011; Ieda et al., 2010; Vierbuchen et al., 2010; Zhou et al., 2008). However, TF-based reprogramming approaches face many challenges in efficiency, safety, and *in vivo* delivery. As a novel and promising solution, small molecules are easy to apply and remove, more efficient and amenable to scale up. Small molecules not only are useful in generating desired cell types *in vitro* for various applications, but also can be further developed as drugs to stimulate patients' endogenous cells to repair and regenerate *in vivo*.

In this review, we will focus on making iPSCs, neurons, cardiomyocytes, hepatocytes and pancreatic β cells by small molecules (Fig. 1). Many interesting small molecules have been identified that can significantly promote cellular reprogramming (Table 1). The rapid and exciting advances in cellular reprogramming by small molecules will undoubtedly advance biomedical studies and clinical translation.

PLURIPOTENT REPROGRAMMING

In 2006, the first generation of iPSCs were derived by using four transcription factors (Oct4, Sox2, Klf4, and c-Myc, termed OSKM) (Takahashi and Yamanaka, 2006). During the last 10 years, substantial efforts have been made to improve the efficiency and safety of iPSC reprogramming by non-integrating viral vectors, synthetic RNAs, recombinant proteins and small molecules (Gonzalez et al., 2011). Particularly, screenings have been applied to identify chemical compounds that can replace individual or multiple

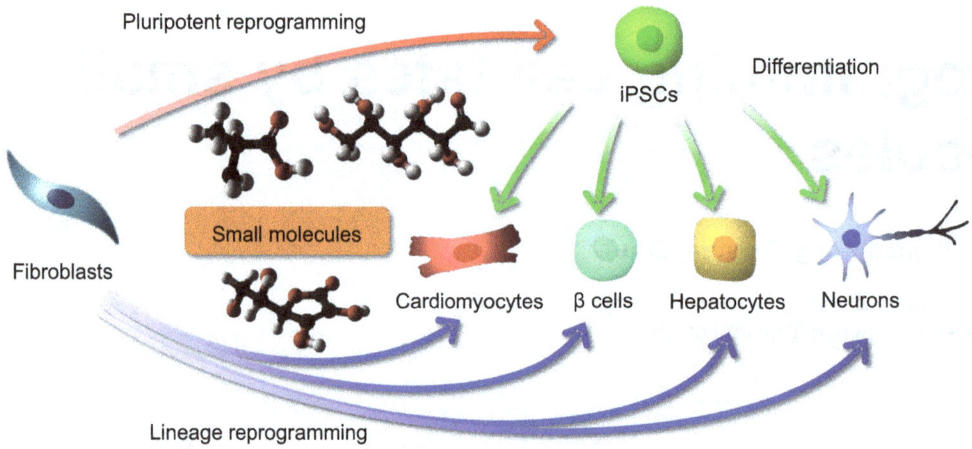

Figure 1. Reprogramming cell fates by small molecules. Chemical approaches can be widely applied to manipulate cell fates and states, including pluripotent reprogramming, directed differentiation, and lineage reprogramming. Small molecules not only are useful in generating functional cell types, such as cardiomyocytes, hepatocytes, pancreatic β cells, and neurons, but also can provide a better understanding of the detailed mechanisms underlying specific reprogramming processes.

Yamanaka factors. Ichida et al. carried out a high-content chemical screening and identified RepSox (also named as E616452, an ALK5 inhibitor) as a chemical replacer of Sox2 (Ichida et al., 2009). The authors suggested that RepSox promoted the completion of reprogramming through induction of *Nanog*. In addition, RepSox possibly promoted mesenchymal-to-epithelial transition (MET) during the early phase of reprogramming (Li et al., 2010). Using a *Nanog*-Luciferase reporter as readout, Kenpaullone was found to functionally replace Klf4 in the presence of OSM (Lyssiotis et al., 2009). Shi et al. found that the combination of BIX01294 and Bayk8644 or BIX01294 and RG108 enabled OK-induced reprogramming of mouse fibroblasts, and Li et al. further reported successful OK-reprogramming of human somatic cells by using CHIR99021 and Tranylcypromine (also named Parnate) (Li et al., 2009; Shi et al., 2008). Huangfu et al. reported that Valproic acid (VPA) could promote OS-induced reprogramming of human fibroblasts (Huangfu et al., 2008). Taken together, these studies demonstrated that epigenetic regulators play important roles in iPSC reprogramming. Subsequently, several groups have focused on Oct4-only reprogramming by employing different chemical cocktails. Chen et al. found that bone morphogenetic proteins (BMPs) could support efficient mouse Oct4-alone reprogramming partly through the promotion of MET (Chen et al., 2011). Interestingly, Yuan et al. reported that AMI-5 and A83-01 enabled Oct4-induced reprogramming of mouse fibroblasts. Remarkably, these Oct4-only iPSCs were truly pluripotent and could give rise to live-born pups by tetraploid complementation assay (Yuan et al., 2011). Furthermore, Li et al. identified a specific chemical combination, including VPA, CHIR99021, E616452 and Tranylcypromine, which was sufficient to reprogramming mouse fibroblasts to iPSCs with Oct4 alone (Li et al., 2011). Subsequently, some interesting chemical substitutes of Oct4 in mouse iPS-reprogramming were identified, including Forskolin, D4476, and 2-Methyl-5-hydroxytryptamine (2-Me-5-HT) (Hou et al., 2013).

All chemical induced pluripotent reprogramming can provide a promising paradigm for cell fate transitions. In 2013, Hou et al. demonstrated successful reprogramming of mouse cells into iPS cells for the first time by using a novel cocktail of seven small molecules, including VPA, CHIR99021, E616452, Tranylcypromine, Forskolin, 3-deazaneplanocin A (DZNep), and TTNPB (Hou et al., 2013). Further refinement of this all-chemical reprogramming led a 1000-fold greater efficiency with additional small molecules, including AM580, EPZ004777, SGC0946, and 5-aza-2-deoxycitidine (5-aza-dC) (Zhao et al., 2015b). Interestingly, chemical induced pluripotent reprogramming process required the early formation of extra-embryonic endoderm (XEN)-like cells and a late transition from XEN-like cells to iPSCs, which fundamentally distinct from the pathway of TF-induced pluripotent reprogramming. However, it is still largely unknown how small molecules activate the endogenous gene regulation network, and gradually

Table 1. Representative small molecules for reprogramming cell fates

Name	Chemical formula	Function	Application
5-Azacytidine (5-aza)	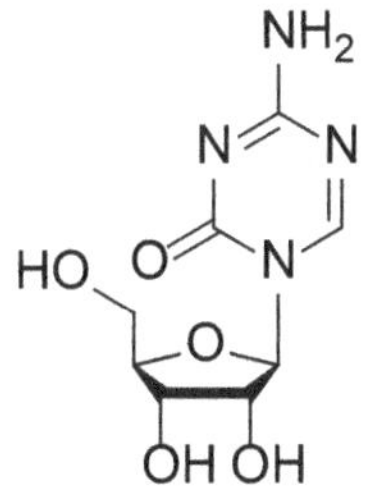	DNA methyltransferase inhibitor	iPSC reprogramming
5-Aza-2'-deoxycitidine (5-aza-dc)	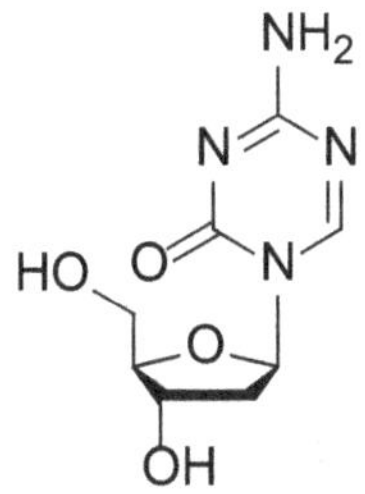	DNA methyltransferase inhibitor	iPSC reprogramming
RepSox (E616452)	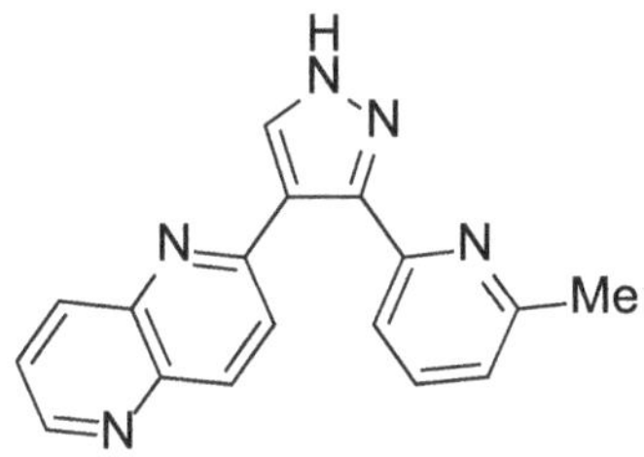	ALK5 inhibitor	iPSC reprogramming
Kenpaullone	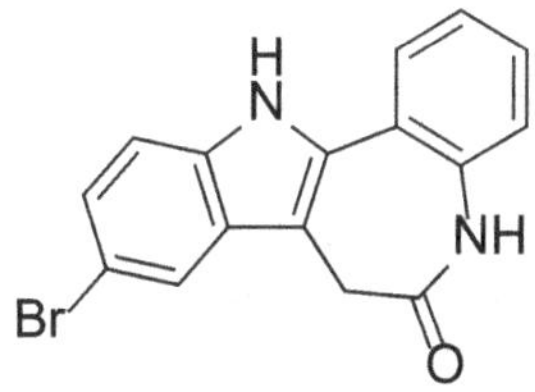	GSK3β inhibitor, CDK inhibitor	iPSC reprogramming
BIX01294	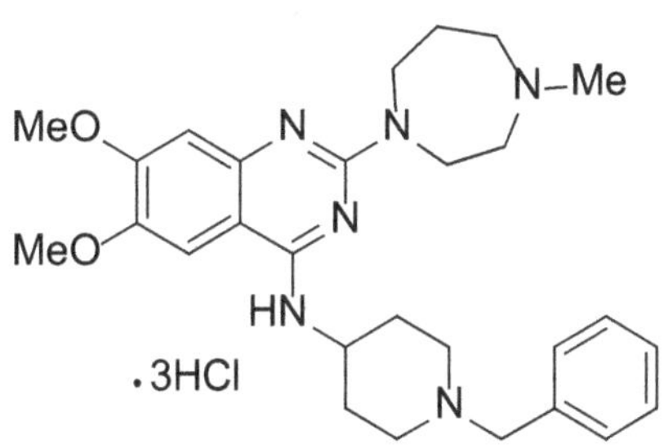	G9a inhibitor	iPSC, iCM, pancreatic β-like cell reprogramming
Bayk8644	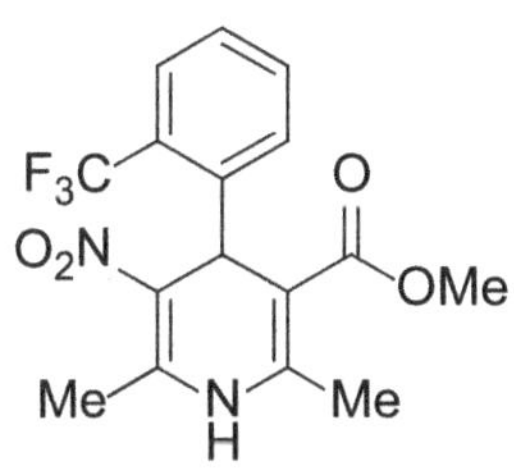	L-type Ca^{2+} channel agonist	iPSC and pancreatic β-like cell reprogramming

Table 1 continued

Name	Chemical formula	Function	Application
RG108	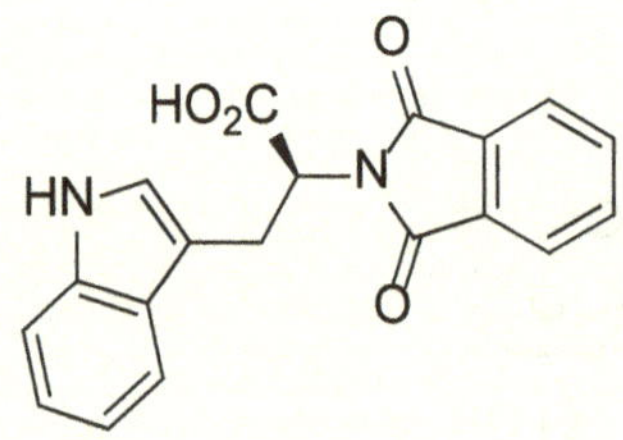	DNA methyltransferase inhibitor	iPSC, iNSC, pancreatic β-like cell reprogramming
CHIR99021	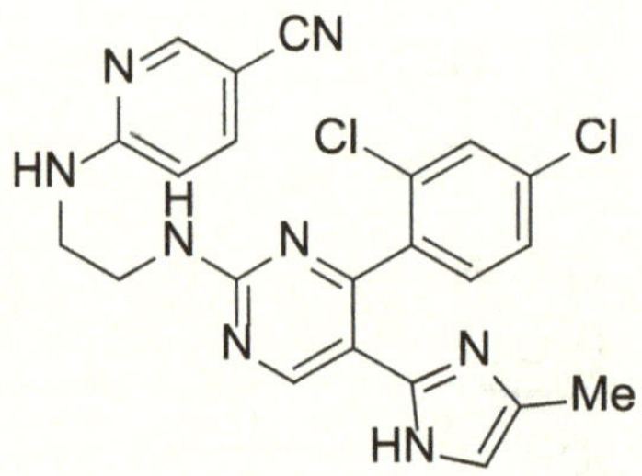	GSK3β inhibitor	iPSC, iN, iNSC, iCM, iCPC, iHep reprogramming
Tranylcypromine (Parnate)	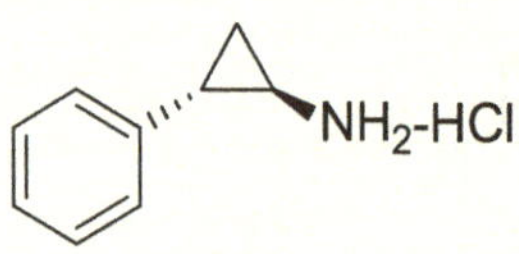	Lysine-specific demethylase 1 inhibitor	iPSC,iNSC, iCM reprogramming
Valproic acid (VPA)	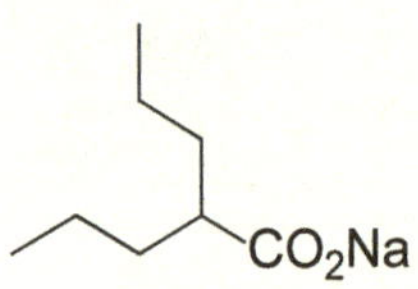	Histone deacetylase inhibitor	iPSC, iNSC, iCM reprogramming
AMI-5	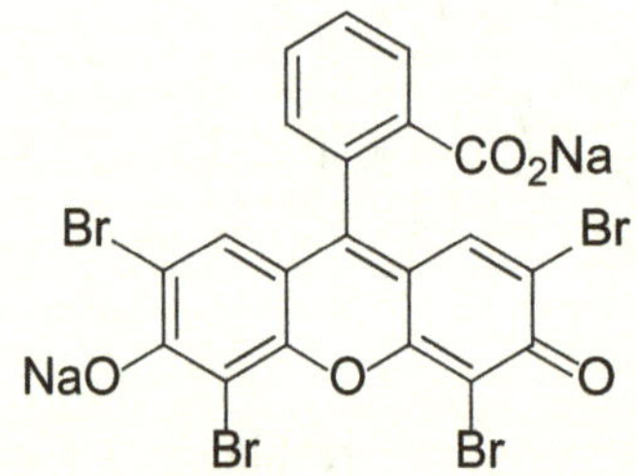	Protein arginine methyltransferase inhibitor	iPSC reprogramming
A83-01	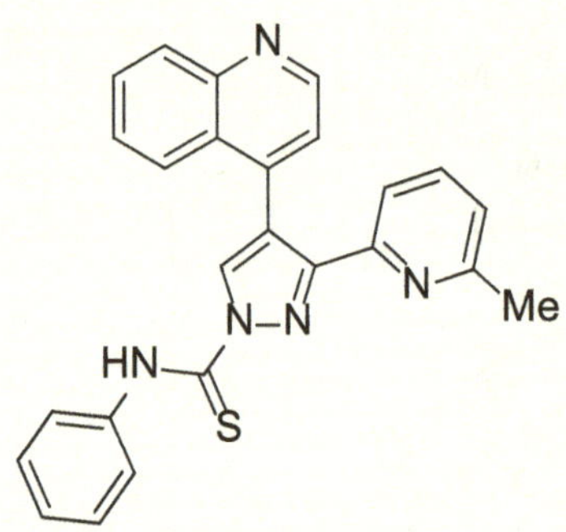	TGFβ signaling pathway inhibitor	iPSC, iNSC, iCM, iHep reprogramming

Table 1 continued

Name	Chemical formula	Function	Application
Forskolin	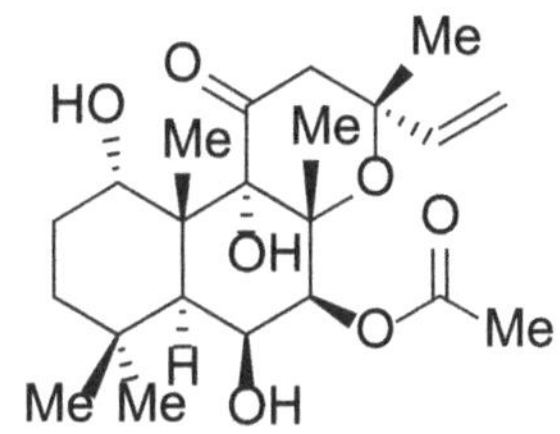	cAMP signaling activator	iPSC, iN, iCM reprogramming
2-Methyl-5-hydroxytryptamine (2-Me-5-HT)	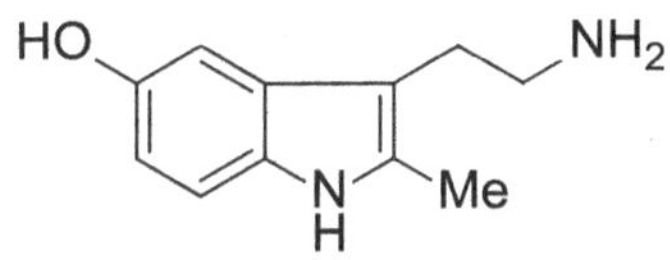	5-HT receptor agonist	iPSC reprogramming
D4476	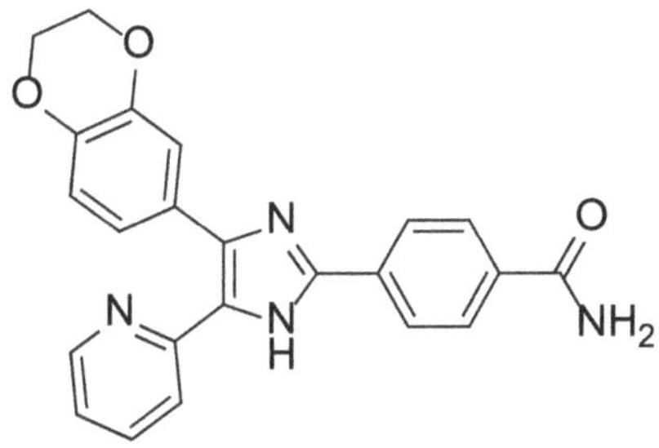	Casein kinase inhibitor	iPSC reprogramming
3-Deazaneplanocin (DZNep)	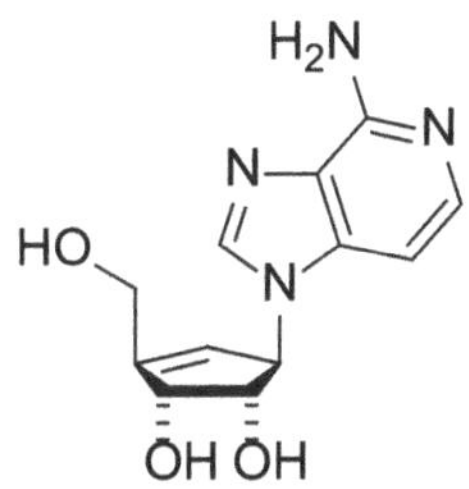	DNA methyltransferase inhibitor	iPSC reprogramming
AM580	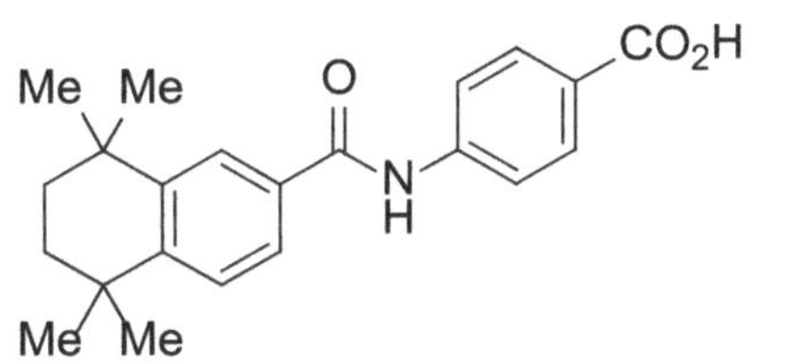	RARα agonist	iPSC reprogramming
EPZ004777	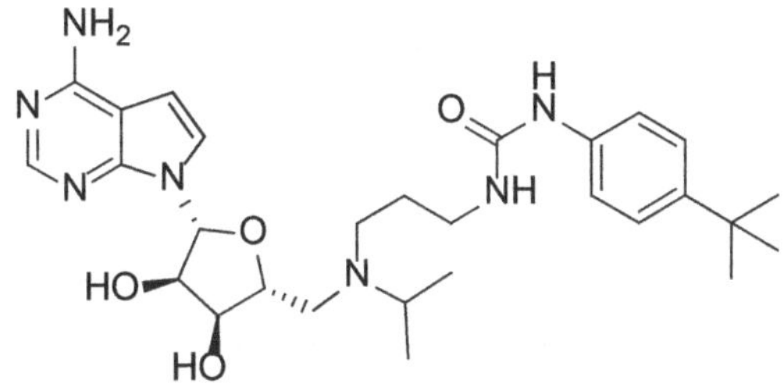	DOT1L inhibitor	iPSC reprogramming

Table 1 continued

Name	Chemical formula	Function	Application
SGC0946	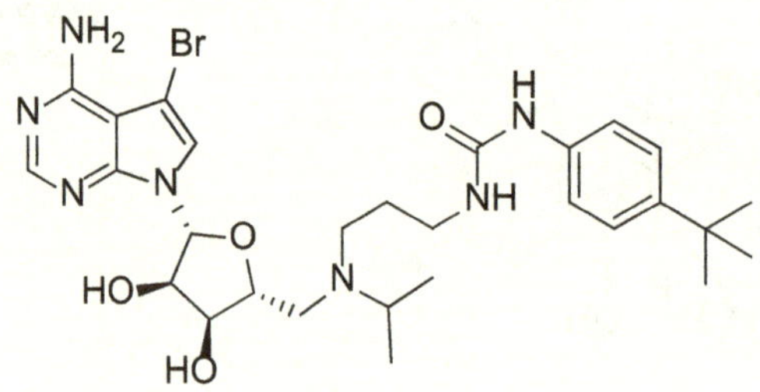	DOT1L inhibitor	iPSC reprogramming
Sodium butyrate (NaB)	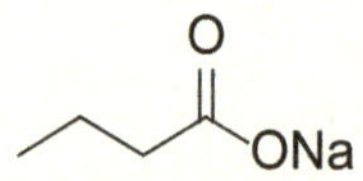	Histone deacetylase inhibitor	iPSC and iNSC reprogramming
PS48	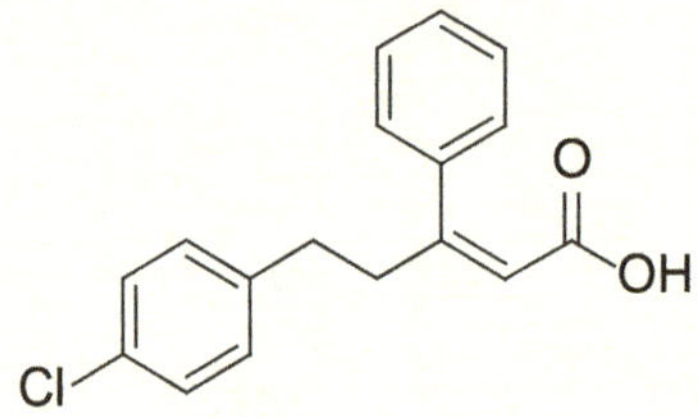	PDK1 activator	iPSC reprogramming
PD0325901	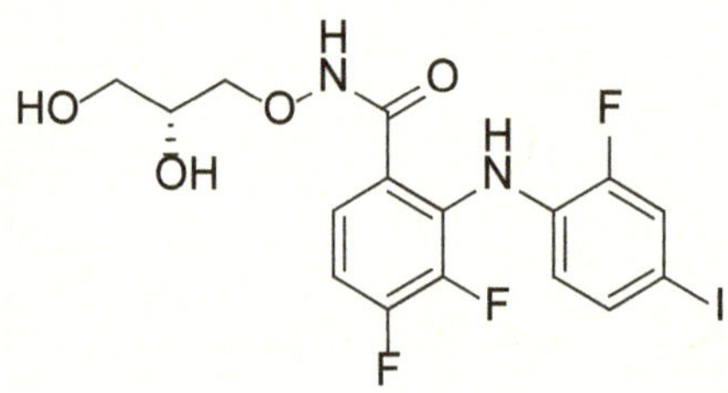	MAPK/ERK signaling pathway inhibitor	iPSC reprogramming
ISX9	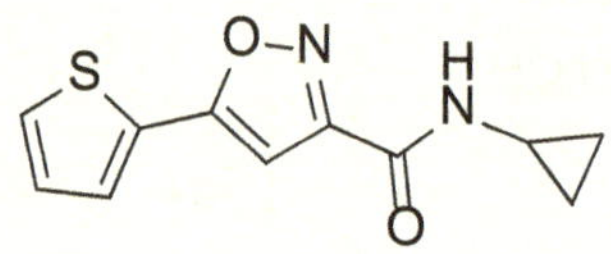	Activates the expression of endogenous neurogenic transcription factors	iN reprogramming
I-BET151	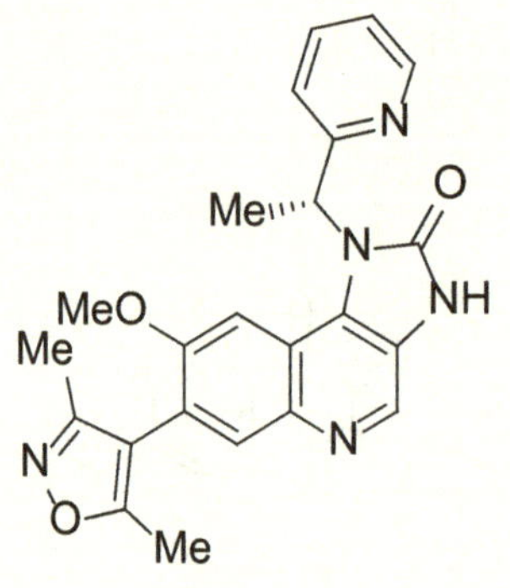	BET family protein inhibitor	iN reprogramming

Table 1 continued

Name	Chemical formula	Function	Application
SP600125	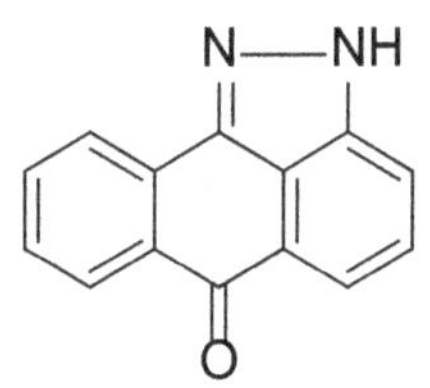	JNK inhibitor	iN and iNSC reprogramming
GO6983	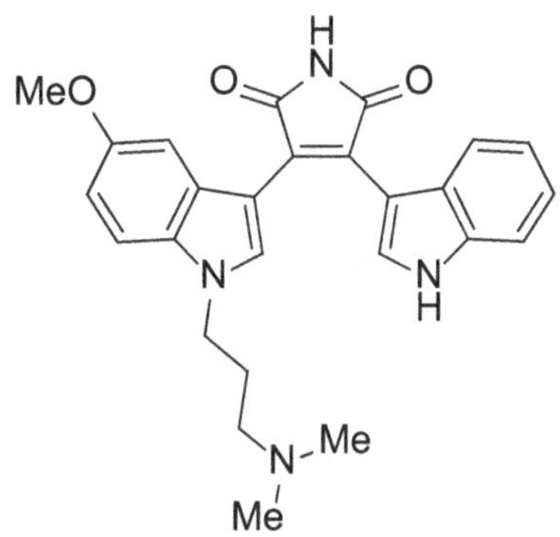	Pan-PKC inhibitor	iN reprogramming
Dorsomorphin	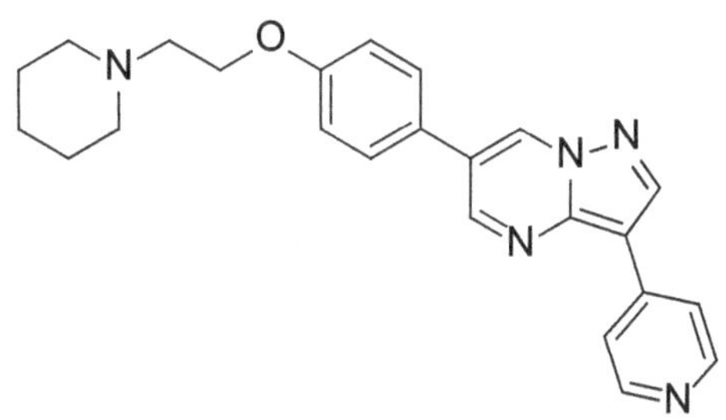	BMP signaling inhibitor	iN reprogramming
LDN193189	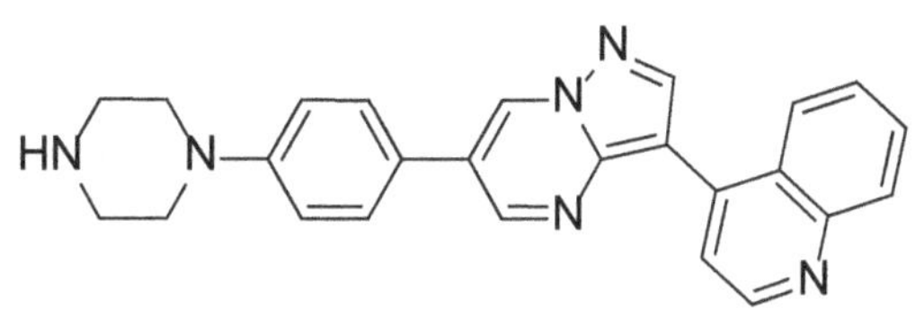	BMP signaling inhibitor	iN, iNSC reprogramming
SB431542	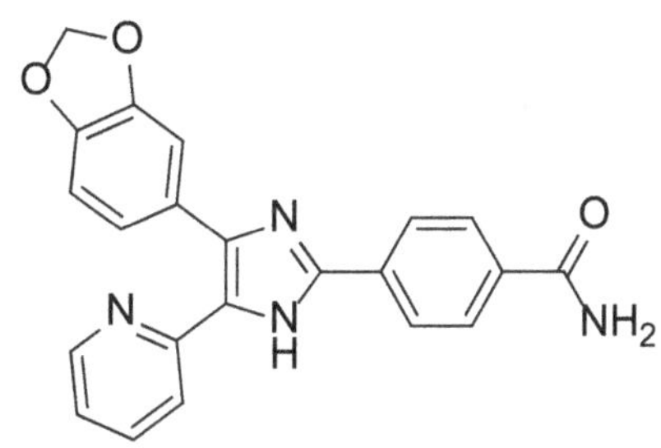	TGFβ signaling pathway inhibitor	iN, iCM, pancreatic β-like cell reprogramming

Table 1 continued

Name	Chemical formula	Function	Application
TTNPB	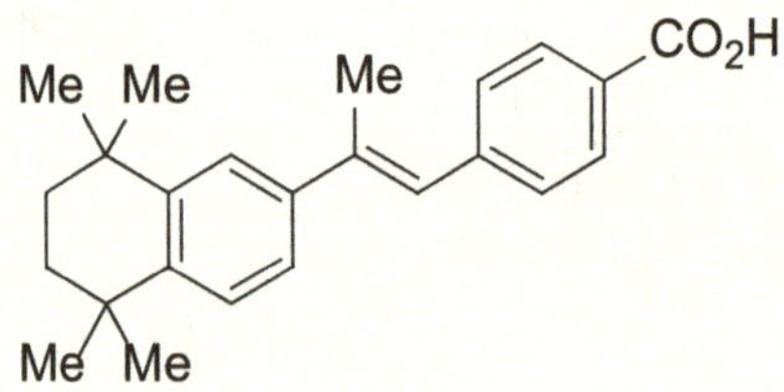	Synthetic retinoic acid receptor ligand	iN reprogramming
Thiazovivin (Tzv)	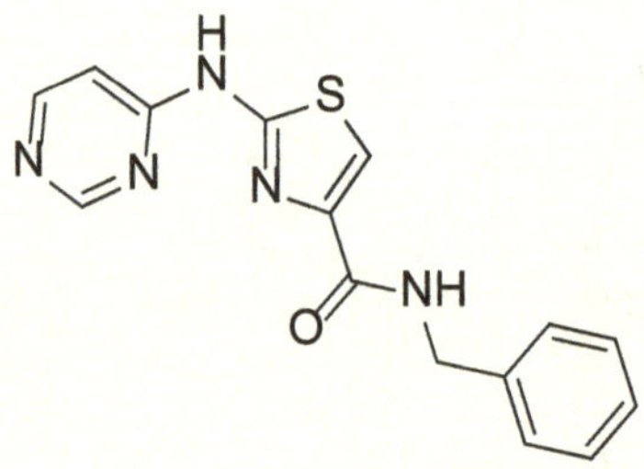	ROCK inhibitor	iPSC, iN reprogramming
DAPT	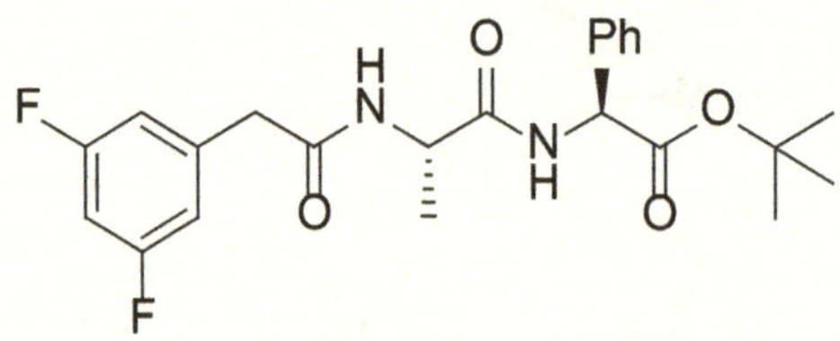	γ-secretase inhibitor	iN reprogramming
Smoothened agonist (SAG)	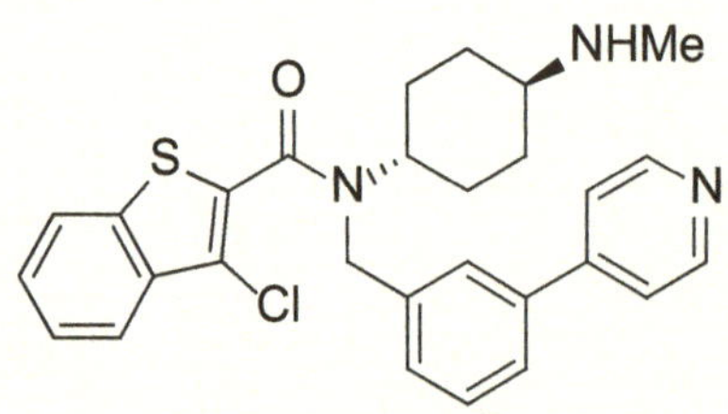	Hedgehog signaling activator	iN reprogramming
Purmorphamine (Purmo)	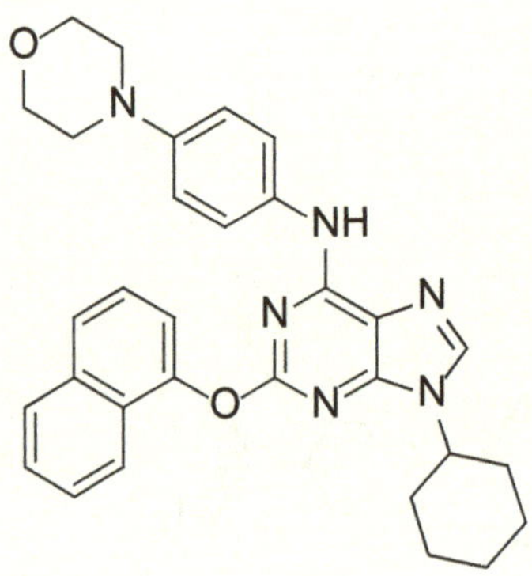	Hedgehog signaling activator	iN reprogramming

Table 1 continued

Name	Chemical formula	Function	Application
Lysophosphatidic acid (LPA)	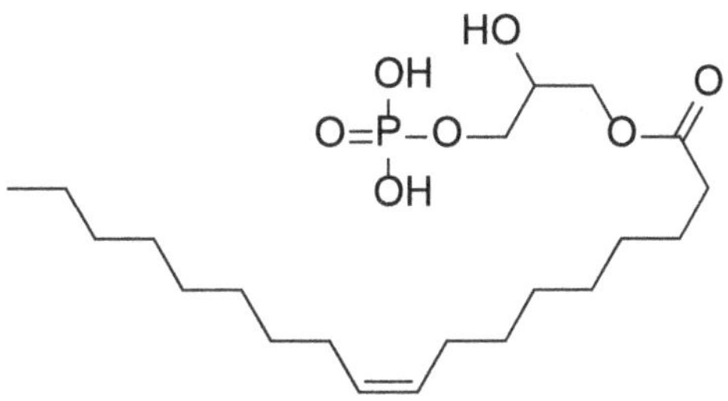	GPCR activator	iNSC reprogramming
Rolipram	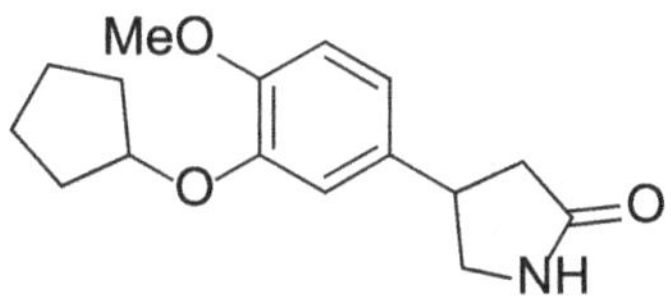	Phosphodiesterase-4 inhibitor	iNSC, iCM reprogramming
Retinoic acid (RA)	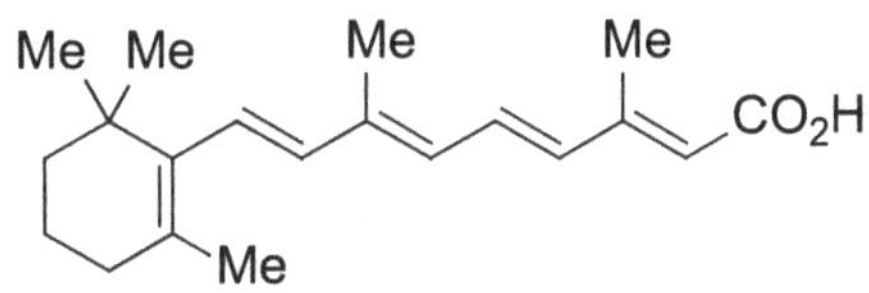	RAR receptor	iNSC reprogramming
Hh-Ag1.5	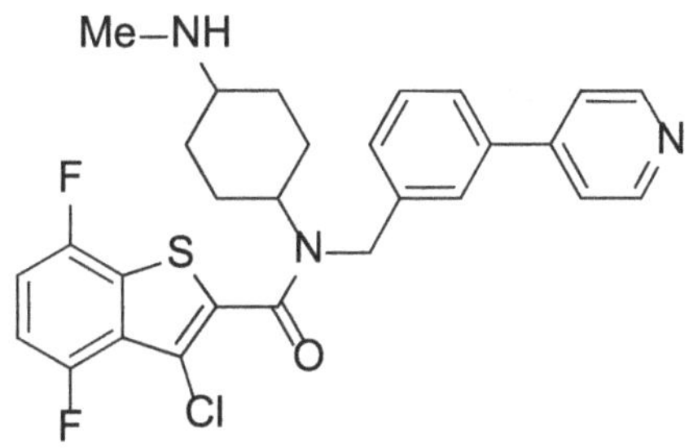	Hedgehog signaling activator	iNSC reprogramming
SMER28	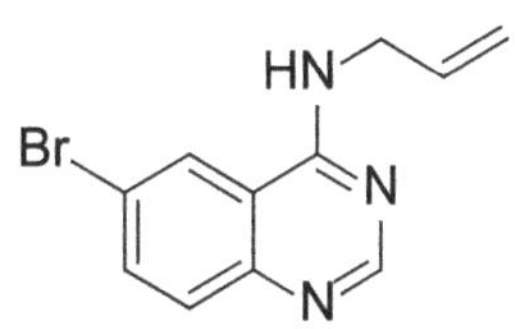	Autophagy enhancer	iNSC reprogramming
Y-27632	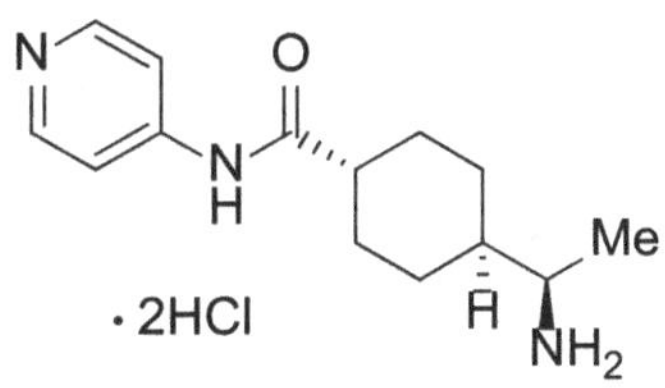	ROCK1 inhibitor	iN, iCM reprogramming

Table 1 continued

Name	Chemical formula	Function	Application
JAK inhibitor I	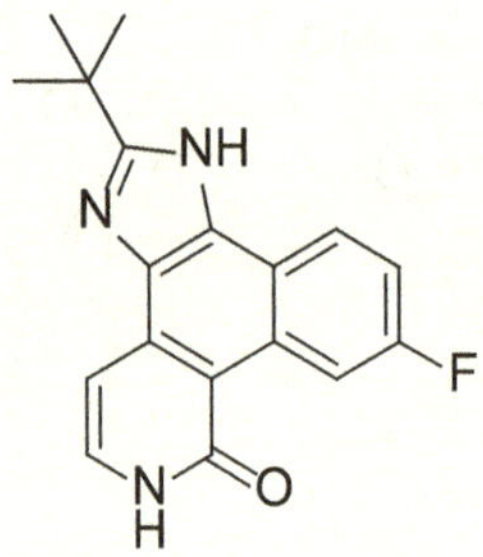	JAK inhibitor	iCM reprogramming
LY-364947	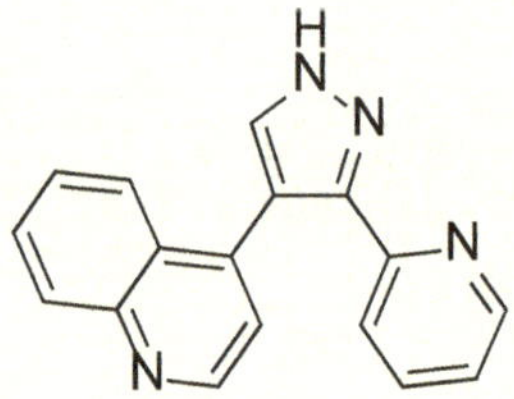	TGFβR-I inhibitor	iCM reprogramming
SD-208	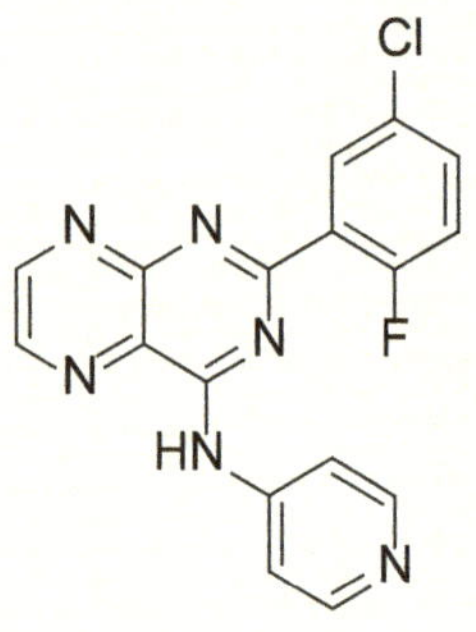	TGFβR-I inhibitor	iCM reprogramming
GW788388	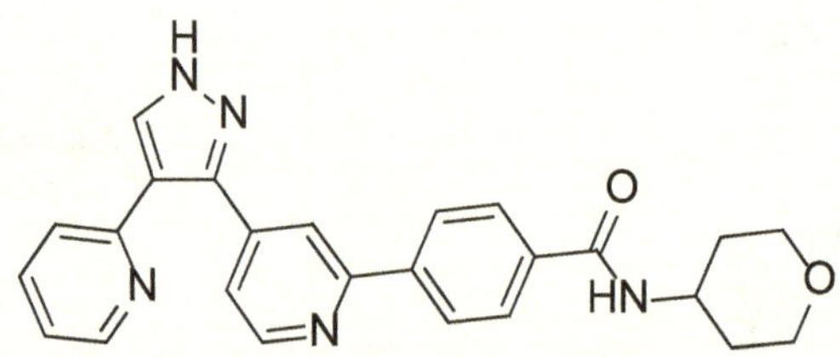	ALK5 inhibitor	iCM reprogramming
ICARIIN	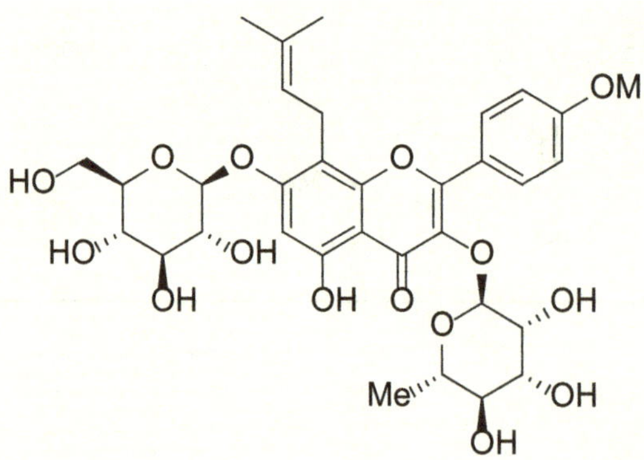	Exhibits a variety of pharmacological actions	iCM reprogramming

Table 1 continued

Name	Chemical formula	Function	Application
PD169316	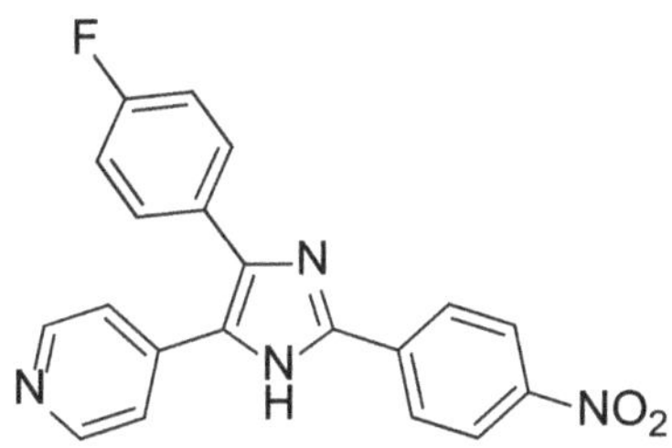	p38 inhibitor	iCM reprogramming
AS8351	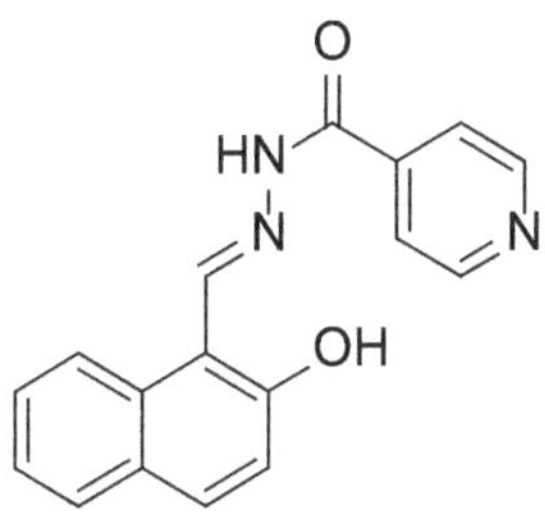	KDM5B inhibitor	iCM reprogramming
SC1	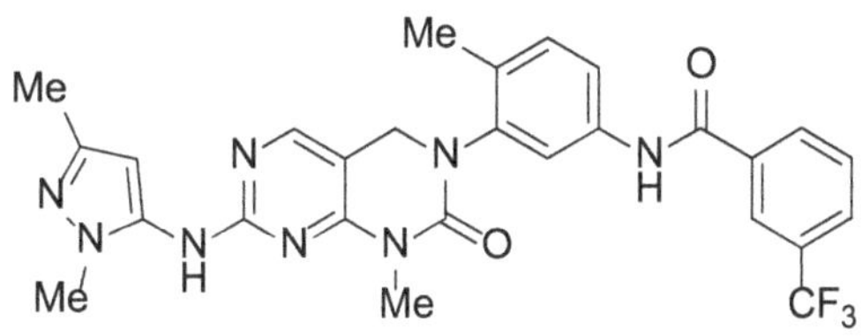	ERK1, RasGAP inhibitor	iCM reprogramming
OAC2	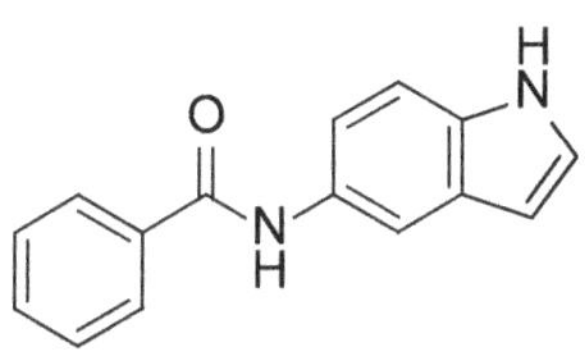	Oct4 activator	iCM reprogramming
SU16F	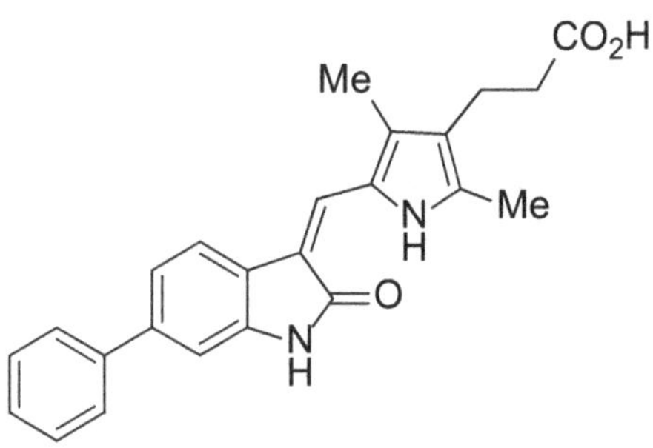	PDGFR inhibitor	iCM reprogramming
JNJ10198409	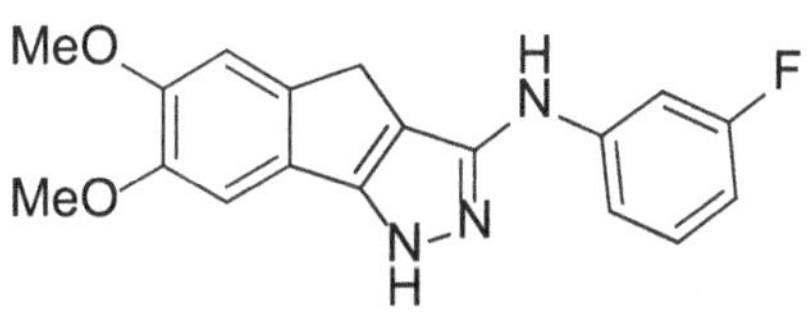	PDGFR inhibitor	iCM reprogramming

Table 1 continued

Name	Chemical formula	Function	Application
SU5402	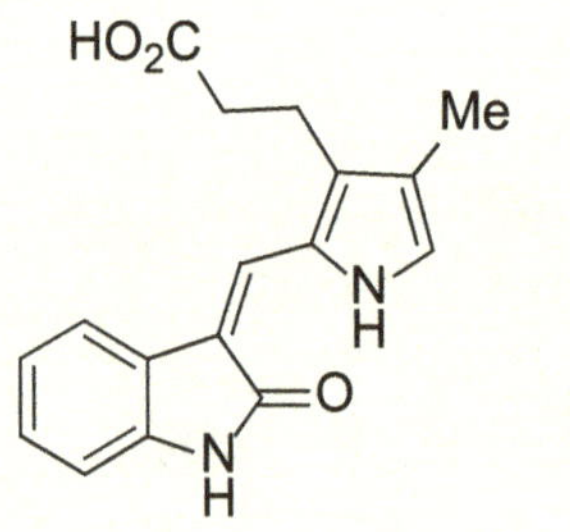	FGFR, VEGFR, and PDGFR inhibitor	iCPC reprogramming
Compound E	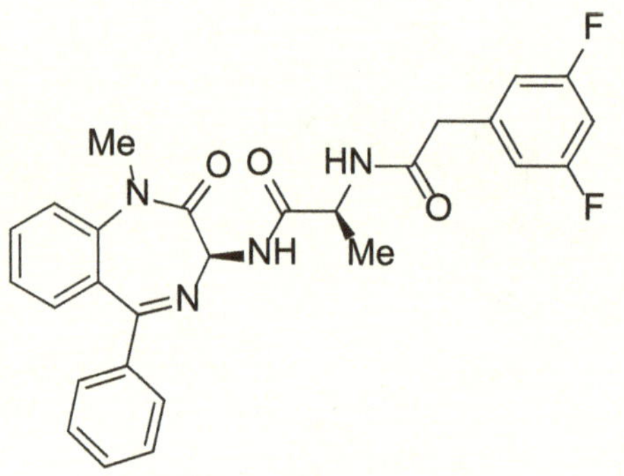	Notch signaling inhibitor	iHep, pancreatic β-like cell reprogramming
Nicotinamide	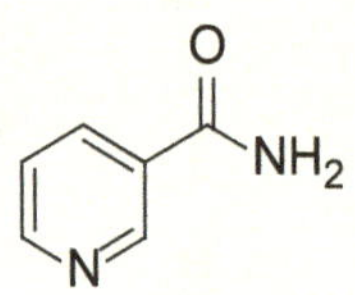	Coenzyme, cofactor	Pancreatic β-cell reprogramming
5'-N-ethylcarboxamidoadenosine (NECA)	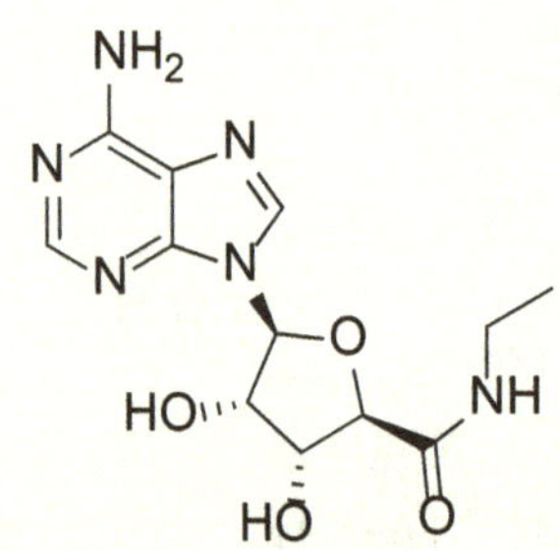	Adenosine receptor agonist	Pancreatic β-like cell reprogramming
BRD7552	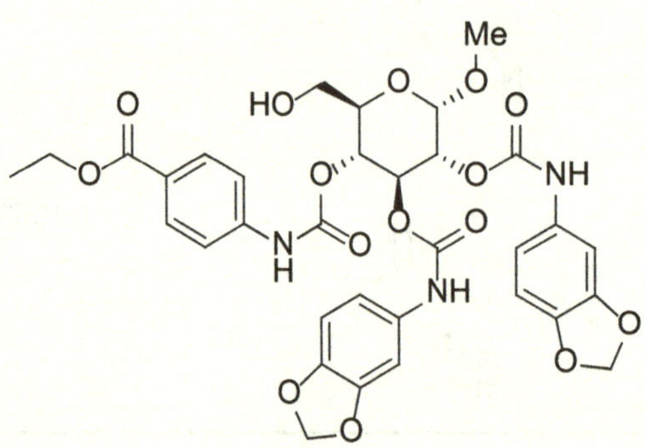	PDX1 expression inducer	Increases insulin expression

Table 1 continued

Name	Chemical formula	Function	Application
Vitamin C	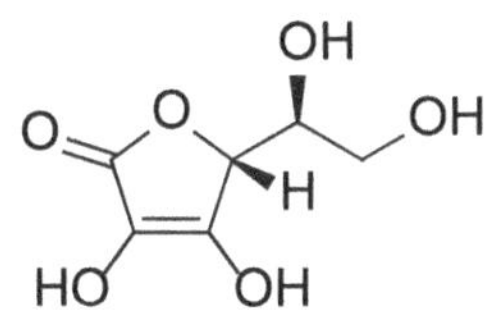	Cofactor of epigenetic modulators	Increases reprogramming efficiency

enable cell fate transitions. The remaining question is obvious: how to achieve chemical induced pluripotent reprogramming of human cells? We previously made some progress and reported that a chemical cocktail of Sodium butyrate (NaB), Parnate, PS48, CHIR99021, A83-01 and PD0325901 enabled OCT4-induced reprogramming of human primary somatic cells (Zhu et al., 2010). These human OCT4-only iPSCs were pluripotent and could give rise to cells of all three germ layers both *in vitro* and *in vivo*. Further mechanistic studies demonstrated that a metabolic switch from oxidative phosphorylation to glycolysis is an important step in iPSC reprogramming. More efforts are required to identify chemical substitutes for OCT4 in human cell reprogramming. The demonstration of human pluripotent reprogramming by an all-chemical approach will greatly benefit the study of stem cell biology and regenerative medicine.

NEURAL REPROGRAMMING

An increasing number of people suffer from neurodegenerative disorders, such as Alzheimer's disease and Parkinson's disease, as life expectancy further extends. It is not feasible to obtain sufficient amounts of patient specific neural cells for disease modeling and drug development. In the last few years, great progress has been made using TF-based direct reprogramming towards induced neurons (iNs) or induced neural stem cells (iNSCs).

In 2010, Vierbuchen et al. firstly reported that mesodermal mouse fibroblasts could be rapidly and efficiently reprogrammed into ectodermal iNs by using three TFs: Brn2, Ascl1, and Myt1 l (BAM) (Vierbuchen et al., 2010). Subsequently, BAM plus NEUROD1 was applied for neuronal reprogramming of human cells (Pang et al., 2011). So far, a number of groups have generated different subtypes of neurons by combining BAM with specific TFs that play an important role in the development of specific neuron subtypes (Caiazzo et al., 2011; Kim, 2011; Pang et al., 2011; Son et al., 2011). Besides TFs, microRNAs, such as miR-9 and miR-124, can also facilitate neuronal reprogramming (Ambasudhan et al., 2011; Yang et al., 2016; Yoo et al., 2011). Obviously, it is desirable to develop non-viral and non-integrative iN reprogramming approaches by small molecules. In 2015, two simultaneous articles reported the successful generation of iN from mouse and human fibroblasts via chemical-only approaches (Hu et al., 2015; Li et al., 2015). Li et al. showed that mouse fibroblasts could be converted into neurons by using an optimal cocktail of four small molecules (Forskolin, ISX9, CHIR99021, and I-BET151, termed FICB) (Li et al., 2015). The authors suggested that I-BET151, a BET family protein inhibitor, suppressed the fibroblast-specific program, and ISX9 activated the expression of endogenous neurogenic transcription factors, which synergistically promoted neuronal conversion (Li et al., 2015). Hu et al. demonstrated the generation of neurons from human fibroblasts using a small molecule cocktail (VPA, CHIR99021, RepSox, Forskolin, SP600125, GO6983, Y-27631, and Dorsomorphin) (Hu et al., 2015). Furthermore, via this chemical cocktail, fibroblasts from familial Alzheimer's disease patients could also be reprogrammed to iNs. Such cells can be used for *in vitro* disease modeling and drug screenings (Hu et al., 2015). Subsequently, Zhang et al. reported that sequential addition of a cocktail of small molecules (LDN193189, SB431542, TTNPB, Thiazovivin (Tzv), CHIR99021, VPA, DAPT, Smoothened agonist (SAG), and Purmorphamine) can reprogram human astrocytes into functional neurons (Zhang et al., 2015). Mechanistically, these small molecules inhibited glial but activated neuronal signaling pathways through epigenetic and transcriptional modulation. Remarkably, these human iNs were functional and could survive more than 5 months under cell culture conditions.

Compared with neurons, expandable and multipotent iNSCs are desirable for downstream applications, like disease modeling and drug screening. In the past few years, many groups reported the generation of iNSCs using neural lineage-specific TFs. These iNSCs are multipotent and can differentiate into functional neurons, astrocytes, and oligodendrocytes both *in vitro* and *in vivo* (Ring et al., 2012; Zhou and Tripathi, 2012). Recently, we achieved both mouse and human iNSC reprogramming by a cell-activation signaling-directed (CASD) strategy (Kim et al., 2011; Zhu et al., 2015). The CASD strategy uses transient exposure of somatic cells to reprogramming factors (Oct4, Sox2, Klf4, and c-Myc) in conjunction with soluble lineage-specific signals to reprogram cells into other cell types, such as

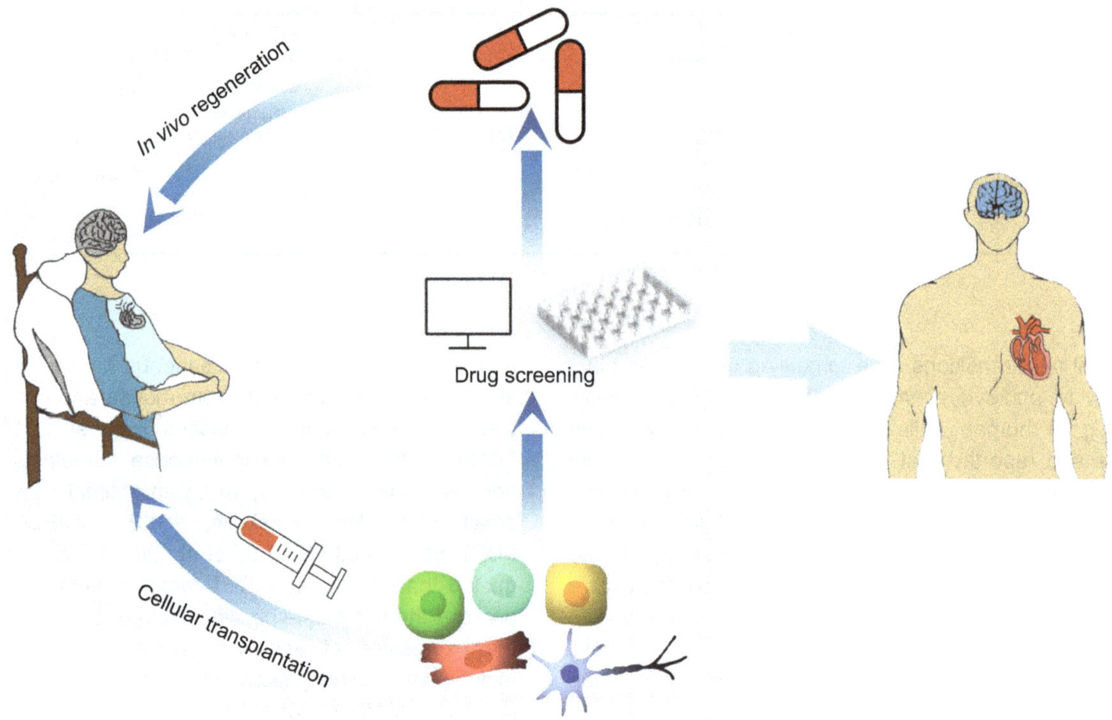

Figure 2. Potential applications of cellular reprogramming. Cellular reprogramming can provide a large number of functional cells, which can be used for cell-based transplantation and high-throughput chemical screenings. This technology will help develop drugs to stimulate patients' endogenous cells to repair and regenerate *in vivo* in the near future. Cellular reprogramming by only small molecules will significantly advance biomedical studies and clinical applications, and realize the long-term goal of curing degenerative diseases, injuries, and aging.

iNSCs. Several interesting small molecules could promote OCT4-mediated iNSC reprogramming process, including A83-01, CHIR99021, NaB, Lysophosphatidic acid (LPA), Rolipram and SP600125 (Zhu et al., 2014a). Furthermore, similar to iN reprogramming, there are also great advances in iNSC reprogramming by using small molecules alone. In 2014, Cheng and colleagues used three small molecules VPA, CHIR99021, and RepSox to derive iNPCs from somatic cells (Cheng et al., 2014). More recently, Zhang et al. achieved more efficient mouse iNSC reprogramming by using a cocktail of nine components (CHIR99021, LDN193189, A83-01, Retinoic acid (RA), Hh-Ag1.5, RG108, Parnate, SMER28, and bFGF) (Zhang et al., 2016a). They provided definitive evidence that these iNSCs could be reprogrammed from fibroblasts using a genetic lineage-tracing system. Interestingly, further mechanistic studies uncovered that these small molecules could gradually and specifically activate key neurogenic regulators, such as Sox2, and then facilitated the neural cell fate transition.

Direct *in vivo* reprogramming will provide a perspective for cell-based clinical regenerative therapy (Chen et al., 2015; Li and Chen, 2016). Glial cells are the most abundant cells in adult brains and several groups have reported the successful TF-based reprogramming of glial cells to neurons or iNPCs. Niu et al. found that delivery of Sox2 could reprogram endogenous astrocytes to proliferating neuroblasts and these neuroblasts further differentiated to functional neurons that integrated into neural networks in the brain (Niu et al., 2013). Guo et al. demonstrated that cortical glial cells activated by injury or disease could be reprogrammed by NeuroD1 *in vivo* (Guo et al., 2014). The further application of knowledge learned from *in vitro* chemical screening and ambitious *in vivo* chemical screening will undoubtedly advance this field.

CARDIAC REPROGRAMMING

The adult mammalian heart possesses little regenerative capacity following injury. Cardiac fibroblasts account for a majority of cells in the heart, and cardiac reprogramming holds great potentials. In 2010, Ieda et al. reported that postnatal cardiac fibroblasts could be directly reprogrammed into induced cardiomyocyte-like cells (iCMs) by transfection with a combination of three TFs (Gata4, Mef2c, Tbx5, termed GMT) (Ieda et al., 2010). Lineage-tracing experiments showed that the cardiac reprogramming with GMT was a direct conversion process. Subsequently, other groups showed that addition of TFs such as Hand2 and Nkx2.5 to GMT promoted the reprogramming efficiency or maturation of iCMs (Addis and Epstein, 2013; Ifkovits et al., 2014; Song et al., 2012). Additionally,miRNAs, such as miR-1 and miR-133, also play important roles in cardiac reprogramming (Ieda, 2016; Jayawardena et al., 2012; Muraoka et al., 2014; Nam et al., 2013; Zhao et al., 2015a). Although the efficiency of cardiac reprogramming has been improved in recent years, the molecular mechanisms of this process are largely unknown. More recently, Zhou et al. carried out a small-scale functional screening and identified that loss of Bmi1 significantly promoted mouse cardiac reprogramming. Mechanistically, Bmi1 blocked cardiac reprogramming through direct interactions with the regulatory regions of many cardiogenic genes (Zhou et al., 2016). Compared with mouse cardiac reprogramming, human cardiac reprogramming is much more challenging to be achieved. Fu et al. showed that GMT plus ESRRG, MESP1, Myocardin, and ZFPM2 could reprogram human fibroblasts to iCMs (Fu et al., 2013). Nam et al. demonstrated that human fibroblasts could be converted to iCMs by introduction of GATA4, HAND2, TBX5, Myocardin, miR-1, and miR-133 under long-term culture (Nam et al., 2013). Wada et al. found that GMT plus MESP1 and Myocardin could reprogram human fibroblasts to iCMs (Rie et al., 2013). Such TF-reprogrammed iCMs exhibited sarcomere formation, calcium transients, and action potentials, but could not contract spontaneously and robustly. These results indicated that further improvements are still required for complete human cardiac reprogramming.

Small molecules can potentially replace TFs and provide a novel approach for cardiac reprogramming. Interestingly, the addition of TGF-β inhibitor SB431542 or A83-01 improved the efficiency of iCM generation (Fu et al., 2013; Zhao et al., 2015a). Other small molecules, including Y-27632, JAK inhibitor I, LY-364947, SD-208 and GW788388, did also enhance cardiac reprogramming (Lalit et al., 2016; Zhao et al., 2015a, b). Tamakawa et al. showed that adding fibroblast growth factor (FGF) 2, FGF10, and vascular endothelial growth factor (VEGF) to GMT enhanced cardiac reprogramming about 100-fold compared with GMT only (Yamakawa et al., 2015). Wang et al. identified a small molecule cocktail, including CHIR99021, SB431542, Parnate, and Forskolin, that enabled efficient conversion of mouse fibroblasts into iCMs with Oct4 alone employing the CASD paradigm. These iCMs spontaneously contracted and exhibited a ventricular phenotype (Wang et al., 2014). Recently, there have been significant achievements in cardiac reprogramming by using small molecules alone. Fu et al. showed that a chemical cocktail (CHIR99021, RepSox, Forskolin, and VPA) could induce beating clusters of cardiac cells from mouse fibroblasts. Other small molecules, including ICARIIN, PD169316 and Rolipram also increased the efficiency of cardiac reprogramming (Fu et al., 2015). Nan et al. showed that human iCMs could be generated from fibroblasts by a specific set of small molecules (Cao et al., 2016). This combination of nine compounds was termed 9C, including CHIR99021, A83-01, BIX01294, AS8351, SC1, Y-27632, OAC2, SU16F, and JNJ10198409. Remarkably, at both transcriptome and epigenetic levels, these human iCMs resembled human cardiomyocytes. The authors observed enrichment of H3K4me3 as well as H3K27ac, and decrease of H3K27me3 on a cohort of heart developmental genes during 9C-induced cardiac reprogramming. Mechanistically, 9C treatment epigenetically and transcriptionally activated key cardiac developmental genes and facilitated the cardiac fate transition. Interestingly, unlike human iCMs from TF-based reprogramming, these human 9C-induced iCMs uniformly contracted. More efforts are required to apply this chemical cardiac reprogramming approach to patient cells.

Expandable and multipotent induced cardiac progenitor cells (iCPCs) are more suitable for basic studies and translational applications. Islas et al. found that human dermal fibroblasts could be converted into cardiac progenitors by ETS2 and MESP1 overexpression (Islas et al., 2012). After a screening of multiple cardiac TFs, Lalit et al. reported that a combination of Mesp1, Tbx5, Gata4, Nkx2.5, and Baf60c could reprogram mouse fibroblasts into iCPCs (Lalit et al., 2016). Meanwhile, Zhang et al. demonstrated that fibroblasts could be induced into expandable iCPCs by the CASD lineage conversion strategy and these iCPCs could be further long-term expanded using a combination of growth factors and small molecules (BMP4, Activin A, CHIR99021 and SU5402) (Zhang et al., 2016b). The activation of Wnt signaling by CHIR99021 not only promoted iCPC proliferation, but also inhibited further differentiation of iCPCs into cardiomyocytes. In addition, treating iCPC with Wnt inhibitor IWP2 facilitated even more efficient cardiac differentiation. Reprogramming mouse and human fibroblasts into expandable cardiac progenitors by small molecules alone could likely be achieved with further screening and modifications in the future.

Cardiac fibroblasts account for a majority of cells in the heart and represent a potential cell source for cardiac reprogramming. Direct *in vivo* cardiac reprogramming has already demonstrated some promising results. Qian et al. converted mouse resident non-myocytes into iCMs *in vivo* by local delivery of GMT after coronary ligation (Qian et al., 2012). Song et al. showed that GMT plus Hand2 could cooperatively reprogram dividing non-cardiomyocytes into functional iCMs in situ (Song et al., 2012). These cardiac reprogramming TF-delivered hearts had decreased infarct size and better cardiac function. Therefore, reprogramming cardiac fibroblasts to iCMs in their native environment by chemical compounds will hold great potentials in cardiac regenerative therapies.

HEPATIC REPROGRAMMING

The liver is a fascinating organ that can regenerate itself; but unfortunately, this remarkable capacity is lost in chronic liver diseases. Currently, there are over 600 million people with liver diseases and 1 million deaths per year worldwide. Because the limitation of organ donors, cell-based therapy may represent an attractive approach for treating life-threatening liver failure, if a sufficient amount of functional hepatocytes can be generated from easily accessible cell sources. Many research groups focus their efforts on the generation of induced hepatocyte-like cells (iHeps) via reprogramming approaches. In 2011, Huang et al. showed that functional iHep cells could be converted from p19 null mouse fibroblasts via transduction of Gata4, Hnf1α, and Foxa3 (Huang et al., 2011). Moreover, these iHep cells restored liver functions *in vivo* and rescued the recipients from death, which provided a promising proof of principle example for the use in regenerative medicine. Meanwhile, Sekiya et al. converted mouse embryonic and adult fibroblasts into iHep cells using a combination of Hnf4α plus Foxa1, Foxa2 or Foxa3 (Sekiya and Suzuki, 2011). Remarkably, after overexpressing Hnf1β and Foxa3, mouse embryonic fibroblasts were converted to induced hepatic stem cells, which had the capability of long-term expansion and bipotential differentiation to hepatocytes and cholangiocytes (Yu et al., 2013). Subsequently, human fibroblasts were successfully reprogrammed into iHep cells by different combinations of TFs (Du et al., 2014; Huang et al., 2014). Huang et al. generated human iHeps from fibroblasts by introduction of FOXA3, HNF1α, and HNF4α (Huang et al., 2014). Meanwhile, Du et al. showed that human fibroblasts could be reprogrammed to functional iHeps by overexpressing HNF1α, HNF4α, HNF6, ATF5, PROX1, and CEBPα (Du et al., 2014). These human iHeps expressed liver-specific markers, and could integrate into mouse liver and demonstrated many metabolic functions of the liver.

Recently, the non-viral and non-integrating transduction systems, including synthetic RNAs, recombinant proteins, and episomal plasmids, become more attractive (Fusaki et al., 2009; Kim et al., 2009, 2015; Warren et al., 2010; Zhou and Freed, 2009). For example, Simeonov et al. used a method of repeating transfection with synthetic modified mRNAs encoding hepatic reprogramming TFs and successfully induced iHep cells from human fibroblasts (Simeonov and Uppal, 2014). Kim et al. applied an oriP/EBNA1-based episomal system to deliver a set of transcription factors, including Gata4, Hnf1α, and Foxa3, to convert fibroblasts into iHep cells (Kim et al., 2015). Small molecules also played a critical role in iHep reprogramming process (Lim et al., 2016; Pournasr et al., 2015). We previously generated induced endodermal progenitor cells by the CASD lineage conversion strategy, and greatly expanded these endodermal progenitor cells by a chemically defined condition, consisting of EGF, bFGF, A83-01, and CHIR99021. Further treatment with soluble growth factors and small molecules, such as A83-01 and Compound E, these induced endodermal progenitor cells efficiently differentiated into functional hepatocytes. After transplantation, human iHeps integrated into immunodeficient mouse livers, expanded extensively, and acquired mature hepatocyte functions (Zhu et al., 2014b). Moreover, Lim et al. reported that direct conversion into iHeps is a stepwise transition involving the sequential erasure of somatic memory, MET transition, and induction of hepatic cell fate (Lim et al., 2016). Through chemical screening, they found that CHIR99021 and A83-01 facilitated one factor Hnf1α-induced hepatic reprogramming. Mechanistically, these small molecules facilitated the robust iHep generation through promoting MET process. In the future, utilizing all-chemical method to induce iHep reprogramming will be a promising way to promote the development of this field.

More recently, *in vivo* hepatic reprogramming was fulfilled by injection of hepatic reprogramming factors *in situ* (Rezvani et al., 2016; Song et al., 2016). Rezvani et al. developed *in vivo* reprogramming of myofibroblasts into hepatocytes using adeno-associated virus (AAV) vectors expressing hepatic transcription factors: *Foxa1*, *Foxa2*, *Foxa3*, *Gata4*, *Hnf1a*, and *Hnf4a*. These iHeps converted from myofibroblsts were functional and reduced liver fibrosis (Rezvani et al., 2016). Song et al. showed that overexpression of *FOXA3*, *GATA4*, *HNF1A*, and *HNF4A* could convert mouse myofibroblasts into iHeps cells in fibrotic mouse livers and also reduced liver fibrosis (Song et al., 2016). More efforts are required to develop better delivery methods and improve the conversion efficiency and cell maturity. Obviously, *in vivo* hepatic reprogramming by small molecules will facilitate the translation of the iHep technology into clinical treatment of chronic liver diseases.

PANCREATIC REPROGRAMMING

Diabetes mellitus represents a global health epidemic and affects more than 300 million people worldwide according to the International Diabetes Federation. Diabetes is caused and developed as a deficiency and/or dysfunction of pancreatic β cells. A key method to study diabetes and treat patients is to obtain unlimited numbers of functional pancreatic β cells. Reprogramming other types of somatic cells to pancreatic β cells is under active investigation. In 2005, Minammi et al. reported that adult mouse pancreatic exocrine cells could be converted into insulin-producing cells *in vitro* by suspension culture with EGF and nicotinamide (Minami et al., 2005). In 2008, Zhou et al. reprogrammed adult pancreatic exocrine cells to β-like cells *in vivo* by introducing three TFs: Pdx1, Ngn3, and Mafa (Zhou et al., 2008). Then, Li et al. developed an improved method for *in vivo* pancreatic conversion (Li et al., 2014b). These induced pancreatic β-like cells formed the islet-like structures and could persist *in vivo* for more than one year (Li et al., 2014a). Additionally, Chen et al. reported that transient intestinal expression of Pdx1, Ngn3, and Mafa promoted rapid conversion of intestinal crypt cells into endocrine cells (Chen et al., 2014). Besides mouse cells, human pancreatic ductal cells could also be converted into pancreatic β-like cells by overexpression of PDX1, NGN3, MAFA, and PAX6 (Lee et al., 2013). These converted pancreatic β-like cells could secrete insulin in response to high concentration of glucose. More interestingly, Ariyachet et al. reported that cells of the antral stomach were ideal cell source for pancreatic reprogramming (Ariyachet et al., 2016). These reprogrammed insulin$^+$ β-like cells performed molecularly and functionally similar to bona fide pancreatic β cells. These results suggested that the initial cell type can affect the reprogramming efficiency and induced cell phenotype and its functional properties.

Introducing exogenous genetic material and altering the genome raise safety concerns (Barrilleaux and Knoepfler, 2011). Therefore, using small molecules to active endogenous TFs is an attractive method for cellular reprogramming. By high-throughput screening, BRD7552 was shown to be an inducer of *PDX1* expression (Yuan et al., 2013). In another report, DNA methyltransferase inhibitor 5-aza-dC was capable of increasing the expression of *Ngn3* in PANC-1 cells (Lefebvre et al., 2010). In our previous study, we found that several specific small molecules could enhance the endodermal conversion process, including CHIR99021, NaB, Parnate, RG108, and 5'-N-ethylcarboxamidoadenosine, and promote pancreatic β-like cell differentiation and maturation, including Compound E, Vitamin C, and Bayk8644 (Zhu et al., 2016). More recently, Wang et al. described the derivation of human induced endodermal progenitor cells from gastrointestinal epithelial cells using a cocktail of defined small molecules, including Bayk8644, BIX01294, RG108, and SB431542, along with support from tissue-specific mesenchymal feeders (Wang et al., 2016). Induced endodermal progenitor cells could subsequently differentiate into more specified cell types, including pancreatic β-like cells. However, the therapeutic potentials of these pancreatic β-like cells have not been investigated so far. Obviously, using small molecules to replace master TFs as well as promote reprogramming to mature pancreatic β cells is a promising therapy for treating diabetes. However, transferring these findings for efficient *in vivo* pancreatic reprogramming, preferably be an all-chemical approach is waiting to be achieved in the future.

PERSPECTIVE

Cell fate transition and regeneration have fascinated biologists for centuries. After the discovery of iPSC technology, a lot of progress has been made in TF-mediated cellular reprogramming into many different defined cell types. Compared with the introduction of TFs by viral methods, small molecules have several advantages, can enhance reprogramming efficiency and improve the quality of the reprogrammed cells. Currently, identification of novel and effective small molecules for cellular reprogramming remains very labor intense. New technologies, such as high-throughput screening (HTS2) platform (Li et al., 2012) and reporter systems generated by CRISPR tools (Hockemeyer and Jaenisch, 2016), can be applied to this field and will accelerate the discovery process. In addition, the molecular mechanisms of most reprogramming processes are largely unknown, and how TFs or small molecules mediate cell fate transitions is still unclear. New technologies, such as single-cell analysis (Wen and Tang, 2016) and CRISPR-based genome-wide screening (Shalem et al., 2014), can provide new insights and help us to better understand the underlying molecular mechanisms. A better understanding of the detailed mechanisms during reprogramming processes can also help to improve the reprogramming efficiency and acquire mature cells with complete functionality for disease modeling, drug development, and cell-based transplantation (Fig. 2). Optimistically, future studies in cellular reprogramming by small molecules will overcome current hurdles, generate new discoveries, and benefit human health.

ACKNOWLEDGEMENTS

We would like to thank Dr. Wanguo Wei at Shanghai Advanced Research Institute, Chinese Academy of Sciences and Dr. Holger A. Russ at the Barbara Davis Center for Childhood Diabetes at the University of Colorado for their generous help. This work was funded by the national key research and development programme of China (2016YFC1305300) and the outstanding youth fund of Zhejiang Province (R17C120002). S.Z. is a scholar of the National 100 Young Talents Program. We apologize to those whose findings are relevant but not cited in this review due to space limitation.

ABBREVIATIONS

2-Me-5-HT, 2-Methyl-5-hydroxytryptamine; AAV, adeno-associated virus; BAM, Brn2, Ascl1, and Myt1 l; BMP, bone morphogenetic protein; CASD, cell-activation signaling-directed; FGF, fibroblast growth factor; iCM, induced cardiomyocyte-like cell; iCPC, induced cardiac progenitor cell; iN, induced neuron; iNSC, induced neural stem cell; iPSC, induced pluripotent stem cell; LPA, Lysophosphatidic acid; NaB, Sodium butyrate; RA, Retinoic acid; SAG, Smoothened agonist; TF, transcription factor; Tzv, Thiazovivin; VEGF, vascular endothelial growth factor; XEN, extra-embryonic endoderm.

COMPLIANCE WITH ETHICS GUIDELINES

Xiaojie Ma, Linghao Kong, and Saiyong Zhu declare that they have no conflict of interest.

REFERENCES

Addis RC, Epstein JA (2013) Induced regeneration–the progress and promise of direct reprogramming for heart repair. Nat Med 19:829–836

Ambasudhan R, Talantova M, Coleman R, Yuan X, Zhu S, Lipton SA, Ding S (2011) Direct reprogramming of adult human fibroblasts to functional neurons under defined conditions. Cell Stem Cell 9:113–118

Ariyachet C, Tovaglieri A, Xiang G, Lu J, Shah MS, Richmond CA, Zhou Q (2016) Reprogrammed stomach tissue as a renewable source of functional beta cells for blood glucose regulation. Cell Stem Cell 18:410–421

Barrilleaux B, Knoepfler P (2011) Transduction of human cells with polymer-complexed ecotropic lentivirus for enhanced biosafety. J Vis Exp 53:1–7

Caiazzo M, Dell'Anno MT, Dvoretskova E, Lazarevic D, Taverna S, Leo D, Broccoli V (2011) Direct generation of functional dopaminergic neurons from mouse and human fibroblasts. Nature 476:224–227

Cao N, Huang Y, Zheng J, Spencer CI, Zhang Y, Fu J, Nie B, Xie M, Zhang M, Wang H, Ma T, Xu T, Shi G, Srivastava D, Ding S (2016) Conversion of human fibroblasts into functional cardiomyocytes by small molecules. Science 352:1216–1220

Chen J, Liu J, Yang J, Chen Y, Chen J, Ni S, Song H, Zeng L, Ding K, Pei D (2011) BMPs functionally replace Klf4 and support efficient reprogramming of mouse fibroblasts by Oct4 alone. Cell Res 21:205–212

Chen Y, Finkbeiner SR, Weinblatt D, Emmett MJ, Tameire F, Yousefi M, Yang C, Maehr R, Zhou Q, Shemer R, Dor Y, Li C, Spence JR, Stanger BZ (2014) De novo formation of insulin-producing "neo-beta cell islets" from intestinal crypts. Cell Rep 6:1046–1058

Chen G, Wernig M, Berninger B, Nakafuku M, Parmar M, Zhang CL (2015) In vivo reprogramming for brain and spinal cord repair. eNeuro 2:e0106-15

Cheng L, Hu W, Qiu B, Zhao J, Yu Y, Guan W, Wang M, Yang W, Pei G (2014) Generation of neural progenitor cells by chemical cocktails and hypoxia. Cell Res 24:665–679

Du Y, Wang J, Jia J, Song N, Xiang C, Xu J, Hou Z, Su X, Liu B, Jiang T, Zhao D, Sun Y, Shu J, Guo Q, Yin M, Sun D, Lu S, Shi Y, Deng H (2014) Human hepatocytes with drug metabolic function induced from fibroblasts by lineage reprogramming. Cell Stem Cell 14:394–403

Fu J, Stone NR, Liu L, Spencer CI, Qian L, Hayashi Y, Delgado-Olguin P, Ding S, Bruneau BG, Srivastava D (2013) Direct reprogramming of human fibroblasts toward a cardiomyocyte-like state. Stem Cell Reports 1:235–247

Fu Y, Huang C, Xu X, Gu H, Ye Y, Jiang C, Qiu Z, Xie X (2015) Direct reprogramming of mouse fibroblasts into cardiomyocytes with chemical cocktails. Cell Res 25:1013–1024

Fusaki N, Ban H, Nishiyama A, Saeki K, Hasegawa M (2009) Efficient induction of transgene-free human pluripotent stem cells using a vector based on Sendai virus, an RNA virus that does not integrate into the host genome. Proce Jpn Acad Ser B 85:348–362

Gonzalez F, Boue S, Izpisua Belmonte JC (2011) Methods for making induced pluripotent stem cells: reprogramming a la carte. Nat Rev Genet 12:231–242

Guo Z, Zhang L, Wu Z, Chen Y, Wang F, Chen G (2014) In vivo direct reprogramming of reactive glial cells into functional neurons after brain injury and in an Alzheimer's disease model. Cell Stem Cell 14:188–202

Gurdon JB (1962) The developmental capacity of nuclei taken from intestinal epithelium cells of feeding tadpoles. J Embryol Exp Morphol 10:622–640

Hockemeyer D, Jaenisch R (2016) Induced pluripotent stem cells meet genome editing. Cell Stem Cell 18:573–586

Hou P, Li Y, Zhang X, Liu C, Guan J, Li H, Zhao T, Ye J, Yang W, Liu K, Ge J, Xu J, Zhang Q, Zhao Y, Deng H (2013) Pluripotent stem cells induced from mouse somatic cells by small-molecule compounds. Science 341:651–654

Hu W, Qiu B, Guan W, Wang Q, Wang M, Li W, Gao L, Shen L, Huang Y, Xie G, Zhao H, Jin Y, Tang B, Yu Y, Zhao J, Pei G (2015) Direct conversion of normal and Alzheimer's disease human fibroblasts into neuronal cells by small molecules. Cell Stem Cell 17:204–212

Huang P, He Z, Ji S, Sun H, Xiang D, Liu C, Hu Y, Wang X, Hui L (2011) Induction of functional hepatocyte-like cells from mouse fibroblasts by defined factors. Nature 475:386–389

Huang P, Zhang L, Gao Y, He Z, Yao D, Wu Z, Cen J, Chen X, Liu C, Hu Y, Lai D, Hu Z, Chen L, Zhang Y, Cheng X, Ma X, Pan G, Wang X, Hui L (2014) Direct reprogramming of human fibroblasts to functional and expandable hepatocytes. Cell Stem Cell 14:370–384

Huangfu D, Maehr R, Guo W, Eijkelenboom A, Snitow M, Chen AE, Melton DA (2008) Induction of pluripotent stem cells by defined factors is greatly improved by small-molecule compounds. Nat Biotechnol 26:795–797

Ichida JK, Blanchard J, Lam K, Son EY, Chung JE, Egli D, Loh KM, Carter AC, Di Giorgio FP, Koszka K, Huangfu D, Akutsu H, Liu DR, Rubin LL, Eggan K (2009) A small-molecule inhibitor of tgf-Beta signaling replaces sox2 in reprogramming by inducing nanog. Cell Stem Cell 5:491–503

Ieda M (2016) Heart development, diseases, and regeneration- new approaches from innervation, fibroblasts, and reprogramming. Circ J 80:2081–2088

Ieda M, Fu J, Delgado-Olguin P, Vedantham V, Hayashi Y, Bruneau BG, Srivastava D (2010) Direct reprogramming of fibroblasts into functional cardiomyocytes by defined factors. Cell 142:375–386

Ifkovits JL, Addis RC, Epstein JA, Gearhart JD (2014) Inhibition of TGFbeta signaling increases direct conversion of fibroblasts to induced cardiomyocytes. PLoS One 9:e89678

Islas JF, Liu Y, Weng K-C, Robertson MJ, Zhang S, Prejusa A, Harger J, Tikhomirova D, Chopra M, Iyer D, Mercola M, Oshima RG, Willerson JT, Potaman VN, Schwartz RJ (2012) Transcription factors ETS2 and MESP1 transdifferentiate human dermal fibroblasts into cardiac progenitors. Proc Natl Acad Sci USA 109:13016–13021

Jayawardena TM, Egemnazarov B, Finch EA, Zhang L, Payne JA, Pandya K, Zhang Z, Rosenberg P, Mirotsou M, Dzau VJ (2012) MicroRNA-mediated in vitro and in vivo direct reprogramming of cardiac fibroblasts to cardiomyocytes. Circ Res 110:1465–1473

Kim KS (2011) Converting human skin cells to neurons: a new tool to study and treat brain disorders? Cell Stem Cell 9:179–181

Kim D, Kim CH, Moon JI, Chung YG, Chang MY, Han BS, Ko S, Yang E, Cha KY, Lanza R, Kim KS (2009) Generation of human induced pluripotent stem cells by direct delivery of reprogramming proteins. Cell Stem Cell 4:472–476

Kim J, Efe JA, Zhu S, Talantova M, Yuan X, Wang S, Lipton SA, Zhang K, Ding S (2011) Direct reprogramming of mouse fibroblasts to neural progenitors. Proc Natl Acad Sci USA 108:7838–7843

Kim J, Kim KP, Lim KT, Lee SC, Yoon J, Song G, Hwang SI, Schöler HR, Cantz T, Han DW (2015) Generation of integration-free induced hepatocyte-like cells from mouse fibroblasts. Sci Rep 5:15706

Lalit PA, Salick MR, Nelson DO, Squirrell JM, Shafer CM, Patel NG, Saeed I, Schmuck EG, Markandeya YS, Wong R, Lea MR, Eliceiri KW, Hacker TA, Crone WC, Kyba M, Garry DJ, Stewart R, Thomson JA, Downs KM, Lyons GE, Kamp TJ (2016) Lineage reprogramming of fibroblasts into proliferative induced cardiac progenitor cells by defined factors. Cell Stem Cell 18:354–367

Lee J, Sugiyama T, Liu Y, Wang J, Gu X, Lei J, Markmann JF, Miyazaki S, Miyazaki J, Szot GL, Bottino R, Kim SK (2013) Expansion and conversion of human pancreatic ductal cells into insulin-secreting endocrine cells. Elife 2:e00940

Lefebvre B, Belaich S, Longue J, Vandewalle B, Oberholzer J, Gmyr V, Pattou F, Kerr-Conte J (2010) 5'-AZA induces Ngn3 expression and endocrine differentiation in the PANC-1 human ductal cell line. Biochem Biophys Res Commun 391:305–309

Li H, Chen G (2016) In vivo reprogramming for CNS repair: regenerating neurons from endogenous glial cells. Neuron 91:728–738

Li W, Zhou H, Abujarour R, Zhu S, Young Joo J, Lin T, Hao E, Schöler HR, Hayek A, Ding S (2009) Generation of human-induced pluripotent stem cells in the absence of exogenous Sox2. Stem Cells 27:2992–3000

Li R, Liang J, Ni S, Zhou T, Qing X, Li H, He W, Chen J, Li F, Zhuang Q, Qin B, Xu J, Li W, Yang J, Gan Y, Qin D, Feng S, Song H, Yang D, Zhang B, Zeng L, Lai L, Esteban MA, Pei D (2010) A mesenchymal-to-epithelial transition initiates and is required for the nuclear reprogramming of mouse fibroblasts. Cell Stem Cell 7:51–63

Li Y, Zhang Q, Yin X, Yang W, Du Y, Hou P, Ge J, Liu C, Zhang W, Zhang X, Wu Y, Li H, Liu K, Wu C, Song Z, Zhao Y, Shi Y, Deng H (2011) Generation of iPSCs from mouse fibroblasts with a single gene, Oct4, and small molecules. Cell Res 21:196–204

Li H, Zhou H, Wang D, Qiu J, Zhou Y, Li X, Rosenfeld MG, Ding S, Fu X (2012) Versatile pathway-centric approach based on high-throughput sequencing to anticancer drug discovery. Proc Natl Acad Sci USA 109:4609–4614

Li W, Cavelti-Weder C, Zhang Y, Clement K, Donovan S, Gonzalez G, Zhu J, Stemann M, Xu K, Hashimoto T, Yamada T, Nakanishi M, Zhang Y, Zeng S, Gifford D, Meissner A, Weir G, Zhou Q (2014a) Long-term persistence and development of induced pancreatic beta cells generated by lineage conversion of acinar cells. Nat Biotechnol 32:1223–1230

Li W, Nakanishi M, Zumsteg A, Shear M, Wright C, Melton DA, Zhou Q (2014b) In vivo reprogramming of pancreatic acinar cells to three islet endocrine subtypes. Elife 3:e01846

Li X, Zuo X, Jing J, Ma Y, Wang J, Liu D, Zhu J, Du X, Xiong L, Du Y, Xu J, Xiao X, Wang J, Chai Z, Zhao Y, Deng H (2015) Small-molecule-driven direct reprogramming of mouse fibroblasts into functional neurons. Cell Stem Cell 17:195–203

Lim KT, Lee SC, Gao Y, Kim KP, Song G, An SY, Adachi K, Jang YJ, Kim J, Oh KJ, Kwak TH, Hwang SI, You JS, Ko K, Koo SH, Sharma AD, Kim JH, Hui L, Cantz T, Schöler HR, Han DW (2016) Small molecules facilitate single factor-mediated hepatic reprogramming. Cell Rep 15:814–829

Lyssiotis CA, Foreman RK, Staerk J, Garcia M, Mathur D, Markoulaki S, Hanna J, Lairson LL, Charette BD, Bouchez LC, Bollong M, Kunick C, Brinker A, Cho CY, Schultz PG, Jaenisch R (2009) Reprogramming of murine fibroblasts to induced pluripotent stem cells with chemical complementation of Klf4. Proc Natl Acad Sci USA 106:8912–8917

Minami K, Okuno M, Miyawaki K, Okumachi A, Ishizaki K, Oyama K, Kawaguchi M, Ishizuka N, Iwanaga T, Seino S (2005) Lineage tracing and characterization of insulin-secreting cells generated from adult pancreatic acinar cells. Proc Natl Acad Sci USA 102:15116–15121

Muraoka N, Yamakawa H, Miyamoto K, Sadahiro T, Umei T, Isomi M, Nakashima H, Akiyama M, Wada R, Inagawa K, Nishiyama T, Kaneda R, Fukuda T, Takeda S, Tohyama S, Hashimoto H, Kawamura Y, Goshima N, Aeba R, Yamagishi H, Fukuda K, Ieda M (2014) MiR-133 promotes cardiac reprogramming by directly repressing Snai1 and silencing fibroblast signatures. EMBO J 33:1565–1581

Nam YJ, Song K, Luo X, Daniel E, Lambeth K, West K, Hill JA, DiMaio JM, Baker LA, Bassel-Duby R, Olson EN (2013) Reprogramming of human fibroblasts toward a cardiac fate. Proc Natl Acad Sci USA 110:5588–5593

Niu W, Zang T, Zou Y, Fang S, Smith DK, Bachoo R, Zhang CL (2013) In vivo reprogramming of astrocytes to neuroblasts in the adult brain. Nat Cell Biol 15:1164–1175

Pang Z, Yang N, Vierbuchen T, Ostermeier A, Fuentes DR, Yang T, Wernig M (2011) Induction of human neuronal cells by defined transcription factors. Nature 476:220–223

Pournasr B, Asghari-Vostikolaee MH, Baharvand H (2015) Transcription factor-mediated reprograming of fibroblasts to hepatocyte-like cells. Eur J Cell Biol 94:603–610

Qian L, Huang Y, Spencer CI, Foley A, Vedantham V, Liu L, Srivastava D (2012) In vivo reprogramming of murine cardiac fibroblasts into induced cardiomyocytes. Nature 485:593–598

Rezvani M, Espanol-Suner R, Malato Y, Dumont L, Grimm AA, Kienle E, Willenbring H (2016) In vivo hepatic reprogramming of myofibroblasts with aav vectors as a therapeutic strategy for liver fibrosis. Cell Stem Cell 18:809–816

Rie W, Muraoka N, Inagawa K, Yamakawa H, Miyamoto K, Sadahiro T, Umei T, Kaneda R, Suzuki T, Kamiya K, Tohyama S (2013) Induction of human cardiomyocyte-like cells from fibroblasts by defined factors. Proc Natl Acad Sci USA 110:12667–12672

Ring KL, Tong LM, Balestra ME, Javier R, Andrews-Zwilling Y, Li G, Walker D, Zhang WR, Kreitzer AC, Huang Y (2012) Direct reprogramming of mouse and human fibroblasts into multipotent neural stem cells with a single factor. Cell Stem Cell 11:100–109

Sekiya S, Suzuki A (2011) Direct conversion of mouse fibroblasts to hepatocyte-like cells by defined factors. Nature 475:390–393

Shalem O, Sanjana NE, Hartenian E, Shi X, Scott DA, Mikkelsen TS, Heckl D, Ebert BL, Root DE, Doench JG, Zhang F (2014) Genome-scale CRISPR-Cas9 knockout screening in human cells. Science 343:84–87

Shi Y, Desponts C, Do JT, Hahm HS, Scholer HR, Ding S (2008) Induction of pluripotent stem cells from mouse embryonic fibroblasts by Oct4 and Klf4 with small-molecule compounds. Cell Stem Cell 3:568–574

Simeonov KP, Uppal H (2014) Direct reprogramming of human fibroblasts to hepatocyte-like cells by synthetic modified mRNAs. PLoS One 9:e100134

Son EY, Ichida JK, Wainger BJ, Toma JS, Rafuse VF, Woolf CJ, Eggan K (2011) Conversion of mouse and human fibroblasts into functional spinal motor neurons. Cell Stem Cell 9:205–218

Song K, Nam YJ, Luo X, Qi X, Tan W, Huang G, Acharya A, Smith CL, Tallquist MD, Neilson EG, Hill JA, Bassel-Duby R, Olson EN (2012) Heart repair by reprogramming non-myocytes with cardiac transcription factors. Nature 485:599–604

Song G, Pacher M, Balakrishnan A, Yuan Q, Tsay HC, Yang D, Reetz J, Brandes S, Dai Z, Pützer BM, Araúzo-Bravo MJ, Steinemann D, Luedde T, Schwabe RF, Manns MP, Schöler HR, Schambach A, Cantz T, Ott M, Sharma AD (2016) Direct reprogramming of hepatic myofibroblasts into hepatocytes in vivo attenuates liver fibrosis. Cell Stem Cell 18:797–808

Tada M, Tada T, Lefebvre L, Barton SC, Surani MA (1997) Embryonic germ cells induce epigenetic reprogramming of somatic nucleus in hybrid cells. EMBO J 16:6510–6520

Takahashi K, Yamanaka S (2006) Induction of pluripotent stem cells from mouse embryonic and adult fibroblast cultures by defined factors. Cell 126:663–676

Takahashi K, Yamanaka S (2016) A decade of transcription factor-mediated reprogramming to pluripotency. Nat Rev Mol Cell Biol 17:183–193

Vierbuchen T, Ostermeier A, Pang Z, Kokubu Y, Sudhof TC, Wernig M (2010) Direct conversion of fibroblasts to functional neurons by defined factors. Nature 463:1035–1041

Wang H, Cao N, Spencer CI, Nie B, Ma T, Xu T, Zhang Y, Wang X, Srivastava D, Ding S (2014) Small molecules enable cardiac reprogramming of mouse fibroblasts with a single factor, Oct4. Cell Rep 6:951–960

Wang Y, Qin J, Wang S, Zhang W, Duan J, Zhang J, Wang X, Yan F, Chang M, Liu X, Feng B, Liu J, Pei X (2016) Conversion of human gastric epithelial cells to multipotent endodermal progenitors using defined small molecules. Cell Stem Cell 19:449–461

Warren L, Manos PD, Ahfeldt T, Loh YH, Li H, Lau F, Ebina W, Mandal PK, Smith ZD, Meissner A, Daley GQ, Brack AS, Collins JJ, Cowan C, Schlaeger TM, Rossi DJ (2010) Highly efficient reprogramming to pluripotency and directed differentiation of human cells with synthetic modified mRNA. Cell Stem Cell 7:618–630

Wen L, Tang F (2016) Single-cell sequencing in stem cell biology. Genome Biol 17:71

Yamakawa H, Muraoka N, Miyamoto K, Sadahiro T, Isomi M, Haginiwa S, Kojima H, Umei T, Akiyama M, Kuishi Y, Kurokawa J, Furukawa T, Fukuda K, Ieda M (2015) Fibroblast growth factors and vascular endothelial growth factor promote cardiac reprogramming under defined conditions. Stem Cell Reports 5:1128–1142

Yang H, Zhang L, An J, Zhang Q, Liu C, He B, Hao D (2016) MicroRNA-mediated reprogramming of somatic cells into neural stem cells or neurons. Mol Neurobiol. doi:10.1007/s12035-016-0115-9

Yoo AS, Sun AX, Li L, Shcheglovitov A, Portmann T, Li Y, Lee-Messer C, Dolmetsch RE, Tsien RW, Crabtree GR (2011) MicroRNA-mediated conversion of human fibroblasts to neurons. Nature 476:228–231

Yu B, He Z, You P, Han Q, Xiang D, Chen F, Wang M, Liu C, Lin X, Borjigin U, Zi X, Li J, Zhu H, Li W, Han C, Wangensteen KJ, Shi Y, Hui L, Wang X, Hu Y (2013) Reprogramming fibroblasts into bipotential hepatic stem cells by defined factors. Cell Stem Cell 13:328–340

Yuan X, Wan H, Zhao X, Zhu S, Zhou Q, Ding S (2011) Brief report: combined chemical treatment enables Oct4-induced reprogramming from mouse embryonic fibroblasts. Stem Cells 29:549–553

Yuan Y, Hartland K, Boskovic Z, Wang Y, Walpita D, Lysy PA, Zhong C, Young DW, Kim YK, Tolliday NJ, Sokal EM, Schreiber SL, Wagner BK (2013) A small-molecule inducer of PDX1 expression identified by high-throughput screening. Chem Biol 20:1513–1522

Zhang L, Yin J, Yeh H, Ma N, Lee G, Chen X, Wang Y, Lin L, Chen L, Jin P, Wu G, Chen G (2015) Small molecules efficiently reprogram human astroglial cells into functional neurons. Cell Stem Cell 17:735–747

Zhang M, Lin Y, Sun Y, Zhu S, Zheng J, Liu K, Cao N, Li K, Huang Y, Ding S (2016a) Pharmacological reprogramming of fibroblasts into neural stem cells by signaling-directed transcriptional activation. Cell Stem Cell 18:653–667

Zhang Y, Cao N, Huang Y, Spencer CI, Fu J, Yu C, Liu K, Nie B, Xu T, Li K, Xu S, Bruneau BG, Srivastava D, Ding S (2016b) Expandable cardiovascular progenitor cells reprogrammed from fibroblasts. Cell Stem Cell 18:368–381

Zhao Y, Londono P, Cao Y, Sharpe EJ, Proenza C, O'Rourke R, Jones KL, Jeong MY, Walker LA, Buttrick PM, McKinsey TA, Song K (2015a) High-efficiency reprogramming of fibroblasts into cardiomyocytes requires suppression of pro-fibrotic signalling. Nat Commun 6:8243

Zhao Y, Zhao T, Guan J, Zhang X, Fu Y, Ye J, Zhu J, Meng G, Ge J, Yang S, Cheng L, Du Y, Zhao C, Wang T, Su L, Yang W, Deng H (2015b) A XEN-like state bridges somatic cells to pluripotency during chemical reprogramming. Cell 163:1678–1691

Zhou W, Freed CR (2009) Adenoviral gene delivery can reprogram human fibroblasts to induced pluripotent stem cells. Stem Cells 27:2667–2674

Zhou Q, Tripathi P (2012) How to remake a fibroblast into a neural stem cell. Cell Stem Cell 10:347–348

Zhou Q, Brown J, Kanarek A, Rajagopal J, Melton DA (2008) In vivo reprogramming of adult pancreatic exocrine cells to beta-cells. Nature 455:627–632

Zhou Y, Wang L, Vaseghi HR, Liu Z, Lu R, Alimohamadi S, Yin C, Fu J, Wang G, Liu J, Qian L (2016) Bmi1 is a key epigenetic barrier to direct cardiac reprogramming. Cell Stem Cell 18:382–395

Zhu S, Li W, Zhou H, Wei W, Ambasudhan R, Lin T, Kim J, Zhang K, Ding S (2010) Reprogramming of human primary somatic cells by OCT4 and chemical compounds. Cell Stem Cell 7:651–655

Zhu S, Ambasudhan R, Sun W, Kim HJ, Talantova M, Wang X, Zhang M, Zhang Y, Laurent T, Parker J, Kim HS, Zaremba JD, Saleem S, Sanz-Blasco S, Masliah E, McKercher SR, Cho YS, Lipton SA, Kim J, Ding S (2014a) Small molecules enable OCT4-mediated direct reprogramming into expandable human neural stem cells. Cell Res 24:126–129

Zhu S, Rezvani M, Harbell J, Mattis AN, Wolfe AR, Benet LZ, Willenbring H, Ding S (2014b) Mouse liver repopulation with hepatocytes generated from human fibroblasts. Nature 508:93–97

Zhu S, Wang H, Ding S (2015) Reprogramming fibroblasts toward cardiomyocytes, neural stem cells and hepatocytes by cell activation and signaling-directed lineage conversion. Nat Protoc 10:959–973

Zhu S, Russ HA, Wang X, Zhang M, Ma T, Xu T, Tang S, Hebrok M, Ding S (2016) Human pancreatic beta-like cells converted from fibroblasts. Nat Commun 7:10080

5

Driver mutations of cancer epigenomes

David M. Roy[1,2], Logan A. Walsh[1], Timothy A. Chan[1,3]✉

[1] Human Oncology and Pathogenesis Program, Memorial Sloan-Kettering Cancer Center, New York, NY 10065, USA
[2] Weill Cornell/Rockefeller/Sloan-Kettering Tri-Institutional MD-PhD Program, New York, NY 10065, USA
[3] Department of Radiation Oncology, Memorial Sloan-Kettering Cancer Center, New York, NY 10065, USA
✉ Correspondence: chant@mskcc.org (T. A. Chan)

ABSTRACT

Epigenetic alterations are associated with all aspects of cancer, from tumor initiation to cancer progression and metastasis. It is now well understood that both losses and gains of DNA methylation as well as altered chromatin organization contribute significantly to cancer-associated phenotypes. More recently, new sequencing technologies have allowed the identification of driver mutations in epigenetic regulators, providing a mechanistic link between the cancer epigenome and genetic alterations. Oncogenic activating mutations are now known to occur in a number of epigenetic modifiers (i.e. *IDH1/2*, *EZH2*, *DNMT3A*), pinpointing epigenetic pathways that are involved in tumorigenesis. Similarly, investigations into the role of inactivating mutations in chromatin modifiers (i.e. *KDM6A*, *CREBBP/EP300*, *SMARCB1*) implicate many of these genes as tumor suppressors. Intriguingly, a number of neoplasms are defined by a plethora of mutations in epigenetic regulators, including renal, bladder, and adenoid cystic carcinomas. Particularly striking is the discovery of frequent histone H3.3 mutations in pediatric glioma, a particularly aggressive neoplasm that has long remained poorly understood. Cancer epigenetics is a relatively new, promising frontier with much potential for improving cancer outcomes. Already, therapies such as 5-azacytidine and decitabine have proven that targeting epigenetic alterations in cancer can lead to tangible benefits. Understanding how genetic alterations give rise to the cancer epigenome will offer new possibilities for developing better prognostic and therapeutic strategies.

KEYWORDS chromatin, cancer, epigenetics, mutations, methylation

INTRODUCTION

Cancer is an evolutionary process through which dysregulation in select cellular mechanisms confers a clonal advantage, leading to tumor growth and eventually, metastasis. In fact, nearly all cancers are defined by several "hallmark" capabilities, including resisting cell death, evading growth suppressors, uncontrolled proliferation, neoangiogenesis, invasion/metastasis, and replicative immortality (Hanahan and Weinberg, 2011). Aberrant control of other processes, such as defective differentiation and DNA damage repair, is also linked to tumor formation. Research efforts have traditionally focused on genetic abnormalities underlying malignant transformation, due to initial technological constraints and our limited understanding of other heritable patterns of gene regulation. Early studies on these genetic alterations—copy number variations, mutations, gene rearrangements—have defined mechanisms of oncogenesis, led to the creation of targeted therapies, and improved patient outcomes for certain cancers. Only recently has it become evident that many genetic alterations in cancer target epigenetic regulators, causing cancer-associated phenotypes via epigenetic dysfunction.

Since the first discovery of cancer-associated loss of DNA methylation, the field of cancer epigenetics has grown remarkably and helped elucidate aspects of cancer biology where genetic explanations alone are insufficient (Feinberg and Vogelstein, 1983). Epigenetics is the process by which cells encode non-genetic, heritable information through alterations that do not change the DNA sequence. Generally, chromatin exists in two main forms—condensed, transcriptional silent heterochromatin and euchromatin, which is transcriptionally active. The functional unit of chromatin is the nucleosome, which is an octameric structure composed of two histones each of H2A, H2B, H3, and H4 encircled by 147 bp of DNA (Margueron and Reinberg, 2010). Regulation

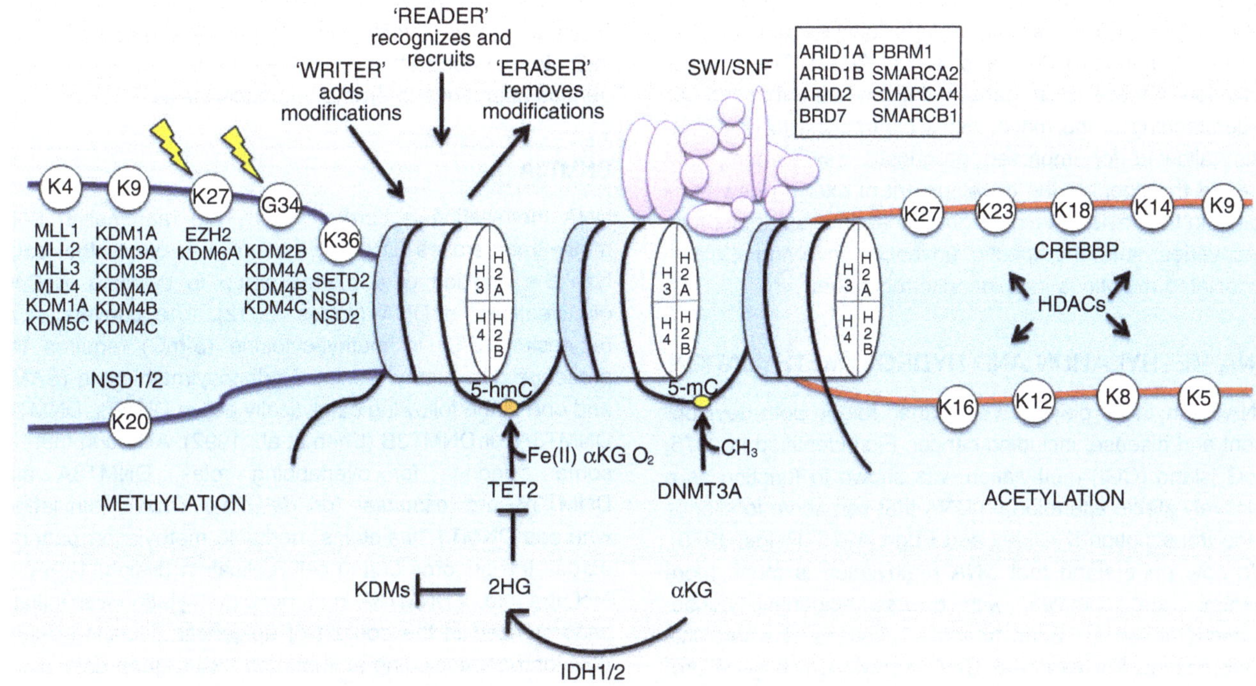

Figure 1. Epigenetic regulation of gene expression. Epigenetic processes include DNA methylation, covalent histone modifications (e.g. methylation, acetylation), and chromatin remodeling (SWI/SNF complex). Modifier proteins with frequent driver mutations in cancer are shown by specific function and target site. Green and red font colors represent histone "writers" and "erasers," respectively. Lightning bolts represent cancer-associated mutations in histones H3.1 (K27) and H3.3 (K27, G34).

of the chromatin state is achieved through DNA methylation, chromatin remodeling, and/or covalent histone modifications, such as methylation, acetylation, phosphorylation, ubiquitination, and sumoylation. The major effectors of these modifications are the chromatin modifier enzymes, highly specific proteins that catalyze the addition or removal of functional groups to DNA or histone tails. These modifications alter chromatin structure through noncovalent interactions within and between nucleosomes, leading to changes in macromolecular organization and promoter accessibility. In addition, these chromatin "marks" serve as signals to other specialized proteins involved in chromatin organization, gene transcription, genome maintenance, and replication (Kouzarides, 2007; Sharma et al., 2010). As a result, aberrations in one or more of these modifiers can have profound effects on normal cell physiology and are now well-documented in many diseases, including cancer (Fig. 1).

It has been known for decades that epigenetic dysregulation occurs in cancer. Aberrant DNA hyper- and hypomethylation was one of the first molecular features noted to be present in cancer cells (Feinberg and Vogelstein, 1983; Gama-Sosa et al., 1983; Rubery and Newton, 1973; Timp and Feinberg, 2013). However, the nature of epigenetic alterations and the molecular underpinnings of these changes are just beginning to be appreciated. The recent accelerated search for mutations in cancers, facilitated through next generation sequencing technologies, has greatly accelerated our understanding of the mechanisms underlying epigenetic dysfunction in cancers.

Genome-wide and exomic sequencing data from recent years have demonstrated that mutations in epigenetic modifiers comprise a large portion of all genetic events in many cancers, including tumors such as renal carcinoma, adenoid cystic carcinoma (ACC), and transitional cell carcinoma (TCC) of the bladder (Dalgliesh et al., 2010; Gui et al., 2011; Ho et al., 2013). In fact, the main objective of these recent genomic analyses is the identification of *bona fide* "driver" mutations in cancer genes. These mutations are defined by their ability to promote or "drive" tumorigenesis and are therefore positively selected for in the development of cancer. In contrast, "passenger" mutations, which comprise the majority of mutations identified, represent genetic events that have no direct or indirect effect on the selective growth advantage of the cell in which it occurred (Stratton et al., 2009; Vogelstein et al., 2013). Candidate epigenetic driver genes have been identified as either mutated across many cancers (e.g. *KDM6A* in over 12 cancers), having highly recurrent mutations (e.g. *IDH1* R132), or having highly prevalent mutations in select tumor histologies (e.g. *MLL2* in follicular lymphoma) (You and Jones, 2012). The characterization of these driver gene mutations has enhanced our understanding of the mechanisms contributing to oncogenesis, allowed for improved prognostic assessment, and opened the door for the development of exciting new therapies. In this review, we highlight recent developments, new discoveries, and therapeutic advances involving cancer-associated mutations in epigenetic regulators.

DNA METHYLATION AND HYDROXYMETHYLATION

DNA methylation plays a well-defined role in both development and disease, including cancer. First identified in 1975, CpG island (CGI) methylation was shown to function as a relatively stable alteration on DNA that can serve to silence gene transcription (Holliday and Pugh, 1975; Riggs, 1975). We now understand that DNA methylation is much more dynamic and complex, with diverse epigenetic consequences linked to varied genomic locations of where this mark occurs. For example, DNA methylation at gene promoter CGIs potently blocks the initiation of transcription, whereas methylation within CpG-poor gene bodies may actually facilitate elongation and influence patterns of alternate splicing. In addition, DNA methylation is frequently found in repeat-rich areas of the genome and is vital for both chromosomal and genomic stability, possibly through the repression of retroviral transposons (Jones, 2012). Still, the role for this epigenetic mark at other regulatory regions, such as enhancers and insulators, has yet to be determined. Regardless, aberrant methylation in human cancer is a defining feature, with global promoter CGI hypermethylation and non-CGI hypomethylation widely reported (Ehrlich, 2002; Sharma et al., 2010). Furthermore, local variations in methylation at only several key loci have been shown to be sufficient for tumorigenesis (Lee et al., 2008; Poage et al., 2011). Importantly, these altered patterns of DNA epigenetic marks (e.g. 5-mC, 5-hmC) are frequently accompanied by a critical imbalance in transcriptional programs involving differentiation and stem cell maintenance, thereby initiating tumorigenesis and sustaining growth (Jones and Baylin, 2007).

DNA methylation can function to silence tumor suppressor genes along with genetic mutations (Herman and Baylin, 2003). For example, in the case of hereditary gastric cancer, methylation of *CDH1* (which encodes the E-cadherin tumor suppressor) can function as a "second hit" and cause gastric cancer when the first allele is mutated (Grady et al., 2000). In sporadic cancers, tumor suppressor genes that are mutated in hereditary versions of the disease are frequently silenced by DNA methylation instead (Esteller et al., 2001). For example, in hereditary nonpolyposis colon cancer (HNPCC), *MLH1* inactivation via mutation can lead to microsatellite instability (MSI) and tumorigenesis, whereas in sporadic colon cancers, *MLH1* is frequently silenced by methylation (Kane et al., 1997; Veigl et al., 1998). These data and others indicate that aberrant DNA methylation can work along with genetic alterations to promote tumorigenesis.

DNMT3A

DNA methylation is carried out by the mammalian DNA methyltransferases (DNMTs), essential enzymes that catalyze the addition of a methyl group to cytosine in CpG dinucleotides in DNA (Jones, 2012). The conversion of 5-cytosine (5-C) to methyl-cytosine (5-mC) requires the presence of a methyl donor, S-adenosylmethionine (SAM), and one of the following catalytically active DNMTs: DNMT1, DNMT3A, or DNMT3B (Shen et al., 1992). Although there is some evidence for overlapping roles, DNMT3A and DNMT3B are essential for *de novo* DNA methylation, whereas DNMT1 "maintains" heritable methylation patterns across the genome during cell replication (Hsieh, 1999). In fact, the role of DNMT3A in *de novo* methylation was initially characterized in the context of epigenetic silencing during development, including at imprinted loci in germ cells (Kaneda et al., 2004; Okano et al., 1999). The DNMTs have long been suspected to play a role in oncogenesis, as well. Aberrations in DNA methylation—both hyper- and hypomethylation—have been well-documented in patient tumors and cell lines for decades (El-Osta, 2004). Additionally, Oka and colleagues first showed that in some cases, DNMT3A and DNMT3B, not DNMT1, mediate the cytotoxic effects of 5-aza-dC, a therapeutic mainstay in the treatment of several hematopoietic malignancies (Oka et al., 2005; Plimack et al., 2007). Not long after, DNMT3A and DNMT3B were further implicated in both hematopoietic stem cell (HSC) renewal and differentiation, two tightly regulated processes whose perturbation can lead to carcinogenesis (Challen et al., 2012; Tadokoro et al., 2007).

DNMT3A somatic mutations were first discovered by Yamashita and colleagues following sequencing of tissue from adult patients with *de novo* acute myeloid leukemia (AML) (Yamashita et al., 2010). Soon after, *DNMT3A* mutations were reported in AML cohorts from others, with frequencies as high as 22.1% (Ley et al., 2010). The majority of mutations occur at R882 (60%–64%), and almost all are heterozygous (Ley et al., 2010; Thol et al., 2011a). *DNMT3A* mutations are enriched in AML patients with intermediate-risk cytogenetics and normal karyotype (Lin et al., 2011; Patel et al., 2012). They are also associated with increased age, M4 and M5 AML subtypes, worse overall survival (OS) and relapse-free survival (RFS), and increased blasts at diagnosis (Hou et al., 2012; Ley et al., 2010; Lin et al., 2011; Marcucci et al., 2012; Thol et al., 2011a; Yan et al., 2011). There is also evidence that *DNTM3A* mutations may be a prognostic marker for decreased time to treatment failure (TTF), duration of complete remission (CR), and disease-free survival (DFS), at least in cytogenetically normal AML (Marcucci et al., 2012; Thol et al., 2011a; Yan et al., 2011). These studies have also identified several frequently co-occurring mutations, which include *NPM1*, *FLT3*, *IDH1/2*, and less commonly, *TET2* (Hou et al., 2012; Ley et al., 2010; Lin et al., 2011; Marcucci et al., 2012; Renneville et al., 2012; Thol et al., 2011a; Yan et al., 2011). As alterations in epigenetic regulators frequently lead to genomic instability, *DNMT3A* mutation may further drive progression of the disease itself (Wakita et al., 2013). Further, recent experimental work has demonstrated that DNMT3A mutations at R882 are only found in major clones, suggesting this genetic alteration may be an initiating lesion in AML (Bisling et al., 2013).

DNMT3A mutations have also been described, albeit at lower frequency, in other myeloid malignancies, such as myelodysplastic syndromes (MDS; 3%–8%) and myeloproliferative neoplasms (MPNs; 2%–10%), as well as early T-cell precursor acute lymphoblastic leukemia (ETP-ALL; 16%–18%) (Brecqueville et al., 2011; Ewalt et al., 2011; Grossmann et al., 2013; Neumann et al., 2013; Stegelmann et al., 2011; Thol et al., 2011b; Traina et al., 2013; Walter et al., 2011). Similar to what is observed in *de novo* AML, R882 is most frequently targeted for mutation (60%) in these neoplasms (Thol et al., 2011b). Clinically, *DNMT3A* mutations also correlated with increased age and predicted prognosis in all types, including worse OS, event-free survival (EFS), and AML-free survival (Lin et al., 2011; Neumann et al., 2013; Renneville et al., 2012; Thol et al., 2011b; Walter et al., 2011). Paradoxically, univariate and multivariate analysis of 92 patients with MDS revealed *DNMT3A* mutations were correlated with better overall response rate (ORR) and progression-free survival (PFS) (Traina et al., 2013). Interestingly, similar to AML, MPNs showed an association between *DNMT3A* alterations and mutation in *JAK2*, *IDH1/2*, and *ASXL1* but not *TET2* (Stegelmann et al., 2011). However, studies in MDS and ETP-ALL have found no association between *DNMT3A* mutations and other known leukemogenic drivers, including *FLT3* (Neumann et al., 2013; Thol et al., 2011b).

DNMT3A is a 102 kDa protein with three highly conserved functional domains: An N-terminal PWWP domain, a cysteine-rich PHD zinc finger domain, and a C-terminal catalytic domain (Hermann et al., 2004). *DNMT3A* mutations in cancer have been reported in all three domains, with most occurring in the catalytic domain, including the R882 mutation (60%) (Ley et al., 2010). Still, many cancer-specific mutations occur in non-catalytic domains. The PWWP domain is essential in localizing DNMT3A to heterochromatic regions of DNA during interphase, though it is unclear if this is related to its reported ability to bind DNA directly (Bachman et al., 2001; Ge et al., 2004; Purdy et al., 2010; Suetake et al., 2011). Alternatively, the PHD domain has also been shown to mediate regional specificity and repression through its interactions with transcriptional repressor RP58, HP1beta, histone deacetylatases (HDACs), and SUV39H1 (Datta et al., 2003; Fuks et al., 2001; Fuks et al., 2003). Exactly how these mutations disrupt protein function is of great interest, although a single unifying mechanism is unlikely to exist. Likewise, a recent clinical study found that R882 mutations confer poor prognosis in older populations, whereas non-R882 mutations confer poor prognosis in younger patients (Marcucci et al., 2012).

Whether these reported mutations are loss-of-function, gain-of-function, or act via a dominant-negative mechanism has also been debated. A strong case can be made that *DNMT3A* is an oncogene, as it is overexpressed in several cancers, depletion results in decreased proliferation and metastasis, and 5-aza-dC causes apoptosis through direct inhibition of DNMT3A (Oka et al., 2005). In regards to mutations themselves, the R882H DNMT3A mutant was sufficient to promote tumorigenesis in an IL-3 dependent transformation assay in leukemic 32D cells (Yan et al., 2011). Alternatively, some studies have demonstrated that mutant DNMT3A (R882H) has a markedly reduced catalytic ability (~50%) in methyltransferase assays and decreased DNA-binding capacity *in vitro*, implying a possible loss-of-function phenotype via a dominant-negative mechanism (Jia et al., 2007; Ley et al., 2010; Yamashita et al., 2010). Clearly, additional studies are necessary to understand the exact nature by which these cancer-associated mutations are transforming.

Characterizing larger DNMT3A-induced changes in the cancer methylome has proven quite challenging. Ley and colleagues showed no changes in genome-wide methylation according to *DNMT3A* status in AML, and although 182 specific regions showed increased hypomethylation in mutant samples, this did not correlate with gene expression (Ley et al., 2010). However, in a more recent cohort, a total of 3878 genomic regions were found to have significantly different methylation patterns using MeDIP-chip and differences of expression levels in 889 of 20,723 annotated genes was observed via an Affymetrix microarray (Yan et al., 2011). Further, this group found during RT-PCR validation that the expression of several HOX family genes significantly increased in DNMT3A mutant samples compared to wild-type.

Recent experiments may offer some insight behind the conflicting methylation and functional data. Protein binding at the DNMT3A tetramerization interface is important for methylation patterning, inducing processive methylation of clustered sites (Jia et al., 2007). Most mutations, including R882, occur within this tetramer interface. Therefore, differences between oligomerization states can explain how *DNMT3A* mutations alter epigenetic silencing and lead to transformation, without global changes in DNA methylation (Holz-Schietinger et al., 2011; Holz-Schietinger et al., 2012). Although the commonly occurring R882H mutation does not disrupt DNMT3A association with required cofactor DNMT3L *in vitro*, the latter is only expressed in early development (Chedin et al., 2002; Jia et al., 2007; Webster et al., 2005;

Yamashita et al., 2010). However, other binding proteins, such as thymine-DNA glycosylase and ecotropic viral integration site 1, can adhere to sites within the DNMT3A catalytic domain and may explain altered mutant DNMT3A activity (Li et al., 2007; Senyuk et al., 2011). In addition, noncatalytic mutations may disrupt protein binding to other domains, as described above. Although the exact mechanism remains elusive, these mutations result in decreased methylation processivity and altered localization, possibly to euchromatic regions of DNA (Jurkowska et al., 2011). This may help explain how *DNMT3A* mutations drive tumor formation in hematopoietic malignancies, even in the absence of larger global methylation changes.

TET2

The ten-eleven translocation (TET) family proteins were first discovered in cancer with the fusion of TET1 to MLL in select AML patients with t(10;11)(q22;q23) (Lorsbach et al., 2003). Mechanistic studies then showed the TET proteins are dioxygenases that depend on 2-oxoglutarate, oxygen, Fe(II), and ascorbic acid to catalyze the conversion of 5-mC to 5-hydroxymethylcytosine (5-hmC) at CpG regions in DNA (Blaschke et al., 2013; Ito et al., 2010; Minor et al., 2013; Tahiliani et al., 2009; Yin et al., 2013). TET enzymes may be responsible for DNA demethylation through both passive and active means. For example, CpG dinucleotides that are "marked" with 5-hmC are not recognized by DNMT1 and therefore, methylation is passively lost at these sites through repeated cycles of cell division (Valinluck and Sowers, 2007). Alternatively, active demethylation can proceed following placement of 5-hmC via the activation-induced cytidine deaminase (AID)-APOBEC DNA repair pathway (Guo et al., 2011). More recently, an even greater role for TET enzymes in active demethylation was shown *in vitro*, with TET enzymes proving sufficient for converting 5-mC to 5-hmC, 5-hmC to 5-formylcytosine (5-fC), and finally 5-fC to 5-carboxylcytosine (5-caC). 5-caC is then targeted by base excision repair enzymes to complete the demethylation process (He et al., 2011; Ito et al., 2011).

TET2 was first suspected to have a role in cancer when six patients with either secondary AML (sAML) or MDS were noted to have minimal deletions via FISH on chromosome 4q24 (Viguie et al., 2005). Soon after, the first *TET2* somatic mutations were identified in 25 patients (14%) with MPNs (Delhommeau et al., 2008). Delhommeau and colleages then sequenced *TET2* in patient tumor samples, becoming the first group to identify *TET2* mutations in multiple myeloid neoplasms, including MDS (19%), MPNs (12%–14%), and sAML (24%). They concluded that *TET2* was a novel *bona fide* tumor suppressor, noting that the majority of mutations are heterozygous (55%) and that *TET2* defects precede the well-known JAK2 V617F driver mutation in MPN HSCs (Delhommeau et al., 2009). More recently, Schaub and colleagues disputed this result using colony formation assays to show that *TET2* mutations can either precede (4 of 8 patients), follow (2 of 8), or occur independently (2 of 8) of JAK2 V617F mutations in MPN patient samples (Schaub et al., 2010). Although the temporal relationship between *TET2* mutations and other leukemogenic drivers is still unclear, the frequency and ubiquitous nature of these mutations in cancer is quite revealing. In addition to MDS, sAML, and MPNs, *TET2* mutations have now been described in other myeloid neoplasms, such as *de novo* AML (12%) and chronic myelomonocytic leukemia (CMML; 42%–46%) (Abdel-Wahab et al., 2009; Smith et al., 2010).

Unlike *DNMT3A* mutations, *TET2* alterations seem to hold limited prognostic utility in leukemia. The overwhelming majority of studies published to date have found no change in OS or any other prognostic tools between patients harboring *TET2* mutations and those who are not. However, one study in a cohort of 96 MDS patients reported that *TET2* mutations conferred an OS, EFS, and AML-free survival advantage (Kosmider et al., 2009). Paradoxically, Abdel-Wahab and colleagues found that *TET2* mutations were linked to worse OS in 93 patients with *de novo* AML, and a recent whole-exome sequencing study reported worse EFS in AML patients with *TET2* mutation (Abdel-Wahab et al., 2009; Weissmann et al., 2012). Further, Nibourel and colleagues reported no association with OS or DFS in their cohort of *de novo* AML patients, though the prevalence of *TET2* mutations in patients who failed to achieve complete remission (CR) trended higher (27% vs. 17%) (Nibourel et al., 2010). Despite questionable association to patient outcomes, *TET2* mutations are linked with other clinical features, including monocytosis, leukocytosis, and advanced age at diagnosis (Jankowska et al., 2009; Smith et al., 2010; Tefferi et al., 2009a; Tefferi et al., 2009c). Although *TET2* mutations show little association to known myeloid leukemogenic drivers—*FLT3-ITD, RUNX1, CEBPA*—they do associate with *NPM1* and *ASXL1* mutations and infrequently co-occur with *IDH1* or *IDH2* mutations (Chou et al., 2011a; Weissmann et al., 2012). Lastly, despite some data indicating no association between cytogenetics and *TET2* status in MPN, *TET2* mutations occur more often in the presence of normal karyotype and intermediate-risk AML (Hussein et al., 2010; Weissmann et al., 2012). In this cytogenetic setting, *TET2* mutations do predict significantly worse OS in AML (Chou et al., 2011a).

Although few *TET2* recurrent mutations have been reported, many mutations result in a frameshift or early stop codon and are therefore inactivating (Tefferi et al., 2009b). In fact, the largest proportion of nonsense mutations occur in exon 3, resulting in a truncated protein lacking the C-terminal catalytic domain (Moran-Crusio et al., 2011). Additionally, several missense mutations have been characterized as loss-of-function, with Ko and colleagues reporting impaired hydroxylation of 5-mC when mutant TET2 was overexpressed in HEK-293T cells. Furthermore, *TET2* mutation status is significantly correlated with decreased global 5-hmC in myeloid tumors (Ko et al., 2010; Konstandin et al., 2011). Functional studies manipulating *TET2* have also been able to recapitulate phenotypes that are characteristic of myeloid neoplasms, suggesting that *TET2* loss may be a key event in leukemic transformation. For example, a conditional mouse model for *TET2* loss in the hematopoietic compartment resulted in increased HSC self-renewal and myeloproliferation including splenomegaly, monocytosis, and extramedullary hematopoiesis (Moran-Crusio et al., 2011). This is consistent with other studies showing that *TET2* inactivation leads to decreased 5-hmC in HSCs, amplification of the stem cell population, and may skew HSCs toward a myeloid lineage (Pronier et al., 2011; Quivoron et al., 2011).

The true effect of *TET2* mutations on DNA methylation status has been difficult to ascertain. Despite an expected increase in 5-mC following *TET2* inactivation, several studies have reported a global decrease in methylation (Ko et al., 2010; Yamazaki et al., 2012). However, analysis of specific gene promoters shows mixed results in TET2 mutant samples, frequently exhibiting promoter-specific hypermethylation in spite of global hypomethylation (Perez et al., 2012; Wu et al., 2011; Yamazaki et al., 2012). Still, Ko and colleagues noted that several AML patients with wild-type TET2 had 5-hmC levels very similar to those patients with mutant TET2 (Ko et al., 2010). This suggests a more complex relationship between *TET2*, DNA methylation status, and malignant transformation.

In the past year, exciting new evidence has emerged to suggest a more diverse role for *TET2* in epigenetic regulation. In addition to known associations with polycomb repressive complex (PRC) regulator SIN3A and NURD complex member MBD3 (Wu et al., 2011; Yildirim et al., 2011), TET2 was recently identified as a direct binding partner with O-linked beta-N-acetylglucosamine transferase (OGT), an enzyme that marks histone H2B S112 at active transcription start sites (TSS) (Chen et al., 2013b). Although OGT doesn't influence TET2 activity in functional assays, TET2 seems to actively target OGT to unmethylated promoters and activate transcription via other means (Vella et al., 2013). Furthermore, Deplus and colleagues showed that TET2 and OGT co-localize at active promoters marked by H3K4me3 through a direct interaction with host-cell factor 1 (HCF1) and that knockdown of *TET2* leads to global decreases of H3K4me3 and GlcNAcylation (Deplus et al., 2013). Another direct interaction has been described between TET2 and EBF1, a transcription factor that is associated with transcriptional activation and "poised" chromatin (Guilhamon et al., 2013; Treiber et al., 2010). Additional binding partners have also been reported within the past year, including NANOG and IDAX (Costa et al., 2013; Ko et al., 2013). Lastly, several novel miRNAs were discovered to negatively regulate *TET2* expression, offering a possible explanation for *TET*-associated transformation in the absence of any genomic alterations (Cheng et al., 2013; Fu et al., 2013; Song et al., 2013). Collectively, evidence is mounting that *TET2* inactivation in cancer may alter more than just DNA methylation; in fact, transformation may result considerably from disrupted interactions with other epigenetic regulators and development-associated transcription factors.

IDH1/2

Isocitrate dehydrogenase 1 (IDH1) and isocitrate dehydrogenase 2 (IDH2) are two homodimeric metabolic enzymes that convert isocitrate to α-ketoglutarate (α-KG) while reducing $NADP^+$ to NADPH. IDH1 is present in the cytosol and peroxisomes whereas IDH2 is located exclusively in mitochondria (Geisbrecht and Gould, 1999; Winkler et al., 1986; Xu et al., 2004). Frequent recurrent mutations in *IDH1* were initially discovered in GBM (12%) following whole-exome sequencing of 22 patient tumor samples (Parsons et al., 2008). Further sequencing efforts revealed that mutations are most prevalent in WHO grade II/III gliomas (71%) and secondary GBMs (88%) but less common in primary GBMs (7%) (Balss et al., 2008). Subsequent studies have shown that *IDH2* mutations are also enriched in WHO grade II/III gliomas, albeit less frequently, and that *IDH1/2* mutations occur in a mutually exclusive manner (Hartmann et al., 2009). These data indicate *IDH* mutation is an early event in glioma oncogenesis, frequently preceding known alterations like *TP53* mutation and 1p/19q loss (Watanabe et al., 2009). Interestingly, recent data suggest *IDH1* and *IDH2* mutations may actually differentially associate with astrocytoma and oligodendrogliomas, respectively (Hartmann et al., 2009). *IDH* mutations are associated with *MGMT* promoter hypermethylation, *TP53* mutation, 1p/19q codeletion, *ATRX* inactivation, younger age, and improved prognosis while being inversely correlated with *EGFR* amplification in glioma (Chou et al., 2010; Wiestler et al., 2013; Yan et al., 2009; Zou et al., 2013). Further, although early studies could not find any *IDH* mutations in other types of solid tumors (Bleeker et al., 2009; Kang et al., 2009), recurrent mutations have since been identified in chondrosarcoma (56%), cholangiocarcinoma (23%), melanoma (10%), and prostate cancer (2%) (Amary et al., 2011; Borger et al., 2012; Ghiam et al., 2012; Shibata et al., 2011).

Soon after the discovery of *IDH* mutations in glioma, recurrent mutations of *IDH* were also identified in AML (Green and Beer, 2010; Mardis et al., 2009). Similar to glioma, *IDH1* and *IDH2* mutations are mutually exclusive, though the mutational frequency of *IDH* in AML is much lower (23%) (Chou et al., 2011b; Ward et al., 2010). In contrast, the utility of *IDH* mutation status as an independent prognostic marker in AML is less clear. In a convincing cohort of 493 adult patients with AML, Chou and colleagues found that *IDH1* mutation had no impact on OS (Chou et al., 2010). Still, other studies have suggested a more disparate role, with mutation in *IDH1* and *IDH2* conferring poor and improved prognosis, respectively (Chou et al., 2011b;

Schnittger et al., 2010). Notably, although another large cohort of 805 patients found that *IDH* mutation did not correlate with prognosis, a subset of patients with *IDH* mutation and CN-AML, *NPM1* mutation, and no *FLT3-ITD* did show significantly reduced OS and RFS (Paschka et al., 2010). Therefore, it is likely that the use of prognostic subsets—*IDH* status along with other genetic markers—may improve the utility of *IDH* status as a biomarker in AML. Other features that correlate with *IDH* mutation status include normal karyotype, intermediate-risk cytogenetics, *NPM1* mutation, and M1 AML subtype (Chou et al., 2010; Schnittger et al., 2010).

To date, practically all *IDH* mutations found in cancer are heterozygous and highly recurrent. Amino acid substitutions at residues IDH1-R132, IDH2-R172 and IDH2-R140Q are the most common, with considerable variability at R132 (R/H, R/C, R/S, R/G, R/L, R/V) and R172 (R/G, R/M, R/K) (Balss et al., 2008; Yan et al., 2009). The remarkable absence of any frameshift or nonsense mutations, deletions, or epigenetic silencing provided early evidence that *IDH* mutations were activating (Flanagan et al., 2012; Zhao et al., 2009). The R132 residue is located in the active site of IDH1 where it forms 2 hydrogen bonds with α- and β- carboxylate of isocitrate, its substrate (Nekrutenko et al., 1998; Xu et al., 2004). Initially, it was believed that these mutations may be loss-of-function or dominant-negative, as mutant IDH showed a reduced affinity for isocitrate and produced markedly less α-KG and NADPH *in vitro* (Yan et al., 2009; Zhao et al., 2009). However, an *in vitro* metabolite screen revealed that *IDH* mutations are neomorphic, producing the novel oncometabolite 2-hydroxyglutarate (2-HG) through heterodimer formation with the remaining wild-type IDH1. This was also verified in patient samples, with a strong correlation between the amount of 2-HG in tumor tissue and *IDH1/2* mutation status (Dang et al., 2009; Ward et al., 2010). In fact, 2-HG levels are increased 10–100 fold in patient sera and can be used to reliably diagnose *IDH* status and monitor response to therapy, though this application may be restricted to myeloid neoplasms (Capper et al., 2012; DiNardo et al., 2013; Ward et al., 2010).

The effects of mutant IDH are pleiotropic and affect numerous cell processes including DNA methylation, histone methylation, HIF1a signaling, collagen synthesis, and metabolism (Cairns and Mak, 2013). Remarkably, α-KG levels are unchanged in mutant *IDH* AML and glioma (Dang et al., 2009; Gross et al., 2010), and it is now clear that 2-HG-mediated inhibition of 2-OG-dependent dioxygenases is the dominant mechanism by which *IDH* mutations are oncogenic (Xu et al., 2011). Early data from the Cancer Genome Atlas project first identified the glioma hypermethylator phenotype (G-CIMP) in GBM and its association with *IDH* mutations (Laffaire et al., 2011; Noushmehr et al., 2010). Following this, Figueroa and colleagues demonstrated that 2-HG inhibition of the α-KG-dependent enzyme TET2 actively generates the hypermethylator phenotype in AML. Further, they showed that *TET2* and *IDH* mutations are mutually exclusive in AML, result in overlapping methylation signatures, and impair HSC differentiation in 32D myeloid cells (Figueroa et al., 2010). Similarly, work from our lab demonstrated that *IDH1* mutation directly causes the G-CIMP phenotype, reduces global 5-hmC through TET2 inhibition, results in hypermethylation of the repressive histone marks H3K9 and H3K27, and blocks differentiation (Turcan et al., 2012). It is important to highlight that widespread loss of 5-hmC is an additional epigenetic hallmark in *IDH* or *TET2* mutated cancers, including melanoma, and that reestablishment of the 5-hmC landscape can suppress tumor invasion and growth in both melanoma cells and a zebrafish model (Lian et al., 2012). Interestingly, *IDH*-associated increases in histone methylation are likely due to 2-HG-mediated inhibition of the Jumonji C (JmjC)-domain-containing histone demethylases (Lu et al., 2012). Still, others have proposed additional mechanisms in *IDH*-mutated cancers such as HIF1a stabilization through PHD inhibition, altered ECM structure due to decreased hydroxylation of collagen, and possible metabolic shifts in NADP/NADPH ratio (Sasaki et al., 2012; Zhao et al., 2009).

Recently, several exciting studies have shed light on novel mechanisms by which *IDH* mutations initiate malignant transformation and how underlying mechanisms may be exploited for therapeutic gain. In mouse models of leukemia and melanoma, IDH mutants accelerated cell cycle transition by activation of the MAPK/ERK pathway and repression of tumor suppressors *CDKN2A* and *CDKN2B* (Chaturvedi et al., 2013; Shibata et al., 2011). Although several studies have noted that *IDH* mutations cause increased colony formation in soft agar and enhanced proliferation, two mouse models for leukemia found that *IDH* mutation primes cells by inducing an MDS- or MPN-like state. However, combining *IDH1* mutants with HOXA9, or *IDH2* mutants with FLT3 or NRAS, was sufficient to initiate transformation (Chaturvedi et al., 2013; Chen et al., 2013a; Xu et al., 2011). This may be cancer-specific though, as mutant IDH2 alone was recently shown to be sufficient to induce sarcoma formation in mice, at least in one model system (Lu et al., 2013). Regardless, the primary mechanism underlying IDH-induced oncogenesis in several model systems is a block in cell differentiation (Pirozzi et al., 2013). Both groups showed restoration of differentiation and increased apoptosis following treatment with IDH inhibitor HMS-101 or Brd4 inhibitor JQ1. Similarly, Losman and colleagues showed that IDH mutant leukemic transformation is specific to the (R)-enantiomer of 2-HG, which can independently promote cytokine independence and block differentiation. Again, this transformation was reversible with IDH inhibitor AGI-5198 (Losman et al., 2013). Lastly, IDH inhibition can reverse novel EMT-associated expression patterns, though a lengthy delay to phenotypic change suggests more stable epigenetic alterations may be to blame (Grassian et al., 2012).

Despite the targeted nature of IDH inhibitors, *IDH* mutation likely unleashes epigenetic marks that are themselves selectable, such as DNA methylation. There is thus interest in the therapeutic potential of 5-azacytidine (5-aza) and decitabine (DAC), due to the complex downstream effects of *IDH* mutations on the cancer methylome. In order to determine if broader epigenetic therapies could "unlock" glioma initiating cells (GIC) from a dedifferentiated state, we treated both wild-type and mutant IDH cell lines with DAC and/or IDH inhibitor AGI-5198. Along with Borodovsky and colleagues, our group found that hypomethylating agents potently induce differentiation, impair colony formation, and suppress *in vivo* growth in IDH mutant cells only (Borodovsky et al., 2013; Turcan et al., 2013). This demonstrates that mutant IDH-induced DNA methylation likely plays a role in maintaining the self-renewal capacity of glioma tumor initiating cells. Similar effectiveness has also been seen in leukemia and chondrosarcoma (Chaturvedi et al., 2013; Chen et al., 2013a; Lu et al., 2013). Interestingly, IDH inhibition using the mutant IDH1 inhibitor AGI-5198 was not nearly as effective, which suggests that broader epigenetic therapies may be necessary to reverse more permanent changes induced by long exposure to mutant IDH. In addition, combination therapy with other IDH affected processes such as histone hypermethylation may have a role and warrant further investigation.

HISTONE METHYLATION

Histone methylation is a reversible process that takes place at the amino acid side chains of lysine, arginine, and histidine residues. Lysine methylation on histones H3 and H4 is the best characterized and catalyzed by the lysine methyltransferases (KMTs) through the required methyl group donor SAM. All of the KMTs except DOTL1/KMT4 have a catalytically active SET domain and are highly specific to both histone residue and degree of methylation (mono- vs. di- vs. tri-methylation) (Feng et al., 2002; Rea et al., 2000). Generally, methylation at H3K4, H3K36, and H3K79 corresponds to euchromatic or transcriptionally active regions of the genome, whereas methylation at H3K9, H3K27, and H4K20, is associated with heterochromatic regions and gene silencing. In addition, each residue is capable of four methylated states: unmethylated or mono-/di-/tri-methylated. This provides further regulatory diversity in the histone code. For example, H3K4me2/3 is found at TSSs of active genes, whereas H3K4me1 tends to localize to enhancer regions (Greer and Shi, 2012; Kampranis and Tsichlis, 2009).

In contrast, the lysine-specific demethylases (KDMs) work in opposition to the KMTs through the catalytic removal of methylation marks on histone tails. The two families of KDMs responsible are the (FAD)-dependent amine oxidases and the larger JmjC-containing family of α-KG/Fe(II)-ion dependent oxygenases (Shi and Whetstine, 2007). KDM1A/B (LSD1/2) and KDM5A-D (JARID1A-D) catalyze demethylation at H3K4, whereas KDM2A/B (JHDM1A/B) and KDM4A-C (JMJD2A-C) target H3K36, leading to repressed gene transcription at these sites. Alternatively, transcriptional activation is induced in part by demethylation at H3K9 and H3K27 by KDM1A, KDM3A-C (JHDM2A-C), or KDM4A-D and KDM6A/B (UTX/JMJD3), respectively (Varier and Timmers, 2011). Due to the broad and essential nature of these epigenetic marks across the genome, genetic aberrations of histone modifiers have powerful effects on vital cellular processes such as differentiation and cell cycle control, among others.

Writers (KMTs)

EZH2

EZH2/KMT6 is the enzymatic component of the polycomb repressor complex 2 (PRC2), which is responsible for methylation at H3K27 and subsequent gene silencing (Kirmizis et al., 2004). Other essential subunits of the PRC2 complex through which EZH2 interacts include embryonic ectoderm development (EED), suppressor of zeste 12 homologue (SUZ12), and RBAP48/RBBP4. Collectively, these polycomb group (PcG) proteins have been shown to regulate vital cellular processes including differentiation, cell identity, stem-cell plasticity, and proliferation (Margueron and Reinberg, 2011; Shih et al., 2012). As a result, aberrations in any PRC2 component can have powerful physiologic consequences on the cell.

Alterations in *EZH2* were first discovered in breast and prostate cancer, where amplification and overexpression first implied it may function as an oncogene (Kleer et al., 2003; Varambally et al., 2002; Yang and Yu, 2013). This finding was validated both *in vitro* and *in vivo*, with *EZH2* overexpression proving sufficient to drive proliferation in cancer cells and transform primary fibroblasts (Bracken et al., 2003; Croonquist and Van Ness, 2005). Recent sequencing studies have identified numerous mutations of *EZH2* in a variety of leukemias and lymphomas, including follicular lymphoma (FL; 7%–22%), diffuse large B-cell lymphoma (DLBCL; 14%–21.7%), high grade B-cell lymphoma (18%), MDS/MPN (6%–13%), CMML (11.1%), T-ALL, and AML (Abdel-Wahab et al., 2011; Bodor et al., 2011; Capello et al., 2011; Ernst et al., 2010; Grossmann et al., 2011; Lohr et al., 2012; Makishima et al., 2010; Morin et al., 2010; Nikoloski et al., 2010; Ryan et al., 2011; Zhang et al., 2012). Interestingly, frequent missense and truncating mutations were observed, which generated some confusion in the field about whether *EZH2* could possess both pro- and anti-oncogenic functions. Clinically, *EZH2* mutations seem to commonly predict poor prognosis—worse OS/leukemia-free survival (LFS), high-risk IPSS score—especially in myeloid malignancies (Guglielmelli et al., 2011; Khan et al., 2013; Nikoloski et al., 2010). Additionally, *EZH2* mutation associates with *BCL2* rearrangement in FL and germinal center B-cell like DLBCL (GCB-DLBCL) and is notably absent from activated B-cell like DLBCL (ABC-DLBCL) (Beguelin et al., 2013; Morin et al., 2010; Ryan et al., 2011).

One landmark finding that emerged from the wealth of recent sequencing data was the identification of the highly recurrent heterozygous Y641 mutation (Y/F, Y/N, Y/H, Y/C) in FL, DLBCL, and other lymphoid neoplasms (Bodor et al., 2013; Bodor et al., 2011; Morin et al., 2010). Initially thought to be inactivating due to reduced catalytic ability against a short H3-like peptide *in vitro*, Y641 mutant EZH2 exhibited a powerful gain-of-function phenotype when incubated against the entire nucleosomal unit (Sneeringer et al., 2010). In addition, Sneeringer and others have observed a powerful synergy between the wild-type and Y641 forms of EZH2 both *in vitro* and *in vivo*. Whereas EZH2 is very efficient at catalyzing monomethylation of H3K27 but not di-/tri-methylation, EZH2 Y641 shows enhanced ability for di-/tri- methylation at H3K27 (Ryan et al., 2011; Wigle et al., 2011; Yap et al., 2011). Similar findings have also been observed with A677G and A687V mutant EZH2, though these are far less prevalent in cancer (Majer et al., 2012; McCabe et al., 2012a). Since almost all *EZH2* gain-of-function mutations are heterozygous, the overall consequence of these mutations is a cooperative and thereby efficient silencing of genes associated with the repressive H3K27 mark.

Recent studies have shown that mutant EZH2-driven tumors can be effectively targeted with small molecule inhibitors. Knutson and colleagues were the first to describe potent phenotypic effects in lymphoma cell lines, following treatment with the SAM-competitive EZH2 inhibitor (EPZ005687) (Knutson et al., 2012). This inhibitor was highly selective, inducing cell death in mutant EZH2-expressing cells only. As expected, these cells showed global reduction of the H3K27me2/me3 histone mark and significant enrichment of cell cycle gene sets by GSEA. The more recent EZH2 inhibitor GSK126 was also highly selective for mutant EZH2 lymphoma cells *in vivo* and led to increased activation of known EZH2 target genes, such as *TXNIP* and *TNFRSF21* (McCabe et al., 2012b). Although *EZH2* mutation alone may be insufficient to induce development of B-cell lymphoma, new evidence suggests it functions as a master regulator of GCB phenotype through repression of *CDKN1A*, *IRF4*, and *PRDM1* (Beguelin et al., 2013). Due to frequent activation of *EZH2* in lymphoma, these new targeted therapies hold exciting promise in the clinic.

SETD2

The major KMT responsible for H3K36 trimethylation is SETD2/KMT3A, which is a novel candidate tumor suppressor gene (TSG) (Edmunds et al., 2008). Gene deletions in clear cell renal cell carcinoma (ccRCC)-derived cell lines are common, reduced expression is seen in breast tumors, and loss is associated with decreased H3K36 trimethylation (Duns et al., 2012; Duns et al., 2010; Newbold and Mokbel, 2010). *SETD2* mutations are quite common in ccRCC (7.4%–11.6%), pediatric high-grade glioma (HGG; 15%), and adult HGG (8%) (Cancer Genome Atlas Research Network, 2013; Dalgliesh et al., 2010; Duns et al., 2010; Fontebasso et al., 2013; Hakimi et al., 2013b; Varela et al., 2011). Almost all of the mutations characterized so far are frameshift or nonsense and therefore truncating, further suggesting *SETD2* may be an important TSG in select malignancies (Hakimi et al., 2013a). Furthermore, Hakimi and colleagues found that *SETD2* mutations were significantly associated with worse cancer-specific survival (CSS) in ccRCC (Hakimi et al., 2013b). Though phenotypic effects of *SETD2* inactivation have not been clarified, recent research showed that *SETD2* loss triggers MSI and can increase genome-wide mutation rates through alterations in H3K36 methylation (Li et al., 2013a; Schmidt and Jackson, 2013).

MLLs

The mammalian mixed lineage leukemia (MLL) family of genes encodes a series of active (MLL1–4/KMT2A–D) and inactive (MLL5/KMT2E) KMTs, which have all been implicated in cancer. MLL1–4 are responsible for methylation at H3K4 and share a common core formed by WDR5, RbBP5, Dpy-30, and Ash2L (Varier and Timmers, 2011). Notably, MLL1–2 form a complex along with the menin (MEN1) tumor suppressor and recent evidence shows that H3K27 demethylase KDM6A/UTX can complex with MLL2–4 (Hughes et al., 2004; Yokoyama and Cleary, 2008; Yokoyama et al., 2005).

The earliest known alterations in *MLL* family genes involved frequent rearrangements of *MLL1* at 11q23, with recombination involving more than 40 different partner genes and occurring in 60%–80% of infants with ALL or AML (Dimartino and Cleary, 1999; Pais et al., 2005; Thirman et al., 1993). Since then, several missense and truncating mutations have been identified in *MLL1* in bladder, lung, and breast cancer (Gui et al., 2011; Kan et al., 2010). Although *MLL1* acts as a dominant oncogene in liquid tumors, these new discoveries suggest a different recessive role for *MLL1* in some solid tumors (Krivtsov and Armstrong, 2007). Infrequent truncating mutations have also been noted in *MLL4* in medulloblastoma and head and neck squamous cell carcinoma (HNSCC), suggesting a minor but alternative role for this family member as well (Pugh et al., 2012; Stransky et al., 2011).

Notably, recently, many new mutations have been identified in both *MLL2* and *MLL3*, showing diversity of both mutation and tumor type. *MLL2* and *MLL3* mutations are frequently nonsense or frameshift, resulting in a truncated protein lacking the active SET domain (Morin et al., 2011; Parsons et al., 2011; Pasqualucci et al., 2011b; Pugh et al., 2012; Stransky et al., 2011). Along with *MLL1/4*, *MLL2/3* mutations suggest a dual role for MLL family proteins in oncogenesis, which may depend heavily on cellular context. In addition, mutations have been discovered in numerous cancers, occasionally at high frequency, including colon (*MLL3*, 14%–17%), DLBCL (*MLL2*, 24%–32%), FL (*MLL2*,

89%), AML, breast, GBM, RCC, prostate, pancreatic, bladder, medulloblastoma, and HNSCC (Balakrishnan et al., 2007; Dalgliesh et al., 2010; Gui et al., 2011; Li et al., 2013b; Lindberg et al., 2013b; Mann et al., 2012; Morin et al., 2011; Parsons et al., 2008; Parsons et al., 2011; Pasqualucci et al., 2011b; Pugh et al., 2012; Sjoblom et al., 2006; Stransky et al., 2011; Vakoc et al., 2009; Watanabe et al., 2011). Unfortunately, functional data remains sparse and the importance of these mutations has yet to be characterized. Watanabe and colleagues made the interesting observation that in colorectal carcinoma, *MLL3* mutations were associated with increased MSI, though no mechanism has been proposed (Watanabe et al., 2011). Regardless, they are intriguing candidates for further study in cancer, especially since MLL family proteins are important regulators of HOX proteins and differentiation (Wang et al., 2009a).

NSD1/2

Histone-lysine N-methyltransferase *NSD2/MMSET* was first implicated in cancer as a target for rearrangement [t(4;14)(p16.3;q32)] in 15%–20% of multiple myeloma (MM) patients (Chesi et al., 1998). This translocation results in aberrant upregulation of *NSD2*, which first suggested that it may be an oncogene. Subsequent work has shown that knockdown in MM KMS11 cells leads to apoptosis and re-expression of wild-type NSD2 causes increased proliferation (Martinez-Garcia et al., 2011). Furthermore, overexpression of wild-type *NSD2* is sufficient to transform *NSD2*$^{-/-}$ cancer cells *in vivo* and in mouse embryonic fibroblasts (MEFs) (Kuo et al., 2011). Functionally, interactions with HDAC1/2 and catalytic activity at H3K4 and H4K20 have been proposed, though these may be minor (Marango et al., 2008). It is now clear that the NSD2-catalyzed conversion of unmethylated H3K36 to mono- or di-methylated forms, with concomitant decreases in H3K27me3, is the dominant mechanism driving *NSD2*-associated oncogenic reprogramming (Kuo et al., 2011; Li et al., 2009). In fact, Kuo and colleagues demonstrated that NSD2 SET catalytic activity is required for transcriptional activation at several oncogenic loci (*TGFA*, *MET*, *PAK1*, *RRAS2*). Pathway analyses have also identified the following as significantly altered in mutant NSD2 tumors: TP53 pathway, cell cycle, DNA repair, focal adhesion, and Wnt (Kuo et al., 2011; Martinez-Garcia et al., 2011).

Recently, sequencing projects revealed the presence of a highly recurrent mutation in *NSD2* (E1099K), which is present in 7.5% of pediatric B-ALL and other lymphoid neoplasms (Jaffe et al., 2013; Oyer et al., 2013). These studies showed that NSD2 E1099K leads to enhanced colony formation in soft agar and expected increases and decreases in H3K36me2 and H3K27me3, respectively. This new discovery has exciting therapeutic potential—similar to the activating mutations in *EZH2* described above. Additionally, the related KMT *NSD1* was recently discovered to harbor point mutations in multiple cancers, including HNSCC and AML (Dolnik et al., 2012; Yan et al., 2011). If these mutations prove to be similarly activating, both NSD members will represent completely novel areas of epigenetic regulation through which small molecule targeted inhibition could be useful.

Erasers (KDMs)

KDM6A

Among the first cancer-associated mutations in KDMs that were identified were those in *KDM6A* following sequencing of 1,390 patient tumor samples (van Haaften et al., 2009). Remarkably, *KDM6A* mutations were found to be widespread across both solid and liquid tumors, including AML, chronic myelogenous leukemia (CML), T-ALL, MM, Hodgkin's lymphoma (HL), TCC, breast, colon, esophageal, pancreas, endometrial, GBM, small cell lung cancer (SCLC), non-small cell lung cancer (NSCLC), and RCC (Dalgliesh et al., 2010; Gui et al., 2011; Mann et al., 2012; Mar et al., 2012; Ross et al., 2013; van Haaften et al., 2009). Since then, *KDM6A* mutations have also been discovered in other tumors such as prostate cancer, medulloblastoma, and adenoid cystic carcinoma (Ho et al., 2013; Lindberg et al., 2013a; Robinson et al., 2012). Although these mutations occur at low frequency in most cancers, *KDM6A* mutations in bladder carcinoma are quite common (20%–29%) (Gui et al., 2011; Poon et al., 2013; Ross et al., 2013). Furthermore, *KDM6A* mutations in bladder carcinoma associate with earlier grade and are inversely correlated with stage (Gui et al., 2011). Therefore, *KDM6A* inactivation may be a powerful driver and early event in bladder oncogenesis. However, whether *KDM6A* mutation status holds significant prognostic value in cancer is yet to be determined.

Despite infrequent inactivation in many cancers, mutation and functional data have established that *KDM6A* is a *bona fide* tumor suppressor gene. KDM6A is a 1401 amino acid protein with several N-terminal tetratricopeptide-repeat (TPR) domains and a single C-terminal Jumonji C (JmjC) domain (Shpargel et al., 2012). Early sequencing efforts revealed that a majority of *KDM6A* mutations are either frameshift or nonsense, and since most occur before the active JmjC demethylase domain, they are most likely inactivating. To test this hypothesis, van Haaften and colleagues re-expressed wild-type *KDM6A* in *KDM6A*-deleted cell lines and observed markedly reduced proliferation (van Haaften et al., 2009). Similarly, we recently showed that cancer-specific missense mutations in the JmjC domain can abrogate this growth suppressive effect and may even contribute to a dominant proliferative phenotype. Furthermore, following overexpression, these tumor-specific mutants exhibited reduced demethylase activity at the repressive chromatin mark H3K27me3 (Ho et al., 2013). Dysregulation of methylation at H3K27 may have important consequences in cancer, as demethylation of H3K27 at HOX genes is required for proper differentiation. Interestingly, it has also been reported that KDM6A binds directly to the HOXB1 locus and that its activation requires KDM6A catalytic activity (Christensen et al., 2007; Morales Torres et al., 2013). In addition to regulation through HOX gene targets, KDM6A catalytic activity also activates RB pathway genes through HBP1 to further influence differentiation and cell cycle control (Herz et al., 2010; Wang et al., 2010).

In addition to contributing to tumor suppression via its catalytic domain, KDM6A also has important demethylase-independent roles in cancer. A recent study found that conditional inactivation of *KDM6A* in a mouse model did not change global levels of H3K27me3, though it did contribute to a MDS-like phenotype and reduced migration of HSCs (Thieme et al., 2013). Furthermore, it is also known that KDM6A can bind to KMTs MLL2–4 to promote H3K4 methylation independent of its catalytic domain and that demethylase-inactive KDM6A is sufficient to induce differentiation (Cho et al., 2007; Issaeva et al., 2007; Lee et al., 2007; Morales Torres et al., 2013). In fact, Wang and colleagues mapped the chromatin occupancy of KDM6A, H3K4me2, and H3K27me3 in primary human fibroblasts and discovered that 62% of KDM6A target genes are enriched for univalent H3K4me2 (Wang et al., 2010). Therefore, KDM6A has at least two independent, yet complimentary, mechanisms for shaping the epigenetic landscape in cancer. Indeed, one study found that inactivating mutations in the catalytic JmjC domain caused increased growth yet also led to simultaneous increases and decreases in H3K27me3 and H3K4me1, respectively (Herz et al., 2010). As a result, it is likely that the *KDM6A*-associated phenotypes in cancer are diverse and are linked to the type and location of each driving mutation.

Other KDMs

Although *KDM6A* mutations are the most prevalent and best characterized among the KDMs, several others have been identified as significantly mutated across cancer, albeit at low frequency (Cerami et al., 2012; Gao et al., 2013; Parsons et al., 2011; Pasqualucci et al., 2011b). Uniquely, ccRCC harbors mutations in many of the KDMs, including *KDM1A*, *KDM2B*, *KDM3A*, *KDM3B*, *KDM4A/B*, and *KDM5C* (Dalgliesh et al., 2010; Hakimi et al., 2013a; Larkin et al., 2012; Shi et al., 2011). The natural function of these KDMs is still being determined, but these studies and others suggest *KDM1A*, *KDM4A–C*, and *KDM5B* are putative oncogenes, whereas *KDM6A/B* and *KDM3B/C* are tumor suppressors. Also, *KDM2A/B* and *KDM5A* seem likely to be both pro- and anti-oncogenic, depending on context (Rotili and Mai, 2011). More recently, Niu and colleagues provided the first *in vivo* evidence that *KDM5C* serves as a tumor suppressor following VHL loss in ccRCC and that cancer-specific mutations were inactivating. Furthermore, they demonstrated that HIF2a binds directly to KDM5C, targeting KDM5C to demethylate H3K4me3 at HIF-repressed gene loci (Niu et al., 2012). It will be exciting to determine if some of these mutations are true oncogenic drivers. Additionally, if gain-of-function mutants are identified in KDM oncoproteins, these may be prime candidates for existing KDM inhibitors or new targeted therapies (Rotili and Mai, 2011).

HISTONE ACETYLATION

Lysine residues on histone tails may also undergo another form of covalent modification through the addition of an acetyl functional group. This process uniquely results in the neutralization of charge normally associated with lysine residues, which weakens the electrostatic interaction between histones and negatively charged DNA. As a result, it is believed that histone acetylation primarily results in a more "open" chromatin configuration, serving as a "mark" of active gene transcription. Several ChIP-seq studies have now confirmed this, showing localization of acetylated histones at enhancers, promoters, and even throughout the transcribed region of active genes (Dawson and Kouzarides, 2012; Heintzman et al., 2007; Wang et al., 2008). In addition to altering the chromatin state directly, these specific histone "marks" further act to recruit other remodelers containing "reader" bromodomains and tandem plant homeodomain (PHD) fingers (Taverna et al., 2007).

The process of histone acetylation is carried out by the lysine acetyltransferase (KAT) enzymes, of which there are two major classes: Type-A, which are usually found in the nucleus and Type-B, which are cytoplasmic and act on free histones. Dynamic regulation of acetylation is also catalyzed by the histone deacetylase (HDAC) enzymes, which oppose the actions of KATs and remove acetyl groups from histone tails. Interestingly, these enzymes are capable of modifying other non-histone proteins—including p53, Rb, and MYC—and have additional roles as transcriptional cofactors, which has led to many challenges in determining their specific roles in cancer and other disease processes (Dawson and Kouzarides, 2012; Iyer et al., 2004).

Writers (CREBBP and EP300)

CREB-binding protein (CREBBP) and E1A binding protein p300 (EP300) are structurally distinct from other KATs and have unique broad substrate specificity, including the ability to acetylate all four histones *in vitro*. In fact, both proteins are highly conserved, with 75% similarity across their entire length and 63% homology at the amino-acid level (Iyer et al., 2004; Shiama, 1997). Not surprisingly, many functional similarities exist. Both proteins engage in several diverse functions, including chromatin remodeling via KAT activity, acetylation of association proteins (p53, Rb, E2F), and the ability to act as scaffolds for transcription factors and other transcriptional machinery (Bannister and Kouzarides, 1996; Gu and Roeder, 1997; Nakajima et al., 1997). Despite being some of the earliest epigenetic modifiers identified, their roles in both normal physiology and disease are just beginning to be appreciated.

Both *CREBBP* and *EP300* have long been linked to cancer, though the specific roles they play have been harder to elucidate. Suspicion that *CREBBP* may be a tumor suppressor first arose in the mid-1990s, when heterozygous germline mutations were identified in the setting of Rubinstein-Taybi syndrome, a developmental disorder with an increased prevalence of cancer, including leukemia and lymphoma (Petrij et al., 1995). Around the same time, *EP300* was first shown to bind to the E1A viral oncoprotein, suggesting it may also function as a tumor suppressor (Eckner et al., 1994). Soon after these discoveries, the first genetic alteration of *CREBBP* in cancer was identified in M4/M5 AML subtypes, albeit a rare t(8,16)(p11,p13) translocation that fuses the *MOZ* gene with the N-terminus of *CREBBP* (Borrow et al., 1996; Panagopoulos et al., 2001). Interestingly, reports of *MOZ-EP300* translocations do exist, though these events may be exceedingly rare (Lai et al., 1985). Though a few early studies identified low frequency *EP300* mutations in colorectal carcinoma, breast cancer, and gastric cancer, the full spectrum of mutational inactivation of *CREBBP* and *EP300* would not be fully evident until the genomics era.

Genomic and exomic sequencing data from the past several years have revealed that *CREBBP* and *EP300* inactivation via mutation is more widespread and frequent than previously thought. For example, *CREBBP* mutations have now been described in NHL (21%), DLBCL (29%), FL (32.6%), TCC (13%), ACC (7%), and relapsed ALL (18.3%), with *EP300* mutations occurring slightly less frequently in NHL (7%), DLBCL (10%), FL (8.7%), TCC (13%), ACC, and relapsed ALL (Gui et al., 2011; Ho et al., 2013; Morin et al., 2011; Mullighan et al., 2011; Pasqualucci et al., 2011a). Additionally, *CREBBP* and *EP300* are collectively mutated in up to 18% of SCLC (Peifer et al., 2012). Interestingly, mutations in both genes are mutually exclusive, suggesting functional equivalency, at least in part. Additionally, the majority of mutations are heterozygous, indicating that both genes most likely function as haploinsufficient tumor suppressors. In line with this, an earlier mouse study showed that heterozygous *CREBBP* loss led to increased neoplasia over wild-type mice (Kung et al., 2000). In almost all of these studies, mutations strongly clustered in the catalytic KAT domain, several of which exhibit reduced acetyltransferase ability *in vitro* at H3K18 and in the non-histone substrates Bcl6 and p53 (Mullighan et al., 2011; Pasqualucci et al., 2011a; Peifer et al., 2012). Though the targets of CREBBP/p300 are diverse, it seems likely that disruption of acetyltransferase ability can be a main contributor to tumor formation.

Despite the identification and initial functional characterization of these tumor-specific mutations, the physiologic consequences leading to oncogenesis remain obscure. For example, in SCLC, *CREBBP/EP300* mutations do not lead to any obvious concerted shifts in gene expression, even though several reported mutations are inactivating *in vitro* (Peifer et al., 2012). Due to the diverse roles of CREBBP and p300, it is possible that some inactivating mutations exert phenotypic consequences through other non-epigenetic mechanisms. For example, a more recent discovery showed that both proteins exhibit acetyltransferase specificity for histones H3 and H4 at double-strand break (DSB) sites, facilitating the recruitment of the SWI/SNF chromatin remodeling complex (Ogiwara et al., 2011). Also, *CREBBP* KAT mutations favor constitutive activity of the Bcl6 oncogene over p53, which may be alone sufficient to promote tumorigenesis. Lastly, paradoxical roles for *CREBBP/EP300* have been described recently. For example, *EP300* is actually upregulated in melanoma cell lines, and inhibition of KAT function *in vitro* reduces melanoma tumor cell growth (Yan et al., 2013). Additional experiments will be necessary to help identify functional consequences of these mutations and in what specific contexts these two KATs function as tumor suppressors.

Erasers (HDACs)

In contrast to acetyltransferases, the 18 member HDAC family is responsible for the removal of acetyl groups from lysine residues on histone tails. Similar to KATs, HDACs have a wide range of protein targets, and are also known to deacetylate nonhistone substrates (Ellis et al., 2009). The HDAC family has four major classes, with class I (nucleus), II (nucleus and cytoplasm), and IV requiring the zinc ion for catalytic activity. Class III (sirtuins) are catalytically active in the absence of zinc and share almost no homology with the other HDACs (New et al., 2012). Notably, several HDACs have been implicated in cancer. Specifically, functional experiments have revealed that these HDACs are pro-oncogenic, with increased apoptosis and reduced proliferation (Class I) or reduced angiogenesis and cell migration (Class II) following specific HDAC knockdown (Ellis et al., 2009). Most importantly, this role is congruent with their acetyltransferase counterparts, the KATs, which have been characterized as tumor suppressors (see above).

In the past decade, much work has been done to develop targeted inhibitors of HDACs, since they are now established oncogenes in many cellular contexts. These HDAC inhibitors (HDACi) work to prevent histone deacetylation, facilitating a more open chromatin configuration and leading to increased gene transcription. Vorinostat, the first FDA approved HDACi, is already in use for select neoplasms, and other more selective HDACi show promising efficacy in early trials at reducing cell growth and increasing apoptosis (Witt et al., 2009). Despite the substantial therapeutic potential for HDACi, there is a need to determine biomarkers for treatment response and resistance.

Sequencing data from the past several years has identified inactivating mutations in several HDACs, which may influence the effectiveness of HDAC inhibitor therapy and predict overall response in patients (You and Jones, 2012). Specifically, mutations have now been reported in *HDAC2* (colon), *HDAC4* (breast), and *HDAC9* (prostate) (Berger et al., 2011; Ropero et al., 2006; Sjoblom et al., 2006). Of these, the recurrent frameshift mutation of *HDAC2* in exon 1 has been the most extensively characterized. This mutation is incredibly common in colon cancer (21%), resulting in a premature stop codon and loss of measurable *HDAC2* expression in 83% of mutant tumors (Ropero et al., 2006). Further, *HDAC2* mutations are enriched in MSI colon cancers (43%) (Hanigan et al., 2008). Importantly, *in vitro* work showed that *HDAC2*-deficient cells were resistant to HDACi via trichostatin A, exhibiting no hyperacetylation at target histones H3 and H4 and no reduction in proliferation compared to wild-type *HDAC2*-expressing cells (Ropero et al., 2006). Recent work has identified the pro-apoptotic gene *APAF1* as a likely target for repression by HDAC2, providing a specific mechanism for both HDACi efficacy and resistance in HDAC2 mutant cells, where HDACi does not appreciably alter *APAF1* levels (Hanigan et al., 2008). Lastly, although there is an important role for using HDAC mutation status as a predictor of HDACi treatment response, Ropero and colleagues have further shown that the *HDAC2* mutation itself can cause changes in gene expression, actively leading to increased levels of multiple pro-tumorigenic proteins (Ropero et al., 2008).

CHROMATIN REMODELING

In addition to gene regulation via covalent histone tail modifications, the ATP-dependent chromatin remodelers also shape chromatin structure and thereby affect gene expression patterns. Several multi-unit effectors share this responsibility, including SWI/SNF, ISWI, INO80, SWR1, and NURD/Mi2/CDH complexes. In the past several years, protein components of the SWI/SNF complex have been found to be frequently inactivated in cancer, and subsequent work has solidified their status as *bona fide* epigenetic tumor suppressors (Wilson and Roberts, 2011).

SWI/SNF complex

The SWI/SNF complex consists of one or two mutually exclusive catalytic ATPases (SMARCA2/BRM or SMARCA4/BRG1), a group of conserved core subunits (SMARCB1/SNF5, SMARCC1/BAF155, SMARCC2/BAF170), and other variant subunits (Wilson and Roberts, 2011). Two important SWI/SNF complexes implicated in cancer are the BAF and PBAF complexes, which contain the mutually exclusive ARID1A or ARID1B subunits and PBRM1 or BRD7 subunits, respectively (Reisman et al., 2009; Wang et al., 2013). Collectively, SWI/SNF complexes remodel chromatin through the mobilization of nucleosomes both by sliding and by the ejection/insertion of histone octomers (Saha et al., 2006). Through these mechanisms, the SWI/SNF complexes have powerful effects on transcriptional regulation, serving an important role in development through the coordinate activation and repression of critical gene expression programs. Importantly, specificity is most likely achieved through the unique combinatorial assembly of the SWI/SNF complex, facilitated by the sheer size and diversity of the protein subunit repertoire (Wang et al., 1996).

ARIDs

The AT-rich interactive-containing domain (ARID) gene superfamily consists of seven members (ARID1–5), of which the following have now been implicated in cancer: *ARID1A/BAF250a*, *ARID1B/250b*, and *ARID2/BAF200*. Mutations in *ARID1A* are the most widely reported in the literature, with remarkable frequency, first reported in ovarian clear cell carcinoma (OCCC; 50%) and endometrioid carcinoma (30%) (Bosse et al., 2013; Jones et al., 2010; Wiegand et al., 2010). Mutations in other cancers exist, including medulloblastoma, breast, lung adenocarcinoma, ACC, hepatocellular carcinoma (HCC), gastric, pancreatic, and neuroblastoma (Fujimoto et al., 2012; Ho et al., 2013; Sausen et al., 2013; Wu and Roberts, 2013; Zang et al., 2012). Interestingly, the majority of mutations are heterozygous, truncating, and evenly spread along the protein, suggesting a possible role as a haploinsufficient tumor suppressor. Functional studies have confirmed this, noting increased proliferation and colony formation, impaired differentiation, and decreased apoptosis following partial *ARID1A* knockdown (Gao et al., 2008; Luo et al., 2008; Nagl et al., 2007; Zang et al., 2012). Correspondingly, re-expression of *ARID1A* decreases cell proliferation (Zang et al., 2012). In addition, a role in differentiation seems likely, though conflicting data on the specific consequences of *ARID1A* inactivation has been complicated through varying technical approaches and model systems (Wu and Roberts, 2013).

Little is currently known about how *ARID1A* inactivation leads to malignant transformation through SWI/SNF chromatin remodeling, though several intriguing possibilities exist. Both ARID1A and ARID1B provide unique and mutually exclusive specificities for SWI/SNF recruitment to chromatin (Wilson and Roberts, 2011). Interestingly, with ARID1A, this process is at least partially independent of its ARID domain, which binds DNA in a non-specific manner only (Dallas et al., 2000). Instead, ARID1A likely contributes to specific recruitment of SWI/SNF by binding transcription factors and transcriptional coactivator/corepressor complexes, including nuclear hormone receptors (Nie et al., 2000; Trotter and Archer, 2004). In fact, Inoue and colleagues showed that re-expression of *ARID1A* in a breast cancer cell line augments transcriptional activation through glucocorticoid receptors, estrogen receptor, and androgen receptor (Inoue et al., 2002). This specificity is likely due to the presence of several nuclear hormone receptor binding sites near the C-terminus of ARID1A (Nie et al., 2000). In light of this, the impressive prevalence of *ARID1A* mutations in hormone-responsive neoplasms (ovary, breast) is likely to be more than just coincidence.

The clinical utility of *ARID1A* mutation status is currently unclear, due to a limited number of sufficiently powered studies. However, Sausen and colleagues recently found that *ARID1A* or *ARID1B* mutations associate with worse OS in patients with neuroblastoma, and a previous breast cancer study noted that decreased *ARID1A* expression can confer worse prognosis (Mamo et al., 2012; Sausen et al., 2013). Several other intriguing observations have been noted, including a tendency toward co-occurrence between mutations in *ARID1A* and *CTNNB1* (β-catenin) or PI3K-Akt pathway alterations, as well as a mutual exclusivity between *ARID1A* and *TP53* mutations (Bosse et al., 2013). Whether *ARID1A*-associated phenotypes arise through additional epigenetic means, such as reported binding partners HDAC1/2, its ubiquitin ligase activity at H2B K120, or by increased MSI, remains to be determined (Bosse et al., 2013; Li et al., 2010).

Mutations in other ARID family members *ARID1B* and *ARID2* have now been reported, though investigation into their clinical and functional relevance is in its infancy. *ARID1B* mutations occur at moderate frequency in neuroblastoma (7%) and HCC (6.7%), with sporadic mutations identified in breast, gastric, and pancreatic cancers too (Fujimoto et al., 2012; Sausen et al., 2013; Shain et al., 2012; Wang et al., 2011). These mutations are usually frameshift and hemizygous, suggesting it may be a tumor suppressor, like *ARID1A*. More intriguing has been the discovery of *ARID2* mutations in HCC (5.8%–6.5%), with strong enrichment in HCV-associated HCC (14%) (Fujimoto et al., 2012; Zhao et al., 2011). *ARID2* seems to mediate antiproliferative signaling by binding to IFN-inducible promoters to remodel chromatin in response to interferon signaling. As such, mutations in *ARID2* may render IFN-related immune processes incapacitated in the setting of HCV infection, leading to accelerated tumorigenesis. More recently, mutations in *ARID2* were identified in NSCLC (7.3%), making this gene one of the most frequently mutated in this type of cancer (Manceau et al., 2013).

SMARCs

SWI/SNF-related matrix-associated actin-dependent regulator of chromatin (SMARC) genes, also known as BRG1-associated factors, are among the most frequently altered and best characterized chromatin remodelers in cancer. These genes encode several SWI/SNF proteins including the catalytic ATPase subunits (either SMARCA2 or SMARCA4), a group of conserved core subunits (e.g. SMARCB1, SMARCC1, and SMARCC2), and variant subunits (e.g. SMARCE1) (Wilson and Roberts, 2011). Tumor-specific mutations in the following SMARCs are especially common: *SMARCB1/SNF5*, *SMARCA2/BRM*, and *SMARCA4/BRG1* (Shain et al., 2012). Alterations in these critical chromatin remodelers have profound effects on vital processes, such as differentiation, cell proliferation, and metastasis.

SMARCB1 is inactivated via biallelic alterations—deletion and mutation (truncating, missense)—in nearly all rhabdoid tumors (RTs; 98%), an especially lethal cancer that predominantly affects young children (Sievert et al., 2009; Versteege et al., 1998). Recent exomic sequencing data has identified mutations in other tumors and pre-malignant lesions, such as sarcoma, gastric carcinoma, schwannomatosis, meningioma, chordoma, and hepatoblastoma (Christiaans et al., 2011; Hulsebos et al., 2007; Kim et al., 2013; Kreiger et al., 2009; Mobley et al., 2010; Trobaugh-Lotrario et al., 2009). Mouse models for *SMARCB1* loss have verified a tumor suppressive role for this protein, with homozygous loss leading to frequent formation of lymphomas and RT-like tumors at only 11 weeks, which is notably faster than *TP53*-inactivated models of sarcoma (Guidi et al., 2001; Roberts et al., 2000; Wilson and Roberts, 2011). Interestingly, *SMARCB1*-deleted mice do not develop tumors in the absence of SWI/SNF ATPase SMARCA4 (Wang et al., 2009b), suggesting that oncogenesis following *SMARCB1* inactivation is due to aberrant residual activity of SMARCA4-containing SWI/SNF complexes. Additionally, global gains of H3K27me3 following loss of *SMARCB1* may directly contribute to tumorigenesis, as simultaneous *EZH2* loss protects against transformation. In fact, EZH2 elevation occurs immediately following *SMARCAB1* inactivation, and re-expression of *SMARCB1* evicts PcG proteins from the *CDKN2A* tumor suppressor locus, activating transcription of this important tumor suppressor (Kia et al., 2008; Wilson et al., 2010). This evidence provides both a necessary mechanism and a possible therapeutic target (EZH2) in *SMARCB1*-inactivated tumors.

Further mechanistic insights have yielded many other potential therapeutic avenues in *SMARCB1*-mutated tumors. For example, *SMARCB1* plays a critical role in cell cycle control. Mouse models have shown that coinactivation with Rb or p16 does not accelerate tumorigenesis, and simultaneous loss of cyclin D1 protects against tumor formation (Isakoff et al., 2005; Oruetxebarria et al., 2004; Tsikitis et al., 2005). In this vein, cyclin D1 inhibitors may have surprising effectiveness in these patients. Furthermore, direct binding to Myc has also been reported, along with increased Myc expression following *SMARCB1* loss (Cheng et al., 1999; Tsikitis et al., 2005). Though the efficacy of synthetic lethal approaches in targeting Myc-driven tumorigenesis has yet to be established, these strategies may be worthwhile therapeutic candidates. In addition to cell cycle dysregulation, *SMARCB1* inactivation leads to alterations in cell differentiation, specifically by increasing hedgehog signaling through the transcription factor Gli1 (Jagani et al., 2010). Thus, inhibitors of Gli1 provide a fourth potential target for therapy in *SMARCB1*-inactivated tumors. Lastly, increased RHOA activity and subsequent cell migration was observed following knockdown of *SMARCB1*, though therapeutic possibilities of this mechanism remain unexplored (Caramel et al., 2008).

SMARCA4, one of two SWI/SNF ATPase subunits, was first identified as a potential tumor suppressor in NSCLC, where expression is lost in 15%–50% of all tumors (Fukuoka et al., 2004; Reisman et al., 2003). *SMARCA4* mutations have also been noted in 35% of NSCLC cell lines, as well as in medulloblastoma, Burkitt lymphoma, melanoma, HCC, ccRCC, HNSCC, RT, pancreatic, breast, and prostate cancer (Cancer Genome Atlas Research Network, 2013; Endo et al., 2013; Love et al., 2012; Medina et al., 2008; Oike et al., 2013; Shain and Pollack, 2013). The majority of mutations and deletions are homozygous, though loss of only one allele may be sufficient to drive tumorigenesis. For instance, 10% of mice heterozygous for *SMARCA4* develop mammary tumors without spontaneous loss of the second allele (Bultman et al., 2000; Bultman et al., 2008). Functional studies have so far identified dual roles for SMARCA4 in both differentiation and cell adhesion/migration. In embryonic stem cells (ESCs), inactivation of *SMARCA4* leads to defective self-renewal and promotes differentiation, while overexpression enhances the epigenetic reprogramming of fibroblasts into induced pluripotent stem (iPS) cells, possibly through increased OCT4 binding to target genes (Singhal et al., 2010). In addition, SMARCA4 has been shown to promote osteoblast differentiation (Flowers et al., 2009). Alternatively, *SMARCA4* overexpression in cervical cancer cell lines leads to increased ROCK1 and stress fiber formation, which is reversible upon *SMARCA4* knockdown (Asp et al., 2002). Through this latter mechanism, SMARCA4 may serve to modulate cell migration and reduce the potential for invasion.

SMARCA2, the other ATPase subunit of the SWI/SNF complex, is also a *bona fide* tumor suppressor with frequent loss of expression via epigenetic silencing and more recently, mutational inactivation (Oike et al., 2013). Mouse models have shown that *SMARCA2* deficiency results in proliferative abnormalities, including increased overall weight and tissue-specific increased growth (Reyes et al., 1998). These mice exhibit increased cell proliferation in the prostate as well as androgen independence (Shen et al., 2008). In a similar lung model, heterozygous or homozygous loss of *SMARCA2* led to increased tumor formation (Glaros et al., 2007). Other studies have proposed another mechanism for enhanced tumorigenicity following *SMARCA2* or *SMARCA4* loss, specifically through the induction of an epithelial-mesenchymal transition (EMT) phenotype via the transcription factor ZEB1 (Matsubara et al., 2013; Sanchez-Tillo et al., 2010). Clinically, *SMARCA2* inactivation may predict poor prognosis, as loss of expression in both NSCLC and HCC correlates with worse OS (Endo et al., 2013; Fukuoka et al., 2004; Reisman et al., 2003). Recently, we identified somatic nonsynonymous mutations in *SMARCA2* (5%) following whole-exome and whole-genome sequencing of 60 patients with ACC (Ho et al., 2013). All of these mutations were located in a region encoding the Helicase C domain, which is involved in regulating gene transcription. Other *SMARCA2* mutations have also been reported in another cohort of patients with melanoma (Nikolaev et al., 2012). Despite epigenetic silencing being the dominant mechanism for *SMARCA2* inactivation, specific mutations may also prove important in the pathogenesis of certain cancers.

PBRM1

Mutations in chromatin state regulators are very common in ccRCC, with *PBRM1/BAF180* mutations noted in 29%–41% of all ccRCCs (Hakimi et al., 2013a; Varela et al., 2011). In fact, the only gene altered more frequently in this malignancy is *VHL*, a well-characterized driver of these cancers. In addition, lower frequency mutation of *PBRM1* has also been described in many other neoplasms, including DLBCL, HNSCC, chronic lymphoid leukemia (CLL), gastric, pancreatic, and breast cancer (Shain and Pollack, 2013; Xia et al., 2008). Many of these mutations are truncating—frameshift and nonsense—though a sizable number are also missense, evenly distributed across the protein (Hakimi et al., 2013a). It is most likely that these mutations are inactivating, as loss of *PBRM1* expression and gene deletion are noted in several cancers and cell lines (Wang et al., 2012). Further, functional studies have shown that re-expression of *PBRM1* in *PBRM1*-deficient cells induces cell growth arrest in G_1 phase, specifically through binding of the *CDKN1A* promoter, leading to induction of p21 (Xia et al., 2008). PBRM1 contains six tandem bromodomains that bind acetylated histone, two BAH domains that mediate protein-protein interactions, and a HMG domain to bind nucleosomal DNA, all of which provide functional specificity to SWI/SNF remodelers (Wilson and Roberts, 2011).

Exactly how these mutations affect epigenetic reprogramming has yet to be determined, though they are likely pleiotropic, diverse, and dependent on location and type of mutation. It is clear that *PBRM1* inactivation does have clinical consequences, at least in ccRCC. For example, *PBRM1* mutations have been shown to correlate to worse clinical outcomes (Pawlowski et al., 2013). Also, patients with *PBRM1* inactivation are more likely to present with stage III or IV disease, increased tumor size, low differentiation grade, increased perinephric or lymphatic invasion, and are less likely to have an "incidentaloma" at diagnosis (da Costa et al., 2013; Hakimi et al., 2013a; Pawlowski et al., 2013). In addition, smaller tumors (<4 cm) with *PBRM1* mutations are more likely to exhibit stage II pathologic features. Most interestingly, *PBRM1* and *BAP1* mutations are mutually exclusive in ccRCC, and in a side-by-side comparison, *PBRM1* mutation correlates with improved OS over *BAP1* alterations (Kapur et al., 2013; Pena-Llopis et al., 2013).

BRD7

BRD7 is a PBAF-specific chromatin remodeler that is frequently deleted in breast cancer. Recently, low frequency mutations in this gene have been identified in several cancers (Cerami et al., 2012; Forbes et al., 2011). For some time, *BRD7* was believed to be an important SWI/SNF tumor suppressor that functions mainly as a cofactor for p53 to activate oncogene-induced senescence (OIS) (Burrows et al., 2010). In fact, *BRD7* was first identified in a loss-of-function screen for genes required for p53-dependent OIS. Drost and colleagues showed that p53 required BRD7 for OIS—and vice versa—and that p53 directly interacts with the N-terminus of BRD7, upstream of its bromodomain. Furthermore, BRD7 was required for induction of several p53 target genes, including *CDKN1A* and *MDM2*, and was capable of direct binding to their associated promoters (Drost et al., 2010). This mechanism likely explains the reported ability of BRD7 to inhibit cell cycle progression from G_1 to S phase, though direct binding at the *E2F3* promoter may also facilitate this (Peng et al., 2006).

Though this mechanism is likely important in certain cancers, the diversity of *BRD7* phenotypes may be dependent on the unique molecular architecture found within different tissues. For example, a recent study focusing on epithelial ovarian carcinoma showed that *BRD7* can act as a tumor suppressor independent of p53 activity, possibly by sequestering β-catenin in the cytoplasm. These data were further supported by a β-catenin responsive TCF-reporter assay *in vitro* and the presence of decreased *BRD7* expression in high-grade epithelial ovarian serous carcinoma clinical specimens (Bae et al., 2013). Regardless of the mechanism, overexpression of *BRD7* has powerful tumor suppressive effects in cancer cell lines, resulting in decreased cell viability and reduced invasion/migration, independent of p53 status. Lastly, BRD7 may also influence covalent histone modifications themselves, with knockdown resulting in decreased acetylation at H3K9 and direct binding observed *in vitro* at histone residue H3K14 (Bae et al., 2013; Peng et al., 2006). Future studies are necessary to determine which mutations have functional consequences and whether any associated therapeutic vulnerabilities can be exploited.

HISTONES

An exciting development in the past few years has been the identification of recurrent mutations in genes encoding the histones themselves. Histone mutations in cancer were first discovered following massive exomic sequencing efforts in colon cancer (*HIST1H1B*, 4%) and NHL (*HIST1H1C*, 7%), though frequencies were low and no recurrent mutations were identified (Morin et al., 2011; Sjoblom et al., 2006). However, a recent pair of pediatric glioma studies revealed highly recurrent heterozygous mutations in *HIST1H3B* and *H3F3A*, which encode histones H3.1 and H3.3, respectively. Collectively, Wu et al. and Schwartzentruber et al. reported common *H3F3A* mutations in diffuse intrinsic pontine glioma (DIPG; 78%) and pediatric GBM (22%–31%), with *HIST1H3B* mutations almost exclusive to DIPG (18%). Most striking is the location of these mutations at amino acid residues K27M and G34 (G/R, G/V), which are at or near critical sites for post-translational modification via methylation and acetylation on the histone tail (e.g. H3K27, H3K36) (Schwartzentruber et al., 2012; Wu et al., 2012). This observation, along with the mutually exclusive and monoallelic nature of the K27 and G34 mutations, led to subsequent investigations into the clinical and biologic consequences of these potentially gain-of-function mutations.

Currently, K27M and G34R/V mutations have been found almost exclusively in pediatric HGGs, including CNS-PNET (11%, G34R), though sporadic mutations at these residues have been recently noted in osteosarcoma (G34W) and MDS (K27N) (Attieh et al., 2013; Gessi et al., 2013; Joseph et al., 2013). In fact, the remarkable prevalence of these mutations in pediatric HGGs has enabled a thorough examination into the clinical significance of these events. For example, despite the mutually exclusive nature of K27M and G34 mutations, which implies a common pathway or biological process, there are important differences. First, several studies have noted that the K27M mutation predicts both younger age and worse OS as compared to G34 and wild-type H3.3 (Chan et al., 2013a; Khuong-Quang et al., 2012; Sturm et al., 2012). In one study, G34 mutations were actually associated with improved OS when compared to wild-type H3.3 (Sturm et al., 2012). Second, *H3F3A* mutations are frequently exclusive with *IDH1* mutations and commonly co-occur with *ATRX/DAXX*, *PDGFRA*, and *TP53* alterations, especially in G34 mutant tumors (Schwartzentruber et al., 2012; Sturm et al., 2012). Overlap between alterations in *H3F3A* and *ATRX/DAXX* is particularly interesting, since both have been linked to increased CNVs and alternative lengthening of telomeres (ALT) in tumor specimens (Yuen and Knoepfler, 2013). Third, *H3F3A* mutations predict tumor anatomic location—K27M and G34 mutant GBMs occur in midline and hemispheric areas of the brain, respectively (Chan et al., 2013a; Schwartzentruber et al., 2012). Likewise, a groundbreaking study by Sturm and colleagues revealed distinct epigenetic subtypes of GBM, with K27 and G34 mutant tumor samples exhibiting independent gene expression and methylation profiles (Bjerke et al., 2013; Schwartzentruber et al., 2012; Sturm et al., 2012). Interestingly, these profiles may also explain the anatomic preferences; the G34 mutant transcriptomic signature resembles embryonic regions of the neocortex and striatum, whereas the K27M signature is more similar to embryonic regions of the striatum and thalamus. Lastly, these epigenetic signatures harbor important differences in marker genes, with DNA hypermethylation and associated silencing at *FOXG1* (K27M), *MICA* (K27M), or *OLIG1/2* (G34) and elevated expression of *PDGFRA* (K27M) (Bender et al.,

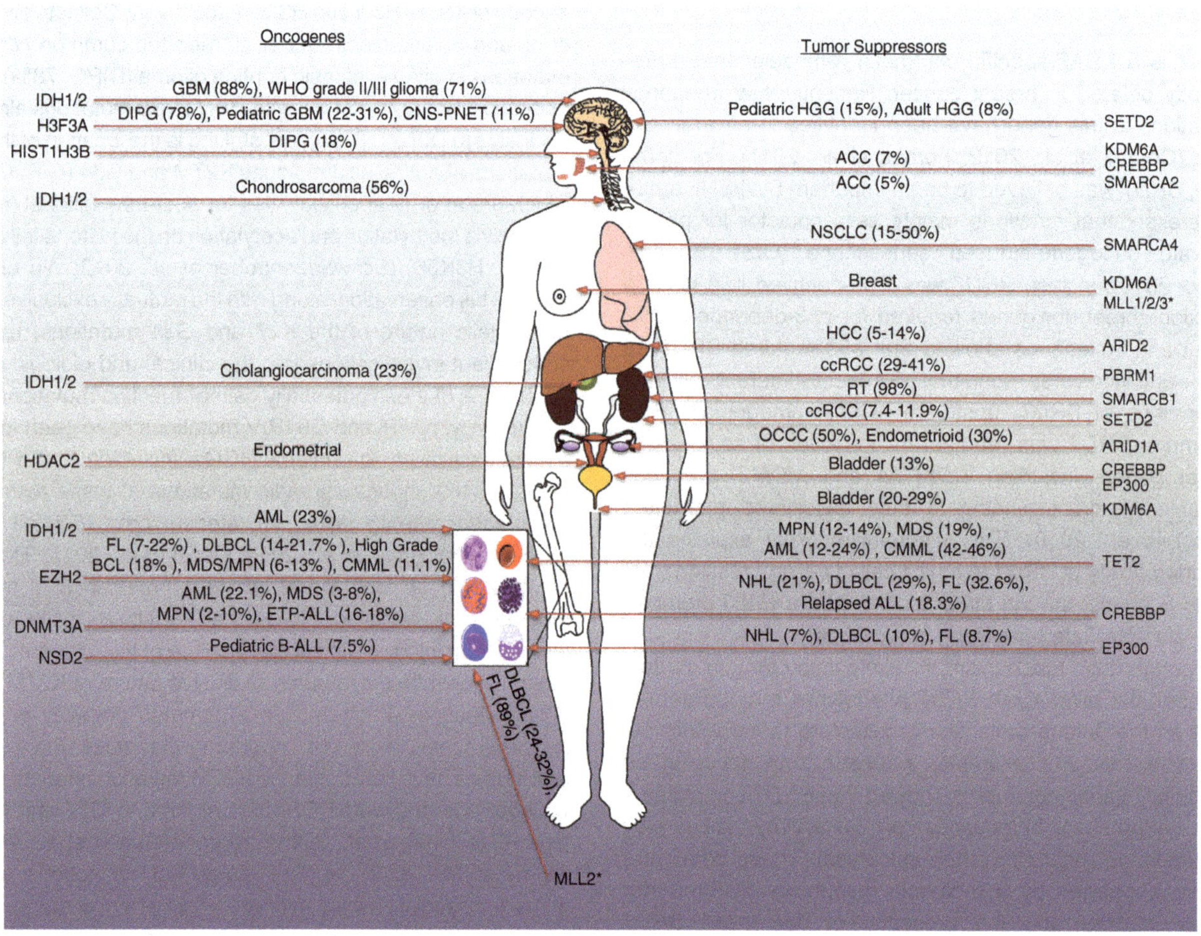

Figure 2. Frequently mutated epigenetic regulators in human cancer. Select oncogenes and tumor suppressors implicated in altered cancer epigenomes are shown, along with mutation frequency by tumor histology. *Role in oncogenesis (oncogene vs. tumor suppressor) is either mixed or undetermined.

2013; Sturm et al., 2012). Collectively, these changes in the methylome and transcriptome define the different H3.3 mutant GBM subtypes, correlate to anatomic tumor locations, and indicate potentially different cell origins or initiation events in tumorigenesis. Further, they provide mechanistic insight into *H3F3A*-induced oncogenesis.

Global DNA hypomethylation and decreases in genome-wide histone methylation suggest that mutant H3.3 functions in a dominant-negative manner (Bender et al., 2013; Chan et al., 2013a; Lewis et al., 2013; Sturm et al., 2012). Several groups have now shown that K27M mutant patient samples and cell lines exhibit globally reduced H3K27 di-/tri-methylation and at many loci, DNA hypomethylation (Bender et al., 2013; Chan et al., 2013a; Lewis et al., 2013). Further, forced overexpression of H3.3 K27M in isogenic GBM cells, 293T cells, and MEFs causes reduced H3K27me2/me3 (but not changes in monomethylation) as well as modest increases in H3K27ac, two features of transcriptionally active chromatin (Bender et al., 2013; Chan et al., 2013b). The major mechanism for loss of histone di-/tri-methylation is K27M-mediated inhibition of the PRC2 complex through direct binding of KMT EZH2 at the catalytic site, as well as binding to PRC2 cofactor SUZ12 (Bender et al., 2013; Chan et al., 2013a; Lewis et al., 2013). In fact, *in vitro* KMT assays revealed that K27M reduces KMT activity by 40%–70% and EZH2 catalytic activity up to 85%, with no loss of PRC2-associated KDM activity (e.g KDM6A/B) (Bender et al., 2013). Somewhat contradictory to this finding, ChIP-seq has revealed that hundreds of gene loci are also enriched for H3K27me3 and EZH2 in K27M mutant tumors or following overexpression of K27M, and analysis of differential expression shows increased gene transcription at these loci (Bender et al., 2013; Bjerke et al., 2013; Chan et al., 2013a). The mechanism underlying these paradoxical changes in H3K27 methylation and associated transcription changes is currently unknown (Chan et al., 2013b).

In contrast to the K27M mutation, broad changes in DNA methylation and histone marks underlying G34R/V induced

Table 1. Selected inhibitors of epigenetic regulators

Substance	Target	Highest clinical status
5-Azacytidine	DNMT	Approved
Decitabine	DNMT	Approved
SGI-110	DNMT	Phase I/II
MG98	DNMT	Phase I
RG108	DNMT	Preclinical
SGI-1027	DNMT	Preclinical
Zebularine	DNMT	Preclinical
EPZ-6438	EZH2	Phase I/II
GSK126	EZH2	Preclinical
GSK343	EZH2	Preclinical
EI1	EZH2	Preclinical
EPZ005687	EZH2	Preclinical
UNC1999	EZH2	Preclinical
Pivanex	HDAC	Phase I/II
Romidepsin	HDAC (class I)	Approved
Vorinostat	HDAC (pan)	Approved
Panobinostat	HDAC (class I/II)	Phase III
Abexinostat	HDAC (class I/II)	Phase II
Belinostat	HDAC (pan)	Phase II
Butyrate	HDAC (class I/IIa)	Phase II
Entinostat	HDAC (class I)	Phase II
Givinostat	HDAC (pan)	Phase II
Mocetinostat	HDAC (class I)	Phase II
Resminostat	HDAC (pan)	Phase II
SB939	HDAC (pan)	Phase II
Valproate	HDAC (class I/IIa)	Phase II
ACY-1215	HDAC6	Phase I/II
PCI-24781	HDAC (class I/II)	Phase I/II
CUDC-101	HDAC (class I/II)	Phase Ib
4SC-202	HDAC (class I)	Phase I
AR-42	HDAC (class I/II)	Phase I
CG200745	HDAC (pan)	Phase I
Tranylcypromine	KDM1A/LSD1	Phase II
ORY-1001	KDM1A/LSD1	Preclinical
AGI-5198	Mutant IDH1	Preclinical
AGI-6780	Mutant IDH2	Preclinical

tumorigenesis are less clear. Further, though investigators have shown that G34R/V mutations disrupt methylation at nearby amino acid residue K36, our understanding of the mechanisms leading to malignant transformation is limited. Despite this, analysis of differential gene expression between H3.3 G34R/V mutant and wild-type tumors has already identified some interesting markers (e.g. OLIG1/2), as stated above. As such, the defining feature of H3.3 mutant tumors is the aberrant activation and repression of numerous genes that uniquely contribute to oncogenesis. Understanding the constellation of changes unique to H3.3 mutants should allow identification of additional subtypes and driver genes. Most exciting is the recent identification of G34-induced activation of *MYCN* in pediatric GBM cell line KNS42 (Bjerke et al., 2013). This well-characterized oncogene is potentially druggable through synthetic lethal means, via aurora kinase A (AURKA), or JQ1 inhibition (Huang and Weiss, 2013). Implementing this strategy in patients harboring *H3F3A* G34 mutations could have immense therapeutic benefit. Similarly, identifying and characterizing other differentially expressed genes in *H3F3A* mutant pediatric gliomas could present new ways of treating this very difficult disease.

SUMMARY AND PERSPECTIVES

Understanding the role of altered epigenetic states has long been a fundamental goal in cancer research. It is now well-documented that distinct chromatin states are both necessary and sufficient to drive tumor formation, sustain increased cellular growth, and encourage metastatic dissemination (Baylin et al., 2001; Jones and Baylin, 2002). However, the link between classical genetics and cancer epigenomics has only recently been explored. In recent years, a plethora of mutations have been discovered in chromatin modifier genes. Studies now show that these alterations have profound effects on the cancer epigenome, leading to potent oncogenic transcriptional programs.

It has become abundantly clear that mutations in epigenetic modifiers are both incredibly diverse and ubiquitous in cancer. In fact, sequencing data has revealed that mutations exist in genes involved in nearly all aspects of epigenetics, including DNA methylation, covalent histone modification, and chromatin remodeling (Fig. 2). Several neoplasms rely on few, yet powerful, mutations in select genes to drive tumorigenesis through an altered epigenome, at least in part (e.g. *IDH1/2* in LGG, *MLL2* in FL, *SMARCB1* in RT). Remarkably, some cancers have such a wide range and high prevalence of alterations in chromatin modifiers (ACC, ccRCC, TCC) that they may be driven primarily through epigenetic means. Interestingly, mutations in regulators of DNA methylation have a strong cancer-specific prevalence (glioma, leukemia), whereas genetic alterations in histone modifiers (e.g. *KDM6A*) are widespread across many cancers, though the significance of this is incompletely understood.

Although the field of epigenomics in cancer is relatively new, the identification of driver mutations in epigenetic regulator genes has already led to new prognostic and therapeutic advances. New subtypes of cancer have been identified, or at least attributed to previously unknown genetic alterations, allowing a more nuanced approach to treatment. Further, several new targeted therapies are currently in development and show great promise at reversing phenotypes caused by activating mutations in certain genes (e.g. *IDH1*, *EZH2*). In addition, understanding the epigenetic mechanisms that underlie these alterations has also provided new therapeutic potential, even if the driver mutations cannot be targeted directly (e.g. HDACi, DNMT inhibitors) (Table 1) (Arrowsmith et al., 2012; Foulks et al., 2012; Helin and Dhanak, 2013; New et al., 2012; Plass et al., 2013).

There is no doubt that we are in the midst of an exciting era in both cancer genomics and epigenetics. Investigations focused on the intersection of these two fields have provided remarkable clarity into numerous aspects of cancer biology and novel mechanisms of oncogenesis. Already, great strides have been made to translate these early discoveries into clinical practice, and the characterization of novel mutations in chromatin modifier genes is rapidly accelerating. With the ongoing development of new epigenetic therapies, it is possible that much improved outcomes, at least in certain cancers, may be possible in the near future. Further, an enhanced understanding of the epigenomic alterations associated with these driver gene mutations could lead to the discovery of novel pathways involved in tumorigenesis, which may themselves possess unique vulnerabilities for therapeutic intervention.

ACKNOWLEDGEMENTS

We apologize to the many authors whose work we could not cite because of space constraints. We would like to thank Deepa Ramaswami for help in reviewing the manuscript. D.M.R. was supported by the HHMI Medical Research Fellows Program and NIH MSTP grant T32GM007739. L.A.W. was supported by The Canadian Institutes of Health Research PDF. T.A.C. is a Department of Defense Era of Hope Scholar and supported in part by the Frederick Adler Fund.

ABBREVIATIONS

ACC, adenoid cystic carcinoma; AML, acute myeloid leukemia; CGI, CpG island; HNPCC, hereditary nonpolyposis colon cancer; HDACs, histone deacetylatases; MSI, microsatellite instability; TCC, transitional cell carcinoma.

COMPLIANCE WITH ETHICS GUIDELINES

Authors David Roy, Logan Walsh, and Timothy Chan declare that they have no conflict of interest.

REFERENCES

Abdel-Wahab O, Mullally A, Hedvat C, Garcia-Manero G, Patel J, Wadleigh M, Malinge S, Yao J, Kilpivaara O, Bhat R et al (2009) Genetic characterization of TET1, TET2, and TET3 alterations in myeloid malignancies. Blood 114:144–147

Abdel-Wahab O, Pardanani A, Patel J, Wadleigh M, Lasho T, Heguy A, Beran M, Gilliland DG, Levine RL, Tefferi A (2011) Concomitant analysis of EZH2 and ASXL1 mutations in myelofibrosis, chronic myelomonocytic leukemia and blast-phase myeloproliferative neoplasms. Leukemia 25:1200–1202

Amary MF, Bacsi K, Maggiani F, Damato S, Halai D, Berisha F, Pollock R, O'Donnell P, Grigoriadis A, Diss T et al (2011) IDH1 and IDH2 mutations are frequent events in central chondrosarcoma and central and periosteal chondromas but not in other mesenchymal tumours. J Pathol 224:334–343

Arrowsmith CH, Bountra C, Fish PV, Lee K, Schapira M (2012) Epigenetic protein families: a new frontier for drug discovery. Nat Rev Drug Discov 11:384–400

Asp P, Wihlborg M, Karlen M, Farrants AK (2002) Expression of BRG1, a human SWI/SNF component, affects the organisation of actin filaments through the RhoA signalling pathway. J Cell Sci 115:2735–2746

Attieh Y, Geng QR, Dinardo CD, Zheng H, Jia Y, Fang ZH, Ganan-Gomez I, Yang H, Wei Y, Kantarjian H et al (2013) Low frequency of H3.3 mutations and upregulated DAXX expression in MDS. Blood 121:4009–4011

Bachman KE, Rountree MR, Baylin SB (2001) Dnmt3a and Dnmt3b are transcriptional repressors that exhibit unique localization properties to heterochromatin. J Biol Chem 276:32282–32287

Bae DS, Park YA, Lee JW, Kim HS, Lee YY, Kim TJ, Choi C, Choi JJ, Jeon HK, Cho YJ et al (2013) Tumor suppressive effects of bromodomain containing protein 7 (BRD7) in epithelial ovarian carcinoma. Clin Cancer Res

Balakrishnan A, Bleeker FE, Lamba S, Rodolfo M, Daniotti M, Scarpa A, van Tilborg AA, Leenstra S, Zanon C, Bardelli A (2007) Novel somatic and germline mutations in cancer candidate genes in glioblastoma, melanoma, and pancreatic carcinoma. Cancer Res 67:3545–3550

Balss J, Meyer J, Mueller W, Korshunov A, Hartmann C, von Deimling A (2008) Analysis of the IDH1 codon 132 mutation in brain tumors. Acta Neuropathol 116:597–602

Bannister AJ, Kouzarides T (1996) The CBP co-activator is a histone acetyltransferase. Nature 384:641–643

Baylin SB, Esteller M, Rountree MR, Bachman KE, Schuebel K, Herman JG (2001) Aberrant patterns of DNA methylation, chromatin formation and gene expression in cancer. Hum Mol Genet 10:687–692

Beguelin W, Popovic R, Teater M, Jiang Y, Bunting KL, Rosen M, Shen H, Yang SN, Wang L, Ezponda T et al (2013) EZH2 is required for germinal center formation and somatic EZH2 mutations promote lymphoid transformation. Cancer Cell 23:677–692

Bender S, Tang Y, Lindroth AM, Hovestadt V, Jones DT, Kool M, Zapatka M, Northcott PA, Sturm D, Wang W et al (2013) Reduced H3K27me3 and DNA hypomethylation are major drivers of gene expression in K27M mutant pediatric high-grade gliomas. Cancer Cell 24:660–672

Berger MF, Lawrence MS, Demichelis F, Drier Y, Cibulskis K, Sivachenko AY, Sboner A, Esgueva R, Pflueger D, Sougnez C et al (2011) The genomic complexity of primary human prostate cancer. Nature 470:214–220

Bisling KE, Brewin JN, McGovern AP, Horne GA, Rider T, Stewart HJ, Ramsahoye BH, Chevassut TJ (2013) DNMT3A mutations at R882 hotspot are only found in major clones of acute myeloid leukemia. Leuk Lymphoma 1:4

Bjerke L, Mackay A, Nandhabalan M, Burford A, Jury A, Popov S, Bax DA, Carvalho D, Taylor KR, Vinci M et al (2013) Histone H3.3 mutations drive pediatric glioblastoma through upregulation of MYCN. Cancer Discov 3:512–519

Blaschke K, Ebata KT, Karimi MM, Zepeda-Martinez JA, Goyal P, Mahapatra S, Tam A, Laird DJ, Hirst M, Rao A et al (2013) Vitamin C induces Tet-dependent DNA demethylation and a blastocyst-like state in ES cells. Nature 500:222–226

Bleeker FE, Lamba S, Leenstra S, Troost D, Hulsebos T, Vandertop WP, Frattini M, Molinari F, Knowles M, Cerrato A et al (2009) IDH1 mutations at residue p. R132 (IDH1(R132)) occur frequently in high-grade gliomas but not in other solid tumors. Hum Mutat 30:7–11

Bodor C, O'Riain C, Wrench D, Matthews J, Iyengar S, Tayyib H, Calaminici M, Clear A, Iqbal S, Quentmeier H et al (2011) EZH2 Y641 mutations in follicular lymphoma. Leukemia 25:726–729

Bodor C, Grossmann V, Popov N, Okosun J, O'Riain C, Tan K, Marzec J, Araf S, Wang J, Lee AM et al (2013) EZH2 mutations are frequent and represent an early event in follicular lymphoma. Blood 122:3165–3168

Borger DR, Tanabe KK, Fan KC, Lopez HU, Fantin VR, Straley KS, Schenkein DP, Hezel AF, Ancukiewicz M, Liebman HM et al (2012) Frequent mutation of isocitrate dehydrogenase (IDH)1 and IDH2 in cholangiocarcinoma identified through broad-based tumor genotyping. Oncologist 17:72–79

Borodovsky A, Salmasi V, Turcan S, Fabius AW, Baia GS, Eberhart CG, Weingart JD, Gallia GL, Baylin SB, Chan TA et al (2013) 5-azacytidine reduces methylation, promotes differentiation and induces tumor regression in a patient-derived IDH1 mutant glioma xenograft. Oncotarget 4:1737–1747

Borrow J, Stanton VP Jr, Andresen JM, Becher R, Behm FG, Chaganti RS, Civin CI, Disteche C, Dube I, Frischauf AM et al (1996) The translocation t(8;16)(p11;p13) of acute myeloid leukaemia fuses a putative acetyltransferase to the CREB-binding protein. Nat Genet 14:33–41

Bosse T, Ter Haar NT, Seeber LM, Diest PJ, Hes FJ, Vasen HF, Nout RA, Creutzberg CL, Morreau H, Smit VT (2013) Loss of ARID1A expression and its relationship with PI3K-Akt pathway alterations, TP53 and microsatellite instability in endometrial cancer. Mod Pathol 26:1525–1535

Bracken AP, Pasini D, Capra M, Prosperini E, Colli E, Helin K (2003) EZH2 is downstream of the pRB-E2F pathway, essential for proliferation and amplified in cancer. EMBO J 22:5323–5335

Brecqueville M, Cervera N, Gelsi-Boyer V, Murati A, Adelaide J, Chaffanet M, Rey J, Vey N, Mozziconacci MJ, Birnbaum D (2011) Rare mutations in DNMT3A in myeloproliferative neoplasms and myelodysplastic syndromes. Blood Cancer J 1:e18

Bultman S, Gebuhr T, Yee D, La Mantia C, Nicholson J, Gilliam A, Randazzo F, Metzger D, Chambon P, Crabtree G et al (2000) A Brg1 null mutation in the mouse reveals functional differences among mammalian SWI/SNF complexes. Mol Cell 6:1287–1295

Bultman SJ, Herschkowitz JI, Godfrey V, Gebuhr TC, Yaniv M, Perou CM, Magnuson T (2008) Characterization of mammary tumors from Brg1 heterozygous mice. Oncogene 27:460–468

Burrows AE, Smogorzewska A, Elledge SJ (2010) Polybromo-associated BRG1-associated factor components BRD7 and BAF180 are critical regulators of p53 required for induction of replicative senescence. Proc Natl Acad Sci USA 107:14280–14285

Cairns RA, Mak TW (2013) Oncogenic isocitrate dehydrogenase mutations: mechanisms, models, and clinical opportunities. Cancer Discov 3:730–741

Cancer Genome Atlas Research Network (2013) Comprehensive molecular characterization of clear cell renal cell carcinoma. Nature 499:43–49

Capello D, Gloghini A, Martini M, Spina M, Tirelli U, Bertoni F, Rinaldi A, Morra E, Rambaldi A, Sinigaglia F et al (2011) Mutations of CD79A, CD79B and EZH2 genes in immunodeficiency-related non-Hodgkin lymphomas. Br J Haematol 152:777–780

Capper D, Simon M, Langhans CD, Okun JG, Tonn JC, Weller M, von Deimling A, Hartmann C (2012) 2-Hydroxyglutarate concentration in serum from patients with gliomas does not correlate with IDH1/2 mutation status or tumor size. Int J Cancer 131:766–768

Caramel J, Quignon F, Delattre O (2008) RhoA-dependent regulation of cell migration by the tumor suppressor hSNF5/INI1. Cancer Res 68:6154–6161

Cerami E, Gao J, Dogrusoz U, Gross BE, Sumer SO, Aksoy BA, Jacobsen A, Byrne CJ, Heuer ML, Larsson E et al (2012) The cBio cancer genomics portal: an open platform for exploring multidimensional cancer genomics data. Cancer Discov 2:401–404

Challen GA, Sun D, Jeong M, Luo M, Jelinek J, Berg JS, Bock C, Vasanthakumar A, Gu H, Xi Y et al (2012) Dnmt3a is essential for hematopoietic stem cell differentiation. Nat Genet 44:23–31

Chan KM, Fang D, Gan H, Hashizume R, Yu C, Schroeder M, Gupta N, Mueller S, James CD, Jenkins R et al (2013a) The histone H3.3K27M mutation in pediatric glioma reprograms H3K27 methylation and gene expression. Genes Dev 27:985–990

Chan KM, Han J, Fang D, Gan H, Zhang Z (2013b) A lesson learned from the H3.3K27M mutation found in pediatric glioma: a new approach to the study of the function of histone modifications in vivo? Cell Cycle 12:2546–2552

Chaturvedi A, Araujo Cruz MM, Jyotsana N, Sharma A, Yun H, Gorlich K, Wichmann M, Schwarzer A, Preller M, Thol F et al (2013) Mutant IDH1 promotes leukemogenesis in vivo and can be specifically targeted in human AML. Blood 122:2877–2887

Chedin F, Lieber MR, Hsieh CL (2002) The DNA methyltransferase-like protein DNMT3L stimulates de novo methylation by Dnmt3a. Proc Natl Acad Sci USA 99:16916–16921

Chen C, Liu Y, Lu C, Cross JR, Morris JPt, Shroff AS, Ward PS, Bradner JE, Thompson C, Lowe SW (2013a) Cancer-associated IDH2 mutants drive an acute myeloid leukemia that is susceptible to Brd4 inhibition. Genes Dev 27:1974–1985

Chen Q, Chen Y, Bian C, Fujiki R, Yu X (2013b) TET2 promotes histone O-GlcNAcylation during gene transcription. Nature 493:561–564

Cheng SW, Davies KP, Yung E, Beltran RJ, Yu J, Kalpana GV (1999) c-MYC interacts with INI1/hSNF5 and requires the SWI/SNF complex for transactivation function. Nat Genet 22:102–105

Cheng J, Guo S, Chen S, Mastriano SJ, Liu C, D'Alessio AC, Hysolli E, Guo Y, Yao H, Megyola CM et al (2013) An extensive network of TET2-targeting microRNAs regulates malignant hematopoiesis. Cell Rep 5:471–481

Chesi M, Nardini E, Lim RS, Smith KD, Kuehl WM, Bergsagel PL (1998) The t(4;14) translocation in myeloma dysregulates both FGFR3 and a novel gene, MMSET, resulting in IgH/MMSET hybrid transcripts. Blood 92:3025–3034

Cho YW, Hong T, Hong S, Guo H, Yu H, Kim D, Guszczynski T, Dressler GR, Copeland TD, Kalkum M et al (2007) PTIP associates with MLL3- and MLL4-containing histone H3 lysine 4 methyltransferase complex. J Biol Chem 282:20395–20406

Chou WC, Hou HA, Chen CY, Tang JL, Yao M, Tsay W, Ko BS, Wu SJ, Huang SY, Hsu SC et al (2010) Distinct clinical and biologic characteristics in adult acute myeloid leukemia bearing the isocitrate dehydrogenase 1 mutation. Blood 115:2749–2754

Chou WC, Chou SC, Liu CY, Chen CY, Hou HA, Kuo YY, Lee MC, Ko BS, Tang JL, Yao M et al (2011a) TET2 mutation is an unfavorable prognostic factor in acute myeloid leukemia patients with intermediate-risk cytogenetics. Blood 118:3803–3810

Chou WC, Lei WC, Ko BS, Hou HA, Chen CY, Tang JL, Yao M, Tsay W, Wu SJ, Huang SY et al (2011b) The prognostic impact and stability of isocitrate dehydrogenase 2 mutation in adult patients with acute myeloid leukemia. Leukemia 25:246–253

Christensen J, Agger K, Cloos PA, Pasini D, Rose S, Sennels L, Rappsilber J, Hansen KH, Salcini AE, Helin K (2007) RBP2 belongs to a family of demethylases, specific for tri-and dimethylated lysine 4 on histone 3. Cell 128:1063–1076

Christiaans I, Kenter SB, Brink HC, van Os TA, Baas F, van den Munckhof P, Kidd AM, Hulsebos TJ (2011) Germline SMARCB1 mutation and somatic NF2 mutations in familial multiple meningiomas. J Med Genet 48:93–97

Costa Y, Ding J, Theunissen TW, Faiola F, Hore TA, Shliaha PV, Fidalgo M, Saunders A, Lawrence M, Dietmann S et al (2013) NANOG-dependent function of TET1 and TET2 in establishment of pluripotency. Nature 495:370–374

Croonquist PA, Van Ness B (2005) The polycomb group protein enhancer of zeste homolog 2 (EZH 2) is an oncogene that influences myeloma cell growth and the mutant ras phenotype. Oncogene 24:6269–6280

da Costa WH, Rezende M, Rocha RM, da Cunha IW, Carraro DM, Guimaraes GC, de Cassio Zequi S (2013) PBRM1, a SWI/SNF complex subunit is a prognostic marker in clear cell renal cell carcinoma. BJU Int

Dalgliesh GL, Furge K, Greenman C, Chen L, Bignell G, Butler A, Davies H, Edkins S, Hardy C, Latimer C et al (2010) Systematic sequencing of renal carcinoma reveals inactivation of histone modifying genes. Nature 463:360–363

Dallas PB, Pacchione S, Wilsker D, Bowrin V, Kobayashi R, Moran E (2000) The human SWI-SNF complex protein p270 is an ARID family member with non-sequence-specific DNA binding activity. Mol Cell Biol 20:3137–3146

Dang L, White DW, Gross S, Bennett BD, Bittinger MA, Driggers EM, Fantin VR, Jang HG, Jin S, Keenan MC et al (2009) Cancer-associated IDH1 mutations produce 2-hydroxyglutarate. Nature 462:739–744

Datta J, Ghoshal K, Sharma SM, Tajima S, Jacob ST (2003) Biochemical fractionation reveals association of DNA methyltransferase (Dnmt) 3b with Dnmt1 and that of Dnmt 3a with a histone H3 methyltransferase and Hdac1. J Cell Biochem 88:855–864

Dawson MA, Kouzarides T (2012) Cancer epigenetics: from mechanism to therapy. Cell 150:12–27

Delhommeau F, Dupont S, James C, Masse A, le Couedic J, Valle V (2008) TET2 is a novel tumor suppressor gene inactivated in myeloproliferative neoplasms: identification of a pre-JAK2 V617F event. Blood 112:11

Delhommeau F, Dupont S, Della Valle V, James C, Trannoy S, Masse A, Kosmider O, Le Couedic JP, Robert F, Alberdi A et al (2009) Mutation in TET2 in myeloid cancers. N Engl J Med 360:2289–2301

Deplus R, Delatte B, Schwinn MK, Defrance M, Mendez J, Murphy N, Dawson MA, Volkmar M, Putmans P, Calonne E et al (2013) TET2 and TET3 regulate GlcNAcylation and H3K4 methylation through OGT and SET1/COMPASS. EMBO J 32:645–655

Dimartino JF, Cleary ML (1999) Mll rearrangements in haematological malignancies: lessons from clinical and biological studies. Br J Haematol 106:614–626

DiNardo CD, Propert KJ, Loren AW, Paietta E, Sun Z, Levine RL, Straley KS, Yen K, Patel JP, Agresta S et al (2013) Serum 2-hydroxyglutarate levels predict isocitrate dehydrogenase mutations and clinical outcome in acute myeloid leukemia. Blood 121:4917–4924

Dolnik A, Engelmann JC, Scharfenberger-Schmeer M, Mauch J, Kelkenberg-Schade S, Haldemann B, Fries T, Kronke J, Kuhn MW, Paschka P et al (2012) Commonly altered genomic regions in acute myeloid leukemia are enriched for somatic mutations involved in chromatin remodeling and splicing. Blood 120:e83–e92

Drost J, Mantovani F, Tocco F, Elkon R, Comel A, Holstege H, Kerkhoven R, Jonkers J, Voorhoeve PM, Agami R et al (2010) BRD7 is a candidate tumour suppressor gene required for p53 function. Nat Cell Biol 12:380–389

Duns G, van den Berg E, van Duivenbode I, Osinga J, Hollema H, Hofstra RM, Kok K (2010) Histone methyltransferase gene SETD2 is a novel tumor suppressor gene in clear cell renal cell carcinoma. Cancer Res 70:4287–4291

Duns G, Hofstra RM, Sietzema JG, Hollema H, van Duivenbode I, Kuik A, Giezen C, Jan O, Bergsma JJ, Bijnen H et al (2012) Targeted exome sequencing in clear cell renal cell carcinoma tumors suggests aberrant chromatin regulation as a crucial step in ccRCC development. Hum Mutat 33:1059–1062

Eckner R, Ewen ME, Newsome D, Gerdes M, DeCaprio JA, Lawrence JB, Livingston DM (1994) Molecular cloning and functional analysis of the adenovirus E1A-associated 300-kD protein (p300) reveals a protein with properties of a transcriptional adaptor. Genes Dev 8:869–884

Edmunds JW, Mahadevan LC, Clayton AL (2008) Dynamic histone H3 methylation during gene induction: HYPB/Setd2 mediates all H3K36 trimethylation. EMBO J 27:406–420

Ehrlich M (2002) DNA methylation in cancer: too much, but also too little. Oncogene 21:5400–5413

Ellis L, Atadja PW, Johnstone RW (2009) Epigenetics in cancer: targeting chromatin modifications. Mol Cancer Ther 8:1409–1420

El-Osta A (2004) The rise and fall of genomic methylation in cancer. Leukemia 18:233–237

Endo M, Yasui K, Zen Y, Gen Y, Zen K, Tsuji K, Dohi O, Mitsuyoshi H, Tanaka S, Taniwaki M et al (2013) Alterations of the SWI/SNF chromatin remodelling subunit-BRG1 and BRM in hepatocellular carcinoma. Liver Int 33:105–117

Ernst T, Chase AJ, Score J, Hidalgo-Curtis CE, Bryant C, Jones AV, Waghorn K, Zoi K, Ross FM, Reiter A et al (2010) Inactivating mutations of the histone methyltransferase gene EZH2 in myeloid disorders. Nat Genet 42:722–726

Esteller M, Fraga MF, Guo M, Garcia-Foncillas J, Hedenfalk I, Godwin AK, Trojan J, Vaurs-Barriere C, Bignon YJ, Ramus S et al (2001) DNA methylation patterns in hereditary human cancers mimic sporadic tumorigenesis. Hum Mol Genet 10:3001–3007

Ewalt M, Galili NG, Mumtaz M, Churchill M, Rivera S, Borot F, Raza A, Mukherjee S (2011) DNMT3a mutations in high-risk myelodysplastic syndrome parallel those found in acute myeloid leukemia. Blood Cancer J 1:e9

Feinberg AP, Vogelstein B (1983) Hypomethylation distinguishes genes of some human cancers from their normal counterparts. Nature 301:89–92

Feng Q, Wang H, Ng HH, Erdjument-Bromage H, Tempst P, Struhl K, Zhang Y (2002) Methylation of H3-lysine 79 is mediated by a new family of HMTases without a SET domain. Curr Biol 12:1052–1058

Figueroa ME, Abdel-Wahab O, Lu C, Ward PS, Patel J, Shih A, Li Y, Bhagwat N, Vasanthakumar A, Fernandez HF et al (2010) Leukemic IDH1 and IDH2 mutations result in a hypermethylation phenotype, disrupt TET2 function, and impair hematopoietic differentiation. Cancer Cell 18:553–567

Flanagan S, Lee M, Li CC, Suter CM, Buckland ME (2012) Promoter methylation analysis of IDH genes in human gliomas. Front Oncol 2:193

Flowers S, Nagl NG Jr, Beck GR Jr, Moran E (2009) Antagonistic roles for BRM and BRG1 SWI/SNF complexes in differentiation. J Biol Chem 284:10067–10075

Fontebasso AM, Schwartzentruber J, Khuong-Quang DA, Liu XY, Sturm D, Korshunov A, Jones DT, Witt H, Kool M, Albrecht S et al (2013) Mutations in SETD2 and genes affecting histone H3K36 methylation target hemispheric high-grade gliomas. Acta Neuropathol 125:659–669

Forbes SA, Bindal N, Bamford S, Cole C, Kok CY, Beare D, Jia M, Shepherd R, Leung K, Menzies A et al (2011) COSMIC: mining complete cancer genomes in the catalogue of somatic mutations in cancer. Nucleic Acids Res 39:D945–D950

Foulks JM, Parnell KM, Nix RN, Chau S, Swierczek K, Saunders M, Wright K, Hendrickson TF, Ho KK, McCullar MV et al (2012) Epigenetic drug discovery: targeting DNA methyltransferases. J Biomol Screen 17:2–17

Fu X, Jin L, Wang X, Luo A, Hu J, Zheng X, Tsark WM, Riggs AD, Ku HT, Huang W (2013) MicroRNA-26a targets ten eleven translocation enzymes and is regulated during pancreatic cell differentiation. Proc Natl Acad Sci USA 110:17892–17897

Fujimoto A, Totoki Y, Abe T, Boroevich KA, Hosoda F, Nguyen HH, Aoki M, Hosono N, Kubo M, Miya F et al (2012) Whole-genome sequencing of liver cancers identifies etiological influences on mutation patterns and recurrent mutations in chromatin regulators. Nat Genet 44:760–764

Fuks F, Burgers WA, Godin N, Kasai M, Kouzarides T (2001) Dnmt3a binds deacetylases and is recruited by a sequence-specific repressor to silence transcription. EMBO J 20:2536–2544

Fuks F, Hurd PJ, Deplus R, Kouzarides T (2003) The DNA methyltransferases associate with HP1 and the SUV39H1 histone methyltransferase. Nucleic Acids Res 31:2305–2312

Fukuoka J, Fujii T, Shih JH, Dracheva T, Meerzaman D, Player A, Hong K, Settnek S, Gupta A, Buetow K et al (2004) Chromatin remodeling factors and BRM/BRG1 expression as prognostic indicators in non-small cell lung cancer. Clin Cancer Res 10:4314–4324

Gama-Sosa MA, Slagel VA, Trewyn RW, Oxenhandler R, Kuo KC, Gehrke CW, Ehrlich M (1983) The 5-methylcytosine content of DNA from human tumors. Nucleic Acids Res 11:6883–6894

Gao X, Tate P, Hu P, Tjian R, Skarnes WC, Wang Z (2008) ES cell pluripotency and germ-layer formation require the SWI/SNF chromatin remodeling component BAF250a. Proc Natl Acad Sci USA 105:6656–6661

Gao J, Aksoy BA, Dogrusoz U, Dresdner G, Gross B, Sumer SO, Sun Y, Jacobsen A, Sinha R, Larsson E et al (2013) Integrative analysis of complex cancer genomics and clinical profiles using the cBioPortal. Sci Signal 6:pl1

Ge YZ, Pu MT, Gowher H, Wu HP, Ding JP, Jeltsch A, Xu GL (2004) Chromatin targeting of de novo DNA methyltransferases by the PWWP domain. J Biol Chem 279:25447–25454

Geisbrecht BV, Gould SJ (1999) The human PICD gene encodes a cytoplasmic and peroxisomal NADP(+)-dependent isocitrate dehydrogenase. J Biol Chem 274:30527–30533

Gessi M, Gielen GH, Hammes J, Dorner E, Muhlen AZ, Waha A, Pietsch T (2013) H3.3 G34R mutations in pediatric primitive neuroectodermal tumors of central nervous system (CNS-PNET) and pediatric glioblastomas: possible diagnostic and therapeutic implications? J Neurooncol 112:67–72

Ghiam AF, Cairns RA, Thoms J, Dal Pra A, Ahmed O, Meng A, Mak TW, Bristow RG (2012) IDH mutation status in prostate cancer. Oncogene 31:3826

Glaros S, Cirrincione GM, Muchardt C, Kleer CG, Michael CW, Reisman D (2007) The reversible epigenetic silencing of BRM: implications for clinical targeted therapy. Oncogene 26:7058–7066

Grady WM, Willis J, Guilford PJ, Dunbier AK, Toro TT, Lynch H, Wiesner G, Ferguson K, Eng C, Park JG et al (2000) Methylation of the CDH1 promoter as the second genetic hit in hereditary diffuse gastric cancer. Nat Genet 26:16–17

Grassian AR, Lin F, Barrett R, Liu Y, Jiang W, Korpal M, Astley H, Gitterman D, Henley T, Howes R et al (2012) Isocitrate dehydrogenase (IDH) mutations promote a reversible ZEB1/microRNA (miR)-200-dependent epithelial-mesenchymal transition (EMT). J Biol Chem 287:42180–42194

Green A, Beer P (2010) Somatic mutations of IDH1 and IDH2 in the leukemic transformation of myeloproliferative neoplasms. N Engl J Med 362:369–370

Greer EL, Shi Y (2012) Histone methylation: a dynamic mark in health, disease and inheritance. Nat Rev Genet 13:343–357

Gross S, Cairns RA, Minden MD, Driggers EM, Bittinger MA, Jang HG, Sasaki M, Jin S, Schenkein DP, Su SM et al (2010) Cancer-associated metabolite 2-hydroxyglutarate accumulates in acute myelogenous leukemia with isocitrate dehydrogenase 1 and 2 mutations. J Exp Med 207:339–344

Grossmann V, Kohlmann A, Eder C, Haferlach C, Kern W, Cross NC, Haferlach T, Schnittger S (2011) Molecular profiling of chronic myelomonocytic leukemia reveals diverse mutations in >80% of patients with TET2 and EZH2 being of high prognostic relevance. Leukemia 25:877–879

Grossmann V, Haferlach C, Weissmann S, Roller A, Schindela S, Poetzinger F, Stadler K, Bellos F, Kern W, Haferlach T et al (2013) The molecular profile of adult T-cell acute lymphoblastic leukemia: mutations in RUNX1 and DNMT3A are associated with poor prognosis in T-ALL. Genes Chromosom Cancer 52:410–422

Gu W, Roeder RG (1997) Activation of p53 sequence-specific DNA binding by acetylation of the p53 C-terminal domain. Cell 90:595–606

Guglielmelli P, Biamonte F, Score J, Hidalgo-Curtis C, Cervantes F, Maffioli M, Fanelli T, Ernst T, Winkelman N, Jones AV et al (2011) EZH2 mutational status predicts poor survival in myelofibrosis. Blood 118:5227–5234

Gui Y, Guo G, Huang Y, Hu X, Tang A, Gao S, Wu R, Chen C, Li X, Zhou L et al (2011) Frequent mutations of chromatin remodeling genes in transitional cell carcinoma of the bladder. Nat Genet 43:875–878

Guidi CJ, Sands AT, Zambrowicz BP, Turner TK, Demers DA, Webster W, Smith TW, Imbalzano AN, Jones SN (2001) Disruption of Ini1 leads to peri-implantation lethality and tumorigenesis in mice. Mol Cell Biol 21:3598–3603

Guilhamon P, Eskandarpour M, Halai D, Wilson GA, Feber A, Teschendorff AE, Gomez V, Hergovich A, Tirabosco R, Fernanda Amary M et al (2013) Meta-analysis of IDH-mutant cancers identifies EBF1 as an interaction partner for TET2. Nat Commun 4:2166

Guo JU, Su Y, Zhong C, Ming GL, Song H (2011) Hydroxylation of 5-methylcytosine by TET1 promotes active DNA demethylation in the adult brain. Cell 145:423–434

Hakimi AA, Chen YB, Wren J, Gonen M, Abdel-Wahab O, Heguy A, Liu H, Takeda S, Tickoo SK, Reuter VE et al (2013a) Clinical and pathologic impact of select chromatin-modulating tumor suppressors in clear cell renal cell carcinoma. Eur Urol 63:848–854

Hakimi AA, Ostrovnaya I, Reva B, Schultz N, Chen YB, Gonen M, Liu H, Takeda S, Voss MH, Tickoo SK et al (2013b) Adverse outcomes in clear cell renal cell carcinoma with mutations of 3p21 epigenetic regulators BAP1 and SETD2: a report by MSKCC and the KIRC TCGA research network. Clin Cancer Res 19:3259–3267

Hanahan D, Weinberg RA (2011) Hallmarks of cancer: the next generation. Cell 144:646–674

Hanigan CL, Van Engeland M, De Bruine AP, Wouters KA, Weijenberg MP, Eshleman JR, Herman JG (2008) An inactivating mutation in HDAC2 leads to dysregulation of apoptosis mediated by APAF1. Gastroenterology 135(1654–1664):e1652

Hartmann C, Meyer J, Balss J, Capper D, Mueller W, Christians A, Felsberg J, Wolter M, Mawrin C, Wick W et al (2009) Type and frequency of IDH1 and IDH2 mutations are related to astrocytic and oligodendroglial differentiation and age: a study of 1,010 diffuse gliomas. Acta Neuropathol 118:469–474

He YF, Li BZ, Li Z, Liu P, Wang Y, Tang Q, Ding J, Jia Y, Chen Z, Li L et al (2011) Tet-mediated formation of 5-carboxylcytosine and its excision by TDG in mammalian DNA. Science 333:1303–1307

Heintzman ND, Stuart RK, Hon G, Fu Y, Ching CW, Hawkins RD, Barrera LO, Van Calcar S, Qu C, Ching KA et al (2007) Distinct and predictive chromatin signatures of transcriptional promoters and enhancers in the human genome. Nat Genet 39:311–318

Helin K, Dhanak D (2013) Chromatin proteins and modifications as drug targets. Nature 502:480–488

Herman JG, Baylin SB (2003) Gene silencing in cancer in association with promoter hypermethylation. N Engl J Med 349:2042–2054

Hermann A, Gowher H, Jeltsch A (2004) Biochemistry and biology of mammalian DNA methyltransferases. CMLS 61:2571–2587

Herz HM, Madden LD, Chen Z, Bolduc C, Buff E, Gupta R, Davuluri R, Shilatifard A, Hariharan IK, Bergmann A (2010) The H3K27me3 demethylase dUTX is a suppressor of Notch- and Rb-dependent tumors in Drosophila. Mol Cell Biol 30:2485–2497

Ho AS, Kannan K, Roy DM, Morris LG, Ganly I, Katabi N, Ramaswami D, Walsh LA, Eng S, Huse JT et al (2013) The mutational landscape of adenoid cystic carcinoma. Nat Genet 45:791–798

Holliday R, Pugh JE (1975) DNA modification mechanisms and gene activity during development. Science 187:226–232

Holz-Schietinger C, Matje DM, Harrison MF, Reich NO (2011) Oligomerization of DNMT3A controls the mechanism of de novo DNA methylation. J Biol Chem 286:41479–41488

Holz-Schietinger C, Matje DM, Reich NO (2012) Mutations in DNA methyltransferase (DNMT3A) observed in acute myeloid leukemia patients disrupt processive methylation. J Biol Chem 287:30941–30951

Hou HA, Kuo YY, Liu CY, Chou WC, Lee MC, Chen CY, Lin LI, Tseng MH, Huang CF, Chiang YC et al (2012) DNMT3A mutations in acute myeloid leukemia: stability during disease evolution and clinical implications. Blood 119:559–568

Hsieh CL (1999) In vivo activity of murine de novo methyltransferases, Dnmt3a and Dnmt3b. Mol Cell Biol 19:8211–8218

Huang M, Weiss WA (2013) G34, another connection between MYCN and a pediatric tumor. Cancer Discov 3:484–486

Hughes CM, Rozenblatt-Rosen O, Milne TA, Copeland TD, Levine SS, Lee JC, Hayes DN, Shanmugam KS, Bhattacharjee A, Biondi CA et al (2004) Menin associates with a trithorax family histone methyltransferase complex and with the hoxc8 locus. Mol Cell 13:587–597

Hulsebos TJ, Plomp AS, Wolterman RA, Robanus-Maandag EC, Baas F, Wesseling P (2007) Germline mutation of INI1/SMARCB1 in familial schwannomatosis. Am J Hum Genet 80:805–810

Hussein K, Abdel-Wahab O, Lasho TL, Van Dyke DL, Levine RL, Hanson CA, Pardanani A, Tefferi A (2010) Cytogenetic correlates of TET2 mutations in 199 patients with myeloproliferative neoplasms. Am J Hematol 85:81–83

Inoue H, Furukawa T, Giannakopoulos S, Zhou S, King DS, Tanese N (2002) Largest subunits of the human SWI/SNF chromatin-remodeling complex promote transcriptional activation by steroid hormone receptors. J Biol Chem 277:41674–41685

Isakoff MS, Sansam CG, Tamayo P, Subramanian A, Evans JA, Fillmore CM, Wang X, Biegel JA, Pomeroy SL, Mesirov JP et al (2005) Inactivation of the Snf5 tumor suppressor stimulates cell cycle progression and cooperates with p53 loss in oncogenic transformation. Proc Natl Acad Sci USA 102:17745–17750

Issaeva I, Zonis Y, Rozovskaia T, Orlovsky K, Croce CM, Nakamura T, Mazo A, Eisenbach L, Canaani E (2007) Knockdown of ALR (MLL2) reveals ALR target genes and leads to alterations in cell adhesion and growth. Mol Cell Biol 27:1889–1903

Ito S, D'Alessio AC, Taranova OV, Hong K, Sowers LC, Zhang Y (2010) Role of Tet proteins in 5mC to 5hmC conversion, ES-cell self-renewal and inner cell mass specification. Nature 466:1129–1133

Ito S, Shen L, Dai Q, Wu SC, Collins LB, Swenberg JA, He C, Zhang Y (2011) Tet proteins can convert 5-methylcytosine to 5-formylcytosine and 5-carboxylcytosine. Science 333:1300–1303

Iyer NG, Ozdag H, Caldas C (2004) p300/CBP and cancer. Oncogene 23:4225–4231

Jaffe JD, Wang Y, Chan HM, Zhang J, Huether R, Kryukov GV, Bhang HE, Taylor JE, Hu M, Englund NP et al (2013) Global chromatin profiling reveals NSD2 mutations in pediatric acute lymphoblastic leukemia. Nat Genet 45:1386–1391

Jagani Z, Mora-Blanco EL, Sansam CG, McKenna ES, Wilson B, Chen D, Klekota J, Tamayo P, Nguyen PT, Tolstorukov M et al (2010) Loss of the tumor suppressor Snf5 leads to aberrant activation of the Hedgehog-Gli pathway. Nat Med 16:1429–1433

Jankowska AM, Szpurka H, Tiu RV, Makishima H, Afable M, Huh J, O'Keefe CL, Ganetzky R, McDevitt MA, Maciejewski JP (2009) Loss of heterozygosity 4q24 and TET2 mutations associated with myelodysplastic/myeloproliferative neoplasms. Blood 113:6403–6410

Jia D, Jurkowska RZ, Zhang X, Jeltsch A, Cheng X (2007) Structure of Dnmt3a bound to Dnmt3L suggests a model for de novo DNA methylation. Nature 449:248–251

Jones PA (2012) Functions of DNA methylation: islands, start sites, gene bodies and beyond. Nat Rev Genet 13:484–492

Jones PA, Baylin SB (2002) The fundamental role of epigenetic events in cancer. Nat Rev Genet 3:415–428

Jones PA, Baylin SB (2007) The epigenomics of cancer. Cell 128:683–692

Jones S, Wang TL, Shih le M, Mao TL, Nakayama K, Roden R, Glas R, Slamon D, Diaz LA Jr, Vogelstein B et al (2010) Frequent mutations of chromatin remodeling gene ARID1A in ovarian clear cell carcinoma. Science 330:228–231

Joseph CG, Hwang H, Jiao Y, Wood LD, Kinde I, Wu J, Mandahl N, Luo J, Hruban RH, Diaz LA Jr et al (2013) Exomic analysis of myxoid liposarcomas, synovial sarcomas, and osteosarcomas. Genes Chromosom Cancer 53:15–24

Jurkowska RZ, Rajavelu A, Anspach N, Urbanke C, Jankevicius G, Ragozin S, Nellen W, Jeltsch A (2011) Oligomerization and binding of the Dnmt3a DNA methyltransferase to parallel DNA molecules: heterochromatic localization and role of Dnmt3L. J Biol Chem 286:24200–24207

Kampranis SC, Tsichlis PN (2009) Histone demethylases and cancer. Adv Cancer Res 102:103–169

Kan Z, Jaiswal BS, Stinson J, Janakiraman V, Bhatt D, Stern HM, Yue P, Haverty PM, Bourgon R, Zheng J et al (2010) Diverse somatic mutation patterns and pathway alterations in human cancers. Nature 466:869–873

Kane MF, Loda M, Gaida GM, Lipman J, Mishra R, Goldman H, Jessup JM, Kolodner R (1997) Methylation of the hMLH1 promoter correlates with lack of expression of hMLH1 in sporadic colon tumors and mismatch repair-defective human tumor cell lines. Cancer Res 57:808–811

Kaneda M, Okano M, Hata K, Sado T, Tsujimoto N, Li E, Sasaki H (2004) Essential role for de novo DNA methyltransferase Dnmt3a in paternal and maternal imprinting. Nature 429:900–903

Kang MR, Kim MS, Oh JE, Kim YR, Song SY, Seo SI, Lee JY, Yoo NJ, Lee SH (2009) Mutational analysis of IDH1 codon 132 in glioblastomas and other common cancers. Int J Cancer 125:353–355

Kapur P, Pena-Llopis S, Christie A, Zhrebker L, Pavia-Jimenez A, Rathmell WK, Xie XJ, Brugarolas J (2013) Effects on survival of BAP1 and PBRM1 mutations in sporadic clear-cell renal-cell carcinoma: a retrospective analysis with independent validation. Lancet Oncol 14:159–167

Khan SN, Jankowska AM, Mahfouz R, Dunbar AJ, Sugimoto Y, Hosono N, Hu Z, Cheriyath V, Vatolin S, Przychodzen B et al (2013) Multiple mechanisms deregulate EZH2 and histone H3 lysine 27 epigenetic changes in myeloid malignancies. Leukemia 27:1301–1309

Khuong-Quang DA, Buczkowicz P, Rakopoulos P, Liu XY, Fontebasso AM, Bouffet E, Bartels U, Albrecht S, Schwartzentruber J, Letourneau L et al (2012) K27M mutation in histone H3.3 defines clinically and biologically distinct subgroups of pediatric diffuse intrinsic pontine gliomas. Acta Neuropathol 124:439–447

Kia SK, Gorski MM, Giannakopoulos S, Verrijzer CP (2008) SWI/SNF mediates polycomb eviction and epigenetic reprogramming of the INK4b-ARF-INK4a locus. Mol Cell Biol 28:3457–3464

Kim JG, Takeshima H, Niwa T, Rehnberg E, Shigematsu Y, Yoda Y, Yamashita S, Kushima R, Maekita T, Ichinose M et al (2013) Comprehensive DNA methylation and extensive mutation analyses reveal an association between the CpG island methylator phenotype and oncogenic mutations in gastric cancers. Cancer Lett 330:33–40

Kirmizis A, Bartley SM, Kuzmichev A, Margueron R, Reinberg D, Green R, Farnham PJ (2004) Silencing of human polycomb target genes is associated with methylation of histone H3 Lys 27. Genes Dev 18:1592–1605

Kleer CG, Cao Q, Varambally S, Shen R, Ota I, Tomlins SA, Ghosh D, Sewalt RG, Otte AP, Hayes DF et al (2003) EZH2 is a marker of aggressive breast cancer and promotes neoplastic transformation of breast epithelial cells. Proc Natl Acad Sci USA 100:11606–11611

Knutson SK, Wigle TJ, Warholic NM, Sneeringer CJ, Allain CJ, Klaus CR, Sacks JD, Raimondi A, Majer CR, Song J et al (2012) A selective inhibitor of EZH2 blocks H3K27 methylation and kills mutant lymphoma cells. Nat Chem Biol 8:890–896

Ko M, Huang Y, Jankowska AM, Pape UJ, Tahiliani M, Bandukwala HS, An J, Lamperti ED, Koh KP, Ganetzky R et al (2010) Impaired hydroxylation of 5-methylcytosine in myeloid cancers with mutant TET2. Nature 468:839–843

Ko M, An J, Bandukwala HS, Chavez L, Aijo T, Pastor WA, Segal MF, Li H, Koh KP, Lahdesmaki H et al (2013) Modulation of TET2 expression and 5-methylcytosine oxidation by the CXXC domain protein IDAX. Nature 497:122–126

Konstandin N, Bultmann S, Szwagierczak A, Dufour A, Ksienzyk B, Schneider F, Herold T, Mulaw M, Kakadia PM, Schneider S et al (2011) Genomic 5-hydroxymethylcytosine levels correlate with TET2 mutations and a distinct global gene expression pattern in secondary acute myeloid leukemia. Leukemia 25:1649–1652

Kosmider O, Gelsi-Boyer V, Cheok M, Grabar S, Della-Valle V, Picard F, Viguie F, Quesnel B, Beyne-Rauzy O, Solary E et al (2009) TET2 mutation is an independent favorable prognostic factor in myelodysplastic syndromes (MDSs). Blood 114:3285–3291

Kouzarides T (2007) Chromatin modifications and their function. Cell 128:693–705

Kreiger PA, Judkins AR, Russo PA, Biegel JA, Lestini BJ, Assanasen C, Pawel BR (2009) Loss of INI1 expression defines a unique subset of pediatric undifferentiated soft tissue sarcomas. Mod Pathol 22:142–150

Krivtsov AV, Armstrong SA (2007) MLL translocations, histone modifications and leukaemia stem-cell development. Nat Rev Cancer 7:823–833

Kung AL, Rebel VI, Bronson RT, Ch'ng LE, Sieff CA, Livingston DM, Yao TP (2000) Gene dose-dependent control of hematopoiesis and hematologic tumor suppression by CBP. Genes Dev 14:272–277

Kuo AJ, Cheung P, Chen K, Zee BM, Kioi M, Lauring J, Xi Y, Park BH, Shi X, Garcia BA et al (2011) NSD2 links dimethylation of histone H3 at lysine 36 to oncogenic programming. Mol Cell 44:609–620

Laffaire J, Everhard S, Idbaih A, Criniere E, Marie Y, de Reynies A, Schiappa R, Mokhtari K, Hoang-Xuan K, Sanson M et al (2011) Methylation profiling identifies 2 groups of gliomas according to their tumorigenesis. Neuro-oncology 13:84–98

Lai JL, Jouet JP, Bauters F, Deminatti M (1985) Chronic myelogenous leukemia with translocation (8;22): report of a new case. Cancer Genet Cytogenet 17:365–366

Larkin J, Goh XY, Vetter M, Pickering L, Swanton C (2012) Epigenetic regulation in RCC: opportunities for therapeutic intervention? Nat Rev Urol 9:147–155

Lee MG, Villa R, Trojer P, Norman J, Yan KP, Reinberg D, Di Croce L, Shiekhattar R (2007) Demethylation of H3K27 regulates polycomb recruitment and H2A ubiquitination. Science 318:447–450

Lee JJ, Geli J, Larsson C, Wallin G, Karimi M, Zedenius J, Hoog A, Foukakis T (2008) Gene-specific promoter hypermethylation without global hypomethylation in follicular thyroid cancer. Int J Oncol 33:861–869

Lewis PW, Muller MM, Koletsky MS, Cordero F, Lin S, Banaszynski LA, Garcia BA, Muir TW, Becher OJ, Allis CD (2013) Inhibition of PRC2 activity by a gain-of-function H3 mutation found in pediatric glioblastoma. Science 340:857–861

Ley TJ, Ding L, Walter MJ, McLellan MD, Lamprecht T, Larson DE, Kandoth C, Payton JE, Baty J, Welch J et al (2010) DNMT3A mutations in acute myeloid leukemia. N Engl J Med 363:2424–2433

Li YQ, Zhou PZ, Zheng XD, Walsh CP, Xu GL (2007) Association of Dnmt3a and thymine DNA glycosylase links DNA methylation with base-excision repair. Nucleic Acids Res 35:390–400

Li Y, Trojer P, Xu CF, Cheung P, Kuo A, Drury WJ 3rd, Qiao Q, Neubert TA, Xu RM, Gozani O et al (2009) The target of the NSD family of histone lysine methyltransferases depends on the nature of the substrate. J Biol Chem 284:34283–34295

Li XS, Trojer P, Matsumura T, Treisman JE, Tanese N (2010) Mammalian SWI/SNF–a subunit BAF250/ARID1 is an E3 ubiquitin ligase that targets histone H2B. Mol Cell Biol 30:1673–1688

Li F, Mao G, Tong D, Huang J, Gu L, Yang W, Li GM (2013a) The histone mark H3K36me3 regulates human DNA mismatch repair through its interaction with MutSalpha. Cell 153:590–600

Li WD, Li QR, Xu SN, Wei FJ, Ye ZJ, Cheng JK, Chen JP (2013b) Exome sequencing identifies an MLL3 gene germ line mutation in a pedigree of colorectal cancer and acute myeloid leukemia. Blood 121:1478–1479

Lian CG, Xu Y, Ceol C, Wu F, Larson A, Dresser K, Xu W, Tan L, Hu Y, Zhan Q et al (2012) Loss of 5-hydroxymethylcytosine is an epigenetic hallmark of melanoma. Cell 150:1135–1146

Lin J, Yao DM, Qian J, Chen Q, Qian W, Li Y, Yang J, Wang CZ, Chai HY, Qian Z et al (2011) Recurrent DNMT3A R882 mutations in Chinese patients with acute myeloid leukemia and myelodysplastic syndrome. PLoS ONE 6:e26906

Lindberg J, Klevebring D, Liu W, Neiman M, Xu J, Wiklund P, Wiklund F, Mills IG, Egevad L, Gronberg H (2013a) Exome sequencing of prostate cancer supports the hypothesis of independent tumour origins. Eur Urol 63:347–353

Lindberg J, Mills IG, Klevebring D, Liu W, Neiman M, Xu J, Wikstrom P, Wiklund P, Wiklund F, Egevad L et al (2013b) The mitochondrial and autosomal mutation landscapes of prostate cancer. Eur Urol 63:702–708

Lohr JG, Stojanov P, Lawrence MS, Auclair D, Chapuy B, Sougnez C, Cruz-Gordillo P, Knoechel B, Asmann YW, Slager SL et al (2012) Discovery and prioritization of somatic mutations in diffuse large B-cell lymphoma (DLBCL) by whole-exome sequencing. Proc Natl Acad Sci USA 109:3879–3884

Lorsbach RB, Moore J, Mathew S, Raimondi SC, Mukatira ST, Downing JR (2003) TET1, a member of a novel protein family, is fused to MLL in acute myeloid leukemia containing the t(10;11)(q22;q23). Leukemia 17:637–641

Losman JA, Looper RE, Koivunen P, Lee S, Schneider RK, McMahon C, Cowley GS, Root DE, Ebert BL, Kaelin WG Jr (2013) (R)-2-hydroxyglutarate is sufficient to promote leukemogenesis and its effects are reversible. Science 339:1621–1625

Love C, Sun Z, Jima D, Li G, Zhang J, Miles R, Richards KL, Dunphy CH, Choi WW, Srivastava G et al (2012) The genetic landscape of mutations in Burkitt lymphoma. Nat Genet 44:1321–1325

Lu C, Ward PS, Kapoor GS, Rohle D, Turcan S, Abdel-Wahab O, Edwards CR, Khanin R, Figueroa ME, Melnick A et al (2012) IDH mutation impairs histone demethylation and results in a block to cell differentiation. Nature 483:474–478

Lu C, Venneti S, Akalin A, Fang F, Ward PS, Dematteo RG, Intlekofer AM, Chen C, Ye J, Hameed M et al (2013) Induction of sarcomas by mutant IDH2. Genes Dev 27:1986–1998

Luo B, Cheung HW, Subramanian A, Sharifnia T, Okamoto M, Yang X, Hinkle G, Boehm JS, Beroukhim R, Weir BA et al (2008) Highly parallel identification of essential genes in cancer cells. Proc Natl Acad Sci USA 105:20380–20385

Majer CR, Jin L, Scott MP, Knutson SK, Kuntz KW, Keilhack H, Smith JJ, Moyer MP, Richon VM, Copeland RA et al (2012) A687V EZH2 is a gain-of-function mutation found in lymphoma patients. FEBS Lett 586:3448–3451

Makishima H, Jankowska AM, Tiu RV, Szpurka H, Sugimoto Y, Hu Z, Saunthararajah Y, Guinta K, Keddache MA, Putnam P et al (2010) Novel homo- and hemizygous mutations in EZH2 in myeloid malignancies. Leukemia 24:1799–1804

Mamo A, Cavallone L, Tuzmen S, Chabot C, Ferrario C, Hassan S, Edgren H, Kallioniemi O, Aleynikova O, Przybytkowski E et al (2012) An integrated genomic approach identifies ARID1A as a candidate tumor-suppressor gene in breast cancer. Oncogene 31:2090–2100

Manceau G, Letouze E, Guichard C, Didelot A, Cazes A, Corte H, Fabre E, Pallier K, Imbeaud S, Le Pimpec-Barthes F et al (2013) Recurrent inactivating mutations of ARID2 in non-small cell lung carcinoma. Int J Cancer 132:2217–2221

Mann KM, Ward JM, Yew CC, Kovochich A, Dawson DW, Black MA, Brett BT, Sheetz TE, Dupuy AJ, Chang DK et al (2012) Sleeping beauty mutagenesis reveals cooperating mutations and pathways in pancreatic adenocarcinoma. Proc Natl Acad Sci USA 109:5934–5941

Mar BG, Bullinger L, Basu E, Schlis K, Silverman LB, Dohner K, Armstrong SA (2012) Sequencing histone-modifying enzymes identifies UTX mutations in acute lymphoblastic leukemia. Leukemia 26:1881–1883

Marango J, Shimoyama M, Nishio H, Meyer JA, Min DJ, Sirulnik A, Martinez-Martinez Y, Chesi M, Bergsagel PL, Zhou MM et al (2008) The MMSET protein is a histone methyltransferase with characteristics of a transcriptional corepressor. Blood 111:3145–3154

Marcucci G, Metzeler KH, Schwind S, Becker H, Maharry K, Mrozek K, Radmacher MD, Kohlschmidt J, Nicolet D, Whitman SP et al (2012) Age-related prognostic impact of different types of DNMT3A mutations in adults with primary cytogenetically normal acute myeloid leukemia. J Clin Oncol 30:742–750

Mardis ER, Ding L, Dooling DJ, Larson DE, McLellan MD, Chen K, Koboldt DC, Fulton RS, Delehaunty KD, McGrath SD et al (2009) Recurring mutations found by sequencing an acute myeloid leukemia genome. N Engl J Med 361:1058–1066

Margueron R, Reinberg D (2010) Chromatin structure and the inheritance of epigenetic information. Nat Rev Genet 11:285–296

Margueron R, Reinberg D (2011) The polycomb complex PRC2 and its mark in life. Nature 469:343–349

Martinez-Garcia E, Popovic R, Min DJ, Sweet SM, Thomas PM, Zamdborg L, Heffner A, Will C, Lamy L, Staudt LM et al (2011) The MMSET histone methyl transferase switches global histone methylation and alters gene expression in t(4;14) multiple myeloma cells. Blood 117:211–220

Matsubara D, Kishaba Y, Ishikawa S, Sakatani T, Oguni S, Tamura T, Hoshino H, Sugiyama Y, Endo S, Murakami Y et al (2013) Lung cancer with loss of BRG1/BRM, shows epithelial mesenchymal transition phenotype and distinct histologic and genetic features. Cancer Sci 104:266–273

McCabe MT, Graves AP, Ganji G, Diaz E, Halsey WS, Jiang Y, Smitheman KN, Ott HM, Pappalardi MB, Allen KE et al (2012a) Mutation of A677 in histone methyltransferase EZH2 in human B-cell lymphoma promotes hypertrimethylation of histone H3 on lysine 27 (H3K27). Proc Natl Acad Sci USA 109:2989–2994

McCabe MT, Ott HM, Ganji G, Korenchuk S, Thompson C, Van Aller GS, Liu Y, Graves AP, Della Pietra A 3rd, Diaz E et al (2012b) EZH2 inhibition as a therapeutic strategy for lymphoma with EZH2-activating mutations. Nature 492:108–112

Medina PP, Romero OA, Kohno T, Montuenga LM, Pio R, Yokota J, Sanchez-Cespedes M (2008) Frequent BRG1/SMARCA4-inactivating mutations in human lung cancer cell lines. Hum Mutat 29:617–622

Minor EA, Court BL, Young JI, Wang G (2013) Ascorbate induces ten-eleven translocation (Tet) methylcytosine dioxygenase-mediated generation of 5-hydroxymethylcytosine. J Biol Chem 288:13669–13674

Mobley BC, McKenney JK, Bangs CD, Callahan K, Yeom KW, Schneppenheim R, Hayden MG, Cherry AM, Gokden M, Edwards MS et al (2010) Loss of SMARCB1/INI1 expression in poorly differentiated chordomas. Acta Neuropathol 120:745–753

Morales Torres C, Laugesen A, Helin K (2013) Utx is required for proper induction of ectoderm and mesoderm during differentiation of embryonic stem cells. PLoS ONE 8:e60020

Moran-Crusio K, Reavie L, Shih A, Abdel-Wahab O, Ndiaye-Lobry D, Lobry C, Figueroa ME, Vasanthakumar A, Patel J, Zhao X et al (2011) Tet2 loss leads to increased hematopoietic stem cell self-renewal and myeloid transformation. Cancer Cell 20:11–24

Morin RD, Johnson NA, Severson TM, Mungall AJ, An J, Goya R, Paul JE, Boyle M, Woolcock BW, Kuchenbauer F et al (2010) Somatic mutations altering EZH2 (Tyr641) in follicular and diffuse large B-cell lymphomas of germinal-center origin. Nat Genet 42:181–185

Morin RD, Mendez-Lago M, Mungall AJ, Goya R, Mungall KL, Corbett RD, Johnson NA, Severson TM, Chiu R, Field M et al (2011) Frequent mutation of histone-modifying genes in non-Hodgkin lymphoma. Nature 476:298–303

Mullighan CG, Zhang J, Kasper LH, Lerach S, Payne-Turner D, Phillips LA, Heatley SL, Holmfeldt L, Collins-Underwood JR, Ma J et al (2011) CREBBP mutations in relapsed acute lymphoblastic leukaemia. Nature 471:235–239

Nagl NG Jr, Wang X, Patsialou A, Van Scoy M, Moran E (2007) Distinct mammalian SWI/SNF chromatin remodeling complexes with opposing roles in cell-cycle control. EMBO J 26:752–763

Nakajima T, Uchida C, Anderson SF, Parvin JD, Montminy M (1997) Analysis of a cAMP-responsive activator reveals a two-component mechanism for transcriptional induction via signal-dependent factors. Genes Dev 11:738–747

Nekrutenko A, Hillis DM, Patton JC, Bradley RD, Baker RJ (1998) Cytosolic isocitrate dehydrogenase in humans, mice, and voles and phylogenetic analysis of the enzyme family. Mol Biol Evol 15:1674–1684

Neumann M, Heesch S, Schlee C, Schwartz S, Gokbuget N, Hoelzer D, Konstandin NP, Ksienzyk B, Vosberg S, Graf A et al (2013) Whole-exome sequencing in adult ETP-ALL reveals a high rate of DNMT3A mutations. Blood 121:4749–4752

New M, Olzscha H, La Thangue NB (2012) HDAC inhibitor-based therapies: can we interpret the code? Mol Oncol 6:637–656

Newbold RF, Mokbel K (2010) Evidence for a tumour suppressor function of SETD2 in human breast cancer: a new hypothesis. Anticancer Res 30:3309–3311

Nibourel O, Kosmider O, Cheok M, Boissel N, Renneville A, Philippe N, Dombret H, Dreyfus F, Quesnel B, Geffroy S et al (2010) Incidence and prognostic value of TET2 alterations in de novo acute myeloid leukemia achieving complete remission. Blood 116:1132–1135

Nie Z, Xue Y, Yang D, Zhou S, Deroo BJ, Archer TK, Wang W (2000) A specificity and targeting subunit of a human SWI/SNF family-related chromatin-remodeling complex. Mol Cell Biol 20:8879–8888

Nikolaev SI, Rimoldi D, Iseli C, Valsesia A, Robyr D, Gehrig C, Harshman K, Guipponi M, Bukach O, Zoete V et al (2012) Exome sequencing identifies recurrent somatic MAP2K1 and MAP2K2 mutations in melanoma. Nat Genet 44:133–139

Nikoloski G, Langemeijer SM, Kuiper RP, Knops R, Massop M, Tonnissen ER, van der Heijden A, Scheele TN, Vandenberghe P, de Witte T et al (2010) Somatic mutations of the histone methyltransferase gene EZH2 in myelodysplastic syndromes. Nat Genet 42:665–667

Niu X, Zhang T, Liao L, Zhou L, Lindner DJ, Zhou M, Rini B, Yan Q, Yang H (2012) The von Hippel-Lindau tumor suppressor protein regulates gene expression and tumor growth through histone demethylase JARID1C. Oncogene 31:776–786

Noushmehr H, Weisenberger DJ, Diefes K, Phillips HS, Pujara K, Berman BP, Pan F, Pelloski CE, Sulman EP, Bhat KP et al (2010) Identification of a CpG island methylator phenotype that defines a distinct subgroup of glioma. Cancer Cell 17:510–522

Ogiwara H, Ui A, Otsuka A, Satoh H, Yokomi I, Nakajima S, Yasui A, Yokota J, Kohno T (2011) Histone acetylation by CBP and p300 at double-strand break sites facilitates SWI/SNF chromatin remodeling and the recruitment of non-homologous end joining factors. Oncogene 30:2135–2146

Oike T, Ogiwara H, Nakano T, Yokota J, Kohno T (2013) Inactivating mutations in SWI/SNF chromatin remodeling genes in human cancer. Jpn J Clin Oncol 43:849–855

Oka M, Meacham AM, Hamazaki T, Rodic N, Chang LJ, Terada N (2005) De novo DNA methyltransferases Dnmt3a and Dnmt3b primarily mediate the cytotoxic effect of 5-aza-2'-deoxycytidine. Oncogene 24:3091–3099

Okano M, Bell DW, Haber DA, Li E (1999) DNA methyltransferases Dnmt3a and Dnmt3b are essential for de novo methylation and mammalian development. Cell 99:247–257

Oruetxebarria I, Venturini F, Kekarainen T, Houweling A, Zuijderduijn LM, Mohd-Sarip A, Vries RG, Hoeben RC, Verrijzer CP (2004) P16INK4a is required for hSNF5 chromatin remodeler-induced cellular senescence in malignant rhabdoid tumor cells. J Biol Chem 279:3807–3816

Oyer JA, Huang X, Zheng Y, Shim J, Ezponda T, Carpenter Z, Allegretta M, Okot-Kotber CI, Patel JP, Melnick A et al (2013) Point mutation E1099K in MMSET/NSD2 enhances its methyltranferase activity and leads to altered global chromatin methylation in lymphoid malignancies. Leukemia

Pais A, Amare Kadam P, Raje G, Sawant M, Kabre S, Jain H, Advani S, Banavali S (2005) Identification of various MLL gene aberrations that lead to MLL gene mutation in patients with acute lymphoblastic leukemia (ALL) and infants with acute leukemia. Leuk Res 29:517–526

Panagopoulos I, Fioretos T, Isaksson M, Samuelsson U, Billstrom R, Strombeck B, Mitelman F, Johansson B (2001) Fusion of the MORF and CBP genes in acute myeloid leukemia with the t (10;16)(q22;p13). Hum Mol Genet 10:395–404

Parsons DW, Jones S, Zhang X, Lin JC, Leary RJ, Angenendt P, Mankoo P, Carter H, Siu IM, Gallia GL et al (2008) An integrated genomic analysis of human glioblastoma multiforme. Science 321:1807–1812

Parsons DW, Li M, Zhang X, Jones S, Leary RJ, Lin JC, Boca SM, Carter H, Samayoa J, Bettegowda C et al (2011) The genetic landscape of the childhood cancer medulloblastoma. Science 331:435–439

Paschka P, Schlenk RF, Gaidzik VI, Habdank M, Kronke J, Bullinger L, Spath D, Kayser S, Zucknick M, Gotze K et al (2010) IDH1 and IDH2 mutations are frequent genetic alterations in acute myeloid leukemia and confer adverse prognosis in cytogenetically normal acute myeloid leukemia with NPM1 mutation without FLT3 internal tandem duplication. J Clin Oncol 28:3636–3643

Pasqualucci L, Dominguez-Sola D, Chiarenza A, Fabbri G, Grunn A, Trifonov V, Kasper LH, Lerach S, Tang H, Ma J et al (2011a) Inactivating mutations of acetyltransferase genes in B-cell lymphoma. Nature 471:189–195

Pasqualucci L, Trifonov V, Fabbri G, Ma J, Rossi D, Chiarenza A, Wells VA, Grunn A, Messina M, Elliot O et al (2011b) Analysis of the coding genome of diffuse large B-cell lymphoma. Nat Genet 43:830–837

Patel JP, Gonen M, Figueroa ME, Fernandez H, Sun Z, Racevskis J, Van Vlierberghe P, Dolgalev I, Thomas S, Aminova O et al (2012) Prognostic relevance of integrated genetic profiling in acute myeloid leukemia. N Engl J Med 366:1079–1089

Pawlowski R, Muhl SM, Sulser T, Krek W, Moch H, Schraml P (2013) Loss of PBRM1 expression is associated with renal cell carcinoma progression. Int J Cancer 132:E11–E17

Peifer M, Fernandez-Cuesta L, Sos ML, George J, Seidel D, Kasper LH, Plenker D, Leenders F, Sun R, Zander T et al (2012) Integrative genome analyses identify key somatic driver mutations of small-cell lung cancer. Nat Genet 44:1104–1110

Pena-Llopis S, Christie A, Xie XJ, Brugarolas J (2013) Cooperation and antagonism among cancer genes: the renal cancer paradigm. Cancer Res 73:4173–4179

Peng C, Zhou J, Liu HY, Zhou M, Wang LL, Zhang QH, Yang YX, Xiong W, Shen SR, Li XL et al (2006) The transcriptional regulation role of BRD7 by binding to acetylated histone through bromodomain. J Cell Biochem 97:882–892

Perez C, Martinez-Calle N, Martin-Subero JI, Segura V, Delabesse E, Fernandez-Mercado M, Garate L, Alvarez S, Rifon J, Varea S et al (2012) TET2 mutations are associated with specific 5-methylcytosine and 5-hydroxymethylcytosine profiles in patients with chronic myelomonocytic leukemia. PLoS ONE 7:e31605

Petrij F, Giles RH, Dauwerse HG, Saris JJ, Hennekam RC, Masuno M, Tommerup N, van Ommen GJ, Goodman RH, Peters DJ et al (1995) Rubinstein-Taybi syndrome caused by mutations in the transcriptional co-activator CBP. Nature 376:348–351

Pirozzi CJ, Reitman ZJ, Yan H (2013) Releasing the block: setting differentiation free with mutant IDH inhibitors. Cancer Cell 23:570–572

Plass C, Pfister SM, Lindroth AM, Bogatyrova O, Claus R, Lichter P (2013) Mutations in regulators of the epigenome and their connections to global chromatin patterns in cancer. Nat Rev Genet 14:765–780

Plimack ER, Kantarjian HM, Issa JP (2007) Decitabine and its role in the treatment of hematopoietic malignancies. Leuk Lymphoma 48:1472–1481

Poage GM, Houseman EA, Christensen BC, Butler RA, Avissar-Whiting M, McClean MD, Waterboer T, Pawlita M, Marsit CJ, Kelsey KT (2011) Global hypomethylation identifies loci targeted for hypermethylation in head and neck cancer. Clin Cancer Res 17:3579–3589

Poon SL, Pang ST, McPherson JR, Yu W, Huang KK, Guan P, Weng WH, Siew EY, Liu Y, Heng HL et al (2013) Genome-wide mutational signatures of aristolochic acid and its application as a screening tool. Sci Transl Med 5:197ra101

Pronier E, Almire C, Mokrani H, Vasanthakumar A, Simon A, da Costa Reis Monte Mor B, Masse A, Le Couedic JP, Pendino F, Carbonne B et al (2011) Inhibition of TET2-mediated conversion of 5-methylcytosine to 5-hydroxymethylcytosine disturbs erythroid and granulomonocytic differentiation of human hematopoietic progenitors. Blood 118:2551–2555

Pugh TJ, Weeraratne SD, Archer TC, Pomeranz Krummel DA, Auclair D, Bochicchio J, Carneiro MO, Carter SL, Cibulskis K, Erlich RL et al (2012) Medulloblastoma exome sequencing uncovers subtype-specific somatic mutations. Nature 488:106–110

Purdy MM, Holz-Schietinger C, Reich NO (2010) Identification of a second DNA binding site in human DNA methyltransferase 3A by substrate inhibition and domain deletion. Arch Biochem Biophys 498:13–22

Quivoron C, Couronne L, Della Valle V, Lopez CK, Plo I, Wagner-Ballon O, Do Cruzeiro M, Delhommeau F, Arnulf B, Stern MH et al (2011) TET2 inactivation results in pleiotropic hematopoietic abnormalities in mouse and is a recurrent event during human lymphomagenesis. Cancer Cell 20:25–38

Rea S, Eisenhaber F, O'Carroll D, Strahl BD, Sun ZW, Schmid M, Opravil S, Mechtler K, Ponting CP, Allis CD et al (2000) Regulation of chromatin structure by site-specific histone H3 methyltransferases. Nature 406:593–599

Reisman DN, Sciarrotta J, Wang W, Funkhouser WK, Weissman BE (2003) Loss of BRG1/BRM in human lung cancer cell lines and primary lung cancers: correlation with poor prognosis. Cancer Res 63:560–566

Reisman D, Glaros S, Thompson EA (2009) The SWI/SNF complex and cancer. Oncogene 28:1653–1668

Renneville A, Boissel N, Nibourel O, Berthon C, Helevaut N, Gardin C, Cayuela JM, Hayette S, Reman O, Contentin N et al (2012) Prognostic significance of DNA methyltransferase 3A mutations in cytogenetically normal acute myeloid leukemia: a study by the Acute Leukemia French Association. Leukemia 26:1247–1254

Reyes JC, Barra J, Muchardt C, Camus A, Babinet C, Yaniv M (1998) Altered control of cellular proliferation in the absence of mammalian brahma (SNF2alpha). EMBO J 17:6979–6991

Riggs AD (1975) X inactivation, differentiation, and DNA methylation. Cytogenet Cell Genet 14:9–25

Roberts CW, Galusha SA, McMenamin ME, Fletcher CD, Orkin SH (2000) Haploinsufficiency of Snf5 (integrase interactor 1) predisposes to malignant rhabdoid tumors in mice. Proc Natl Acad Sci USA 97:13796–13800

Robinson G, Parker M, Kranenburg TA, Lu C, Chen X, Ding L, Phoenix TN, Hedlund E, Wei L, Zhu X et al (2012) Novel mutations target distinct subgroups of medulloblastoma. Nature 488:43–48

Ropero S, Fraga MF, Ballestar E, Hamelin R, Yamamoto H, Boix-Chornet M, Caballero R, Alaminos M, Setien F, Paz MF et al (2006) A truncating mutation of HDAC2 in human cancers confers resistance to histone deacetylase inhibition. Nat Genet 38:566–569

Ropero S, Ballestar E, Alaminos M, Arango D, Schwartz S Jr, Esteller M (2008) Transforming pathways unleashed by a HDAC2 mutation in human cancer. Oncogene 27:4008–4012

Ross JS, Wang K, Al-Rohil RN, Nazeer T, Sheehan CE, Otto GA, He J, Palmer G, Yelensky R, Lipson D et al (2013) Advanced urothelial carcinoma: next-generation sequencing reveals diverse genomic alterations and targets of therapy. Mod Pathol

Rotili D, Mai A (2011) Targeting histone demethylases: a new avenue for the fight against cancer. Genes Cancer 2:663–679

Rubery ED, Newton AA (1973) DNA methylation in normal and tumour virus-transformed cells in tissue culture. I. The level of DNA methylation in BHK21 cells and in BHK21 cells transformed by polyoma virus (PyY cells). Biochim Biophys Acta 324:24–36

Ryan RJ, Nitta M, Borger D, Zukerberg LR, Ferry JA, Harris NL, Iafrate AJ, Bernstein BE, Sohani AR, Le LP (2011) EZH2 codon 641 mutations are common in BCL2-rearranged germinal center B cell lymphomas. PLoS ONE 6:e28585

Saha A, Wittmeyer J, Cairns BR (2006) Chromatin remodelling: the industrial revolution of DNA around histones. Nat Rev Mol Cell Biol 7:437–447

Sanchez-Tillo E, Lazaro A, Torrent R, Cuatrecasas M, Vaquero EC, Castells A, Engel P, Postigo A (2010) ZEB1 represses E-cadherin and induces an EMT by recruiting the SWI/SNF chromatin-remodeling protein BRG1. Oncogene 29:3490–3500

Sasaki M, Knobbe CB, Itsumi M, Elia AJ, Harris IS, Chio II, Cairns RA, McCracken S, Wakeham A, Haight J et al (2012) D-2-hydroxyglutarate produced by mutant IDH1 perturbs collagen maturation and basement membrane function. Genes Dev 26:2038–2049

Sausen M, Leary RJ, Jones S, Wu J, Reynolds CP, Liu X, Blackford A, Parmigiani G, Diaz LA Jr, Papadopoulos N et al (2013) Integrated genomic analyses identify ARID1A and ARID1B alterations in the childhood cancer neuroblastoma. Nat Genet 45:12–17

Schaub FX, Looser R, Li S, Hao-Shen H, Lehmann T, Tichelli A, Skoda RC (2010) Clonal analysis of TET2 and JAK2 mutations suggests that TET2 can be a late event in the progression of myeloproliferative neoplasms. Blood 115:2003–2007

Schmidt CK, Jackson SP (2013) On your mark, get SET(D2), go! H3K36me3 primes DNA mismatch repair. Cell 153:513–515

Schnittger S, Haferlach C, Ulke M, Alpermann T, Kern W, Haferlach T (2010) IDH1 mutations are detected in 6.6% of 1414 AML patients and are associated with intermediate risk karyotype and unfavorable prognosis in adults younger than 60 years and unmutated NPM1 status. Blood 116:5486–5496

Schwartzentruber J, Korshunov A, Liu XY, Jones DT, Pfaff E, Jacob K, Sturm D, Fontebasso AM, Quang DA, Tonjes M et al (2012) Driver mutations in histone H3.3 and chromatin remodelling genes in paediatric glioblastoma. Nature 482:226–231

Senyuk V, Premanand K, Xu P, Qian Z, Nucifora G (2011) The oncoprotein EVI1 and the DNA methyltransferase Dnmt3 co-operate in binding and de novo methylation of target DNA. PLoS ONE 6:e20793

Shain AH, Pollack JR (2013) The spectrum of SWI/SNF mutations, ubiquitous in human cancers. PLoS ONE 8:e55119

Shain AH, Giacomini CP, Matsukuma K, Karikari CA, Bashyam MD, Hidalgo M, Maitra A, Pollack JR (2012) Convergent structural alterations define SWItch/Sucrose NonFermentable (SWI/SNF) chromatin remodeler as a central tumor suppressive complex in pancreatic cancer. Proc Natl Acad Sci USA 109:E252–E259

Sharma S, Kelly TK, Jones PA (2010) Epigenetics in cancer. Carcinogenesis 31:27–36

Shen JC, Rideout WM 3rd, Jones PA (1992) High frequency mutagenesis by a DNA methyltransferase. Cell 71:1073–1080

Shen H, Powers N, Saini N, Comstock CE, Sharma A, Weaver K, Revelo MP, Gerald W, Williams E, Jessen WJ et al (2008) The SWI/SNF ATPase Brm is a gatekeeper of proliferative control in prostate cancer. Cancer Res 68:10154–10162

Shi Y, Whetstine JR (2007) Dynamic regulation of histone lysine methylation by demethylases. Mol Cell 25:1–14

Shi L, Sun L, Li Q, Liang J, Yu W, Yi X, Yang X, Li Y, Han X, Zhang Y et al (2011) Histone demethylase JMJD2B coordinates H3K4/H3K9 methylation and promotes hormonally responsive breast carcinogenesis. Proc Natl Acad Sci USA 108:7541–7546

Shiama N (1997) The p300/CBP family: integrating signals with transcription factors and chromatin. Trends Cell Biol 7:230–236

Shibata T, Kokubu A, Miyamoto M, Sasajima Y, Yamazaki N (2011) Mutant IDH1 confers an in vivo growth in a melanoma cell line with BRAF mutation. Am J Pathol 178:1395–1402

Shih AH, Abdel-Wahab O, Patel JP, Levine RL (2012) The role of mutations in epigenetic regulators in myeloid malignancies. Nat Rev Cancer 12:599–612

Shpargel KB, Sengoku T, Yokoyama S, Magnuson T (2012) UTX and UTY demonstrate histone demethylase-independent function in mouse embryonic development. PLoS Genet 8:e1002964

Sievert AJ, Jackson EM, Gai X, Hakonarson H, Judkins AR, Resnick AC, Sutton LN, Storm PB, Shaikh TH, Biegel JA (2009) Duplication of 7q34 in pediatric low-grade astrocytomas detected by high-density single-nucleotide polymorphism-based genotype arrays results in a novel BRAF fusion gene. Brain Pathol 19:449–458

Singhal N, Graumann J, Wu G, Arauzo-Bravo MJ, Han DW, Greber B, Gentile L, Mann M, Scholer HR (2010) Chromatin-Remodeling Components of the BAF Complex Facilitate Rèprogramming. Cell 141:943–955

Sjoblom T, Jones S, Wood LD, Parsons DW, Lin J, Barber TD, Mandelker D, Leary RJ, Ptak J, Silliman N et al (2006) The consensus coding sequences of human breast and colorectal cancers. Science 314:268–274

Smith AE, Mohamedali AM, Kulasekararaj A, Lim Z, Gaken J, Lea NC, Przychodzen B, Mian SA, Nasser EE, Shooter C et al (2010) Next-generation sequencing of the TET2 gene in 355 MDS and CMML patients reveals low-abundance mutant clones with early origins, but indicates no definite prognostic value. Blood 116:3923–3932

Sneeringer CJ, Scott MP, Kuntz KW, Knutson SK, Pollock RM, Richon VM, Copeland RA (2010) Coordinated activities of wild-type plus mutant EZH2 drive tumor-associated hypertrimethylation of lysine 27 on histone H3 (H3K27) in human B-cell lymphomas. Proc Natl Acad Sci USA 107:20980–20985

Song SJ, Ito K, Ala U, Kats L, Webster K, Sun SM, Jongen-Lavrencic M, Manova-Todorova K, Teruya-Feldstein J, Avigan DE et al (2013) The oncogenic microRNA miR-22 targets the TET2 tumor suppressor to promote hematopoietic stem cell self-renewal and transformation. Cell Stem Cell 13:87–101

Stegelmann F, Bullinger L, Schlenk RF, Paschka P, Griesshammer M, Blersch C, Kuhn S, Schauer S, Dohner H, Dohner K (2011) DNMT3A mutations in myeloproliferative neoplasms. Leukemia 25:1217–1219

Stransky N, Egloff AM, Tward AD, Kostic AD, Cibulskis K, Sivachenko A, Kryukov GV, Lawrence MS, Sougnez C, McKenna A et al (2011) The mutational landscape of head and neck squamous cell carcinoma. Science 333:1157–1160

Stratton MR, Campbell PJ, Futreal PA (2009) The cancer genome. Nature 458:719–724

Sturm D, Witt H, Hovestadt V, Khuong-Quang DA, Jones DT, Konermann C, Pfaff E, Tonjes M, Sill M, Bender S et al (2012) Hotspot mutations in H3F3A and IDH1 define distinct epigenetic and biological subgroups of glioblastoma. Cancer Cell 22:425–437

Suetake I, Mishima Y, Kimura H, Lee YH, Goto Y, Takeshima H, Ikegami T, Tajima S (2011) Characterization of DNA-binding activity in the N-terminal domain of the DNA methyltransferase Dnmt3a. Biochem J 437:141–148

Tadokoro Y, Ema H, Okano M, Li E, Nakauchi H (2007) De novo DNA methyltransferase is essential for self-renewal, but not for differentiation, in hematopoietic stem cells. J Exp Med 204:715–722

Tahiliani M, Koh KP, Shen Y, Pastor WA, Bandukwala H, Brudno Y, Agarwal S, Iyer LM, Liu DR, Aravind L et al (2009) Conversion of 5-methylcytosine to 5-hydroxymethylcytosine in mammalian DNA by MLL partner TET1. Science 324:930–935

Taverna SD, Li H, Ruthenburg AJ, Allis CD, Patel DJ (2007) How chromatin-binding modules interpret histone modifications: lessons from professional pocket pickers. Nat Struct Mol Biol 14:1025–1040

Tefferi A, Levine RL, Lim KH, Abdel-Wahab O, Lasho TL, Patel J, Finke CM, Mullally A, Li CY, Pardanani A et al (2009a) Frequent TET2 mutations in systemic mastocytosis: clinical, KITD816V and FIP1L1-PDGFRA correlates. Leukemia 23:900–904

Tefferi A, Lim KH, Abdel-Wahab O, Lasho TL, Patel J, Patnaik MM, Hanson CA, Pardanani A, Gilliland DG, Levine RL (2009b) Detection of mutant TET2 in myeloid malignancies other than myeloproliferative neoplasms: CMML, MDS, MDS/MPN and AML. Leukemia 23:1343–1345

Tefferi A, Pardanani A, Lim KH, Abdel-Wahab O, Lasho TL, Patel J, Gangat N, Finke CM, Schwager S, Mullally A et al (2009c) TET2 mutations and their clinical correlates in polycythemia vera, essential thrombocythemia and myelofibrosis. Leukemia 23:905–911

Thieme S, Gyarfas T, Richter C, Ozhan G, Fu J, Alexopoulou D, Muders MH, Michalk I, Jakob C, Dahl A et al (2013) The histone demethylase UTX regulates stem cell migration and hematopoiesis. Blood 121:2462–2473

Thirman MJ, Gill HJ, Burnett RC, Mbangkollo D, McCabe NR, Kobayashi H, Ziemin-van der Poel S, Kaneko Y, Morgan R, Sandberg AA et al (1993) Rearrangement of the MLL gene in acute lymphoblastic and acute myeloid leukemias with 11q23 chromosomal translocations. N Engl J Med 329:909–914

Thol F, Damm F, Ludeking A, Winschel C, Wagner K, Morgan M, Yun H, Gohring G, Schlegelberger B, Hoelzer D et al (2011a) Incidence and prognostic influence of DNMT3A mutations in acute myeloid leukemia. J Clin Oncol 29:2889–2896

Thol F, Winschel C, Ludeking A, Yun H, Friesen I, Damm F, Wagner K, Krauter J, Heuser M, Ganser A (2011b) Rare occurrence of DNMT3A mutations in myelodysplastic syndromes. Haematologica 96:1870–1873

Timp W, Feinberg AP (2013) Cancer as a dysregulated epigenome allowing cellular growth advantage at the expense of the host. Nat Rev Cancer 13:497–510

Traina F, Visconte V, Elson P, Tabarroki A, Jankowska AM, Hasrouni E, Sugimoto Y, Szpurka H, Makishima H, O'Keefe CL et al (2013) Impact of molecular mutations on treatment response to DNMT inhibitors in myelodysplasia and related neoplasms. Leukemia 28:78–87

Treiber T, Mandel EM, Pott S, Gyory I, Firner S, Liu ET, Grosschedl R (2010) Early B cell factor 1 regulates B cell gene networks by activation, repression, and transcription-independent poising of chromatin. Immunity 32:714–725

Trobaugh-Lotrario AD, Tomlinson GE, Finegold MJ, Gore L, Feusner JH (2009) Small cell undifferentiated variant of hepatoblastoma: adverse clinical and molecular features similar to rhabdoid tumors. Pediatr Blood Cancer 52:328–334

Trotter KW, Archer TK (2004) Reconstitution of glucocorticoid receptor-dependent transcription in vivo. Mol Cell Biol 24:3347–3358

Tsikitis M, Zhang Z, Edelman W, Zagzag D, Kalpana GV (2005) Genetic ablation of Cyclin D1 abrogates genesis of rhabdoid tumors resulting from Ini1 loss. Proc Natl Acad Sci USA 102:12129–12134

Turcan S, Rohle D, Goenka A, Walsh LA, Fang F, Yilmaz E, Campos C, Fabius AW, Lu C, Ward PS et al (2012) IDH1 mutation is sufficient to establish the glioma hypermethylator phenotype. Nature 483:479–483

Turcan S, Fabius AW, Borodovsky A, Pedraza A, Brennan C, Huse J, Viale A, Riggins GJ, Chan TA (2013) Efficient induction of differentiation and growth inhibition in IDH1 mutant glioma cells by the DNMT inhibitor decitabine. Oncotarget 4:1729–1736

Vakoc CR, Wen YY, Gibbs RA, Johnstone CN, Rustgi AK, Blobel GA (2009) Low frequency of MLL3 mutations in colorectal carcinoma. Cancer Genet Cytogenet 189:140–141

Valinluck V, Sowers LC (2007) Endogenous cytosine damage products alter the site selectivity of human DNA maintenance methyltransferase DNMT1. Cancer Res 67:946–950

van Haaften G, Dalgliesh GL, Davies H, Chen L, Bignell G, Greenman C, Edkins S, Hardy C, O'Meara S, Teague J et al (2009) Somatic mutations of the histone H3K27 demethylase gene UTX in human cancer. Nat Genet 41:521–523

Varambally S, Dhanasekaran SM, Zhou M, Barrette TR, Kumar-Sinha C, Sanda MG, Ghosh D, Pienta KJ, Sewalt RG, Otte AP et al (2002) The polycomb group protein EZH2 is involved in progression of prostate cancer. Nature 419:624–629

Varela I, Tarpey P, Raine K, Huang D, Ong CK, Stephens P, Davies H, Jones D, Lin ML, Teague J et al (2011) Exome sequencing identifies frequent mutation of the SWI/SNF complex gene PBRM1 in renal carcinoma. Nature 469:539–542

Varier RA, Timmers HT (2011) Histone lysine methylation and demethylation pathways in cancer. Biochim Biophys Acta 1815:75–89

Veigl ML, Kasturi L, Olechnowicz J, Ma AH, Lutterbaugh JD, Periyasamy S, Li GM, Drummond J, Modrich PL, Sedwick WD et al (1998) Biallelic inactivation of hMLH1 by epigenetic gene silencing, a novel mechanism causing human MSI cancers. Proc Natl Acad Sci USA 95:8698–8702

Vella P, Scelfo A, Jammula S, Chiacchiera F, Williams K, Cuomo A, Roberto A, Christensen J, Bonaldi T, Helin K et al (2013) Tet proteins connect the O-linked *N*-acetylglucosamine transferase Ogt to chromatin in embryonic stem cells. Mol Cell 49:645–656

Versteege I, Sevenet N, Lange J, Rousseau-Merck MF, Ambros P, Handgretinger R, Aurias A, Delattre O (1998) Truncating mutations of hSNF5/INI1 in aggressive paediatric cancer. Nature 394:203–206

Viguie F, Aboura A, Bouscary D, Ramond S, Delmer A, Tachdjian G, Marie JP, Casadevall N (2005) Common 4q24 deletion in four cases of hematopoietic malignancy: early stem cell involvement? Leukemia 19:1411–1415

Vogelstein B, Papadopoulos N, Velculescu VE, Zhou S, Diaz LA Jr, Kinzler KW (2013) Cancer genome landscapes. Science 339:1546–1558

Wakita S, Yamaguchi H, Omori I, Terada K, Ueda T, Manabe E, Kurosawa S, Iida S, Ibaraki T, Sato Y et al (2013) Mutations of the epigenetics-modifying gene (DNMT3a, TET2, IDH1/2) at diagnosis may induce FLT3-ITD at relapse in de novo acute myeloid leukemia. Leukemia 27:1044–1052

Walter MJ, Ding L, Shen D, Shao J, Grillot M, McLellan M, Fulton R, Schmidt H, Kalicki-Veizer J, O'Laughlin M et al (2011) Recurrent DNMT3A mutations in patients with myelodysplastic syndromes. Leukemia 25:1153–1158

Wang W, Xue Y, Zhou S, Kuo A, Cairns BR, Crabtree GR (1996) Diversity and specialization of mammalian SWI/SNF complexes. Genes Dev 10:2117–2130

Wang Z, Zang C, Rosenfeld JA, Schones DE, Barski A, Cuddapah S, Cui K, Roh TY, Peng W, Zhang MQ et al (2008) Combinatorial patterns of histone acetylations and methylations in the human genome. Nat Genet 40:897–903

Wang P, Lin C, Smith ER, Guo H, Sanderson BW, Wu M, Gogol M, Alexander T, Seidel C, Wiedemann LM et al (2009a) Global analysis of H3K4 methylation defines MLL family member targets and points to a role for MLL1-mediated H3K4 methylation in the regulation of transcriptional initiation by RNA polymerase II. Mol Cell Biol 29:6074–6085

Wang X, Sansam CG, Thom CS, Metzger D, Evans JA, Nguyen PT, Roberts CW (2009b) Oncogenesis caused by loss of the SNF5 tumor suppressor is dependent on activity of BRG1, the ATPase of the SWI/SNF chromatin remodeling complex. Cancer Res 69:8094–8101

Wang JK, Tsai MC, Poulin G, Adler AS, Chen S, Liu H, Shi Y, Chang HY (2010) The histone demethylase UTX enables RB-dependent cell fate control. Genes Dev 24:327–332

Wang K, Kan J, Yuen ST, Shi ST, Chu KM, Law S, Chan TL, Kan Z, Chan AS, Tsui WY et al (2011) Exome sequencing identifies frequent mutation of ARID1A in molecular subtypes of gastric cancer. Nat Genet 43:1219–1223

Wang L, Tsutsumi S, Kawaguchi T, Nagasaki K, Tatsuno K, Yamamoto S, Sang F, Sonoda K, Sugawara M, Saiura A et al (2012) Whole-exome sequencing of human pancreatic cancers and characterization of genomic instability caused by MLH1 haploinsufficiency and complete deficiency. Genome Res 22:208–219

Wang X, Haswell JR, Roberts CW (2013) Molecular pathways: SWI/SNF (BAF) complexes are frequently mutated in cancer-mechanisms and potential therapeutic insights. Clin Cancer Res 20:21–27

Ward PS, Patel J, Wise DR, Abdel-Wahab O, Bennett BD, Coller HA, Cross JR, Fantin VR, Hedvat CV, Perl AE et al (2010) The common feature of leukemia-associated IDH1 and IDH2 mutations is a neomorphic enzyme activity converting alpha-ketoglutarate to 2-hydroxyglutarate. Cancer Cell 17:225–234

Watanabe T, Nobusawa S, Kleihues P, Ohgaki H (2009) IDH1 mutations are early events in the development of astrocytomas and oligodendrogliomas. Am J Pathol 174:1149–1153

Watanabe Y, Castoro RJ, Kim HS, North B, Oikawa R, Hiraishi T, Ahmed SS, Chung W, Cho MY, Toyota M et al (2011) Frequent alteration of MLL3 frameshift mutations in microsatellite deficient colorectal cancer. PLoS ONE 6:e23320

Webster KE, O'Bryan MK, Fletcher S, Crewther PE, Aapola U, Craig J, Harrison DK, Aung H, Phutikanit N, Lyle R et al (2005) Meiotic and epigenetic defects in Dnmt3L-knockout mouse spermatogenesis. Proc Natl Acad Sci USA 102:4068–4073

Weissmann S, Alpermann T, Grossmann V, Kowarsch A, Nadarajah N, Eder C, Dicker F, Fasan A, Haferlach C, Haferlach T et al (2012) Landscape of TET2 mutations in acute myeloid leukemia. Leukemia 26:934–942

Wiegand KC, Shah SP, Al-Agha OM, Zhao Y, Tse K, Zeng T, Senz J, McConechy MK, Anglesio MS, Kalloger SE et al (2010) ARID1A mutations in endometriosis-associated ovarian carcinomas. N Engl J Med 363:1532–1543

Wiestler B, Capper D, Holland-Letz T, Korshunov A, von Deimling A, Pfister SM, Platten M, Weller M, Wick W (2013) ATRX loss refines the classification of anaplastic gliomas and identifies a subgroup of IDH mutant astrocytic tumors with better prognosis. Acta Neuropathol 126:443–451

Wigle TJ, Knutson SK, Jin L, Kuntz KW, Pollock RM, Richon VM, Copeland RA, Scott MP (2011) The Y641C mutation of EZH2 alters substrate specificity for histone H3 lysine 27 methylation states. FEBS Lett 585:3011–3014

Wilson BG, Roberts CW (2011) SWI/SNF nucleosome remodellers and cancer. Nat Rev Cancer 11:481–492

Wilson BG, Wang X, Shen X, McKenna ES, Lemieux ME, Cho YJ, Koellhoffer EC, Pomeroy SL, Orkin SH, Roberts CW (2010) Epigenetic antagonism between polycomb and SWI/SNF complexes during oncogenic transformation. Cancer Cell 18:316–328

Winkler BS, DeSantis N, Solomon F (1986) Multiple NADPH-producing pathways control glutathione (GSH) content in retina. Exp Eye Res 43:829–847

Witt O, Deubzer HE, Milde T, Oehme I (2009) HDAC family: what are the cancer relevant targets? Cancer Lett 277:8–21

Wu JN, Roberts CW (2013) ARID1A mutations in cancer: another epigenetic tumor suppressor? Cancer Discov 3:35–43

Wu H, D'Alessio AC, Ito S, Wang Z, Cui K, Zhao K, Sun YE, Zhang Y (2011) Genome-wide analysis of 5-hydroxymethylcytosine distribution reveals its dual function in transcriptional regulation in mouse embryonic stem cells. Genes Dev 25:679–684

Wu G, Broniscer A, McEachron TA, Lu C, Paugh BS, Becksfort J, Qu C, Ding L, Huether R, Parker M et al (2012) Somatic histone H3 alterations in pediatric diffuse intrinsic pontine gliomas and non-brainstem glioblastomas. Nat Genet 44:251–253

Xia W, Nagase S, Montia AG, Kalachikov SM, Keniry M, Su T, Memeo L, Hibshoosh H, Parsons R (2008) BAF180 is a critical regulator of p21 induction and a tumor suppressor mutated in breast cancer. Cancer Res 68:1667–1674

Xu X, Zhao J, Xu Z, Peng B, Huang Q, Arnold E, Ding J (2004) Structures of human cytosolic NADP-dependent isocitrate dehydrogenase reveal a novel self-regulatory mechanism of activity. J Biol Chem 279:33946–33957

Xu W, Yang H, Liu Y, Yang Y, Wang P, Kim SH, Ito S, Yang C, Wang P, Xiao MT et al (2011) Oncometabolite 2-hydroxyglutarate is a competitive inhibitor of alpha-ketoglutarate-dependent dioxygenases. Cancer Cell 19:17–30

Yamashita Y, Yuan J, Suetake I, Suzuki H, Ishikawa Y, Choi YL, Ueno T, Soda M, Hamada T, Haruta H et al (2010) Array-based genomic resequencing of human leukemia. Oncogene 29:3723–3731

Yamazaki J, Taby R, Vasanthakumar A, Macrae T, Ostler KR, Shen L, Kantarjian HM, Estecio MR, Jelinek J, Godley LA et al (2012) Effects of TET2 mutations on DNA methylation in chronic myelomonocytic leukemia. Epigenetics 7:201–207

Yan H, Parsons DW, Jin G, McLendon R, Rasheed BA, Yuan W, Kos I, Batinic-Haberle I, Jones S, Riggins GJ et al (2009) IDH1 and IDH2 mutations in gliomas. N Engl J Med 360:765–773

Yan XJ, Xu J, Gu ZH, Pan CM, Lu G, Shen Y, Shi JY, Zhu YM, Tang L, Zhang XW et al (2011) Exome sequencing identifies somatic mutations of DNA methyltransferase gene DNMT3A in acute monocytic leukemia. Nat Genet 43:309–315

Yan G, Eller MS, Elm C, Larocca CA, Ryu B, Panova IP, Dancy BM, Bowers EM, Meyers D, Lareau L et al (2013) Selective inhibition of p300 HAT blocks cell cycle progression, induces cellular senescence, and inhibits the DNA damage response in melanoma cells. J Invest Dermatol 133:2444–2452

Yang YA, Yu J (2013) EZH2, an epigenetic driver of prostate cancer. Protein Cell 4:331–341

Yap DB, Chu J, Berg T, Schapira M, Cheng SW, Moradian A, Morin RD, Mungall AJ, Meissner B, Boyle M et al (2011) Somatic mutations at EZH2 Y641 act dominantly through a mechanism of selectively altered PRC2 catalytic activity, to increase H3K27 trimethylation. Blood 117:2451–2459

Yildirim O, Li R, Hung JH, Chen PB, Dong X, Ee LS, Weng Z, Rando OJ, Fazzio TG (2011) Mbd3/NURD complex regulates expression of 5-hydroxymethylcytosine marked genes in embryonic stem cells. Cell 147:1498–1510

Yin R, Mao SQ, Zhao B, Chong Z, Yang Y, Zhao C, Zhang D, Huang H, Gao J, Li Z et al (2013) Ascorbic acid enhances Tet-mediated 5-methylcytosine oxidation and promotes DNA demethylation in mammals. J Am Chem Soc 135:10396–10403

Yokoyama A, Cleary ML (2008) Menin critically links MLL proteins with LEDGF on cancer-associated target genes. Cancer Cell 14:36–46

Yokoyama A, Somervaille TC, Smith KS, Rozenblatt-Rosen O, Meyerson M, Cleary ML (2005) The menin tumor suppressor protein is an essential oncogenic cofactor for MLL-associated leukemogenesis. Cell 123:207–218

You JS, Jones PA (2012) Cancer genetics and epigenetics: two sides of the same coin? Cancer Cell 22:9–20

Yuen BT, Knoepfler PS (2013) Histone h3.3 mutations: a variant path to cancer. Cancer Cell 24:567–574

Zang ZJ, Cutcutache I, Poon SL, Zhang SL, McPherson JR, Tao J, Rajasegaran V, Heng HL, Deng N, Gan A et al (2012) Exome sequencing of gastric adenocarcinoma identifies recurrent somatic mutations in cell adhesion and chromatin remodeling genes. Nat Genet 44:570–574

Zhang J, Ding L, Holmfeldt L, Wu G, Heatley SL, Payne-Turner D, Easton J, Chen X, Wang J, Rusch M et al (2012) The genetic basis of early T-cell precursor acute lymphoblastic leukaemia. Nature 481:157–163

Zhao S, Lin Y, Xu W, Jiang W, Zha Z, Wang P, Yu W, Li Z, Gong L, Peng Y et al (2009) Glioma-derived mutations in IDH1 dominantly inhibit IDH1 catalytic activity and induce HIF-1alpha. Science 324:261–265

Zhao H, Wang J, Han Y, Huang Z, Ying J, Bi X, Zhao J, Fang Y, Zhou H, Zhou J et al (2011) ARID2: a new tumor suppressor gene in hepatocellular carcinoma. Oncotarget 2:886–891

Zou P, Xu H, Chen P, Yan Q, Zhao L, Zhao P, Gu A (2013) IDH1/IDH2 mutations define the prognosis and molecular profiles of patients with gliomas: a meta-analysis. PLoS ONE 8:e68782

6

Human INO80/YY1 chromatin remodeling complex transcriptionally regulates the BRCA2- and CDKN1A-interacting protein (BCCIP) in cells

Jiaming Su[1], Yi Sui[1], Jian Ding[1], Fuqiang Li[1], Shuang Shen[1], Yang Yang[1], Zeming Lu[1], Fei Wang[1], Lingling Cao[1,5], Xiaoxia Liu[2], Jingji Jin[1,3,4]✉, Yong Cai[1,3,4]✉

[1] School of Life Sciences, Jilin University, 2699 Qianjin Street, Changchun 130012, China
[2] Department of Gynecologic Oncology, The First Clinical Hospital of Jilin University, 71 Xinmin Street, Changchun 130021, China
[3] National Engineering Laboratory for AIDS Vaccine, Jilin University, 2699 Qianjin Street, Changchun 130012, China
[4] Key Laboratory for Molecular Enzymology and Engineering, The Ministry of Education, Jilin University, 2699 Qianjin Street, Changchun 130012, China
[5] The first Hospital and Institute of Epigenetic Medicine, Jilin University, Changchun 130061, China
✉ Correspondence: jjjin@jlu.edu.cn (J. Jin), caiyong62@jlu.edu.cn (Y. Cai)

ABSTRACT

The BCCIP (BRCA2- and CDKN1A-interacting protein) is an important cofactor for BRCA2 in tumor suppression. Although the low expression of *BCCIP* is observed in multiple clinically diagnosed primary tumor tissues such as ovarian cancer, renal cell carcinoma and colorectal carcinoma, the mechanism of how BCCIP is regulated in cells is still unclear. The human INO80/YY1 chromatin remodeling complex composed of 15 subunits catalyzes ATP-dependent sliding of nucleosomes along DNA. Here, we first report that *BCCIP* is a novel target gene of the INO80/YY1 complex by presenting a series of experimental evidence. Gene expression studies combined with siRNA knockdown data locked candidate genes including BCCIP of the INO80/YY1 complex. Silencing or over-expressing the subunits of the INO80/YY1 complex regulates the expression level of BCCIP both in mRNA and proteins in cells. Also, the functions of INO80/YY1 complex in regulating the transactivation of BCCIP were confirmed by luciferase reporter assays. Chromatin immunoprecipitation (ChIP) experiments clarify the enrichment of INO80 and YY1 at +0.17 kb downstream of the *BCCIP* transcriptional start site. However, this enrichment is significantly inhibited by either knocking down INO80 or YY1, suggesting the existence of both INO80 and YY1 is required for recruiting the INO80/YY1 complex to *BCCIP* promoter region. Our findings strongly indicate that *BCCIP* is a potential target gene of the INO80/YY1 complex.

KEYWORDS BCCIP, human INO80, chromatin remodeling complex

INTRODUCTION

The BRCA2- and CDKN1A (Cip1, p21)-interacting protein BCCIP is alternatively spliced α and β isoforms (Liu et al. 2001; Ono et al. 2000). Both isoforms share N-terminal acidic domain (NAD, include 258 amino acids) and the internal conserved domain (ICD) (Liu et al. 2001). As an important cofactor for BRCA2 in tumor suppression, BCCIP has been implicated in many important cellular processes with obvious links to cancer. Knocking down BCCIP with siRNA in cells leads to defective DNA damage repair (Lu et al. 2005), abnormal cell cycle (Meng et al. 2004; Meng et al. 2004) and genomic instability (Meng et al. 2007). Recently, other colleagues and we have reported that *BCCIP* was down-regulated in several cancer tissues such as renal cell carcinoma, ovarian cancer and colorectal carcinoma (Meng et al. 2003; Liu et al. 2013). Thus, it is important to clarify the role of BCCIP in tumorigenesis, especially to know how the BCCIP is regulated in cells.

Yin Yang 1 (YY1), a member of the GLI-Krüppel class proteins, was first discovered as a DNA binding protein (Shi et al. 1991; Seto et al. 1991; Park and Atchison 1991; Hariharan et al. 1991). It is a ubiquitously expressed and evolutionarily conserved protein in human cells. Domain research found that YY1 includes not only an activation domain, but also contains a repression domain (Thomas and Seto 1999; Shi et al. 1997). In subsequent studies, YY1 has been clarified as a versatile protein which can either repress or activate gene transcription by recruiting different cofactors such as histone deacetylases (HDACs), methyltransferase enhancer of zeste homolog 2 (Ezh2), CREB-binding protein (CBP), and P300/CBP-associated factor (PCAF) (Yao et al. 2001; Lee et al. 1995).

Using biochemical purification approaches, we previously verified that YY1 is tightly associated with the human INO80 (INO80) chromatin remodeling complex (Jin et al. 2005). All evolutionarily conserved subunits (include actin-related proteins Arp4, Arp5, Arp8, Tip49a and Tip49b AAA$^+$ ATPases, and hIes2 and hIes6) assembled on the conserved helicase-SANT-associated/post-HSA (HSA/PTH) and ATPase domains of INO80 protein (Jin et al. 2005). Both HSA/PTH and ATPase domains in INO80 protein are essential for catalyzing the ATP-dependent nucleosome remodeling activity of the INO80 complex. Based on YY1 with Arp4 and Arp8 together participate in assembling helicase-SANT-associated/post-HSA (HSA/PTH) module, suggesting that like other conserved subunits, YY1 is also essential for maintaining the ATP-dependent nucleosome remodeling activity of the INO80 complex (Chen et al. 2011). Experimental evidence indicates that *yeast* Ino80 is required for proper transcriptional regulation of many target genes (Morrison and Shen 2009), while in *Drosophila*, INO80 facilitates transcriptional repression of ecdysone-regulated genes during prepupal development (Neuman et al. 2014). Recent experimental results demonstrate that INO80 is also required in ESC self-renewal, somatic cell reprogramming, and blastocyst development (Wang et al. 2014). Although genome-wide studies show that INO80 binds many, but not all, loci in the genome (Moshkin et al. 2012), it is unclear whether there is a direct correlation between INO80 binding and target gene expression levels. One possibility is that YY1 as a transcription factor helps INO80 complex to recognize INO80-target genes and recruits the complex to some target genes such as CDC6 and GRP78 (Cai et al. 2007). In this study, using molecular and cell biology approaches, we first show that the BCCIP might be a novel target gene of the INO80 /YY1 chromatin remodeling complex. Our results will provide a strong theoretical basis to clucidate the BCCIP functions in cells.

RESULTS

BCCIP was selected as a candidate target gene of the INO80/YY1 complex from gene expression profiles

Human INO80 complex is one of the most highly conserved chromatin remodelers. Eight core subunits (Arp5, Arp8, TIP49a/b, Ies2, Ies6, Arp4, and YY1) are evolutionary conserved and form an enzyme core including HSA and SNF2 modules. Except for conserved subunits, INO80 complex contains 6 metazoan-specific subunits which all assemble on N-terminus of INO80 protein and form an N-terminal regulatory module (Jin et al. 2005; Chen et al. 2011) (Fig. 1A). Increasing evidence suggests the functions of INO80/YY1 complex in gene transcriptional regulation (Morrison and Shen 2009; Conaway and Conaway 2009), but the precise mechanisms are still unclear. To investigate the target genes of the INO80/YY1 complex, total RNA from HeLa cells with specific siRNA (siINO80, siArp8, Arp5, siIes2 and siIes6) knocked down (Fig. 1B) were sent to EMTD Science and Technology Development Co., Ltd. (Beijing, China) for DNA microarray. As shown in Fig. 1C, a total of 1932, 1445, 1235, 2707 and 2159 genes were differentially expressed among INO80, Arp8, Arp5, hIes6 or hIes2 and NT siRNA knockdown HeLa cells, respectively. Hundreds of overlapping genes are found to be regulated by INO80, Arp8, Arp5, hIes6 and hIes2, which are components of the HSA and SNF2 module. On the other hand, a total of 602 genes were co-regulated by INO80 and Arp8 which participate in assembling HSA module (data not shown). Selected overlapping co-regulated genes (8 down and 6 up) in INO80/YY1 complex knockdown HeLa cells are shown in Table 1. mRNA from INO80- and Arp8-siRNA knockdown cells (Fig. 1E) was measured with RT-qPCR (Fig. 1F). Compared to NT siRNA control, selected gene mRNA including BCCIP, TNFRSF21, and BRMS1L was down-regulated with siRNA knockdown of INO80 and Arp8 in HeLa cells. In contrast, RAF1 and PTTG1IP mRNAs was up-regulated. However, there was no change of mRNA of RAB24, RAB22A, and ST7L. Table 2 compared the results of microarray illumina and RT-qPCR. In addition, differentially expressed genes (DEGs) in INO80/YY1 complex knocked down cells were used for KEGG annotation by using DAVID web annotation tool. DEG-enriched pathways in INO80 complex knockdown cells were shown previously (Chen et al. 2011). Fig. 1D shows the differentially regulated microarray genes in cancer pathway after knocking down the indicated subunits of INO80/YY1 complex.

The expression of BCCIP was regulated by INO80/YY1 complex in HeLa and 293T cells

UCSC (The University of California-Santa Cruz) Genome Browser at http://genome.ucsc.edu provides timely, convenient access to a database of high-quality genome sequence

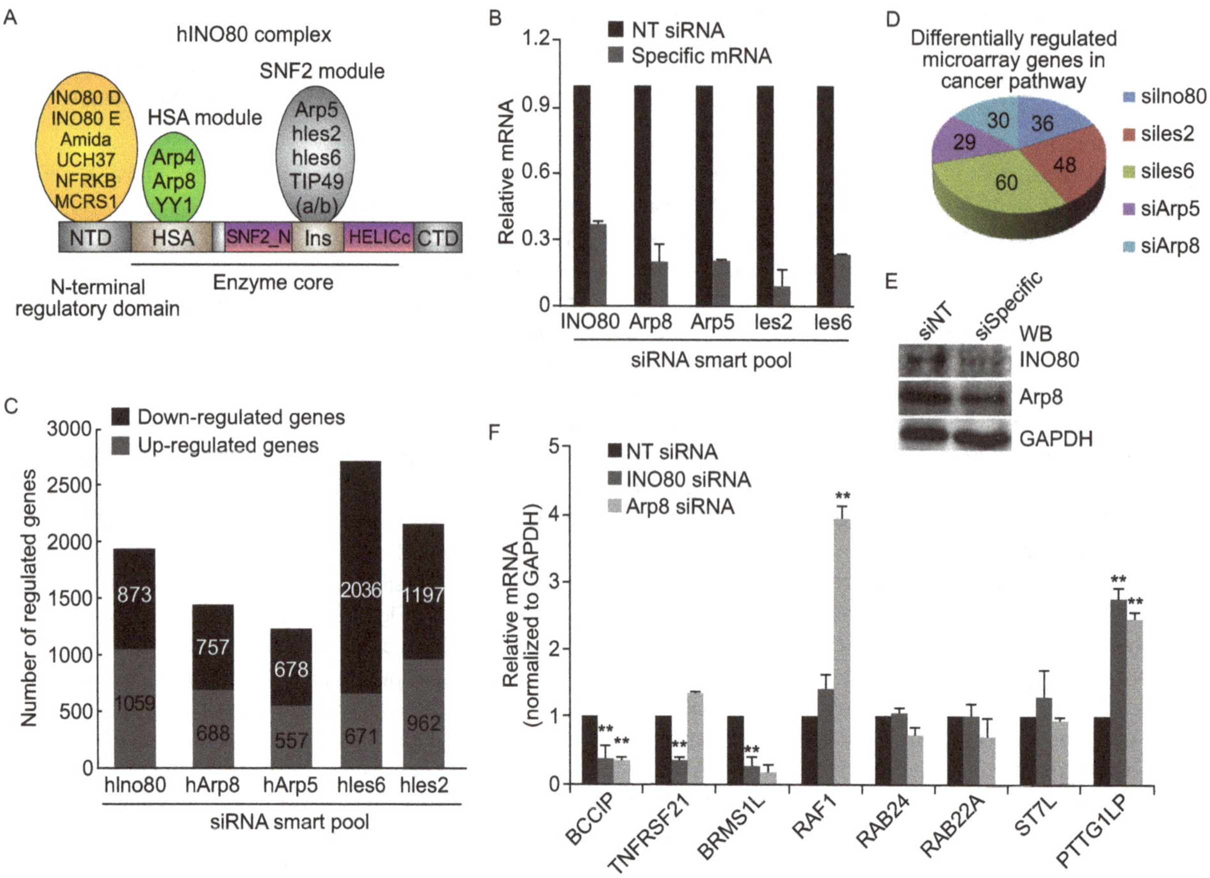

Figure 1. BCCIP is found in overlapping expressed gene profiles in INO80/YY1 complex-knockdown HeLa cells. (A) Functional domains of INO80/YY1 complex. NTD, N-terminal domain; HSA, helicase-SANT-associated domain; SNF2_N, SNF2 family N-terminal domain; Ins, insertion domain; HELICc, Helicase C-terminal domain; CTD, C-terminal domain. (B) Relative mRNA levels in siINO80, siArp8, siArp5, siIes2, and siIes6 knockdown HeLa cells. (C) Gene expression profiles in specific-siRNA knockdown Hela cells. Stacked column chart represents the total number of regulated genes in indicated specific siRNA knockdown HeLa cells. Each bar in the stacked column chart is composed of down- and up-regulated genes. (D) Differentially regulated microarray genes in cancer pathway in INO80/YY1 complex knockdown HeLa cells. (E) Knockdown efficiency. HeLa cells were transfected with 15 pmol INO80- or hArp8-siRNA and siNT (as control). (F) Verification of the mRNA of select genes from gene expression profiles. 48 h after siRNA transfection, RT-qPCR was performed to assess the relative mRNA levlel (n = 3). Bar graphs show ratios of RT-qPCR signals to GAPDH (all signals normalized to siNT). **P < 0.01 in comparison with siNT control (Student t-test).

and annotation (Karolchik et al. 2013). Based on UCSC database (ChIP-Seq) search, we found that YY1 was restricted to the *BCCIP* transcriptional start site proximal region in different types of cell lines including human lung carcinoma type II epithelium-like A549, human liver hepatocellular carcinoma HepG2 and human colon cancer HCT116. To clarify this observation, we first examined the BCCIP expression level in YY1-siRNA knockdown HeLa cells. As shown in Fig. 2A, dose-dependent down-regulation of *BCCIP* (lower panel) in YY1-siRNA knockdown (upper panel) HeLa cells was confirmed by qPCR. Furthermore, low protein levels of BCCIP in YY1-siRNA knockdown HeLa cells were also detected by western blot (Fig. 2B) and immunofluorescence staining (Fig. 2C) approaches, respectively. Given that YY1 is a subunit of the INO80 chromatin remodeling complex, the implication of INO80 complex in regulating BCCIP is speculated. To determine whether BCCIP is a potential target gene of the INO80/YY1 complex, we evaluated the BCCIP expression level in INO80- and Arp8- (evolutionary conserved subunit of the INO80 complex) siRNA knockdown HeLa cells. Declined protein levels of BCCIP by knocking down INO80 and Arp8 in HeLa cells were presented by western blot (Fig. 2E and 2F) and immunofluorescence staining (Fig. 2D), suggesting the involvement of INO80/YY1 complex in regulating the BCCIP expression.

Table 1. Select overlapping co-regulated genes in INO80/YY1 knockdown gene expression profiles

Gene symbol	Gene ID	Gene description	Up or Down	Fold change
AMHR2	NM_020547	Anti-mullerian hormone receptor, type II	Down	−3.58
BCCIP	NM_016567	BRCA2 and CDKN1A interacting protein	Down	−3.62
CDH1	NM_0043609	Cadherin 1, type 1	Down	−4.97
CCL28	NM_019846	Chemokine (C-C motif) ligand 28	Down	−2.61
FGFBP1	NM_005130	Fibroblast growth factor binding protein1	Down	−3.65
GRTP1	NM_024719	Growth hormone regulated TBC protein 1	Down	−3.21
TPD52	NM_001025252	Tumor protein D52	Down	−2.35
WTIP	NM_059037	Wilms tumor1 interacting protein	Down	−3.15
CDKN1A (p21)	NM_078467	Cyclin-dependent kinase inhibitor 1A	Up	6.92
EGFR	NM_005228	Epidermal growth factor receptor	Up	9.5
GAS1	NM_002048	Growth arrest-specifc 1	Up	4.0
MBD1	NM_015845	Methy-CpG binding domain protein 1	Up	5.26
RAF1	NM_002880	v-raf1 murine leukemia viral oncogene homolog1	Up	4.41
TGFBR1	NM_004612	Transforming growth factor, beta receptor 1	Up	4.56

Genes shown in table are the selected overlapping expressed genes in at least three subunits of INO80/YY1 complex knockdown HeLa cells.

Table 2. Comparison of the results of the microarray-illumina and qPCR

Gene symbol	Gene ID	siINO80		siArp8	
		Illumina	qPCR	Illumina	qPCR
BCCIP	NM_016567	Down	Down	Down	Down
TNFRSF21	NM_014452	Down	Down	Down	–
BRMS1L	NM_032352	Down	Down	Down	Down
RAF1	NM_002880	Up	–	Up	Up
RAB24	NM_001031677	Up	–	Up	–
RAB22A	NM_020673	Down	–	Down	–
ST7L	NM_138729	Up	–	Up	–
PTTF1IP	NM_004339	Up	Up	Up	Up

Co-regulated genes selected from different specific subunit of the INO80/YY1 complex knockdown gene expression profiles are tested by quantitative PCR (qPCR). The performed qPCR results were then compared with DNA microarray illumine data.

Human INO80 facilitates the transactivation of BCCIP

To further explore whether the INO80/YY1 chromatin remodeling complex affects the transactivation of BCCIP, three designed BCCIP promoter region was sub-cloned into pGL4 luciferase vector (Fig. 3A). Constructed pGL4-BCCIP-Luc plasmids containing different BCCIP promoter region were first tested in dual luciferase assay (Fig. 3B). Then, the impact of INO80 on BCCIP transactivation was estimated by co-transfection of pGL4-BCCIP-Luc with Flag-INO80 plasmids. As shown in Fig. 3C, in contrast to a basal-level luciforase activity, co-transfection of pGL4-BCCIP-Luc and INO80 dose-dependently increased BCCIP-luciferase activity in all cases, indicating the roles of INO80 complex in regulating the transactivation of BCCIP.

Chromatin remodeling activity of the INO80 complex might be associated with regulation of BCCIP expression

To investigate the effects of INO80 complex on BCCIP-mediated transactivation, pGL4-BCCIP-Luc vector containing 391 bp promoter region (−330 ~+61 bp) was used in subsequent experiments. As shown in Fig. 4A, co-transfection of pGL4-BCCIP-Luc vector with Flag-tagged INO80, Arp5, and Arp8 increased not only luciferase activities (data not shown) but also intracellular BCCIP proteins in a dose-dependent manner. Compared to pcDNA3.1 control group, statistically significant differences were found in INO80 ($P < 0.01$ at 0.2, 0.4, and 0.8 μg of plasmids, respectively), Arp5 ($P < 0.05$ at 0.8 μg of plasmids), or Arp8 ($P < 0.01$ at 0.4 and 0.8 μg of

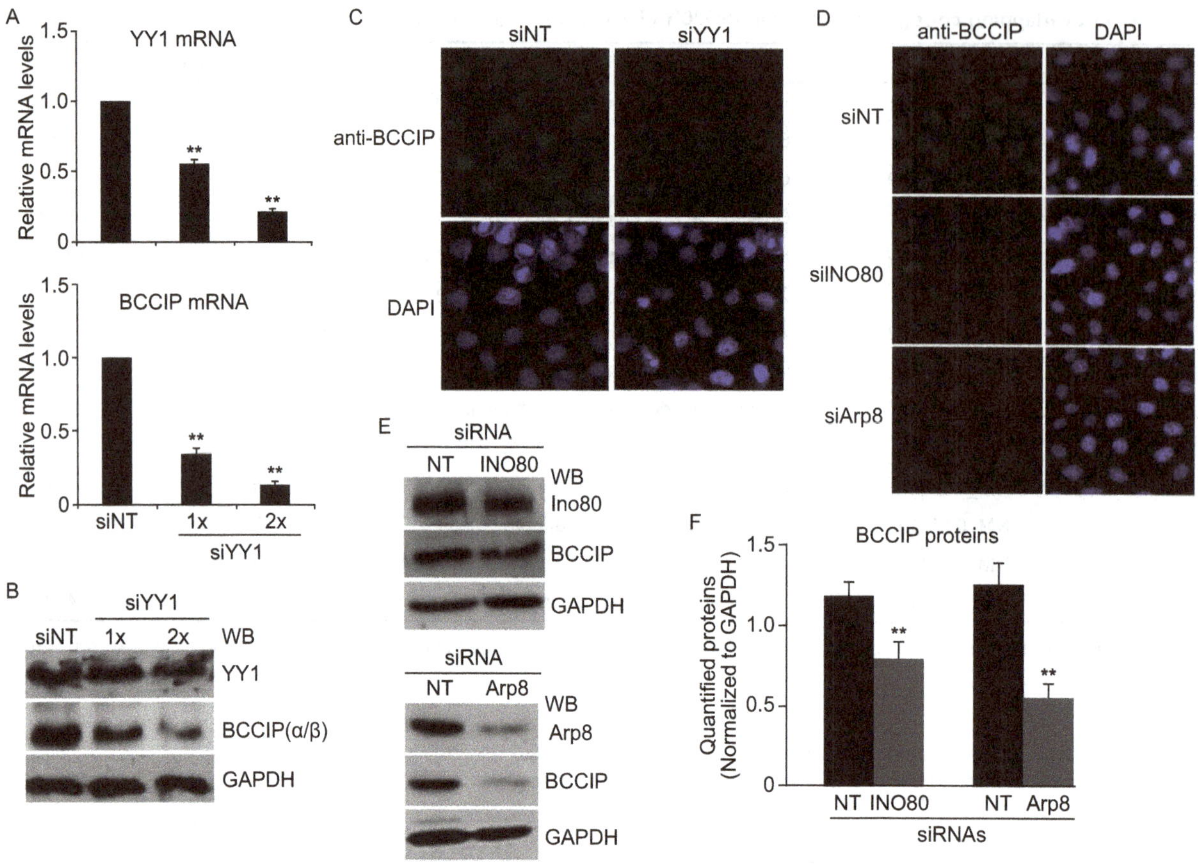

Figure 2. The INO80/YY1 chromatin remodeling complex is implicated in regulating BCCIP expression in HeLa cells. (A) Down regulation of BCCIP mRNA in siYY1 knockdown HeLa cells. Relative mRNA levels of YY1 and BCCIP were measured by RT-qPCR. Each bar represents the mean of 3 independent experiments. ***P* < 0.01 in comparison with siNT-transfected control (Student *t*-test). (B) Reduction of BCCIP protein levels by knocking-down YY1 in HeLa cells. Whole cell lysate from siYY1 knockdown cells was analyzed by western blot. Specific proteins were detected by indicated antibodies (GAPDH as internal control). Anti-BCCIP antibody recognizes both α- and β- isoform of BCCIP. (C and D) Immunofluorescence stained BCCIP proteins. HeLa cells were transfected with YY1-, INO80-, and Arp8 siRNA. 48 h later, immunofluorescence staining was carried out using BCCIP antibody. DAPI staining shows total nuclei. (E) Western blot analysis of BCCIP proteins in INO80 (upper) and Arp8 (down) knocked down HeLa cells. Prepared whole cell lysate from INO80- or Arp8- siRNA knocked down HeLa cells were analyzed by western blot. Quantified protein levels were shown in (F). ***p* < 0.01 in comparison with siNT-transfected control (Student *t*-test).

plasmids, respectively) transfected 293T cells (Fig. 4B), suggesting the importance of enzyme core including HSA and SNF2 modules of the INO80 complex in regulating BCCIP expression. To further confirm this result, point or delete mutations of INO80 protein were designed (Fig. 4C). BCCIP protein levels in overexpressed INO80wt, INO80mt, INO80 N-HSA, and INO80CTD proteins in 293T cells were analyzed by western blot with anti-BCCIP antibody. Consistent with previous results, increased tendency of both BCCIPα and BCCIPβ was detected in INO80wt over-expressed 293T cells (Fig. 4D, left panel). However, this phenomenon is reversed by over-expression of INO80mt (E653Q, a DEAD/H box point mutation) which interfere with INO80 ATPase activity. As a result, BCCIP proteins are gradually decreased by increasing the transfection amount of INO80mt (Fig. 4D, second panel from the left), demonstrating the importance of ATPase activity of the INO80 complex in BCCIP expression. In addition, except for BCCIPα at 0.8 μg transfection of INO80 N-HSA (*P* < 0.05), little changes of BCCIP proteins were observed in INO80 N-HSA transfected 293T cells (Fig. 4D, third panel from the left). Furthermore, co-transfection of pGL4-BCCIP-Luc vector with INO80CTD, decreasing tendency of both BCCIPα and BCCIPβ was represented in a dose dependent manner, and the statistically significant differences were obtained at 0.8 μg transfection of INO80CTD (Fig. 4D, right panel). Above experimental results indicate that enzymatic activities of the INO80 complex might be associated with the regulation of BCCIP expression. Quantified proteins of BCCIPα and BCCIPβ for Fig. 4D are shown in Fig. 4E.

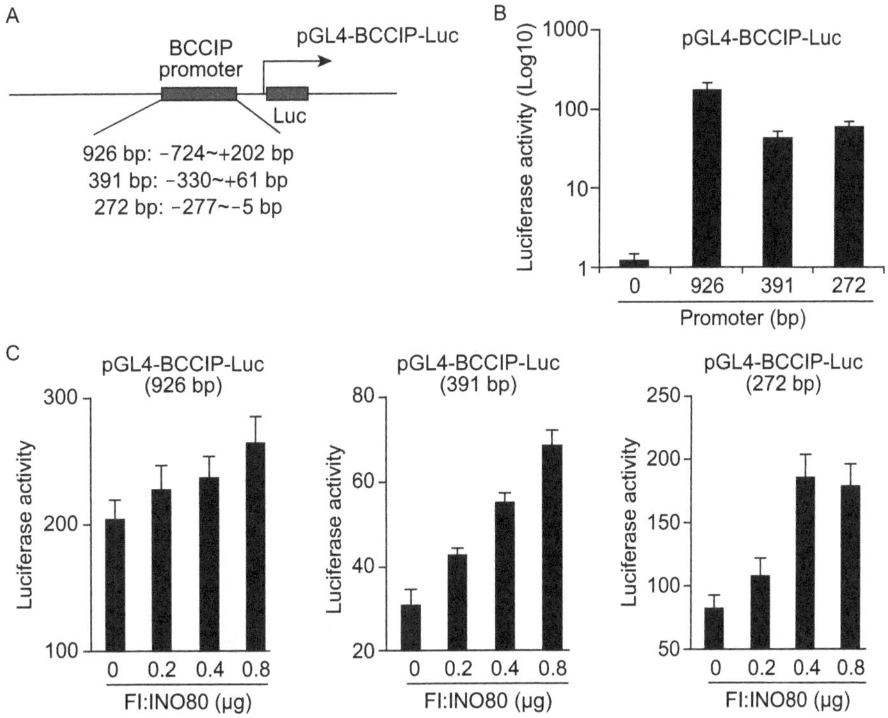

Figure 3. Transactivation of BCCIP is facilitated by INO80/YY1 chromatin remodeling complex. (A) Schematic of construction of pGL4-BCCIP-Luc plasmids. A length of 926 bp (−724 ~+202 bp), 391 bp (−330 ~+61 bp), and 272 bp (−277 ~ −5 bp) region of BCCIP promoter was introduced into pGL4-Luc vector. The luciferase activity of each plasmid was shown in B. (C) A dual luciferase assay. pGL4-BCCIP-Luc vectors and INO80wt plasmid (0.2, 0.4 and 0.8 μg) were co-transfected into 293T cells. 48 h later, luciferase activity in the lysates was determined by a dual luciferase assay. Error bars represent data from three independent experiments.

INO80/YY1 chromatin remodeling complex is restricted to the BCCIP transcriptional start site proximal region

Data in the previous experiment clearly show that BCCIP level in both mRNA and protein, and the transactivation of BCCIP in cells is regulated by INO80/YY1 chromatin remodeling complex. To further determine whether the *BCCIP* is transcriptionally controlled by INO80/YY1 complex, indicated primer sets were designed to amplify ChIP DNA (Fig. 5A, upper panel). Amplified ChIP DNA by each specific primer set for BCCIP promoter proximal region as well as the region far away from the transcriptional start site was confirmed by qPCR, and visualized in 2.5% DNA agarose with Ethidium bromide (Fig. 5A, lower panel). Human INO80 and YY1 specific antibodies were then used in ChIP assays. As shown in Fig. 5B, both INO80 (upper) and YY1 (lower) are recruited at +0.17 kb downstream of the *BCCIP* transcriptional start site, indicating the co-localization of INO80 and YY1 on the *BCCIP* promoter region. Further analysis of performed qPCR products on 2.5% DNA agarose gel acquired a significant enrichment both in INO80 and YY1 ChIP at +0.17 kb downstream of the *BCCIP* transcriptional start site (Fig. 5C). However, knockdown INO80 by transfecting pBS-shINO80 (Fig. 6A) in 293T cells significantly blocked the recruitment of INO80 and YY1 at the +0.17 kb downstream of the *BCCIP* transcriptional start site ($P < 0.01$ in both cases), confirming the binding of INO80/YY1 complex at *BCCIP* promoter region (Fig. 6B). Similarly, silencing YY1 by siRNA (Fig. 6C) in 293T cells inhibited the recruitment of INO80 at the +0.17 kb downstream of the *BCCIP* transcriptional start site ($P < 0.01$), but not at +2.8 bp far away from the downstream of the *BCCIP* transcriptional start site, suggesting the requirement of YY1 for binding of INO80 to *BCCIP* promoter region (Fig. 6D).

DISCUSSION

It has been known that the INO80 complex is highly conserved from yeast to human. Evolutionarily conserved subunits including YY1 form an enzyme core which is composed of two modules (HSA and SNF2) assembled on the conserved HSA/PTH and ATPase domains of the INO80 protein (Fig. 1A). Although the precise functions of those two modules remain to be explored, actin-related protein Arp4 and Arp8 have been reported to bind to DNA and histones, suggesting the contributions of the INO80 complex in recognition of DNA and/or nucleosome substrates (Seeber et al. 2013; Shen et al. 2003). On the other hand, recognition of DNA and/or histones by subunits of INO80 complex is

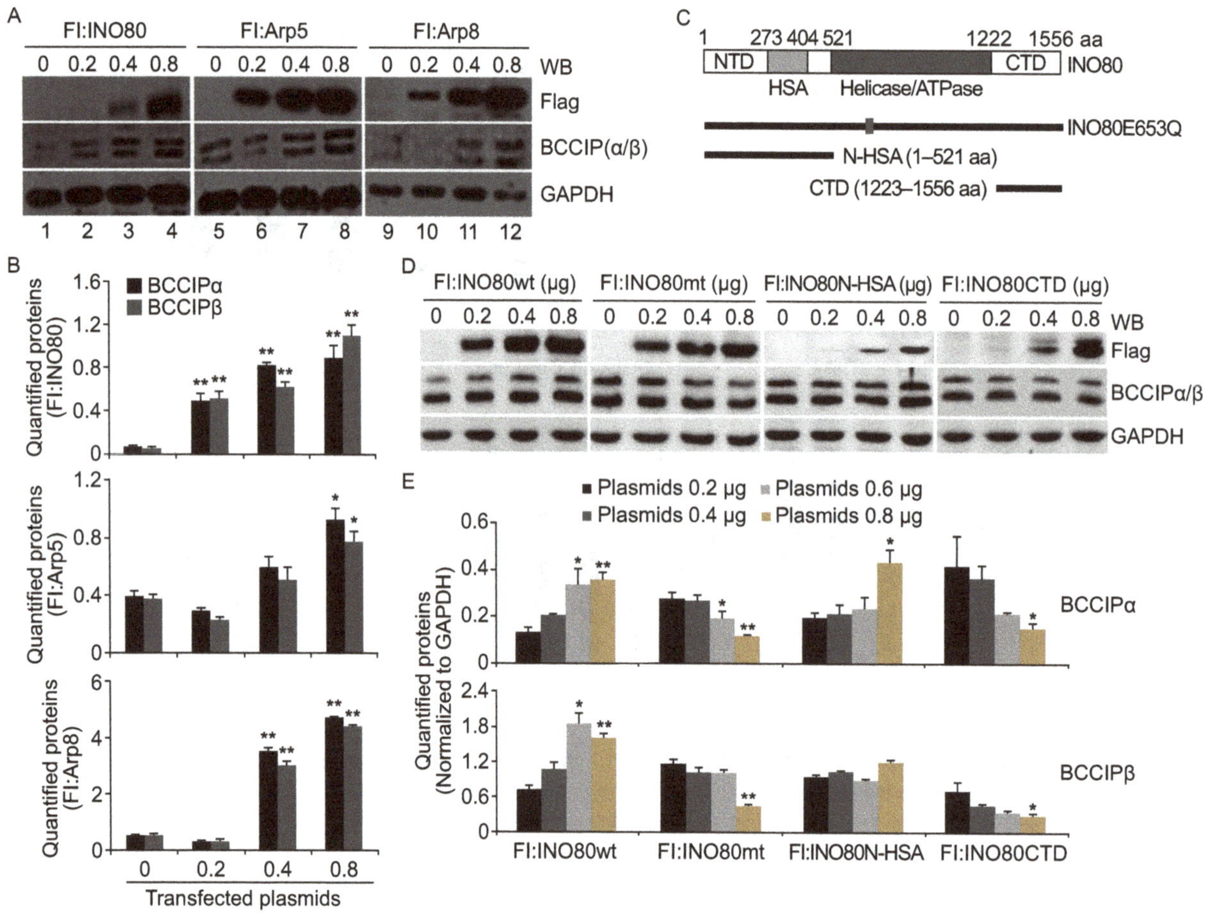

Figure 4. Chromatin remodeling activity of the INO80/YY1 complex might be important for regulating the BCCIP expression. (A) Facilitation of BCCIP expression by INO80/YY1 complex. 293T cells were co-transfected with pGL4-BCCIP (contains 391 bp promoter region) -Luc vector and Flag tagged INO80, Arp5, and Arp8 pcDNA plasmids (0.2, 0.4 and 0.8 μg). 48 h later, Cells were harvested and whole cell lysate was prepared. BCCIP proteins were checked by western blot with anti-BCCIP antibody which can recognize both BCCIPα and BCCIPβ. Here shows representative figures of western blot analysis. (B) Quantified proteins. Western blot images from (A) experiments were quantified with densitometry using Quantity One Basic software (BioRad). Bar graph shows the standard error of the mean of three independent experiments. *P < 0.05, *P < 0.01 in comparison with pcDNA3.1-vector transfected control (Student *t*-test). (C) Point or delete mutation of INO80. (D) Effect of INO80 protein on BCCIP expression. pGL4-BCCIP-Luc (contains 391 bp promoter region)vector was co-transfected with Flag tagged INO80wt, INO80mt (E653Q), INO80 N-HSA, and INO80CTD pcDNA plasmids (0.2, 0.4 and 0.8 μg) into 293T cells. Prepared whole cell lysate was subjected to western blot analysis. (E) Quantified proteins. Western blot images from (D) three independent experiments were scanned and quantified with densitometry using Quantity One Basic software (BioRad). *P < 0.05, *P < 0.01 in comparison with pcDNA3.1-vector transfected control (Student *t*-test).

beneficial to the ATP-dependent nucleosome remodeling activity, thereby regulating gene expression through chromatin structure alteration. Based on the analysis of performed gene expression profiles from INO80-, Arp8-, Arp5-, hIes6- and hIes2-knockdown HeLa cells, we not only clarified that over thousands of genes were regulated in each gene expression profile, but also found that hundreds of genes were co-regulated by silencing the subunits which composed of SNF2 (Arp5, Ies6 and Ies2) and HSA (Arp8) modules (Cao et al. 2015). Also, enrichment of differentially expressed genes in multi-KEGG pathways including pathways in cancer was confirmed by KEGG pathway enrichment analysis, indicating the importance of the INO80 complex in cellular biological processes such as cancer pathway (Fig. 1D) (Chen et al. 2011).

We previously reported that zinc-finger transcription factor YY1 is tightly associated with human INO80 chromatin remodeling complex, and with actin-related proteins (Arp4 and Arp8) together, YY1 assembles HSA module on the HSA/PTH domain of INO80 protein, demonstrating the roles of YY1 in maintaining the nucleosome remodeling activity of the INO80 complex (Chen et al. 2011; Chen et al. 2013).

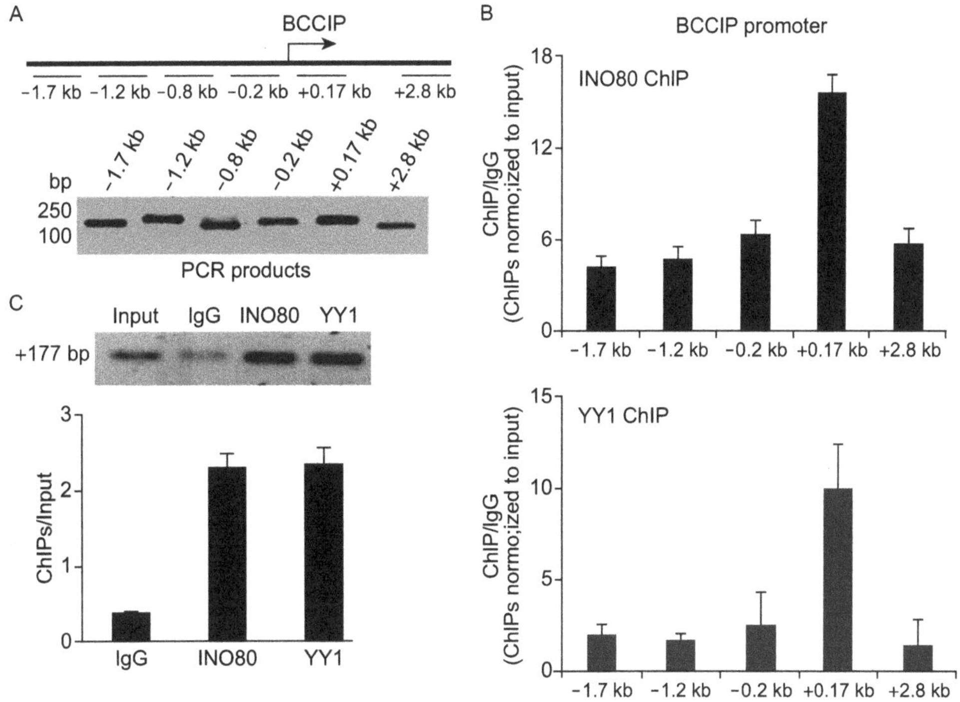

Figure 5. INO80 co-localize with YY1 at the BCCIP promoter. (A) Seven primer sets designed for amplifying ChIP DNA (upper panel). PCR products from each primer set were confirmed using 2.5% agarose gel (lower panel). (B) Co-occupying of INO80 and YY1 at the *BCCIP* promoter region. ChIP assays were performed using INO80 (upper panel) or YY1 (lower panel) antibodies. ChIP DNA was analyzed by qPCR. Bar graph shows the ratios of ChIP DNA signals (normalized to input) to IgG (also normalized to input). Error bars represent the standard error of the mean of three independent experiments. (C) Enrichment of INO80/YY1 at the +0.17 kb downstream of the BCCIP transcriptional start site. qPCR product at +0.17 kp downstream of BCCIP transcriptional start site was visualized by ethidium bromide (upper panel) in 2.5% agarose gel. Quantified PCR signals (Quantity One software) were analyzed by *t*-test (lower panel). Error bars indicate mean ± SE, and the significant difference is expressed as ***P* < 0.01 (Student *t*-test).

More importantly, as an essential co-activator, YY1 can recruit the INO80 complex to some target genes such as CDC6 and GRP78 (Cai et al. 2007). Based on our gene expression profiles and screening experiments, *BCCIP* was chosen as one of the potential candidates of target genes of the INO80 complex. Based on database search in UCSC Genome Browser, we found that YY1 is restricted at the BCCIP transcriptional start site proximal region in different types of cancer cell lines including A549, HepG2 and HCT116, suggesting the functions of YY1 or INO80/YY1 complex in regulating the expression of *BCCIP*.

In this report, we first verified that *BCCIP* is a novel target gene of the INO80 chromatin remodeling complex by presenting a series of experimental evidence. The levels of BCCIP both in mRNA and protein were not only regulated by silencing- or over-expressing- YY1 (Fig. 2), but also were changed by knocking down INO80 and Arp8 or over-expression of INO80, Arp5, and Arp8 HeLa cells (Figs. 2 and 4A), suggesting the roles of INO80/YY1 complex, but not limited to YY1, in regulating the BCCIP expression. ChIP assays further verified that INO80 and YY1 co-localize the *BCCIP* promoter region, and the binding site to BCCIP gene was enriched at +0.17 kb downstream of the transcriptional start site (Fig. 5). However, this enrichment was significantly inhibited in either INO80- or YY1- knocking down 293T cells (Fig. 6), suggesting that (i) INO80/YY1 complex binds to the *BCCIP* promoter region; (ii) binding of YY1 to the *BCCIP* promoter region requires INO80; (iii) INO80 complex is required for YY1 to gain access to target gene promoters and to activate transcription of target genes.

In summary, we first demonstrate that BCCIP might be a novel target gene of the human INO80/YY1 chromatin remodeling complex. This result will provide a theoretical basis for further understanding and elucidating the functions of BCCIP in tumorigenesis and cellular processes.

MATERIALS AND METHODS

Antibodies

Anti-BCCIP (16043-1-AP) antibody was obtained from Proteintech Group (China, Wuhan). Anti-YY1 (sc-1703X) antibody, rabbit IgG (sc-2027) and mouse IgG (sc-2025) were purchased from Santa Cruz Biotechnology (U.S.A.). Anti-Flag (M2)-Agarose and

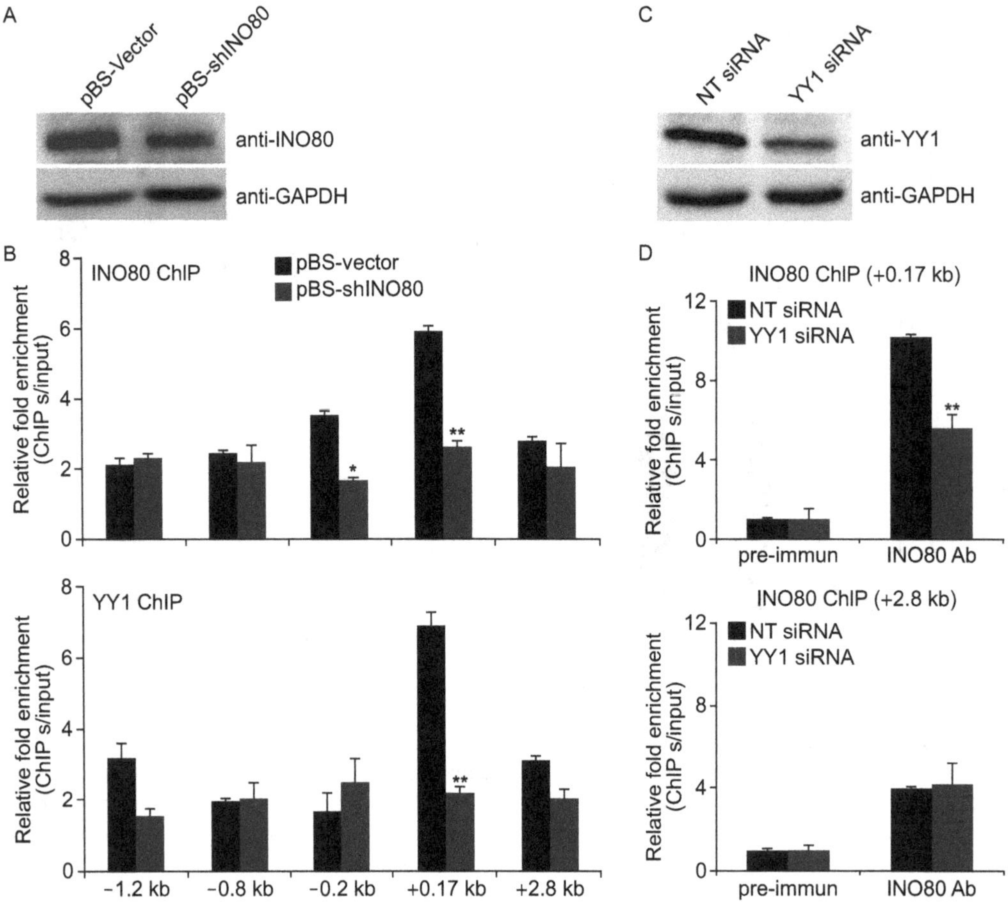

Figure 6. The INO80 chromatin remodeling complex may activate BCCIP gene through YY1. INO80- and YY1- ChIP in siINO80 knockdown 293T cells were performed. INO80 knockdown efficiency was shown as A. ChIP DNA was analyzed by qPCR (B). Y-axis shows the ratio of ChIP DNA signals to IgG. All ChIPs were normalized to input. Error bars represent mean ± SE (n = 3). * $P < 0.05$, ** $P < 0.01$ in comparison with pBS empty vector control (Student t-test). (C) Confirmation of YY1 knockdown. (D) INO80 ChIP in YY1-knockdown 293T cells. INO80 ChIP was carried out in YY1 knocked down 293T cells. +0.17 kb and +2.8 kb promoter primer sets were used in ChIP experiments.

monoclonal antibody (F3165) were from Sigma (U.S.A.). Anti-INO80 (NM_017553, residue 1–526 aa), anti-Arp8 (NM_022899, full length), and anti-GAPDH (NM_002046, full length) rabbit polyclonal antibodies were raised against bacterially expressed proteins (Jilin University).

Cell culture/maintenance

HEK293T (human embryonic kidney) and HeLa (human cervical cancer) cells were grown on plates at 37 °C in DMEM (Dulbecco's modified Eagle's Medium, GIBCO, Life Technologies™, U.S.A.) with 10% fetal bovine serum (KangYuan Biology, China) and 1% Penicillin-Streptomycin mixture (Thermo Fisher Scientific, U.S.A) in the presence of 5% CO_2.

Reverse transcription PCR (RT-PCR)

Total RNA was isolated using RNAiso Plus (Takara, Japan) and reverse transcribed to cDNA using PrimeScript 1st Strand cDNA Synthesis Kit (Takara, Japan). The resulting cDNA was analyzed by quantitative real time PCR (qPCR) with Eco™ Real-Time PCR System (Illumina, Gene Company Limited) (China, Hongkong). All PCR products were synthesized using the following program: initial denaturation step was conducted at 95 °C for 30 s, followed by 35 cycles of denaturation at 95 °C for 5 s, annealing at 60 °C for 30 s and extension at 72 °C for 30 s. The following primer sets were used for qPCR analysis: YY1, 5′-CCCTCATAAAGGCTGCACAA-3′ (forward) and 5′-TGAACCAGTTGGTGTCGTTT-3′ (reverse), yielded a 147-bp product; Arp8, 5′-CCAGGCTGAGAAGGGTGATA-3′ (forward) and 5′-GCAGGAAGAGTGTCTGTGGC-3′ (reverse), yielded a 200-bp product; BCCIP, 5′-TCAAGAGTTGGTTCTACGCTTC-3′ (forward) and 5′- CATGGGCAGAGCGATCTGT-3′ (reverse), yielded a 111-bp product; INO80, 5′-CGGAATCGGCTTTTGCTA-3′ (forward) and 5′-TGTCGGCTGGTCAGTTGG-3′ (reverse), yielded a 292-bp product; and the internal control GAPDH, 5′-ATCACTGCCACCCA-GAAGAC-3′ (forward) and 5′- ATGAGGTCCACCACCCTGTT-3′ (reverse), yielded a 443-bp product.

Transient transfection

cDNAs encoding the INO80 full length (NM_017553), INO80 point mutation (E653Q), deletion mutations including INO80 N-HSA (residue 1–521 aa), INO80CTD (residue 1223–1556 aa), Arp8 (NM_022899), Arp5 (BC038402), and YY1 (BC065366) proteins were sub-cloned with Flag tags into pcDNA3.1(-). 293T cells cultured in 6-well tissue culture plates were transiently transfected with Flag-tagged plasmids (1–2 μg) using PEI (Cat. 23966, Polysciences, U.S.A.) according to the manufacturer's recommendation. The ratio of plasmids (μg) and PEI (μL) is 1:2. 48 h later, the whole cell lysate was prepared and the proteins were isolated using SDS-PAGE gel (10%). The specific proteins were detected by western blot with specific antibodies.

siRNA/shRNA knockdown

HeLa cells were transfected with 10–20 pmol INO80- (D-004176), YY1- (sc-36863), Arp8- (Customized) and non-targeting (NT)- siRNAs (D-001206, as control) for knocking down specific genes. INO80-, Arp8- and non-targeting (NT)- siRNAs were ordered from Dharmacon (U.S.A.). YY1 siRNAs were purchased from Santa Cruz Biotechnology (U.S.A.). 48 h after transfection, total RNA or whole-cell extract was prepared for RT-PCR and western blot analysis. In ChIP assay, 293T cells were transfected with pBS-shINO80 plasmid (10 μg/10 cm culture plate) or YY1 siRNAs for knocking down INO80 and YY1.

Luciferase reporter assay

BCCIP promoter region (926 bp, −724 ~+202 bp; 391 bp, −330 ~+61 bp; 272 bp, −277 ~ −5 bp) was introduced into pGL4-Luc vector. Similar to previously described (Wu et al. 2013), 293T cells grown on 12-well plates were co-transfected with 0.4 μg of pGL4 which encodes firefly luciferase, 0.12 ng of the control plasmid Renilla luciferase vector which encodes renilla luciferase and effector plasmid expressing pGL4-BCCIP using PEI reagent (Polysciences, U.S.A.). Total effector plasmids in each transfection were adjusted to 0.8 μg with empty vector. 48 h after transfection, the transactivation activity of pGL4-BCCIP was determined by measuring firefly and *Renilla* luciferase activities using the Dual-Luciferase Reporter assay kit (Promega, U.S.A.).

Immunofluorescence staining

~30% confluence HeLa cells in 24-well plates containing a coverslips (8D1007, Nest) on each well were transfected with INO80-, Arp8- and YY1- siRNAs using RNAiMAX (Invitrogen). After 48 h, cells were permeabilized with 0.3% TritonX-100 and incubated with BCCIP antibody (1:200) at room temperature, then stained with FITC-conjugated secondary antibody (rabbit/green: 1:300, sc-2012). Cell nuclei were stained with DAPI containing Vectashield (Vector Laboraries, Inc. H-1200). Fluorescence images were observed with Olympus BX40F Microscope (Olympus Corporation).

Chromatin immunoprecipitation (ChIP) assays

1×10^7 of 293T cells grown to 80% of confluence were used for each ChIP. INO80 and YY1 antibodies were used in ChIP assay. ChIP'd DNA was subjected to qPCR. Each experiment was performed 2–3 independent times. Both specific antibody- and IgG-ChIP signal were normalized to total input. The following primer sets were used in qPCR: BCCIP −1.7 kb (−1769 ~ −1603 bp), 5′-GACAGCTTCCTCTCTTCTCCA-3′ (forward) and 5′-ATTTCGGTCTTTCCTGGGAGT-3′ (reverse); −1.2 kb (−1281 ~ −1093 bp), 5′-CCCTCCACTCTTCTCTCCAAA-3′ (forward) and 5′-CCCTCCACTCTTCTCTCCAAA-3′ (reverse); −0.8 kb (−860 ~ −694 bp), 5′-CTCGAACTCCTGACCTCTGGTG-3′ (forward) and 5′-CTGGAGCGCATTTCTTGACC-3′ (reverse); −0.2 kb (−205 ~ −53 bp), 5′-AGCTTTGGGAGGAGTGATGGAG-3′ (forward) and 5′-TAGCCAAGGGGGTGAGGAAC-3′ (reverse); +0.17 kb (+177 ~+353 bp), 5′-TCATTGACGAGGTGAGAAGGAC-3′ (forward) and 5′-GCAGTGTGTTAAAGACCGAGGAG-3′ (reverse); +2.8 kb (+2841 ~+2984 bp), 5′-GCTGCCGTGTCTGGTTTATTTG-3′ (forward) and 5′-GCCAAGATTGCCATCTGTGAAG-3′ (reverse).

DNA microarray

Specific genes (include hINO80, Arp5, Arp8, Ies2 and Ies6) knocked down with siRNAs in HeLa cells and the total RNA sent to EMTD Science and Technology Development Co., Ltd. (Beijing, China) for DNA microarray analysis. The Illumina microarray datasets were accessible from the National Center for Biotechnology Information Gene Expression Omnibus (GEO) data repository (http://www.ncbi.nlm.nih.gov/geo/) using the series accession number GSE68655. The online biological classification tool DAVID (Database for Annotation, Visualization and Integrated Discovery) was used to perform the enrichment analysis (Huang da et al. 2009; Huang da et al. 2009), Gene Ontology (GO) (Ashburner et al. 2000) function analysis and Kyoto Encyclopedia of Genes and Genomes (KEGG) (Ogata et al. 1999) pathway enrichment analysis of differentially expressed genes (DEGs). During the analysis, $P < 0.05$ and FDR < 0.02 genes used in the annotation were defined as statistically significant.

Statistics

Two-tailed student's *t*-test was used in this study. Data shown is mean + SD from at least three independent experiments. Statistical probability is expressed as $^*P < 0.05$, $^{**}P < 0.01$.

ACKNOWLEDGMENTS

This work was supported by the National Natural Science Foundation of China (Grant Nos. 31071131 and 31171245), by the National Laboratory of Biomacromolecules (O5SY02110A and 2012kf04), and by the Project of Jilin Province Science and Technology Development Program (20130413002GH).

ABBREVIATIONS

BCCIP, BRCA2- and CDKN1A (Cip1, p21)-interacting protein; HSA/PTH, helicase-SANT-associated/post-HSA; ChIP, chromatin immunoprecipitation; qPCR, quantitative real time

PCR; DEGs, differentially expressed genes; GO, Gene Ontology; KEGG, Kyoto Encyclopedia of Genes and Genomes; GEO, Gene Expression Omnibus.

COMPLIANCE WITH ETHICAL STANDARDS

Jiaming Su, Yi Sui, Jian Ding, Fuqiang Li, Shuang Shen, Yang Yang, Zeming Lu, Fei Wang, Lingling Cao, Xiaoxia Liu, Jingji Jin and Yong Cai declare that they have no conflict of interest. This article does not contain any studies with human or animal subjects performed by any of the authors.

AUTHOR CONTRIBUTIONS

JS, YS, JD, FL, SS, YY, ZL and LC performed experiments. JJ and FW analyzed data. XL provides reagents. JS, YC and JJ designed experiments and drafted the manuscript. All authors read and approved the final manuscript.

REFERENCES

Ashburner M, Ball CA, Blake JA, Botstein D, Butler H, Cherry JM (2000) Gene ontology: tool for the unification of biology.The gene ontology consortium. Nat Genet 25:25–29

Cai Y, Jin J, Yao T, Gottschalk AJ, Swanson SK, Wu S, Shi Y, Washburn MP, Florens L, Conaway RC, Conaway JW (2007) YY1 functions with INO80 to activate transcription. Nat Struc Mol Biol 14:872–874

Cao L, Ding J, Dong L, Zhao J, Su J, Wang L, Sui Y, Zhao T, Wang F, Jing J, Cai Y (2015) Negative regulation of $p21^{Waf1/Cip1}$ by human INO80 chromatin remodeling complex is implicated in cell cycle phase G2/M arrest and abnormal chromosome stability. PLoS ONE 10(9):e0137411. doi:10.1371/journal.pone.0137411

Chen L, Cai Y, Jin J, Florens L, Swanson SK, Washburn MP, Conaway JW, Conaway RR (2011) Subunit organization of the human INO80 chromatin remodeling complex: an evolutionarily conserved core complex catalyzes ATP-dependent nucleosome remodeling. J Biol Chem 286:11283–11289

Chen L, Conaway RC, Conaway JW (2013) Multiple modes of regulation of the human Ino80 SNF2 ATPase by subunits of the INO80 chromatin remodeling complex. Proc Natl Acad Sci USA 110:20497–20502

Conaway RC, Conaway JW (2009) The INO80 chromatin remodeling complex in transcription, replication and repair. Trends Boichem Sci 34:71–77

da Huang W, Sherman BT, Lempicki RA (2009a) Systematic and integrative analysis of large gene lists using DAVID bioinformatics resources. Nat Protoc 4:44–57

da Huang W, Sherman BT, Lempicki RA (2009b) Bioinformatics enrichment tools: paths toward the comprehensive functional analysis of large gene lists. Nucleic Acids Res 37:1–13

Hariharan N, Kelley DE, Perry RP (1991) Delta a transcription factor that binds to downstream elements in several polymerase II promoters, is a functionally versatile zinc finger protein. Proc Natl Acad Sci USA 88:9799–9803

Jin J, Cai Y, Yao T, Gottschalk AJ, Florens L, Swanson SK, Gutierrez J, Coleman MK, Workman JL, Mushegian A, Washburn MP, Conaway RC, Conaway JW (2005) A mammalian chromatin remodeling complex with similarities to the *yeast* INO80 complex. J Biol Chem 280:41207–41212

Karolchik D, Barber GP, Casper J, Clawson H, Cline MS, Diekhans M, Dreszer TR, Fujita PA, Guruvadoo L, Haeussler M, Harte RA, Heitner S, Hinrichs AS, Learned K, Lee BT, Li CH, Raney BJ, Rhead B, Rosenbloom KR, Sloan CA, Speir M, Zweig AS, Haussler D, Kuhn RM, Kent WJ (2013) The UCSC genome vrowser database: 2014 update. Nucleic Acids Res 1–7

Lee JS, Galvin KM, See RH, Eckner R, Livingston D, Moran E, Shi Y (1995) Relief of YY1 transcriptional repression by adenovirus E1A is mediated by E1A-associated protein p300. Genes Dev 9:1188–1198

Liu J, Yuan Y, Huan J, Shen Z (2001) Inhibition of breast and brain cancer cell growth by BCCIP alpha, an evolutionary conserved nuclear protein that interacts with BRCA2. Oncogene 20:336–345

Liu X, Cao L, Ni J, Liu N, Zhao X, Wang Y, Zhu L, Wang L, Wang J, Yue Y, Cai Y, Jin J (2013) Differential BCCIP gene expression in primary human ovarian cancer, renal cell carcinoma, and colorectal cancer tissues. Int J Oncol 43:1925–1934

Lu H, Guo X, Meng X, Liu J, Allen X, Wray J, Nickoloff JA, Shen Z (2005) The BRCA2-interacting protein BCCIP functions in RAD51 and BRCA2 focus formation and homologous recombinational repair. Mol Cell Biol 25:1949–1957

Meng X, Liu J, Shen Z (2003) Genomic structure of the human BCCIP gene and its expression in cancer. Gene 302:139–146

Meng X, Liu J, Shen Z (2004a) Injibition of G1 to S cell cycle progression by BCCIP beta. Cell Cycle 3:343–348

Meng X, Lu H, Shen Z (2004b) BCCIP functions through p53 to regulate the expression of p21Waf1/Cip1. Cell Cycle 3:1457–1462

Meng X, Fan J, Shen Z (2007) Roles of BCCIP in chromosome stability and cytokinesis. Oncogene 26:6253–6260

Morrison AJ, Shen X (2009) Chromatin remodeling beyond transcription: the INO80 and SWR1 complexes. Nat Rev Mol Biol 10:373–384

Moshkin YM, Chalkley GE, Kan TW, Reddy BA, Ozgur Z, van Ijcken WF, Dekkers DH, Demmers JA, Travers AA, Verrijzer CP (2012) Remodelers organize cellular chromatin by counteracting intrinsic histone-DNA sequence preferences in a class-specific manner. Mol Cell Biol 32:675–688

Neuman SD, Ihry RJ, Gruetzmacher KM, Bashirullah A (2014) INO80-dependent repression of ecdysone-induced transcriptional responses regulates developmental timing in Drosophila. Dev Biol 387:229–239

Ogata H, Goto S, Sato K, Fujibuchi W, Bono H, Kanehisa M (1999) KEGG: kyoto encyclopedia of genes and genomes. Nucleic Acids Res 27:29–34

Ono T, Kitaura H, Ugai H, Murata T, Yokoyama KK, Iguchi-Ariga SM, Ariga H (2000) Tok-1, a novel $p21^{Cip1}$-binding protein that cooperatively enhances p21-dependent inhibitory activity toward CDK2 kinase. J Biol Chem 275:31145–31154

Park K, Atchison ML (1991) Isolation of a candidate repressor/activator, NF-E1 (YY-1, delta), that binds to the immunoglobulin kappa 3′ enhancer and the immunoglobulin heavy-chain mu E1 site. Proc Natl Acad Sci USA 88:9804–9808

Seeber A, Hauer M, Gasser SM (2013) Nucleosome remodelers in double-strand break repair. Curr Opin Genet Dev 23:174–184

Seto E, Shi Y, Shenk T (1991) YY1 is an initiator sequence-binding protein that directs and activates transcription in vitro. Nature 354:241–245

Shen X, Ranallo R, Choi E, Wu C (2003) Involvement of actin-related proteins in ATP-dependent chromatin remodeling. Mol Cell 12:147–155

Shi Y, Seto E, Chang LS, Shenk T (1991) Transcriptional repression by YY1, a human GLI-Kruppel- related protein, and relief of repression by adenovirus E1A protein. Cell 67:377–388

Shi Y, Lee JS, Galvin KM (1997) Everything you have ever wanted to know about Yin Yang 1. Biochim Biophys Acta 1332:F49–F66

Thomas MJ, Seto E (1999) Unlocking the mechanisms of transcription factor YY1: Are chromatin modifying enzymes the key? Gene 236:197–208

Wang L, Du Y, Ward JM, Shimbo T, Lackford B, Zheng X, Miao YL, Zhou B, Han L, Fargo DC, Jothi R, William CJ, Wade PA, Hu G (2014) INO80 facilitates pluripotency gene activation in embryonic stem self-renewal, reprogramming, and blastocyst development. Cell Stem Cell 14:575–591

Wu D, Zhang R, Zhao R, Chen G, Cai Y, Jin J (2013) A novel function of novobiocin: disrupting the interaction of HIF1α and p300/CBP through direct binding to the HIF1α C-ternimal activation domain. PLoS One 8:e62014

Yao YL, Yang WM, Seto E (2001) Regulation of transcription factor YY1 by acetylation and deacetylation. Mol Cell Biol 21:5979–5991

7

CRISPR/Cas9-mediated targeted gene correction in amyotrophic lateral sclerosis patient iPSCs

Lixia Wang[1,2,3], Fei Yi[4], Lina Fu[1,3], Jiping Yang[1,3], Si Wang[1,3], Zhaoxia Wang[5], Keiichiro Suzuki[6,7], Liang Sun[9], Xiuling Xu[1], Yang Yu[8], Jie Qiao[8], Juan Carlos Izpisua Belmonte[6], Ze Yang[9], Yun Yuan[5]✉, Jing Qu[2,3]✉, Guang-Hui Liu[1,3,10,11]✉

[1] National Laboratory of Biomacromolecules, CAS Center for Excellence in Biomacromolecules, Institute of Biophysics, Chinese Academy of Sciences, Beijing 100101, China
[2] State Key Laboratory of Stem Cell and Reproductive Biology, Institute of Zoology, Chinese Academy of Sciences, Beijing 100101, China
[3] University of Chinese Academy of Sciences, Beijing 100049, China
[4] Department of Molecular and Cellular Physiology, Stanford University School of Medicine, Stanford, CA 94305, USA
[5] Department of Neurology, Peking University First Hospital, Beijing 100034, China
[6] Gene Expression Laboratory, Salk Institute for Biological Studies, 10010 North Torrey Pines Road, La Jolla, CA 92037, USA
[7] Universidad Católica San Antonio de Murcia (UCAM), Campus de los Jerónimos, N 135 Guadalupe, 30107 Murcia, Spain
[8] Department of Gynecology and Obstetrics, Peking University Third Hospital, Beijing 100191, China
[9] Beijing Hospital of the Ministry of Health, Beijing 100730, China
[10] Key Laboratory of Regenerative Medicine of Ministry of Education, Institute of Aging and Regenerative Medicine, Jinan University, Guangzhou 510632, China
[11] Beijing Institute for Brain Disorders, Capital Medical University, Beijing 100069, China
✉ Correspondence: yuanyun2002@126.com (Y. Yuan), qujing@ioz.ac.cn (J. Qu), ghliu@ibp.ac.cn (G.-H. Liu)

ABSTRACT

Amyotrophic lateral sclerosis (ALS) is a complex neurodegenerative disease with cellular and molecular mechanisms yet to be fully described. Mutations in a number of genes including *SOD1* and *FUS* are associated with familial ALS. Here we report the generation of induced pluripotent stem cells (iPSCs) from fibroblasts of familial ALS patients bearing *SOD1*$^{+/A272C}$ and *FUS*$^{+/G1566A}$ mutations, respectively. We further generated gene corrected ALS iPSCs using CRISPR/Cas9 system. Genome-wide RNA sequencing (RNA-seq) analysis of motor neurons derived from *SOD1*$^{+/A272C}$ and corrected iPSCs revealed 899 aberrant transcripts. Our work may shed light on discovery of early biomarkers and pathways dysregulated in ALS, as well as provide a basis for novel therapeutic strategies to treat ALS.

KEYWORDS ALS, CRISPR/Cas9, gene correction, iPSC disease modeling

INTRODUCTION

Amyotrophic lateral sclerosis (ALS) are a group of progressive but fatal neurodegenerative diseases due to the selective loss of functional motor neurons in the brain and spinal cord. On average, most ALS patients die within 5 years once symptom appears. Excitotoxic/Inflammatory/Oxidative insults, misfolded proteins and aggregates, aberrant RNA processing, unstable genome, and mitochondria dysfunction have all been implicated in ALS (Pasinelli and Brown, 2006; Kiskinis et al., 2014). Despite extensive studies over the past 40–50 years, the etiology of ALS is still far from clear. ALS is generally classified into two categories: familial ALS (FALS) and sporadic ALS (SALS). Two forms are clinically similar. FALS accounts for 10% of total ALS cases and is mostly dominant inheritance (Robberecht and Philips, 2013). To date, many genes have been implicated in

FALS, such as *SOD1*, *FUS*, *TDP43*, *C9ORF72*, *VAPB*, etc. Gene *SOD1* encoding superoxide dismutase 1 (SOD1) is the first identified gene linked to FALS and accounts for about 20% of FALS cases (Rosen et al., 1993). Mutant SOD1 was found as aberrant misfolded aggregates in FALS (Bruijn et al., 1998; Bosco et al., 2010). *FUS* was initially identified in cancer and encodes a RNA/DNA binding protein (Crozat et al., 1993; Baechtold et al., 1999). A number of mutations in *FUS* gene were identified in ALS cases, like A1564G, C1574T, G1566A, etc. (Drepper et al., 2011; Lai et al., 2011; Lattante et al., 2013). Aggregates containing mutant FUS protein have been found in motor neurons from ALS patients by postmortem analysis (Vance et al., 2009).

Although great progresses have been achieved in studying animal models of ALS, the clinical relevance of many observations in animals is uncertain. For instance, most of the rodent ALS models are transgenic animals carrying extremely high levels of the specific proteins when compared with human ALS pathological tissues (Julien and Kriz, 2006; Turner and Talbot, 2008; Huang et al., 2011; Sharma et al., 2016). Likewise, many early events of neuronal dysfunction and retraction have been well documented in animal models; however, correlated timings of those events in human ALS pathology especially of earlier ones are unclear (Fischer et al., 2004; Casci and Pandey, 2015). Moreover, the anti-excitotoxic agent Riluzole is by far the only compound discovered through ALS animal models, but it only demonstrated moderate therapeutic effect in clinical trials (Zhu et al., 2015). The genetic and developmental differences between human and model species urge more understanding in human system. Recently, vast progresses in using human induced pluripotent stem cells (iPSCs) and gene engineering tools to model human genetic diseases have brought promising prospects (Liu et al., 2011a; Liu et al., 2011b; Egawa et al., 2012; Liu et al., 2012a; Liu et al., 2012b; Yang et al., 2013; Liu et al., 2014; Suzuki et al., 2014; Zhang et al., 2015; Kubben et al., 2016). A few groups reported the establishment of novel ALS disease models by combining iPSCs and gene engineering tool (Chen et al., 2014; Kiskinis et al., 2014; Wainger et al., 2014; Lenzi et al., 2015; Higelin et al., 2016; Ichiyanagi et al., 2016). Other ALS iPSCs disease models were also reported however lacked isogenic gene corrected controls; thus the underlying mechanism may be obscured due to variable genetic backgrounds (Egawa et al., 2012; Li et al., 2015; Liu et al., 2015). More importantly, none of these studies focused on the early molecular events in ALS pathogenesis.

Here, we report the generation of iPSCs from ALS patients carrying heterozygous mutations of $SOD1^{+/A272C}$ and $FUS^{+/G1566A}$, respectively. We also achieved targeted correction of the mutant genes via combing CRISPR/Cas9 and single-stranded oligodeoxynucleotide (ssODN) without leaving any genomic footprint. Using a modified motor neuron differentiation method, we subsequently generated $ISL1^{+}$ motor neurons from isogenic iPSC lines and performed genome-wide RNA sequencing (RNA-seq) analysis. To the end, we identified a group of aberrant transcripts which may be involved in $SOD1^{+/A272C}$-mediated pathogenesis of ALS. Thus, our study provides an isogenic platform to study ALS disease mechanisms and cues to develop novel therapies.

RESULTS

Establishment of $SOD1^{+/A272C}$ and $FUS^{+/G1566A}$ iPSCs

We obtained primary fibroblasts with a heterozygous missense mutation (A272C) in *SOD1* gene and a heterozygous nonsense mutation (G1566A) in *FUS* gene from two ALS patients, respectively. We then generated $SOD1^{+/A272C}$ specific iPSCs ($SOD1^{+/A272C}$ iPSCs) and $FUS^{+/G1566A}$ specific iPSCs ($FUS^{+/G1566A}$ iPSCs) according to a modified reprogramming protocol (Takahashi et al., 2007; Okita et al., 2011; Liu et al., 2014; Ding et al., 2015; Fu et al., 2016) (Fig. 1A). About 20 human embryonic stem cell (hESC)-like colonies were generated from 5×10^5 starting fibroblasts. We did not observed discernible difference in reprogramming efficiency between the two lines of patients-derived fibroblasts (data not shown). The generated iPSCs displayed typical pluripotent stem cell-like morphology, expressed pluripotency markers, such as OCT4, NANOG and SOX2, formed teratomas consisting of three germ-layers *in vivo*, maintained unmethylated CpG islands in the promoter of *OCT4*, and demonstrated normal karyotypes (Fig. 1B–F). DNA sequencing results demonstrated the presence of $SOD1^{+/A272C}$ or $FUS^{+/G1566A}$ mutations in the two lines of ALS iPSCs, respectively (Fig. 1G). These data collectively supported the successful generation of $SOD1^{+/A272C}$ and $FUS^{+/G1566A}$ iPSCs from ALS patients' fibroblasts.

Targeted gene correction of *FUS* and *SOD1* mutations with CRISPR/Cas9 system

In order to circumvent clonal variations of iPSC lines and better elucidate the pathogenic mechanism of the ALS mutations, we sought to generate isogenic wild type iPSC lines by targeted gene correction. Using CRISPR/Cas9 system and a donor plasmid containing wild type coding sequence and homology arms (HA) as the repair template (Fig. S1A), we obtained the gene corrected iPSC clone ($FUS^{+/+}$-HA iPSCs) from $FUS^{+/G1566A}$ iPSCs. After homologous recombination and the subsequent removal of neo-resistance cassette, a flippase recognition target site (FRT footprint) was left in *FUS* intron. Site-specific gene correction was confirmed by genomic DNA PCR and sequencing (Fig. S1B and S1C). Additionally, $FUS^{+/+}$-HA iPSCs demonstrated normal pluripotency (Fig. S1D). Recent studies have shown that linear templates with short (60–70 bases) homology arms, like single-stranded oligodeoxynucleotide (ssODN), could also be used in gene correction instead of plasmid-based construct. This strategy bypasses the selection cassette insertion and leaves no genetic

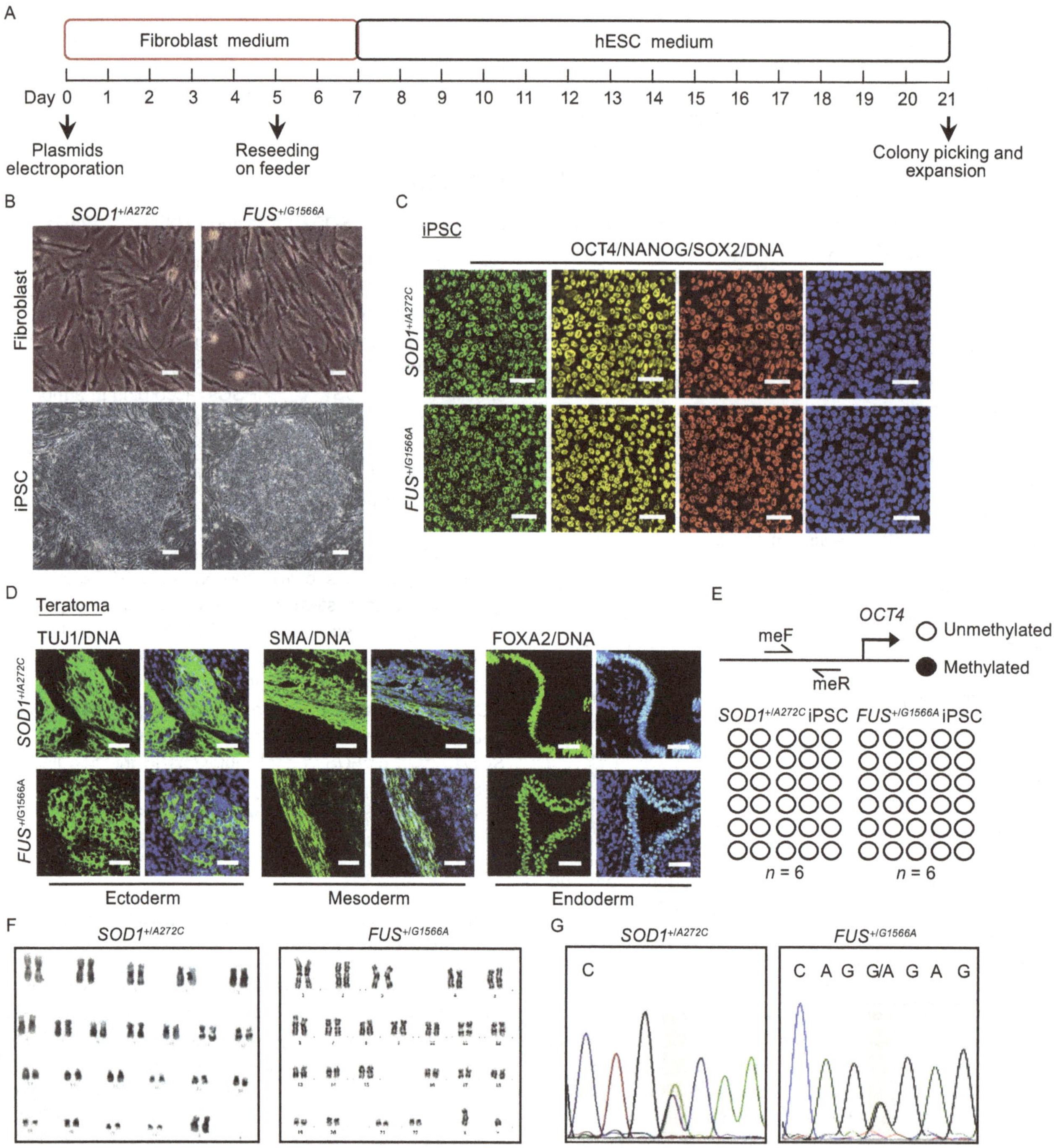

Figure 1. Establishment of *SOD1*$^{+/A272C}$ iPSCs and *FUS*$^{+/G1566A}$ iPSCs. (A) Schematic procedure of generating iPSCs from ALS patient fibroblasts. (B) Phase-contrast images of ALS patient fibroblasts (top panels) and iPSCs (bottom panels). Scale bars = 25 μm (top bottom) and 100 μm (bottom panels). (C) Immunofluorescent staining of pluripotency markers, OCT4, NANOG, and SOX2. Nuclei were stained with Hoechst 33342 (blue). Scale bars = 50 μm. (D) Immunofluorescent staining of TUJ1 (ectoderm), SMA (mesoderm), and FOXA2 (endoderm) in teratomas derived from ALS iPSCs *in vivo*. Nuclei were stained with Hoechst 33342 (blue). Scale bars = 50 μm. (E) DNA methylation analysis of the *OCT4* promoter in ALS iPSCs. A pair of primers used is shown as arrows. Open and closed circles indicate unmethylated and methylated CpG dinucleotides respectively, as indicated. (F) Karyotyping analysis of ALS patient iPSCs. (G) Confirmation of the heterozygous mutation of *SOD1*$^{+/A272C}$ and *FUS*$^{+/G1566A}$ in ALS iPSCs by DNA sequencing.

footprint (Corti et al., 2012; Ding et al., 2013). We next tested this strategy by electroporating an expression vector encoding both mCherry and guide RNA, a plasmid for Cas9-2A-GFP, and the ssODN template all together into patient iPSCs. After fluorescence-activated cell sorting (FACS) for GFP (Cas9) and mCherry (guide RNA) double positive cells, gene corrected clones ($FUS^{+/+}$ iPSCs) were successfully obtained (Fig. 2A). The *Bsa*XI restriction fragment length polymorphism (RFLP) was specific to the corrected allele and DNA sequencing also confirmed that $FUS^{+/G1566A}$ was correctly targeted without any footprint (Figs. 2B, 2C, and S2A.). The gene targeting efficiency at *FUS* gene was about 1% (Fig. S2A). In view of the potential clinical application of gene editing technique (Suzuki et al., 2014; Veres et al., 2014), we further investigated the specificity of gene editing in the system. We carefully examined the top 3 potential off-target sites and found no unwanted off-target effect (Figs. 2D and S3A). Comparing both gene correction platforms, we found no obvious difference regarding targeting efficiency (data not shown). Accordingly, we chose to correct $SOD1^{+/A272C}$ via CRISPR/Cas9 system using ssODN as repair template (Fig. 2E). Correction of $SOD1^{+/A272C}$ was confirmed by genomic DNA sequencing, as well as elimination of *Ape*K1 RFLP (Figs. 2F, 2G, and S2B). The gene targeting efficiency at *SOD1* gene was about 20% (Fig. S2B). We did not find off-target cleavages in $SOD1^{+/+}$ iPSCs by checking the predicted top 3 off-target sites as well (Figs. 2H, S3B, and S4). Additionally, both corrected iPSCs expressed pluripotency markers including OCT4, NANOG, and SOX2 *in vitro,* and formed teratomas composed of three germ layers *in vivo* (Fig. 2I). Taken together, we successfully generated gene corrected ALS iPSCs displaying normal pluripotency.

Figure 2. Targeted gene correction of *FUS* and *SOD1* mutations. (A) Strategy of correcting $FUS^{+/G1566A}$ mutation. The sequence of gRNA is shown with the PAM sequence. Red line represents the mutant allele, and blue line represents the wildtype allele. HR, homologous recombination. ssODN, single-stranded oligodeoxynucleotide. (B) *Bsa*XI restriction digestion of PCR product before and after gene correction. The red letter highlights the corrected base. The mutation G>A eliminates *Bsa*XI restriction site that is present in the corrected line. Primers used are shown as arrows in Fig. 2A (P1, P2). (C) DNA sequencing demonstrates the correction of $FUS^{+/G1566A}$ mutation. The red shadow highlights the corrected base. (D) DNA sequencing of the top 3 potential off-target sites. The off-target sites were predicated at http://crispr.genome-engineering.org/. Sequencing of PCR product shows no off-targets found. FOT, potential off-target site at gRNA targeted *FUS* gene. +1, plus strand. −1, minus strand. √, no off targets. (E) Strategy of correcting $SOD1^{+/A272C}$ mutation. The sequence of gRNA is shown with the PAM sequence. Red line represents the mutant allele, and blue line represents the wildtype allele. HR, homologous recombination. ssODN, single-stranded oligodeoxynucleotide. (F) *Ape*KI restriction digestion of PCR product before and after gene correction. The red letter highlights the mutant base. The mutation A>C creates *Ape*KI restriction site that is absent in the corrected line. Primers used are shown as arrows in Fig. 2E (P3, P4). (G) DNA sequencing demonstrates the correction of $SOD1^{+/A272C}$ mutation. The red shadow highlights the corrected base. (H) DNA sequencing of the top 3 potential off-target sites. The off-target sites were predicated at http://crispr.genome-engineering.org/. Sequencing of PCR product shows no off-targets found. SOT, potential off-target site at gRNA targeted *SOD1* gene. +1, plus strand. −1, minus strand. √, no off targets. (I) Immunofluorescent images of pluripotency markers, OCT4, NANOG, and SOX2 incorrected clone *in vitro* and three-layer markers, TUJ1 (ectoderm), SMA (mesoderm), and FOXA2 (endoderm) in teratomas derived from corrected iPSCs *in vivo*. Nuclei were stained with Hoechst 33342 (blue). Scale bars = 50 μm.

Directed motor neuron differentiation from isogenic pair of iPSC lines

The cardinal symptom of ALS patients is muscle atrophy due to the dysfunction of motor neurons (Al-Chalabi and Hardiman, 2013). To investigate the biological effect of ALS-causing genetic mutations, we generated motor neurons from $SOD1^{+/A272C}$ iPSCs and corresponding isogenic controls using a modified protocol with high differentiation efficiency (Qu et al., 2014; Maury et al., 2015) (Fig. 3A). During the differentiation, four key compounds were sequentially added, including dorsomorphin for neural induction, retinoic acid (RA), and smo agonist (SAG) to activate neural patterning, and DAPT (γ-secretase inhibitor) to accelerate motor neurons maturation. Upon day 12, most cells showed neuronal morphology (Fig. 3B). Nearly all of differentiated cells were $MAP2^+$, about 70%–80% cells were $ISL1^+$, and up to 70% of the cells were $HB9^+$ (Figs. 3C and S5A). Mutation and genetic correction were further confirmed in differentiated motor neurons by sequencing (Fig. 3D). It has been reported that wildtype and SOD1 mutant iPSC lines may have comparable neuronal differentiation capacity (Kiskinis et al., 2014). Similarly, we did not observe significant difference in the efficiency of motor neuron generated from ALS iPSCs and their wildtype isogenic controls (Figs. 3C and S5A).

RNA-seq revealed $SOD1^{+/A272C}$-affected pathways in human motor neurons

In order to better understand the molecular mechanisms of ALS-causing mutations especially those molecular events at early stage of ALS pathogenesis, we performed RNA-seq analysis to compare global gene expression in motor neurons carrying *SOD1* mutation with isogenic wild type controls. Filtering by setting *q*-values less than 0.05, we identified 364 upregulated genes and 535 downregulated

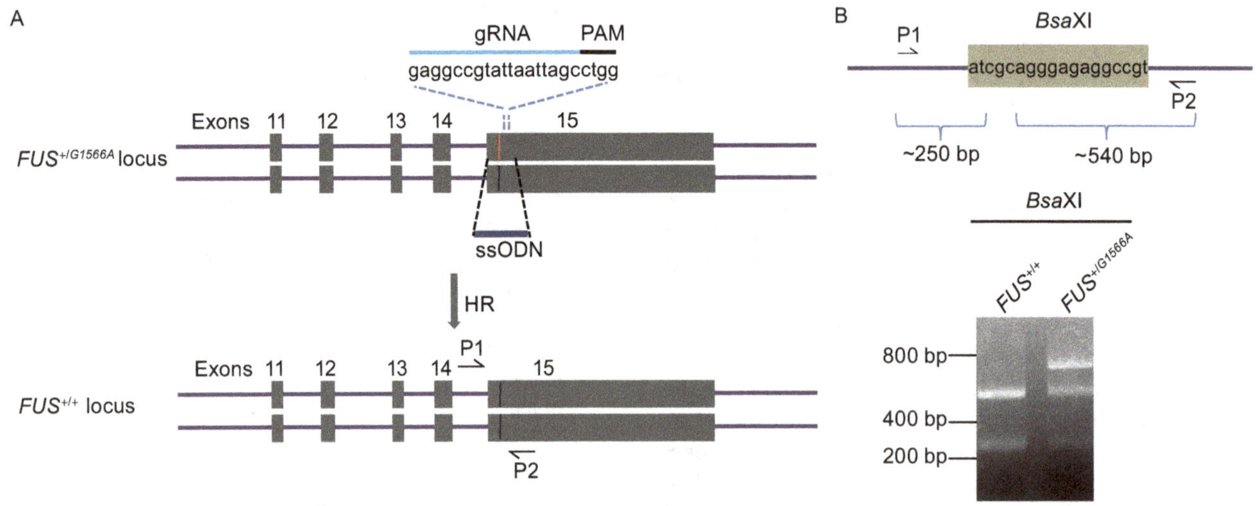

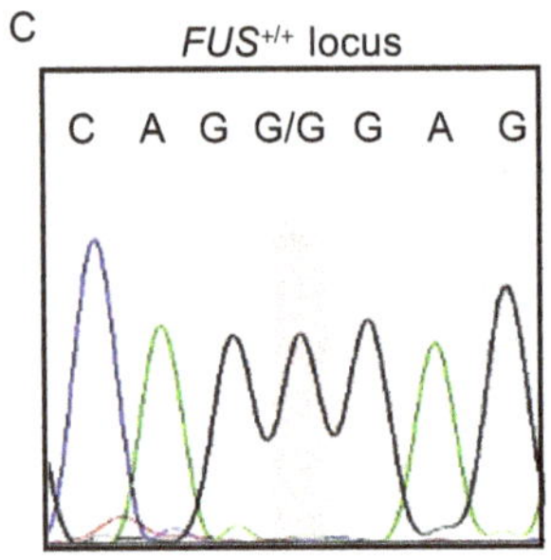

D

Candidate off-targets

	Chromosome	Strand	Position	No. of mismatches	*FUS*$^{+/+}$ iPSC
FOT1	2	−1	176329841	2	√
FOT2	1	+1	82926900	3	√
FOT3	11	−1	128032235	4	√

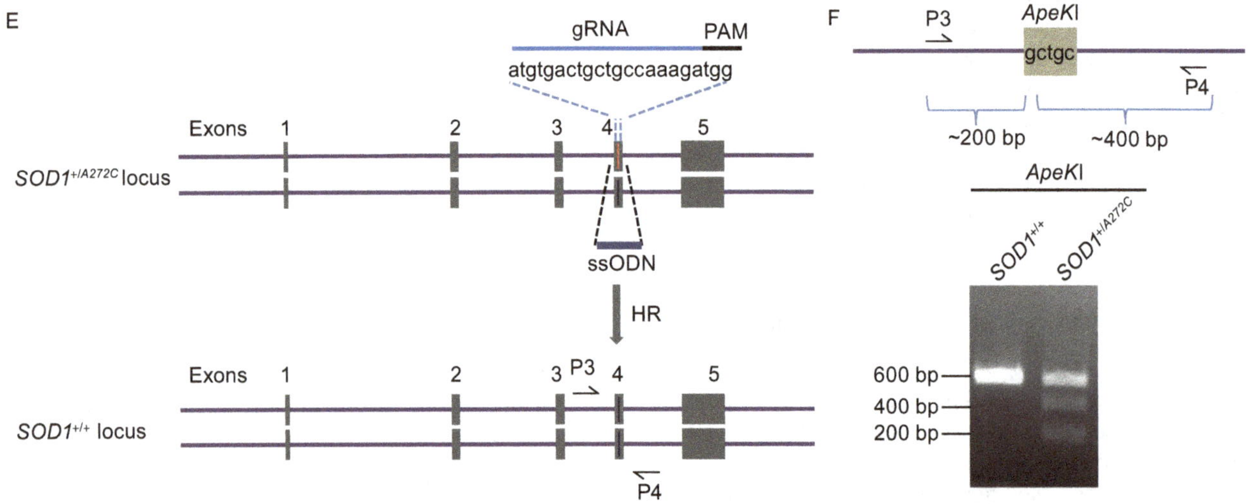

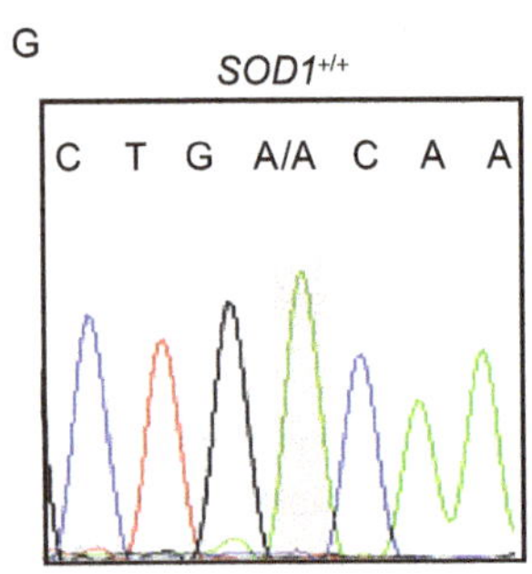

H

Candidate off-targets

	Chromosome	Strand	Position	No. of mismatches	*SOD1*$^{+/+}$ iPSC
SOT1	5	−1	89005769	2	√
SOT2	9	+1	12287385	2	√
SOT3	17	+1	35109248	3	√

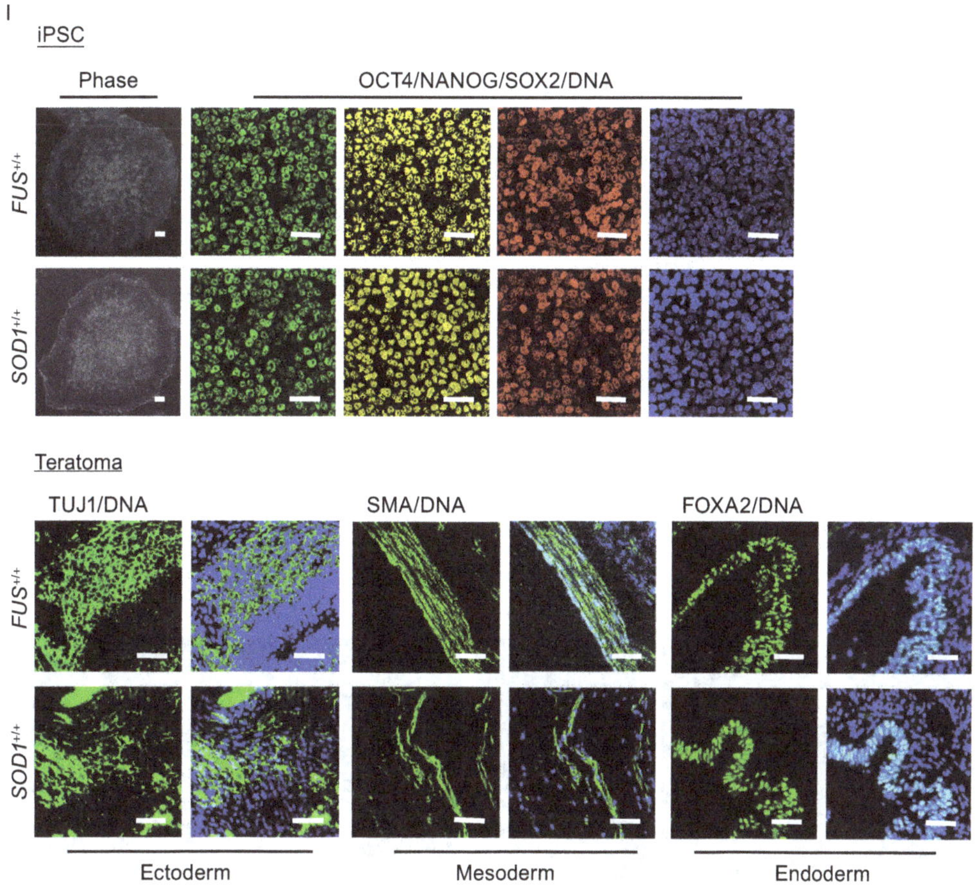

Figure 2. continued.

genes in *SOD1*$^{+/A272C}$ motor neurons (Figs. 4A, S5B, S5C, and Table S1). Gene ontology analysis revealed that upregulated transcripts of *SOD1*$^{+/A272C}$ motor neurons were associated with regulation of nervous system activity and signal transduction, etc. (Fig. S5D, S5E, and Table S2). Downregulated transcripts were associated with the terms of extracellular matrix, calcium homeostasis, and endoplasmic reticulum (ER) homeostasis, etc. (Figs. 4B, 4C, S5E, and Table S2). Key gene expression changes were verified by RT-qPCR (Fig. 4D). These gene expression changes may be implicated in early molecular events accounting for ALS pathogenesis.

DISCUSSION

Basic research in ALS animal models has provided valuable insights into the mechanism of the disease. However, knowledge we learned from animal models were often poorly transferable into clinic. Amongst all variable factors, large genetic variations among different species and strategy of overexpressing mutant genes in animal models may account for major gaps between animal models and clinical cases of ALS. Generating human iPSCs bearing disease mutation together with an efficient neuronal differentiation strategy may provide an unprecedented tool to study cellular and molecular mechanisms of ALS in human neurons. Here, we generated two human iPSC lines from ALS patient-specific fibroblasts bearing heterozygous disease-causing mutations (*FUS*$^{+/G1566A}$ and *SOD1*$^{+/A272C}$), and generated their respective isogenic disease-free iPSCs by CRISPR/Cas9 mediated gene correction.

Previous studies have reported genetic correction of *SOD1*$^{+/A272C}$ mutation and *SOD1*$^{+/C14T}$ mutation in ALS patients derived iPSCs. Notably, these research teams employed TALEN- and ZFN-mediated gene editing together with correction vectors containing drug resistance cassette, resulting in FRT footprint in the engineered genome (Chen et al., 2014; Kiskinis et al., 2014). In our study, we generated isogenic gene-corrected iPSCs using CRISPR/Cas9 system with ssODN as repair template for correcting *SOD1*$^{+/A272C}$ and *FUS*$^{+/G1566A}$ into wildtypes. Using ssODN but not plasmids with resistance cassette in our strategy simplified the gene correction procedure, and avoided introducing any exogenous sequence into the genome. This strategy offers a great potential for generating other human genetic disease models as well.

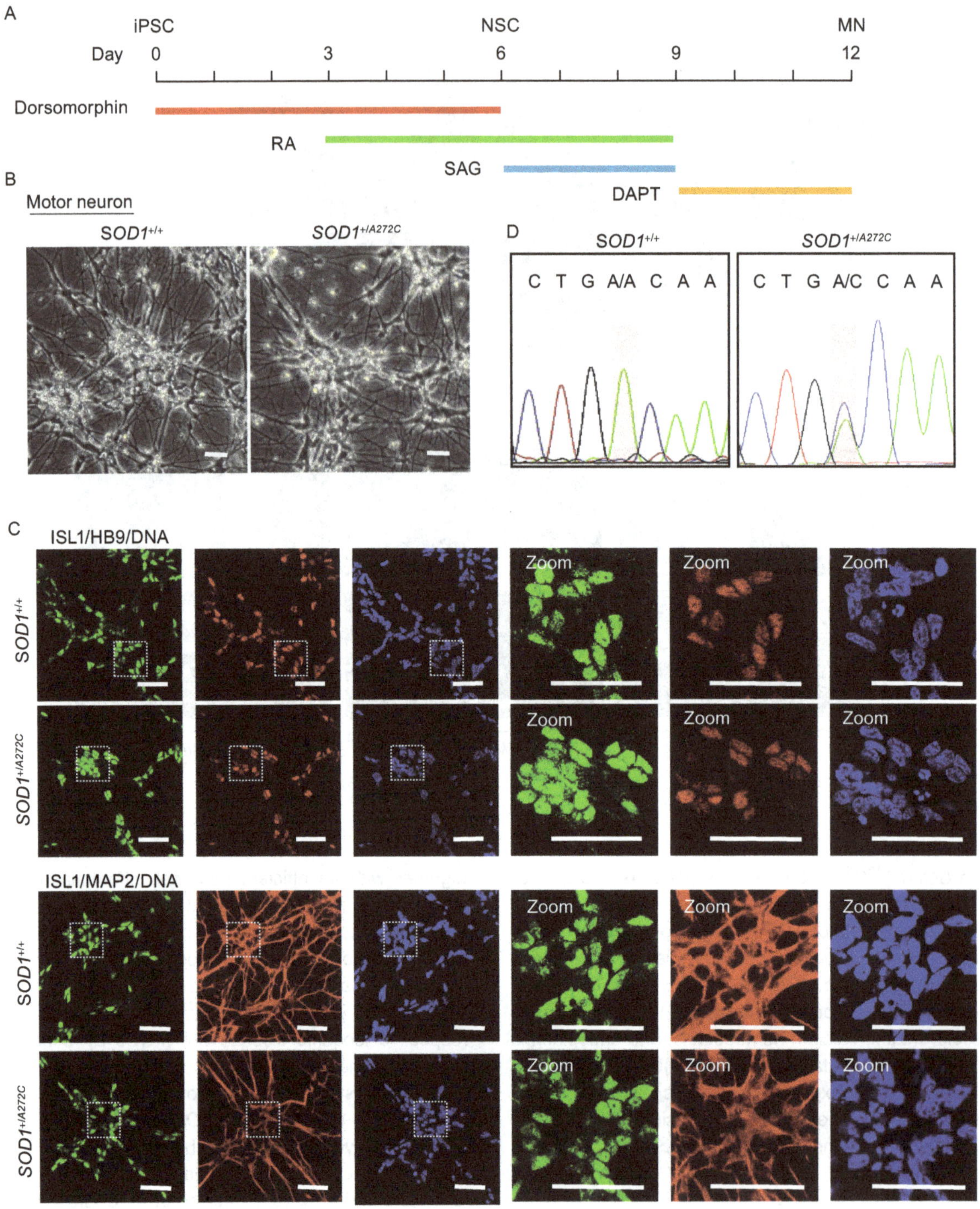

Figure 3. Directed motor neuron differentiation from isogenic pair of iPSC lines. (A) Schematic overview of motor neuron differentiation from iPSCs. Compounds are added at different time points as indicated. NSC, neural stem cell. MN, motor neuron. (B) Phase contrast images of motor neurons derived from *SOD1*$^{+/A272C}$ iPSCs and its isogenic control iPSCs at day 12. Scale bars = 75 μm. (C) Immunofluorescent images of motor neuron markers (ISL1, HB9) and other neural marker (MAP2) at day 12. Nuclei were stained with Hoechst 33342 (blue). Scale bars = 50 μm. (D) DNA Sequencing confirming *SOD1* genotype in motor neurons derived from *SOD1*$^{+/A272C}$ iPSCs and its isogenic control iPSCs at day 12.

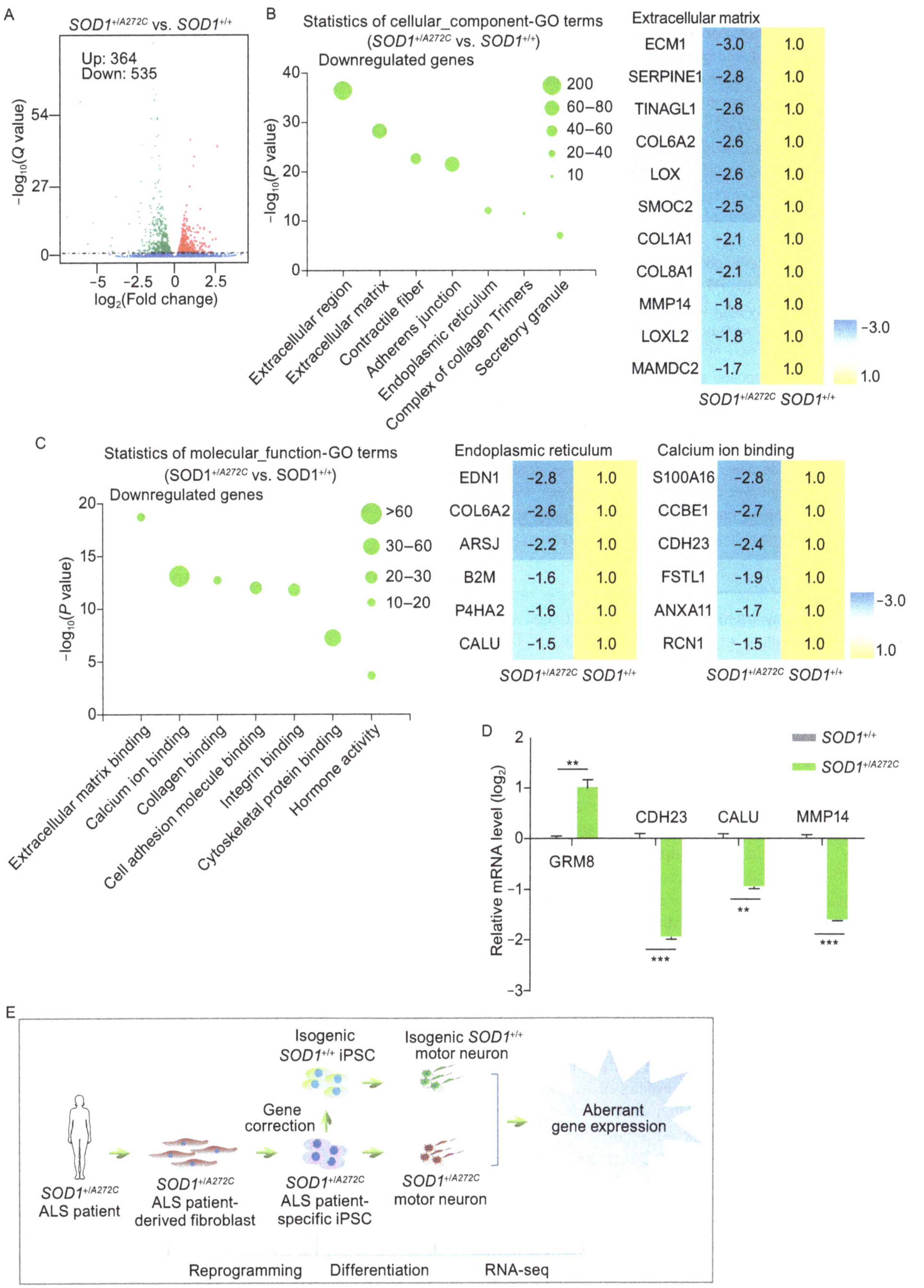
A
SOD1+/A272C vs. SOD1+/+
Up: 364
Down: 535
−log10(Q value)
log2(Fold change)
B
Statistics of cellular_component-GO terms
(SOD1+/A272C vs. SOD1+/+)
Downregulated genes
−log10(P value)
200
60–80
40–60
20–40
10
Extracellular region
Extracellular matrix
Contractile fiber
Adherens junction
Endoplasmic reticulum
Complex of collagen Trimers
Secretory granule
Extracellular matrix
ECM1 -3.0 1.0
SERPINE1 -2.8 1.0
TINAGL1 -2.6 1.0
COL6A2 -2.6 1.0
LOX -2.6 1.0
SMOC2 -2.5 1.0
COL1A1 -2.1 1.0
COL8A1 -2.1 1.0
MMP14 -1.8 1.0
LOXL2 -1.8 1.0
MAMDC2 -1.7 1.0
SOD1+/A272C SOD1+/+
C
Statistics of molecular_function-GO terms
(SOD1+/A272C vs. SOD1+/+)
Downregulated genes
−log10(P value)
>60
30–60
20–30
10–20
Extracellular matrix binding
Calcium ion binding
Collagen binding
Cell adhesion molecule binding
Integrin binding
Cytoskeletal protein binding
Hormone activity
Endoplasmic reticulum
EDN1 -2.8 1.0
COL6A2 -2.6 1.0
ARSJ -2.2 1.0
B2M -1.6 1.0
P4HA2 -1.6 1.0
CALU -1.5 1.0
Calcium ion binding
S100A16 -2.8 1.0
CCBE1 -2.7 1.0
CDH23 -2.4 1.0
FSTL1 -1.9 1.0
ANXA11 -1.7 1.0
RCN1 -1.5 1.0
D
Relative mRNA level (log2)
SOD1+/+
SOD1+/A272C
GRM8
CDH23
CALU
MMP14
E
Isogenic SOD1+/+ iPSC
Isogenic SOD1+/+ motor neuron
Gene correction
Aberrant gene expression
SOD1+/A272C ALS patient
SOD1+/A272C ALS patient-derived fibroblast
SOD1+/A272C ALS patient-specific iPSC
SOD1+/A272C motor neuron
Reprogramming
Differentiation
RNA-seq

◀ **Figure 4. RNA-seq revealed *SOD1*$^{+/A272C}$-affected early pathways in human motor neurons.** (A) Volcano plot analysis showing significantly altered genes (q value < 0.05) between *SOD1*$^{+/A272C}$ and its isogenic control motor neurons. 535 downregulated genes (green) and 364 upregulated genes (red) were found in *SOD1*$^{+/A272C}$ motor neurons compared with its isogenic control motor neurons. (B and C) GO terms based cellular_component and molecular_function enrichment analysis of the significantly downregulated gene sets in *SOD1*$^{+/A272C}$ motor neurons. Heatmap of dysregulated genes between *SOD1*$^{+/A272C}$ motor neurons and its isogenic control motor neurons. Number of altered genes in each GO term is indicated by size of the bubble. (D) RT-qPCR analysis of dysregulated genes in *SOD1*$^{+/A272C}$ motor neurons and its isogenic control motor neurons. Values were normalized against *GAPDH*. Data were presented as mean ± SEM, $n = 4$, *$P < 0.05$, **$P < 0.01$, ***$P < 0.001$. (E) A proposed strategy generating ALS disease model using iPSC, gene editing, and cell differentiation approaches. Fibroblasts are obtained from ALS patient bearing *SOD1*$^{+/A272C}$ mutation and reprogrammed to iPSCs. Isogenic control is created via gene editing. RNA-seq is performed in motor neurons to uncover *SOD1*$^{+/A272C}$-affected early events underlying ALS pathogenesis. RNA-seq analysis shows *SOD1*$^{+/A272C}$ mutation induces aberrant gene expression.

Motor neuron differentiation protocols were reported from embryoid body based strategy to single cell based differentiation (Boulting et al., 2011; Qu et al., 2014; Du et al., 2015). There are apparent limitations regarding the complicacy of differentiation procedure and the purity of neurons generated. By optimizing the combination of chemical molecules and simplifying extracellular matrix, we developed a robust and easy-handle method to produce human ISL1^{+} motor neurons at around 70% purity. As we learned from animal models, biological changes of ALS started early in life however the manifestation of the disease may become clinically apparent much later. For example, the dysfunction of neuromuscular junction (NMJ) and the loss of synapse and axons occur before onset of ALS symptom. The initial alterations are not well studied due to the lack of a precisely-controlled disease model (Conforti et al., 2007; Kanning et al., 2010). Taking advantage of iPSC and directed cell differentiation *in vitro*, it is now feasible to collect different phases of motor neurons bearing ALS-related mutation to comprehensively investigate pathogenesis of ALS including early events. Notably, differences in gene expression profiles of ALS motor neurons without obvious degeneration were carefully explored and directly compared to their gene-corrected controls in our study. Therefore, our ALS iPSC disease model has a great value and advantage in studying early events of ALS progression.

SOD1$^{+/A272C}$ encodes for an aspartic acid to alanine mutation in SOD1 protein, and it is a prevalent mutation in Europe ALS patients. In contrast, *FUS*$^{+/G1566A}$ bears an arginine to arginine nonsense alteration. Although this mutation was reported in clinical cases (Drepper et al., 2011; Lai et al., 2011), the underlying pathogenic mechanism was unknown. Our work demonstrated for the first time that both *SOD1*$^{+/A272C}$ and *FUS*$^{+/G1566A}$ mutations are able to be corrected by CRISPR/Cas9-mediated gene editing technique in ALS patient specific iPSCs. We also developed an isogenic iPSC disease model for studying the pathogenic effect of *SOD1*$^{+/A272C}$ on early stage of motor neuron pathology. Comparing ALS motor neurons with their disease-free controls, we found that several pathways may be essentially implicated in SOD1-associated early ALS pathogenesis. In *SOD1*$^{+/A272C}$ motor neurons, we identified transcriptional changes in a subset of genes involved in signal transduction, organization of extracellular matrix, cellular homeostasis maintenance, and neurogenesis, etc. Dysfunctional cell-cell signaling has been reported as one of the causes contributing to death of motor neurons. For example, toxic factors secreted from astrocytes progressively damage motor neurons in ALS (Nagai et al., 2007). Previous study showed that GRM3 and GRM5, which are glutamate metabotropic receptors belonging to G protein-coupled receptors (GPCR) family, are highly expressed in human ALS pathological specimens. This observation indicates that cell communication might play critical regulatory roles in underlying ALS pathogenesis (Aronica et al., 2001). Our data showed that GRM8, another member of GPCR, was upregulated in *SOD1*$^{+/A272C}$ motor neurons as well. Notably, the role of glutamate-mediated excitotoxicity in ALS pathogenesis has been well studied, and the only FDA-approved ALS-treated drug Riluzole aims at decreasing the level of glutamate (Zhu et al., 2014).

Loss of calcium homeostasis is another factor which may contribute to motor neuron death in ALS cases (Bruijn et al., 2004). Many calcium-binding proteins were downregulated in ALS patient autopsy specimens (Alexianu et al., 1994). In this study, we identified that calcium-binding protein CDH23, a calcium dependent cell-cell adhesion glycoproteins, was reduced in *SOD1*$^{+/A272C}$ motor neurons. In addition, calumenin (CALU), another calcium-binding protein that safeguards ER homeostasis (Lee et al., 2013), decreased in *SOD1*$^{+/A272C}$ motor neurons. Both dysfunction of calcium signaling and increased ER stress have been linked to ALS-associated neurodegeneration at early stages (Lee et al., 2016; Woehlbier et al., 2016) .

Extracellular matrix has been shown to influence neuronal degeneration (Soleman et al., 2013; De Luca and Papa, 2016). However the role of matrix metalloproteinases in neuronal disorder diseases has not been well established (Cirillo et al., 2016). We identified that matrix metalloproteinase 14 (MMP14) decreased in *SOD1*$^{+/A272C}$ motor neurons. Consistently, a previous study identified that MMP14 significantly decreased in SOD1^{G93A} transgenic mice and SALS patients (Kudo et al., 2010). Thus, the role of extracellular matrix in ALS pathogenesis may represent an interesting topic for future study.

Enhanced neurogenesis in ALS transgenic mice model has been recently reported (Chi et al., 2007; Lee et al., 2011). While neural stem cells (NSCs) isolated from presymptomatic ALS transgenic mice proliferate and differentiate well, NSCs from late stage ALS mice lose their functional activities (Lee et al., 2011). Our RNA-seq data also supported that ALS was associated with a tendency of increased neurogenesis at gene expression level, despite that no obvious phenotypic defect was observed.

In summary, our study successfully established valuable isogenic ALS iPSC disease models using CRISPR/Cas9-mediated gene editing. More importantly, we identified potential molecular hallmarks associated with early pathological events of motor neuron with *SOD1*$^{+/A272C}$ ALS mutation. The experimental system established in our report may hold a potential for further mechanistic study, drug screening, and inspire autologous therapy against ALS in the future (Fig. 4E).

MATERIALS AND METHODS

Cell culture

Human ALS patient fibroblasts (ND29149, heterozygous for *SOD1*$^{+/A272C}$; ND29563, heterozygous for *FUS*$^{+/G1566A}$) were obtained from Coriell Institute (http://www.coriell.org). All fibroblasts were maintained in high glucose DMEM (Invitrogen) supplemented with 15% fetal bovine serum (FBS, Hyclone), 1% glutamax (Invitrogen), 1% non-essential amino acids (Invitrogen), 1% penicillin/streptomycin (Invitrogen). iPSCs were maintained on a layer of mitomycin C-inactivated mouse embryonic fibroblast (MEF) with cDF12 medium or on Matrigel (BD Biosciences) with mTeSR1 medium (Stem Cell Technologies) as previously described (Liu et al., 2014; Ding et al., 2015; Fu et al., 2016).

iPSC generation

ALS fibroblasts were reprogrammed using the episomal vectors as described (Okita et al., 2011; Liu et al., 2014; Fu et al., 2016).

Plasmid construction

Guide RNA (gRNA) was designed in http://crispr.mit.edu. gRNAs were cloned into pCAGmCherry-gRNA (Suzuki et al., 2016). pCAG-1BPNLS-Cas9-1BPNLS-2AGFP was used as Cas9 and GFP expression plasmid (Suzuki et al., 2016). For the experiments using CRISPR/Cas9 system together with a plasmid donor including neo, a gRNA_cloning vector (Mali et al., 2013) was employed, and the homology arms were cloned into Neo-pCR2.1 vector (donor plasmid) as previously described (Duan et al., 2015).

Primers for gRNA are listed as follows: *FUS*-gRNA-F, 5′-GAAGGAAAATTAACTCAGG-3′; *FUS*-gRNA-R, 5′-CTGAGTTAATTTTCCTTCC-3′; *FUS*-gRNA-mCherry-F, 5′-GAGGCCGTATTAATTAGCC-3′; *FUS*-gRNA-mCherry-R, 5′-GGCTAATTAATACGGCCTCC-3′; *SOD1*-gRNA-mCherry-F, 5′-ATGTGACTGCTGCCAAAGA-3′; *SOD1*-gRNA-mCherry-R, 5′-CTTTGGCAGCAGTCACATC-3′. Primers for homology arms are listed as follows: *FUS*-LA-F, 5′-ATAGGGCCCTGGTACTGAGGTATGTGCGTGTTTTCCAAAGAA-3′; *FUS*-LA-R, 5′-CCCTCGAGTTAATTTTCCTTCCCTCTCCACTACTGGTTACAAC-3′; *FUS*-RA-F, 5′-CGGGATCCCTCAGGGGGAGTGAATCTGTAGACCCAC-3′; *FUS*-RA-R, 5′-CCCAAGCTTCTCAAGCCCTCTGAGTACAGGCAGGATG-3′. The ssODNs used to repair mutant allele are listed as follows: *FUS*-ssODN, 5′-ACCTGGGGAGCCAGGCTAATTAATACGGCCTCTCCCTGCGATCCTGTCTGTGCTCACCCCTGC-3′; *SOD1*-ssODN, 5′-TCTTCAATAGACACATCGGCCACACCATCTTTGTCAGCAGTCACATTGCCCAAGTCTCCAACAT-3′.

Targeted gene correction in ALS iPSCs via CRISPR/Cas9 system

Correction of *FUS*$^{+/G1566A}$ mutation via homologous recombination was carried out as previously described with some modifications (Duan et al., 2015; Pan et al., 2016).

In brief, *FUS*$^{+/G1566A}$ iPSCs cultured on Matrigel using mTeSR1 medium were pre-treated with ROCK inhibitor Y-27632 (Sigma) overnight and dissociated with TrypLE (Invitrogen). For electroporation, 5 × 10^6 cells were collected. Cells were resuspended in 100 µL Opti-Mem (Gibco) containing 7 µg Cas9, 7 µg gRNA-FUS, and 7 µg donor plasmids. After electroporation, cells were plated on mitomycin C-inactivated DR4-MEF feeder. Two days after electroporation, G418 (50 µg/mL, Gibco) were added to enrich gene-targeted cells. After additional two weeks, G418-resistant clones were picked and verified by genomic PCR and DNA sequencing. The correct clones were expanded and used for further experiments. Primers used for identifying correct clones are listed as followed: P5, 5′-AATGATACCAGTTGCTTGATGGATACTAGGTGCTT-3′; P6, 5′-ACCTTTCTGCTCTTGGGTTAATGTTACGCTCT-3′. Neomycin-resistance cassette was removed from gene-targeted iPSCs as previously described (Duan et al., 2015; Pan et al., 2016).

Gene correction with single-stranded oligodeoxynucleotide (ssODN) as repair template was carried out as previously described with some modifications (Peters et al., 2013). Briefly, 5 × 10^6 iPSCs were resuspended in 100 µL Opti-Mem (Gibco), containing 8 µg Cas9-2A-GFP, 4 µg gRNA-mCherry, and 8 µg ssODN. After electroporation, cells were plated on Matrigel-coated plate and cultured with mTeSR1 medium. 48 hours after electroporation, mCherry$^+$/GFP$^+$ cells were collected by FACS and reseeded on MEF feeder. Two weeks later, clones were picked and screened by RFLP and DNA sequencing. Primers used are listed as follows: P1-*FUS*-S-F, 5′-GAGAAAGTGGTTTCATTTTGAGGGCTAGGTGGA-3′; P2-*FUS*-S-R, 5′-TTGTTTGAGCCTCACCATTAAAAGGGCCAAAAG-3′; P3-*SOD1*-S-F, 5′-CCCATCTTTCTTCCCAGAGCATTAGTGTGTAGACG-3′; P4-*SOD1*-S-R, 5′-ACAAAATGTTCTGTTTAACAAGTGAGAAACCCAATCCT-3′.

Motor neuron differentiation

Motor neuron differentiation was performed as previously described with some modifications (Liu et al., 2014; Maury et al., 2015). Briefly, iPSCs were passaged onto MEF feeder and then motor neuron differentiation was initialized. Medium 1 containing 50% advanced DMEM/F12 (Invitrogen), 50% neurobasal (Invitrogen), 1× N2 (Invitrogen), 1× B27 (Invitrogen), 1% GlutaMAX (Invitrogen), 1% NEAA (Invitrogen), 1% penicillin/streptomycin (Invitrogen), 4 µmol/L

CHIR99021 (Cellagentech), 3 μmol/L SB431542 (Cellagentech), 1 μmol/L dorsomorphin (Sigma), and 0.1 μmol/L compound E (EMD Chemicals), 0.2% heparin (Sigma) was used for the first 3 days. Then, RA (100 nmol/L) was added to the medium for another 3 days (medium 2). On day 6, dorsomorphin was then removed, and SAG (500 nmol/L) was added (medium 3). At day 8, cells were dissociated to single cells with Accumax (Millipore) and reseeding on Matrigel-coated plates. DAPT and laminin were added on day 9 to accelerate MN differentiation.

Teratoma assay

5 × 10^6 iPSCs were injected subcutaneously into NOD-SCID mice (male, 6–8 weeks old). After 8–12 weeks of injection, teratomas were dissected and analyzed by immunostaining (Liu et al., 2014; Fu et al., 2016; Ren et al., 2017). All murine experiments were conducted in compliance with animal protocols approved by the Chinese Academy of Science Institutional Animal Care and Use Committee.

Immunofluorescence

Cells were washed once using PBS, fixed with 4% paraformaldehyde for 30 min at room temperature, permeabilized with 0.4% Triton X-100 in PBS for 20 min and were blocked in donkey serum (10% in PBS) for 60 min. cells were incubated with primary antibodies overnight in 4°C. Then, the cells were stained with secondary antibodies and Hoechst 33342 at room temperature. The antibodies used in immunofluorescence assay are as follows: anti-NANOG (Abcam, 21624, 1:200), anti-OCT3/4 (Santa Cruz, 5279, 1:100), anti-SOX2 (Santa Cruz,17320, 1:100), anti-TUJ1 (Sigma, T2220, 1:500), anti-FOXA2 (CST, 8186, 1:200), anti-SMA (Sigma, A5228, 1:200), anti-ISL1 (Abcam, ab20670 1:250), anti-HB9 (DSHB, 81.5C10, 1:50), anti-MAP2 (Sigma, 4403, 1:500).

Bisulfite sequencing analysis

Bisulfite sequencing of the *OCT4* promoter was carried out as previously described (Liu et al., 2011a; Duan et al., 2015). Primers used are listed as follows: meF-*OCT4*, 5′-ATTTGTTTTTTGGGTAGTTAAAG GT-3′; meR-*OCT4*, 5′-CCAACTATCTTCATCTTAATAACATCC-3′.

Off-target analysis

Off-target analysis was performed following a previous study with modifications (Liang et al., 2015). The off-target sites were predicated in http://crispr.genome-engineering.org/. The top 3 off-target sites were analyzed by DNA sequencing. Primers used are listed as follows: *FUS*-OT1-F, 5′-AGCTTCTGCCAAGATCTGGTTTCTTCGTC-3′; *FUS*-OT1-R, 5′-TTTCAAAGACACACACCACCCTGACCAT-3′; *FUS*-OT2-F, 5′-ACCTGCCATCATAGTCTAGTATCGTTCTCT-3′; *FUS*-OT 2-R, 5′-AACAATCGACCCACTCCCATCATGACC-3′; *FUS*-OT3-F, 5′-A CATTTCTGGCTCAGCTTCAATCATGGT-3′; *FUS*-OT3-R, 5′-ATTC CTTGGCTTGAAGATTATAGGTGAC-3′; *SOD1*-OT1-F, 5′-TTCAATG GCATGTAGGGAAGGACCAAAGTTGAA-3′; *SOD1*-OT1-R, 5′-ATAG CTCTATAAATGCCAGCTGTTGAAGGCAGG-3′; *SOD1*-OT2-F, 5′-AT ATTCAGATGGCCTGAATGTCCAGATGCTGTG-3′; *SOD1*-OT2-R, 5′-CTCACAAGTAGGGTGATAACAGCTGCCATACTG-3′; *SOD1*-OT3-F, 5′-CCCTTCTCCAGCCTTACTCTTTCCATATAGCC-3′; *SOD1*-OT3-R, 5′-TCTGAGCCTGCTTTAAGTCCCAGACACGTT-3′.

RT-qPCR

Total RNA was extracted from cells by TRIzol Reagent (Invitrogen). Around 2 μg total RNA was employed for cDNA synthesis with reverse transcription Master Mix (Promega). RT-qPCR was performed using iTaq Universal SYBR Green Supermix (Bio-Rad) on a CFX384 Real-Time PCR system (Bio-Rad). All data were normalized by *GAPDH* transcript and calculated using the ΔΔCq method. Primers used for RT-qPCR are listed in Table S3.

RNA-seq library construction

Motor neurons were collected for constructing sequencing library. Total RNA was isolated with TRIzol Reagent (Invitrogen). After qualified by Fragment Analyzer (Advanced Analytical), 3 μg RNA was used to construct sequencing library NEBNext® Ultra™ RNA Library Prep Kit for Illumina® (NEB, USA) following manufacturer's recommendations (Li et al., 2016).

RNA-seq data analysis

Clean data (clean reads) were obtained from raw data (raw reads) by removing reads containing adapter, reads containing ploy-N and low quality reads (>50% of reads with Q phred ≤ 20) from raw data. Then reads were mapped to the Homo sapiens hg19 genome (UCSC) by Tophat2 (Trapnell et al., 2009). Differential expression analysis was performed using the DESeq R package (1.18.0) (Anders and Huber, 2010).

Gene Ontology (GO) enrichment analysis of differentially expressed genes was implemented by the GOseq R package, in which gene length bias was corrected. GO terms with corrected *P* value less than 0.05 were considered significantly enriched by differential expressed genes (Anders and Huber, 2010).

Statistical analysis

Results were presented as mean ± SEM. Two tailed Student's *t*-test was conducted using Graph-Pad Prism Software. *P* values < 0.05 were considered statistically significant (*).

Accession number

All of the RNA-seq data have been deposited in GEO under the accession number GSE95089.

ACKNOWLEDGMENTS

This work was supported by the National Basic Research Program of China (973 Program) (Nos. 2015CB964800, 2014CB964600, and 2014CB910503), the National High Technology Research and Development Program of China (2015AA020307), CAS (XDA01020312, KJZDEW-TZ-L05, and CXJJ-16M271), the National Natural Science Foundation of China (Grant Nos. 81625009,

81330008, 31222039, 81371342, 81300261, 81300677, 81271266, 81471414, 81422017, 81401159, 31671429, 81601233, 81671377, 31601109, and 31601158), Beijing Natural Science Foundation (7141005 and 5142016), Program of Beijing Municipal Science and Technology Commission (Z151100003915072), Key Research Program of the Chinese Academy of Sciences (KJZDEW-TZ-L05), the Thousand Young Talents Program of China, and the State Key Laboratory of Stem Cell and Reproductive Biology (2016SRLabKF13). The Lab of J.C.I.B. was supported by UCAM, the G. Harold and Leila Y. Mathers Charitable Foundation, the Leona M. and Harry B. Helmsley Charitable Trust (2012-PG-MED002) and the Moxie Foundation.

ABBREVIATIONS

ALS, amyotrophic lateral sclerosis; CRISPR/Cas9, clustered regularly interspaced short palindromic repeat/Cas9; GO, gene ontology; HR, homologous recombination; iPSC, induced pluripotent stem cell; MN, motor neuron; RNA-seq, genome-wide RNA sequencing; ssODN, single-stranded oligodeoxynucleotide.

COMPLIANCE WITH ETHICS GUIDELINES

Lixia Wang, Fei Yi, Lina Fu, Jiping Yang, Si Wang, Zhaoxia Wang, Keiichiro Suzuki, Liang Sun, Xiuling Xu, Yang Yu, Jie Qiao, Juan Carlos Izpisua Belmonte, Ze Yang, Yun Yuan, Jing Qu, and Guang-Hui Liu declare that they have no conflict of interest. All institutional and national guidelines for the care and use of laboratory animals were followed.

REFERENCES

Al-Chalabi A, Hardiman O (2013) The epidemiology of ALS: a conspiracy of genes, environment and time. Nat Rev Neurol 9:617–623

Alexianu ME, Ho BK, Mohamed AH, La Bella V, Smith RG, Appel SH (1994) The role of calcium-binding proteins in selective motoneuron vulnerability in amyotrophic lateral sclerosis. Ann Neurol 36:846–858

Anders S, Huber W (2010) Differential expression analysis for sequence count data. Genome Biol 11:1–12

Aronica E, Catania MV, Geurts J, Yankaya B, Troost D (2001) Immunohistochemical localization of group I and II metabotropic glutamate receptors in control and amyotrophic lateral sclerosis human spinal cord: upregulation in reactive astrocytes. Neuroscience 105:509–520

Baechtold H, Kuroda M, Sok J, Ron D, , Lopez BS, Akhmedov AT (1999) Human 75-kDa DNA-pairing protein is identical to the pro-oncoprotein TLS/FUS and is able to promote D-loop formation. J Biol Chem 274:34337–34342

Bosco DA, Morfini G, Karabacak NM, Song Y, Gros-Louis F, Pasinelli P, Goolsby H, Fontaine BA, Lemay N, McKenna-Yasek D et al (2010) Wild-type and mutant SOD1 share an aberrant conformation and a common pathogenic pathway in ALS. Nat Neurosci 13:1396–1403

Boulting GL, Kiskinis E, Croft GF, Amoroso MW, Oakley DH, Wainger BJ, Williams DJ, Kahler DJ, Yamaki M, Davidow L, Rodolfa CT et al (2011) A functionally characterized test set of human induced pluripotent stem cells. Nat Biotechnol 29:279–286.

Bruijn LI, Houseweart MK, Kato S, Anderson KL, Anderson SD, Ohama E, Reaume AG, Scott RW, Cleveland DW (1998) Aggregation and motor neuron toxicity of an ALS-linked SOD1 mutant independent from wild-type SOD1. Science 281:1851–1854

Bruijn LI, Miller TM, Cleveland DW (2004) Unraveling the mechanisms involved in motor neuron degeneration in ALS. Annu Rev Neurosci 27:723–749

Casci I, Pandey UB (2015) A fruitful endeavor: modeling ALS in the fruit fly. Brain Res 1607:47–74

Chen H, Qian K, Du Z, Cao J, Petersen A, Liu H, Blackbourn LW, Huang CL, Errigo A, Yin Y et al (2014) Modeling ALS with iPSCs reveals that mutant SOD1 misregulates neurofilament balance in motor neurons. Cell Stem Cell 14:796–809.

Chi L, Gan L, Luo C, Luo C, Lien L, Liu R (2007) Temporal response of neural progenitor cells to disease onset and progression in amyotrophic lateral sclerosis-like transgenic mice. Stem Cells Dev 16:5579–5588

Cirillo G, Colangelo AM, De Luca C, Savarese L, Barillari MR, Alberghina L, Papa M (2016) Modulation of matrix metalloproteinases activity in the ventral horn of the spinal cord restores neuroglial synaptic homeostasis and neurotrophic support following peripheral nerve injury. PLoS ONE 11:152750–152762

Conforti L, Adalbert R, Coleman MP (2007) Neuronal death: where does the end begin? Trends Neurosci 30:159–166

Corti S, Nizzardo M, Simone C, Falcone M, Nardini M, Ronchi D, Donadoni C, Salani S, Riboldi G, Magri F et al (2012) Genetic correction of human induced pluripotent stem cells from patients with spinal muscular atrophy. Sci Transl Med 4:165–197

Crozat A, Aman P, Mandahl N, Mandahl N, Ron D (1993) Fusion of CHOP to a novel RNA-binding protein in human myxoid liposarcoma. Nature 363:640–644

De Luca C, Papa MA (2016) Looking inside the matrix: perineuronal nets in plasticity, maladaptive plasticity and neurological disorders. Neurochem Res 41:1507–1515

Ding Q, Regan SN, Xia Y, Oostrom LA, Cowan CA, Musunuru K (2013) Enhanced efficiency of human pluripotent stem cell genome editing through replacing TALENs with CRISPRs. Cell Stem Cell 12:393–394

Ding Z, Sui L, Ren R, Liu Y, Xu X, Fu L, Bai R, Yuan T, Hao Y, Zhang W et al (2015) A widely adaptable approach to generate integration-free iPSCs from non-invasively acquired human somatic cells. Protein Cell 6:386–389

Drepper C, Herrmann T, Wessig C, Beck M, Sendtner M (2011) C-terminal FUS/TLS mutations in familial and sporadic ALS in Germany. Neurobiol Aging 32:548–552

Du ZW, Chen H, Liu H, Lu J, Qian K, Huang CL, Zhong X, Fan F, Zhang SC (2015) Generation and expansion of highly pure motor neuron progenitors from human pluripotent stem cells. Nat Commun 6:6626–6638

Duan S, Yuan G, Liu X, Ren R, Li J, Zhang W, Wu J, Xu X, Fu L, Li Y et al (2015) PTEN deficiency reprogrammes human neural stem cells towards a glioblastoma stem cell-like phenotype. Nat Commun 6:10068–10082

Egawa N, Kitaoka S, Tsukita K, Naitoh M, Takahashi K, Yamamoto T, Adachi F, Kondo T, Okita K, Asaka I, Aoi T et al (2012) Drug screening for ALS using patient-specific induced pluripotent stem cells. Sci Transl Med 4:145104–145112

Fischer LR, Culver DG, Tennant P, Davis AA, Wang M, Castellano-Sanchez A, Khan J, Polak MA, Glass JD (2004) Amyotrophic lateral sclerosis is a distal axonopathy: evidence in mice and man. Exp Neurol 185:232–240

Fu L, Xu X, Ren R, Wu J, Zhang W, Yang J, Ren X, Wang S, Zhao Y, Sun L et al (2016) Modeling xeroderma pigmentosum associated neurological pathologies with patients-derived iPSCs. Protein Cell 7:210–221

Higelin J, Demestre M, Putz S, Delling JP, Jacob C, Lutz AK, Bausinger J, Huber AK, Klingenstein M, Barbi G et al (2016) FUS mislocalization and vulnerability to DNA damage in ALS patients derived hiPSCs and aging motoneurons. Front Cell Neurosci 10:290–311

Huang C, Zhou H, Tong J, Chen H, Liu YJ, Wang D, Wei X, Xia XG (2011) FUS transgenic rats develop the phenotypes of amyotrophic lateral sclerosis and frontotemporal lobar degeneration. PLoS Genet 7:1–10

Ichiyanagi N, Fujimori K, Yano M, Ishihara-Fujisaki C, Sone T, Akiyama T, Okada Y, Akamatsu W, Matsumoto T, Ishikawa M et al (2016) Establishment of in vitro FUS-associated familial amyotrophic lateral sclerosis model using human induced pluripotent stem cells. Stem Cell Rep 6:496–510

Julien J-P, Kriz J (2006) Transgenic mouse models of amyotrophic lateral sclerosis. Biochim Biophys Acta (BBA) 1762:1013–1024

Kanning KC, Kaplan A, Henderson CE (2010) Motor neuron diversity in development and disease. Annu Rev Neurosci 33:409–440

Kiskinis E, Sandoe J, Williams LA, Boulting GL, Moccia R, Wainger BJ, Han S, Peng T, Thams S, Mikkilineni S et al (2014) Pathways disrupted in human ALS motor neurons identified through genetic correction of mutant SOD1. Cell Stem Cell 14:781–795

Kubben N, Zhang W, Wang L, Voss TC, Yang J, Qu J, Liu GH, Misteli T (2016) Repression of the antioxidant NRF2 pathway in premature aging. Cell 165:1361–1374

Kudo LC, Parfenova L, Vi N, Lau K, Pomakian J, Valdmanis P, Rouleau GA, Vinters HV, Wiedau-Pazos M, Karsten SL (2010) Integrative gene–tissue microarray-based approach for identification of human disease biomarkers: application to amyotrophic lateral sclerosis. Hum Mol Genet 19:3233–3253

Lai SL, Abramzon Y, Schymick JC, Stephan DA, Dunckley T, Dillman A, Cookson M, Calvo A, Battistini S, Giannini F et al (2011) FUS mutations in sporadic amyotrophic lateral sclerosis. Neurobiol Aging 32:551–554

Lattante S, Rouleau GA, Kabashi E (2013) TARDBP and FUS mutations associated with amyotrophic lateral sclerosis: summary and update. Hum Mutat 34:812–826

Lee JC, Jin Y, Jin J, Kang BG, Nam DH, Joo KM, Cha CI (2011) Functional neural stem cell isolation from brains of adult mutant SOD1 (SOD1(G93A)) transgenic amyotrophic lateral sclerosis (ALS) mice. Neurol Res 33:33–37

Lee JH, Kwon DH (2013) Calumenin has a role in the alleviation of ER stress in neonatal rat cardiomyocytes. Biochem Biophys Res Commun 439:327–332

Lee S, Shang Y, Redmond SA, Urisman A, Tang AA, Li KH, Burlingame AL, Pak RA, Jovicic A, Gitler AD et al (2016) Activation of HIPK2 promotes ER stress-mediated neurodegeneration in amyotrophic lateral sclerosis. Neuron 91:41–55

Lenzi J, De Santis R, de Turris V, Morlando M, Laneve P, Calvo A, Caliendo V, Chio A, Rosa A, Bozzoni I (2015) ALS mutant FUS proteins are recruited into stress granules in induced pluripotent stem cell-derived motoneurons. Dis Model Mech 8:755–766

Li Y, Balasubramanian U, Cohen D, Zhang P-W, Mosmiller E, Sattler R, Maragakis NJ, Rothstein JD (2015) A comprehensive library of familial human amyotrophic lateral sclerosis induced pluripotent stem cells. PLOS ONE 10:118266–118279

Li Y, Zhang W, Chang L, Han Y, Sun L, Gong X, Tang H, Liu Z, Deng H, Ye Y et al (2016) Vitamin C alleviates aging defects in a stem cell model for Werner syndrome. Protein Cell 7:478–488

Liang P, Xu Y, Zhang X, Ding C, Huang R, Zhang Z, Lv J, Xie X, Chen Y, Li Y et al (2015) CRISPR/Cas9-mediated gene editing in human tripronuclear zygotes. Protein Cell 6:363–372

Liu G-H, Ding Z, Izpisua Belmonte JC (2012a) iPSC technology to study human aging and aging-related disorders. Curr Opin Cell Biol 24:765–774

Liu GH, Barkho BZ, Ruiz S, Diep D, Qu J, Yang S-L, Panopoulos AD, Suzuki K, Kurian L, Walsh C, Thompson J et al (2011a) Recapitulation of premature ageing with iPSCs from Hutchinson-Gilford progeria syndrome. Nature 472: 221–225

Liu GH, Qu J, Suzuki K, Nivet E, Li M, Montserrat N, Yi F, Xu X, Ruiz S, Zhang W et al (2012b) Progressive degeneration of human neural stem cells caused by pathogenic LRRK2. Nature 491:603–607

Liu GH, Suzuki K, Li M, Qu J, Montserrat N, Tarantino C, Gu Y, Yi F, Xu X, Zhang W et al (2014) Modelling Fanconi anemia pathogenesis and therapeutics using integration-free patient-derived iPSCs. Nat Commun 5:4330–4347

Liu GH, Suzuki K, Qu J, Sancho-Martinez I, Yi F, Li M, Kumar S, Nivet E, Kim J, Soligalla RD et al (2011b) Targeted gene correction of laminopathy-associated LMNA mutations in patient-specific iPSCs. Cell Stem Cell 8:688–694

Liu X, Chen J, Liu W, Li X, Chen Q, Liu T, Gao S, Deng M (2015) The fused in sarcoma protein forms cytoplasmic aggregates in motor neurons derived from integration-free induced pluripotent stem cells generated from a patient with familial amyotrophic lateral sclerosis carrying the FUS-P525L mutation. Neurogenetics 16:223–231

Mali P, Yang L, Esvelt KM, Aach J, Guell M, DiCarlo JE, Norville JE, Church GM (2013). RNA-guided human genome engineering via Cas9. Science 339:823–826

Maury Y, Come J, Piskorowski RA, Salah-Mohellibi N, Chevaleyre V, Peschanski M, Martinat C, Nedelec S (2015) Combinatorial analysis of developmental cues efficiently converts human pluripotent stem cells into multiple neuronal subtypes. Nat Biotechnol 33:89–96

Nagai M, Re DB, Nagata T, Chalazonitis A, Jessell TM, Wichterle H, Przedborski S (2007) Astrocytes expressing ALS-linked mutated SOD1 release factors selectively toxic to motor neurons. Nat Neurosci 10:615–622

Okita K, Matsumura Y, Sato Y, Okada A, Morizane A, Okamoto S, Hong H, Nakagawa M, Tanabe K, Tezuka K et al (2011) A more efficient method to generate integration-free human iPS cells. Nat Methods 8:409–412

Pan H, Guan D, Liu X, Li J, Wang L, Wu J, Zhou J, Zhang W, Ren R, Zhang W et al (2016) SIRT6 safeguards human mesenchymal stem cells from oxidative stress by coactivating NRF2. Cell Research 26:190–205

Pasinelli P, Brown RH (2006) Molecular biology of amyotrophic lateral sclerosis: insights from genetics. Nat Rev Neurosci 7:710–723

Peters DT, Cowan CA, Musunuru K (2013) Genome editing in human pluripotent stem cells. StemBook.

Qu Q, Li D, Louis KR, Li X, Yang H, Sun Q, Crandall SR, Tsang S, Zhou J, Cox CL et al (2014) High-efficiency motor neuron differentiation from human pluripotent stem cells and the function of Islet-1. Nat Commun 5:3449–3462

Ren R, Deng L, Xue Y, Suzuki K, Zhang W, Yu Y, Wu J, Sun L, Gong X, Luan H et al (2017) Visualization of aging-associated chromatin alterations with an engineered TALE system. Cell Res. doi:10.1038/cr.2017.18

Robberecht W, Philips T (2013) The changing scene of amyotrophic lateral sclerosis. Nat Rev Neurosci 14:248–264

Rosen DR, Siddique T, Patterson D, Figlewicz DA, Sapp P, Hentati A, Donaldson D, Goto J, O'Regan JP, Deng H-X et al (1993) Mutations in Cu/Zn superoxide dismutase gene are associated with familial amyotrophic lateral sclerosis. Nature 362:59–62

Sharma A, Lyashchenko AK, Lu L, Nasrabady SE, Elmaleh M, Mendelsohn M, Nemes A, Tapia JC, Mentis GZ, Shneider NA (2016) ALS-associated mutant FUS induces selective motor neuron degeneration through toxic gain of function. Nat Commun 7:10465–10479

Soleman S, Filippov MA, Dityatev A, Fawcett JW (2013) Targeting the neural extracellular matrix in neurological disorders. Neuroscience 253:194–213

Suzuki K, Tsunekawa Y, Hernandez-Benitez R, Wu J, Zhu J, Kim EJ, Hatanaka F, Yamamoto M, Araoka T, Li Z et al (2016) In vivo genome editing via CRISPR/Cas9 mediated homology-independent targeted integration. Nature 540:144–149

Suzuki K, Yu C, Qu J, Li M, Yao X, Yuan T, Goebl A, Tang S, Ren R, Aizawa E et al (2014) Targeted gene correction minimally impacts whole-genome mutational load in human-disease-specific induced pluripotent stem cell clones. Cell Stem Cell 15:31–36

Takahashi K, Tanabe K, Ohnuki M, Narita M, Ichisaka T, Tomoda K, Yamanaka S (2007) Induction of pluripotent stem cells from adult human fibroblasts by defined factors. Cell 131:861–872

Trapnell C, Pachter L, Salzberg SL (2009) TopHat: discovering splice junctions with RNA-Seq. Bioinformatics 25:1105–1111

Turner BJ, Talbot K (2008) Transgenics, toxicity and therapeutics in rodent models of mutant SOD1-mediated familial ALS. Prog Neurobiol 85:94–134

Vance C, Rogelj B, Hortobágyi T, De Vos KJ, Nishimura AL, Sreedharan J, Hu X, Smith B, Ruddy D, Wright P et al (2009) Mutations in FUS, an RNA processing protein, cause familial amyotrophic lateral sclerosis type 6. Science 323:1208–1211

Veres A, Gosis BS, Ding Q, Collins R, Ragavendran A, Brand H, Erdin S, Talkowski ME, Musunuru K (2014) Low incidence of off-target mutations in individual CRISPR-Cas9 and TALEN targeted human stem cell clones detected by whole-genome sequencing. Cell Stem Cell 15:27–30

Wainger Brian J, Kiskinis E, Mellin C, Wiskow O, Han Steve SW, Sandoe J, Perez Numa P, Williams Luis A, Lee S, Boulting G et al (2014) Intrinsic membrane hyperexcitability of amyotrophic lateral sclerosis patient-derived motor neurons. Cell Reports 7:1–11

Woehlbier U, Colombo A, Saaranen MJ, Perez V, Ojeda J, Bustos FJ, Andreu CI, Torres M, Valenzuela V, Medinas DB et al (2016) ALS-linked protein disulfide isomerase variants cause motor dysfunction. EMBO J 35:845–865

Yang Yin M, Gupta Shailesh K, Kim Kevin J, Powers Berit E, Cerqueira A, Wainger Brian J, Ngo Hien D, Rosowski Kathryn A, Schein Pamela A, Ackeifi Courtney A et al (2013) A small molecule screen in stem-cell-derived motor neurons identifies a kinase inhibitor as a candidate therapeutic for ALS. Cell Stem Cell 12:713–726

Zhang W, Li J, Suzuki K, Qu J, Wang P, Zhou J, Liu X, Ren R, Xu X, Ocampo A et al (2015) A Werner syndrome stem cell model unveils heterochromatin alterations as a driver of human aging. Science 348:1160–1163

Zhu Y, Fotinos A, Mao LL, Atassi N, Zhou EW, Ahmad S, Guan Y, Berry JD, Cudkowicz ME, Wang X (2014) Neuroprotective agents target molecular mechanisms of disease in ALS. Drug Discov Today 20:65–75

Zhu Y, Fotinos A, Mao LL, Atassi N, Zhou EW, Ahmad S, Guan Y, Berry JD, Cudkowicz ME, Wang X (2015) Neuroprotective agents target molecular mechanisms of disease in ALS. Drug Discov Today 20:65–75

8

CRISPR/Cas9-mediated gene knockout reveals a guardian role of NF-κB/RelA in maintaining the homeostasis of human vascular cells

Ping Wang[1,4], Zunpeng Liu[2,4], Xiaoqian Zhang[2,4], Jingyi Li[1,4,5], Liang Sun[7], Zhenyu Ju[8], Jian Li[7], Piu Chan[5], Guang-Hui Liu[1,4,5,6,8]✉, Weiqi Zhang[1,4,5]✉, Moshi Song[3,4,6]✉, Jing Qu[2,4,6]✉

[1] National Laboratory of Biomacromolecules, CAS Center for Excellence in Biomacromolecules, Institute of Biophysics, Chinese Academy of Sciences, Beijing 100101, China
[2] State Key Laboratory of Stem Cell and Reproductive Biology, Institute of Zoology, Chinese Academy of Sciences, Beijing 100101, China
[3] State Key Laboratory of Membrane Biology, Institute of Zoology, Chinese Academy of Sciences, Beijing 100101, China
[4] University of Chinese Academy of Sciences, Beijing 100049, China
[5] National Clinical Research Center for Geriatric Disorders, Xuanwu Hospital of Capital Medical University, Beijing 100053, China
[6] Institute of Stem cell and Regeneration, Chinese Academy of Sciences, Beijing 100101, China
[7] The MOH Key Laboratory of Geriatrics, Beijing Hospital, National Center of Gerontology, Beijing 100730, China
[8] Key Laboratory of Regenerative Medicine of Ministry of Education, Institute of Aging and Regenerative Medicine, Jinan University, Guangzhou 510632, China
✉ Correspondence: ghliu@ibp.ac.cn (G.-H. Liu), weiqizhang@aliyun.com (W. Zhang), songmoshi@ioz.ac.cn (M. Song), qujing@ioz.ac.cn (J. Qu)

ABSTRACT

Vascular cell functionality is critical to blood vessel homeostasis. Constitutive NF-κB activation in vascular cells results in chronic vascular inflammation, leading to various cardiovascular diseases. However, how NF-κB regulates human blood vessel homeostasis remains largely elusive. Here, using CRISPR/Cas9-mediated gene editing, we generated *RelA* knockout human embryonic stem cells (hESCs) and differentiated them into various vascular cell derivatives to study how NF-κB modulates human vascular cells under basal and inflammatory conditions. Multi-dimensional phenotypic assessments and transcriptomic analyses revealed that RelA deficiency affected vascular cells via modulating inflammation, survival, vasculogenesis, cell differentiation and extracellular matrix organization in a cell type-specific manner under basal condition, and that RelA protected vascular cells against apoptosis and modulated vascular inflammatory response upon tumor necrosis factor α (TNFα) stimulation. Lastly, further evaluation of gene expression patterns in *IκBα* knockout vascular cells demonstrated that IκBα acted largely independent of RelA signaling. Taken together, our data reveal a protective role of NF-κB/RelA in modulating human blood vessel homeostasis and map the human vascular transcriptomic landscapes for the discovery of novel therapeutic targets.

KEYWORDS NF-κB, RelA, Stem cell, Inflammation, Apoptosis

INTRODUCTION

Blood vessels carry nutrients and oxygen via blood stream throughout the body. Blood vessel homeostasis is critical to health. Disorders of blood vessels impair vascular cell function, resulting in the pathogenesis of severe

Ping Wang, Zunpeng Liu, and Xiaoqian Zhang have contributed equally to this work.

cardiovascular diseases like myocardial infarction, atherosclerosis, and stroke.

Blood vessels consist of three layers, namely the tunica intima, tunica media, and tunica adventitia. The major cell types of these three layers are vascular endothelial cells (VECs), vascular smooth muscle cells (VSMCs), and mesenchymal stem cells (MSCs), respectively. VECs compose a monolayer of cells called endothelium that modulates nutrients transport, vascular tone, host defense, homeostasis and angiogenesis (Galley and Webster, 2004). VSMCs constitute the majority of blood vessels and regulate local blood pressure via vasoconstriction and vasodilation. MSCs are part of the adventitia that modulates vascular inflammatory response, trophic supply, and vessel damage repair (Breitbach et al., 2018; Caplan and Correa, 2011; da Silva Meirelles et al., 2008). Vascular dysfunction manifests various cellular abnormalities like neointimal growth, thrombosis, inflammation, and vascular apoptosis and further leads to ischemia and infarction, thus providing prognostic parameters of cardiovascular diseases (Nedeljkovic et al., 2003; Zhang et al., 2018).

NF-κB is a transcription factor that modulates inflammatory response apart from regulating proliferation and survival (Perkins and Gilmore, 2006). In most cell types, the predominant form of NF-κB is a heterodimer composed of RelA/p65 and p50, which can be activated by various stimuli such as TNFα and IL1β (Tilstra et al., 2011). NF-κB is sequestered in the cytoplasm in an inactive complex bound to the inhibitory proteins, inhibitors of NF-κB (IκBs). Upon stimulations, the degradation of IκB releases NF-κB, which subsequently translocates to the nucleus to modulate genes for the maintenance of cellular homeostasis (Liu et al., 2008; Perkins, 2007). Previous studies have shown that NF-κB activation is present in atherosclerotic lesion, resulting in chemokine secretion, adhesion of circulating monocytes to endothelium and persistent inflammatory response in VECs (Baker et al., 2011; Brand et al., 1996; Hajra et al., 2000). In VSMCs, NF-κB activation promotes extracellular matrix expression, neointimal proliferation, aberrant inflammatory response, and cell apoptosis, leading to vascular restenosis and plaque rupture (Rudijanto, 2007). Conversely, inhibition of NF-κB in vascular cells often exhibits therapeutic effects on cardiovascular diseases, such as the amelioration of vascular atherosclerotic lesion in apolipoprotein E (ApoE)-deficient mice (Chiba et al., 2006; Gareus et al., 2008; Mallavia et al., 2013). However, little has been illustrated how NF-κB regulates the physiological functions of different human vascular cells.

Here, using CRISPR/Cas9-mediated genome editing, we generated RelA-deficient human embryonic stem cells (ESCs) to abolish NF-κB activity. We obtained RelA-deficient VECs, VSMCs and MSCs via directed differentiation and uncovered a protective role of NF-κB/RelA in different human vascular cells under basal and inflammatory conditions. RelA deficiency in VECs resulted in impaired vasculogenesis and monocyte-endothelium adhesion, as well as excessive proliferation. In MSCs, loss of RelA resulted in defective proliferation and dysregulated differentiation potential. Additionally, RelA deficiency promoted TNFα-induced apoptosis in different human vascular cells. Further transcriptomic analysis of RelA-deficient vascular cells revealed the cellular changes in vascular matrix organization, angiogenesis, inflammatory response, cell proliferation and survival. Lastly, analysis of gene expression patterns in various *IκBα* knockout vascular cells showed that IκBα acted largely independent of RelA signaling under basal condition and upon TNFα stimulation. Taken together, our data illustrate a novel protective role of NF-κB/RelA in human vascular cells and provide a platform for the study of NF-κB function in human adult stem cells and somatic cells.

Figure 1. Generation and characterization of *RelA*$^{-/-}$ human ESCs. (A) Schemic diagram of *RelA* knockout strategy via CRISPR/Cas9 in human ESCs. A neomycin-resistant cassette (Neo) was included for positive selection. (B) Genomic PCR verification of *RelA* exon 1 knockout in *RelA*$^{-/-}$ ESCs. Water was used as a negative control (NC). (C) Western blot analysis of RelA protein levels in WT and *RelA*$^{-/-}$ ESCs. β-Actin was used as a loading control. (D) Representative colony morphology and immunostaining of pluripotency markers in WT and *RelA*$^{-/-}$ ESCs. Scale bar, 30 μm. (E) Measurement of the mRNA expression levels of pluripotency markers by semi-quantitative PCR in WT and *RelA*$^{-/-}$ ESCs. *18S* was used as a loading control. (F) Teratoma analysis of WT and *RelA*$^{-/-}$ ESCs with three germ layer markers. Markers were stained in red; DNA was labeled in blue by Hoechst 33342. Scale bar, 100 μm. (G) Karyotype analysis of WT and *RelA*$^{-/-}$ ESCs. (H) Ki67 immunostaining in WT and *RelA*$^{-/-}$ ESCs. Ki67 was stained in red; DNA was labeled by Hoechst 33342. Scale bar, 30 μm. ▶

RESULTS

Generation of RelA-deficient human ESCs

We generated RelA-deficient human ESCs (hESCs) targeting the exon 1 of *RelA* by CRISPR/Cas9-mediated genome editing (Fig. 1A). Successful removal of the targeted exon was verified by PCR (Fig. 1B) and the resulting loss of RelA protein was verified by Western blot (Fig. 1C). The *RelA*$^{-/-}$ ESCs exhibited common pluripotent stem cell features including typical colony morphology, expression of pluripotency markers OCT4, SOX2 and NANOG (Fig. 1D and 1E). The *in vivo* differentiation ability of *RelA*$^{-/-}$ ESCs was validated by teratoma formation assay (Fig. 1F). Furthermore, karyotype and cell proliferation were each normal in *RelA*$^{-/-}$ ESCs when compared to wildtype (WT) controls (Fig. 1G and 1H). These data suggest that the *RelA*$^{-/-}$ ESCs maintained typical hESC features.

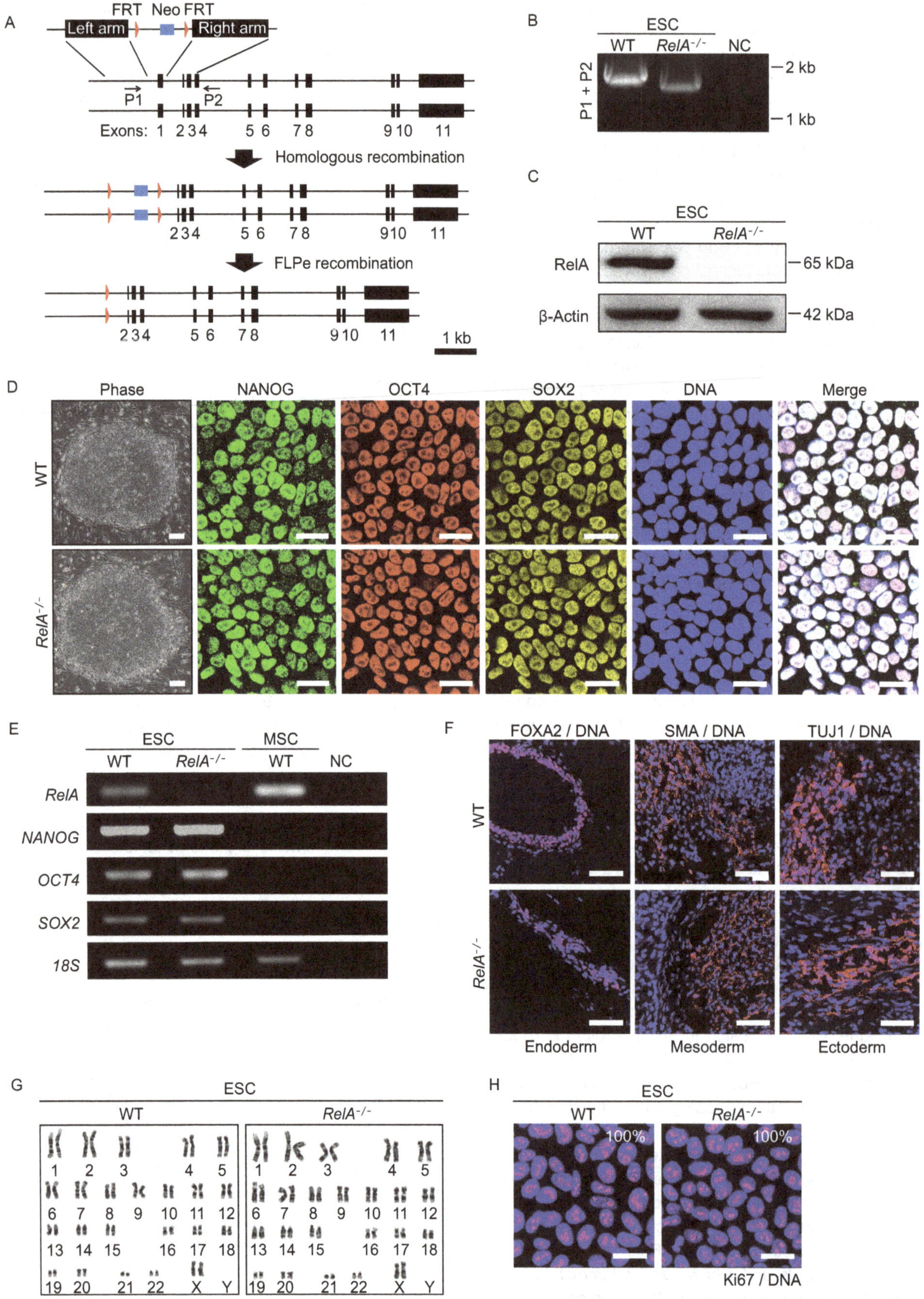
A
FRT Neo FRT
Left arm
Right arm
P1
P2
Exons: 1 2 3 4 5 6 7 8 9 10 11
Homologous recombination
FLPe recombination
1 kb
B
ESC
WT
RelA-/-
NC
P1 + P2
2 kb
1 kb
C
ESC
WT
RelA-/-
RelA
65 kDa
β-Actin
42 kDa
D
Phase
NANOG
OCT4
SOX2
DNA
Merge
WT
RelA-/-
E
ESC
MSC
WT
RelA-/-
WT
NC
RelA
NANOG
OCT4
SOX2
18S
F
FOXA2 / DNA
SMA / DNA
TUJ1 / DNA
WT
RelA-/-
Endoderm
Mesoderm
Ectoderm
G
ESC
WT
RelA-/-
H
ESC
WT
RelA-/-
100%
100%
Ki67 / DNA

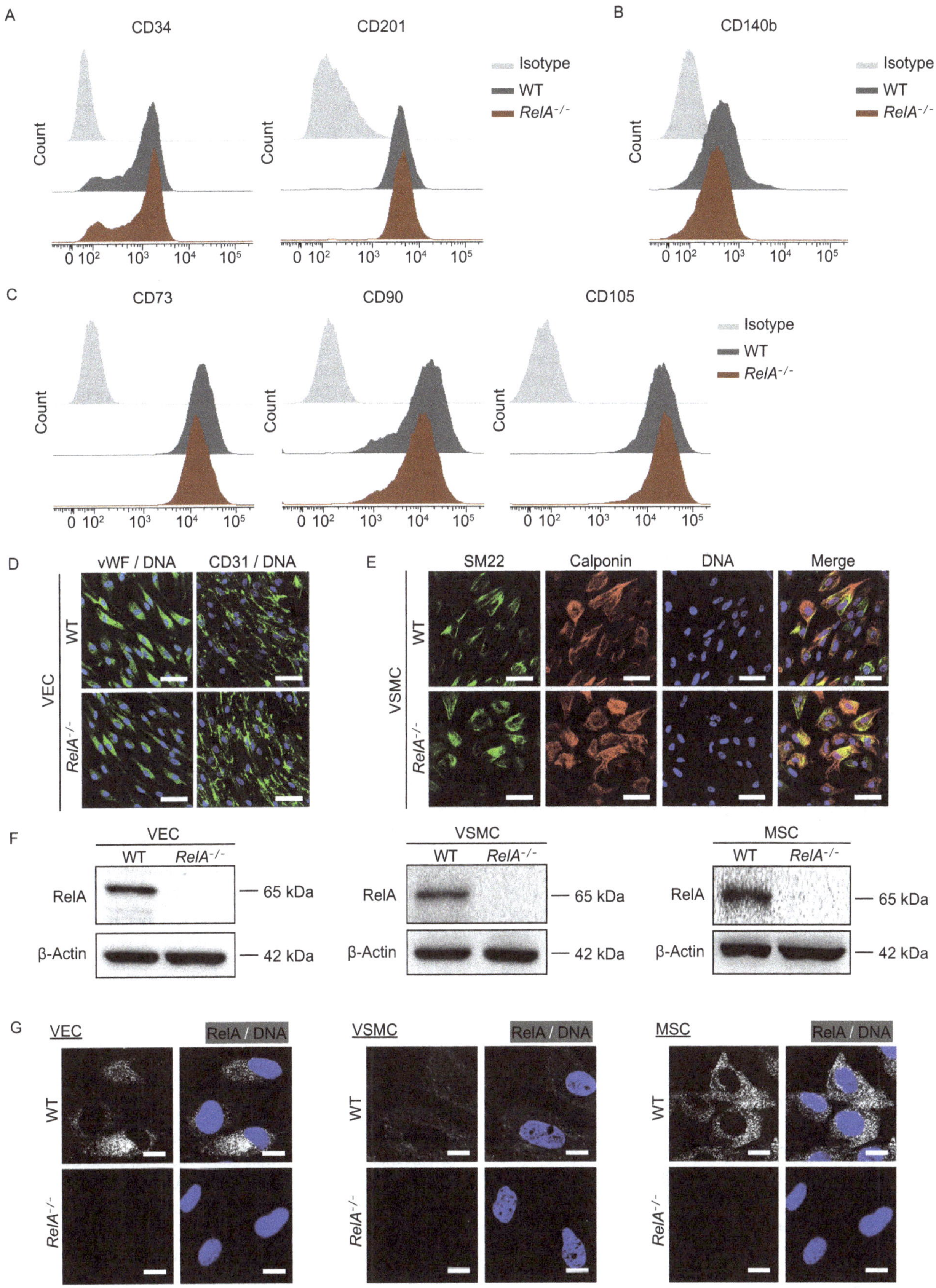
A
CD34
CD201
Isotype
WT
RelA-/-
Count
0 10² 10³ 10⁴ 10⁵
B
CD140b
C
CD73
CD90
CD105
D
vWF / DNA
CD31 / DNA
VEC
WT
RelA-/-
E
SM22
Calponin
DNA
Merge
VSMC
F
VEC
VSMC
MSC
RelA
β-Actin
65 kDa
42 kDa
G
RelA / DNA

◀ **Figure 2. Derivation of *RelA*$^{-/-}$ VECs, VSMCs, and MSCs from *RelA*$^{-/-}$ ESCs.** (A) Flow cytometric analysis of WT and *RelA*$^{-/-}$ VECs with VEC-specific markers CD34 and CD201. IgG-FITC and IgG-PE were used as isotype controls. (B) Flow cytometric analysis of WT and *RelA*$^{-/-}$ VSMCs with VSMC-specific marker, CD140b. IgG-APC was used as an isotype control. (C) Flow cytometric analysis of WT and *RelA*$^{-/-}$ MSCs with MSC-specific markers, CD73, CD90 and CD105. IgG-FITC, IgG-PE and IgG-APC were used as isotype controls. (D) Immunostaining of WT and *RelA*$^{-/-}$ VECs with VEC-specific markers, vWF and CD31. DNA was labeled by Hoechst 33342. Scale bar, 30 μm. (E) Immunostaining of WT and *RelA*$^{-/-}$ VSMCs with VSMC-specific markers, SM22 and Calponin. DNA was labeled by Hoechst 33342. Scale bar, 30 μm. (F) Western blot analysis of RelA protein in WT and *RelA*$^{-/-}$ VECs, VSMCs and MSCs, respectively. β-Actin was used as a loading control. (G) Immunostaining of RelA in WT and *RelA*$^{-/-}$ VECs, VSMCs and MSCs under basal condition. DNA was labeled by Hoechst 33342. Scale bar, 10 μm.

Derivation of different human vascular cells from RelA-deficient hESCs

To study how RelA is involved in human vasculature homeostasis, we generated human VECs, VSMCs and MSCs via directed differentiation of *RelA*$^{-/-}$ and WT ESCs. Cells were purified by fluorescent-activated cell sorting (FACS) using proper cell surface markers (Fig. 2A–C). Cell purity was confirmed by immunofluorescent staining of additional VEC-specific markers, vWF and CD31 (Fig. 2D) and VSMC-specific markers, SM22 and Calponin (Fig. 2E). While RelA was predominantly retained in the cytoplasm of wildtype vascular cells, loss of RelA protein was verified in different types of RelA-deficient vascular cells by western blotting and immunofluorescent staining (Fig. 2F and 2G).

RelA deficiency impaired vasculogenesis in VECs and perturbed differentiation potential in MSCs

We next investigated the functional consequences of RelA deficiency in different vascular cells. Although *RelA*$^{-/-}$ VECs had comparable ability to uptake acetylated low-density lipoprotein (Ac-LDL) compared to that of WT VECs (Fig. 3A), RelA deficiency severely interrupted tube formation of VECs *in vitro* (Fig. 3B), indicative of dysregulated VEC function.

Functional MSCs undergo adipogenesis, osteogenesis and chondrogenesis for regeneration *in vivo* (Uccelli et al., 2008). Here we tested whether RelA deficiency interferes with the differentiation potential of MSCs into adipocytes, osteoblasts and chondrocytes. Adipogenesis was slightly enhanced from *RelA*$^{-/-}$ MSCs, evidenced by an increase in oil red O staining (Fig. 3C) and upregulation of adipocyte-specific genes like *FABP4*, *PPARG* and *LPL*, which were dramatically enriched in WT adipocytes relative to WT MSCs (Fig. 3D). Despite of increased osteoblast-specific gene expression, there were less calcium deposits stained by Von Kossa in derived *RelA*$^{-/-}$ osteoblasts (Fig. 3E and 3F), indicative of aberrant osteogenesis from *RelA*$^{-/-}$ MSCs. Moreover, RelA deficiency resulted in defective chondrogenesis with less condensate structure stained by toluidine blue dye (Fig. 3G). These data indicate that RelA was required for maintaining the homeostasis of VECs and MSCs.

RelA deficiency impeded inflammatory response in human vascular cells

RelA is a well-known component of NF-κB heterodimer, which is a major mediator to regulate inflammation (Barnes and Karin, 1997; Salminen et al., 2008). Accordingly, we measured the mRNA levels of several NF-κB target genes involved in inflammation, including vascular cell adhesion molecule 1 (*VCAM1*), monocyte chemoattractant protein 1 (*MCP1*), *IL6*, and *IL8* in human VECs, VSMCs, and MSCs. Loss of RelA led to reduced mRNA levels of these NF-κB target genes in rested VECs, VSMCs and MSCs (Fig. 4A). As expected, while RelA was translocated from cytosol to nucleus in WT VECs, VSMCs and MSCs upon TNFα treatment, no RelA immunofluorescence signal was observed in *RelA*$^{-/-}$ cells (Fig. 4B).

Given that monocyte recruitment to sites of vessel injury and subsequent adhesion to endothelium is critical for inflammation to repair damage and maintain blood vessel homeostasis (Kirton and Xu, 2010; Schober and Weber, 2005), we tested how RelA deficiency affected the adherence of monocytes to endothelium. *RelA*$^{-/-}$ VECs exhibited compromised ability in recruiting monocytes compared to WT VECs under basal condition, and this defect was even evident upon TNFα stimulation (Fig. 4C). Furthermore, RelA deficiency resulted in reduced expression of genes implicated in monocyte adhesion to vascular endothelium including *Selectin E*, *VCAM1* and intercellular adhesion molecule 1 (*ICAM1*) upon TNFα treatment (Fig. 4D and 4E). Thus, our data demonstrate that RelA deficiency impeded inflammatory response in human vascular cells.

RelA deficiency altered proliferative ability in human vascular cells

We next checked how RelA affected the proliferative ability of various human vascular cells. RelA deficiency enhanced proliferation in VECs, but not in VSMCs (Fig. 5A–D). In MSCs, however, RelA deficiency led to decreased proliferative ability at early passages (Passage 3) (Fig. 5E and 5F). Given that MSCs as adult stem cells have the self-renewal ability, we further studied how RelA affected MSC proliferation after serial passaging and found even worse proliferative ability in late-passage *RelA*$^{-/-}$ MSCs (Passage 7) (Fig. 5G and 5H). Therefore, RelA demonstrated distinct regulatory activities towards the proliferation of VECs, VSMCs and MSCs.

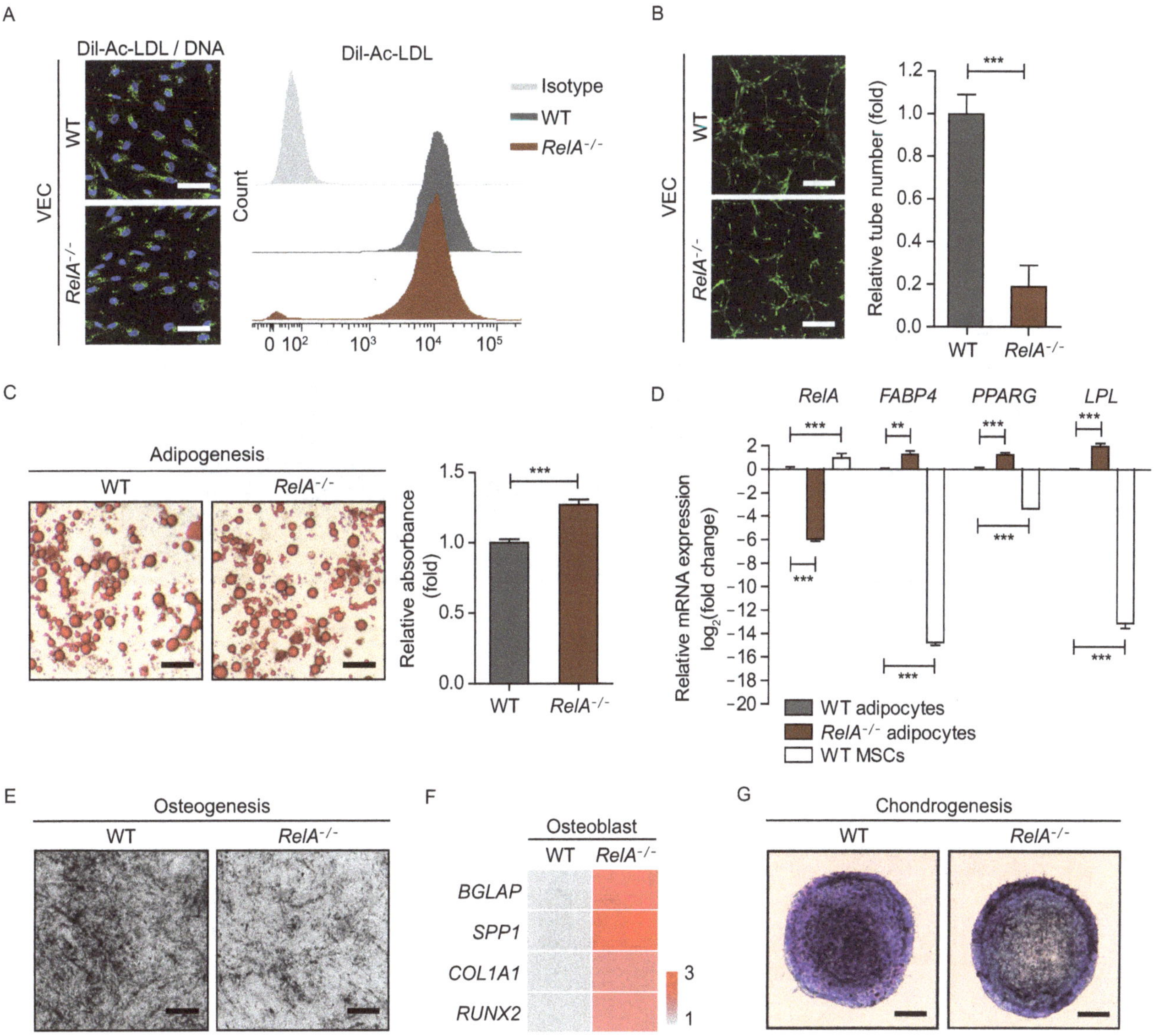

Figure 3. RelA deficiency affected vascular cell homeostasis. (A) Immunostaining and flow cytometry analysis of the Dil-Ac-LDL uptake capacity in WT and *RelA*$^{-/-}$ VECs. DNA was labeled by Hoechst 33342. Scale bar, 30 μm. (B) Representative micrographs of matrigel tubes formed by WT and *RelA*$^{-/-}$ VECs *in vitro* ($n = 3$). Scale bar, 3 mm. (C) Oil red O staining of WT and *RelA*$^{-/-}$ adipocytes derived from MSCs, respectively. The quantification of adipocytes was measured by absorbance at 510 nm ($n = 4$). *** $P < 0.001$. Scale bar, 3 mm. (D) Transcriptional expression of adipocyte-specific genes in WT and *RelA*$^{-/-}$ adipocytes via RT-qPCR detection ($n = 4$). WT MSCs were used as a negative control. *18S* was used as a loading control. * $P < 0.05$. ** $P < 0.01$. *** $P < 0.001$. (E) Representative micrographs of WT and *RelA*$^{-/-}$ osteoblasts by Von Kossa staining. Scale bar, 3 mm. (F) Transcriptional levels of osteoblast-specific gene expression in WT and *RelA*$^{-/-}$ osteoblasts via RT-qPCR detection ($n = 4$). *18S* was used as a loading control. (G) Representative toluidine blue staining images of WT and *RelA*$^{-/-}$ chondrocytes. Scale bar, 3 mm.

RelA deficiency promoted TNFα-induced apoptosis in human vascular cells

NF-κB modulates the expression of anti-apoptotic genes, thus promoting cell survival (Kucharczak et al., 2003). Here we tested how RelA affected the survival of human vascular cells under basal and inflammatory conditions. Consistently, the protective effects of NF-κB on inhibiting TNFα-induced apoptosis were impaired by RelA deficiency in various vascular cells. Interestingly, at baseline RelA deficiency exhibited an anti-apoptotic effect in VSMCs, but not in VECs or MSCs (Fig. 6A and 6B). These data indicate that RelA protected human vascular cells against TNFα-induced cell apoptosis.

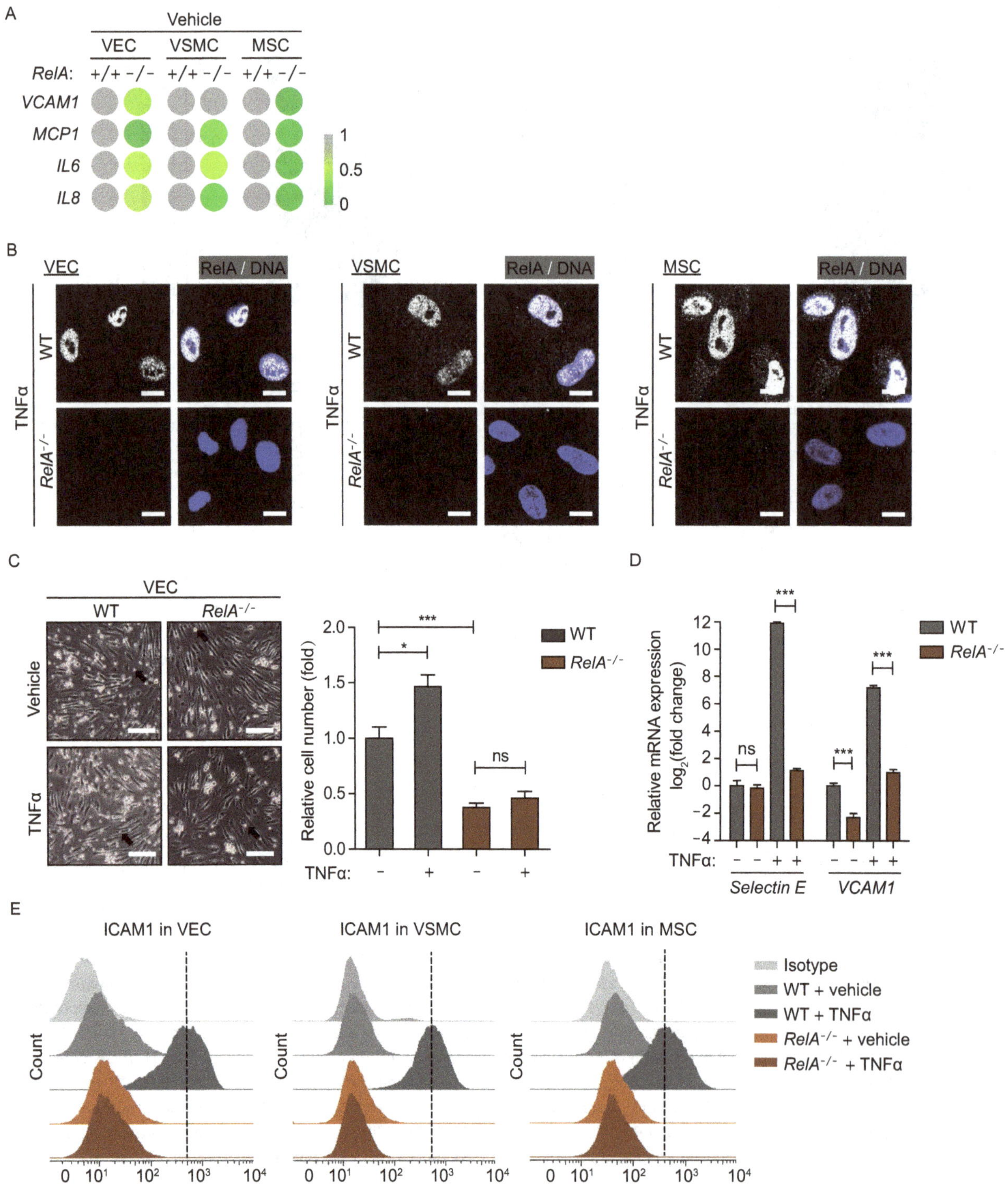

Figure 4. Inflammatory response defect in *RelA*$^{-/-}$ vascular cells. (A) Transcriptional expression of NF-κB target genes in WT and *RelA*$^{-/-}$ VECs, VSMCs, and MSCs under basal condition via RT-qPCR detection (n = 4). (B) Immunostaining of RelA in WT and *RelA*$^{-/-}$ VECs, VSMCs and MSCs upon 10 ng/mL TNFα treatment. Scale bar, 10 μm. (C) Monocyte adhesion on WT and *RelA*$^{-/-}$ endothelium under basal and 10 ng/mL TNFα-induced inflammatory conditions (n = 3). Black arrows indicate monocytes. Random fields were selected and the numbers of monocytes were counted by ImageJ software. ns, not significant; * $P < 0.05$; *** $P < 0.001$. (D) Transcriptional expression of adhesion molecules in WT and *RelA*$^{-/-}$ VECs under basal and 10 ng/mL TNFα-induced inflammatory conditions via RT-qPCR detection (n = 4). ns, not significant. *** $P < 0.001$. (E) Flow cytometric analysis of ICAM1 in WT and *RelA*$^{-/-}$ VECs, VSMCs and MSCs under basal and 10 ng/mL TNFα-induced inflammatory conditions. The black dotted line represents the mean ICAM1 protein level upon TNFα treatment. IgG-PE was used as an isotype control.

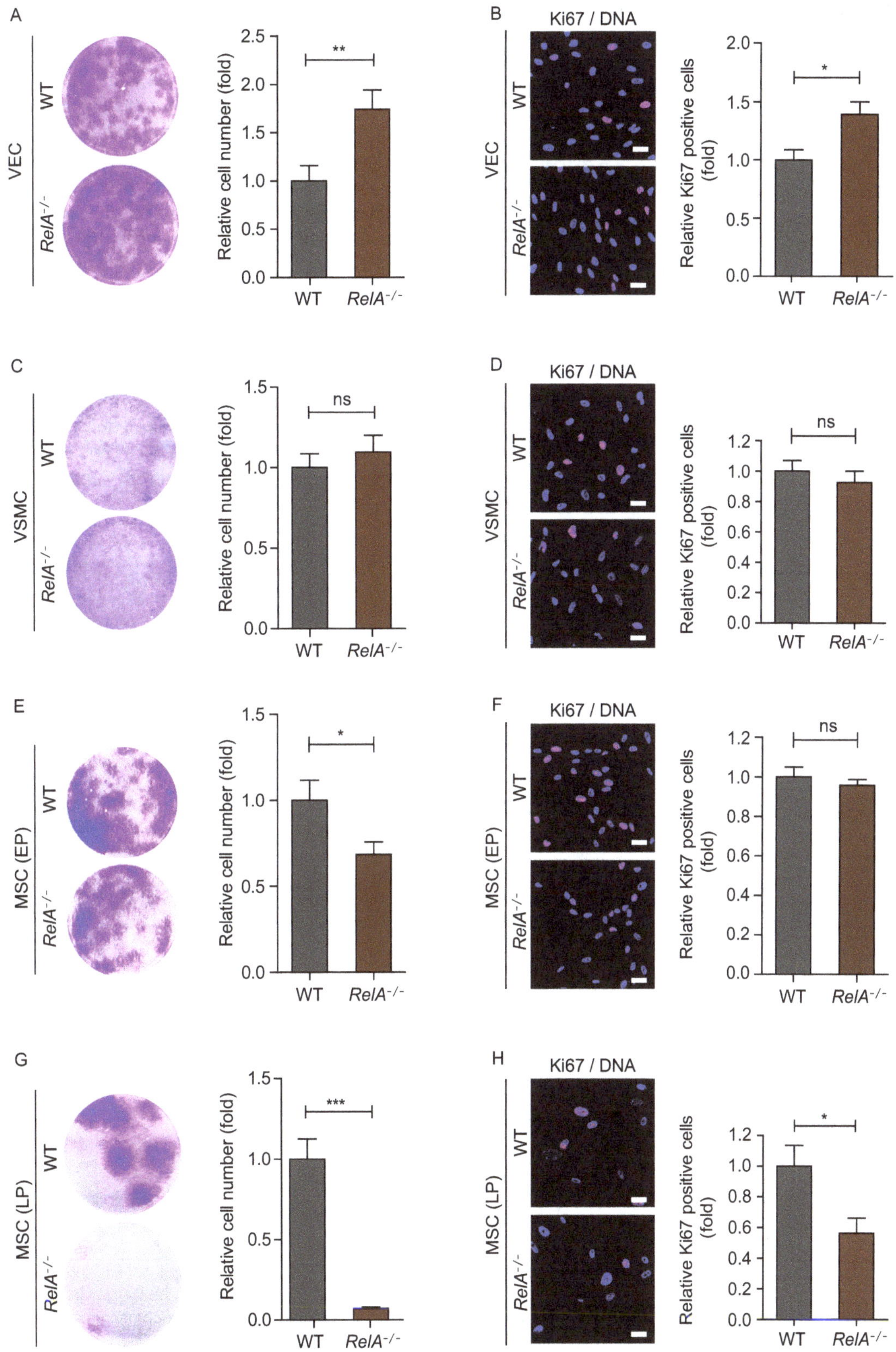

A
VEC
WT
RelA-/-
Relative cell number (fold)
**
WT
RelA-/-
B
Ki67 / DNA
VEC
Relative Ki67 positive cells (fold)
*
C
VSMC
ns
D
E
MSC (EP)
*
F
G
MSC (LP)
***
H

◄ **Figure 5. RelA deficiency resulted in distinct proliferative activity in various vascular cells.** (A) Colony formation of WT and *RelA*$^{-/-}$ VECs (n = 3). ** P < 0.01. (B) Ki67 immunostaining of WT and *RelA*$^{-/-}$ VECs (n = 3). DNA was labeled by Hoechst 33342. * P < 0.05. Scale bar, 30 μm. (C) Colony formation of WT and *RelA*$^{-/-}$ VSMCs (n = 3). ns, not significant. (D) Ki67 immunostaining of WT and *RelA*$^{-/-}$ VSMCs (n = 3). DNA was labeled by Hoechst 33342. ns, not significant. Scale bar, 30 μm. (E) Colony formation of WT and *RelA*$^{-/-}$ MSCs at passage 3 (early passage, EP) (n = 3). * P < 0.05. (F) Ki67 immunostaining of WT and *RelA*$^{-/-}$ MSCs at passage 3 (early passage, EP) (n = 3). DNA was labeled by Hoechst 33342. ns, not significant. Scale bar, 30 μm. (G) Colony formation of WT and *RelA*$^{-/-}$ MSCs at passage 7 (late passage, LP) (n = 3). *** P < 0.001. (H) Ki67 immunostaining of WT and *RelA*$^{-/-}$ MSCs at passage 7 (late passage, LP) (n = 3). DNA was labeled by Hoechst 33342. * P < 0.05. Scale bar, 30 μm.

MSCs exhibited similar transcriptional signatures upon TNFα or IL1β stimulation

Inflammatory cytokines such as TNFα and IL1β are strong activators of NF-κB signaling pathway (Kida et al., 2005; Osborn et al., 1989). To explore how these cytokines modulate human vascular cells via NF-κB signaling pathway, we measured the mRNA levels of two NF-κB target genes, *ICAM1* and *VCAM1*, upon TNFα or IL1β stimulation in a dose- and time-dependent manner. Upon TNFα induction, the mRNA levels of *ICAM1* and *VCAM1* were both increased with peak expression at 20 ng/mL and by 8 h of treatment (Fig. 7A). By comparison, *ICAM1* and *VCAM1* were upregulated even at low IL1β concentrations with peak expression by 4 h of treatment (Fig. 7B).

We next mapped the transcriptomic landscapes induced by TNFα or IL1β in WT and RelA-deficient MSCs by genome-wide RNA-seq (Fig. 7C). The changes in gene expression patterns were similar upon TNFα or IL1β

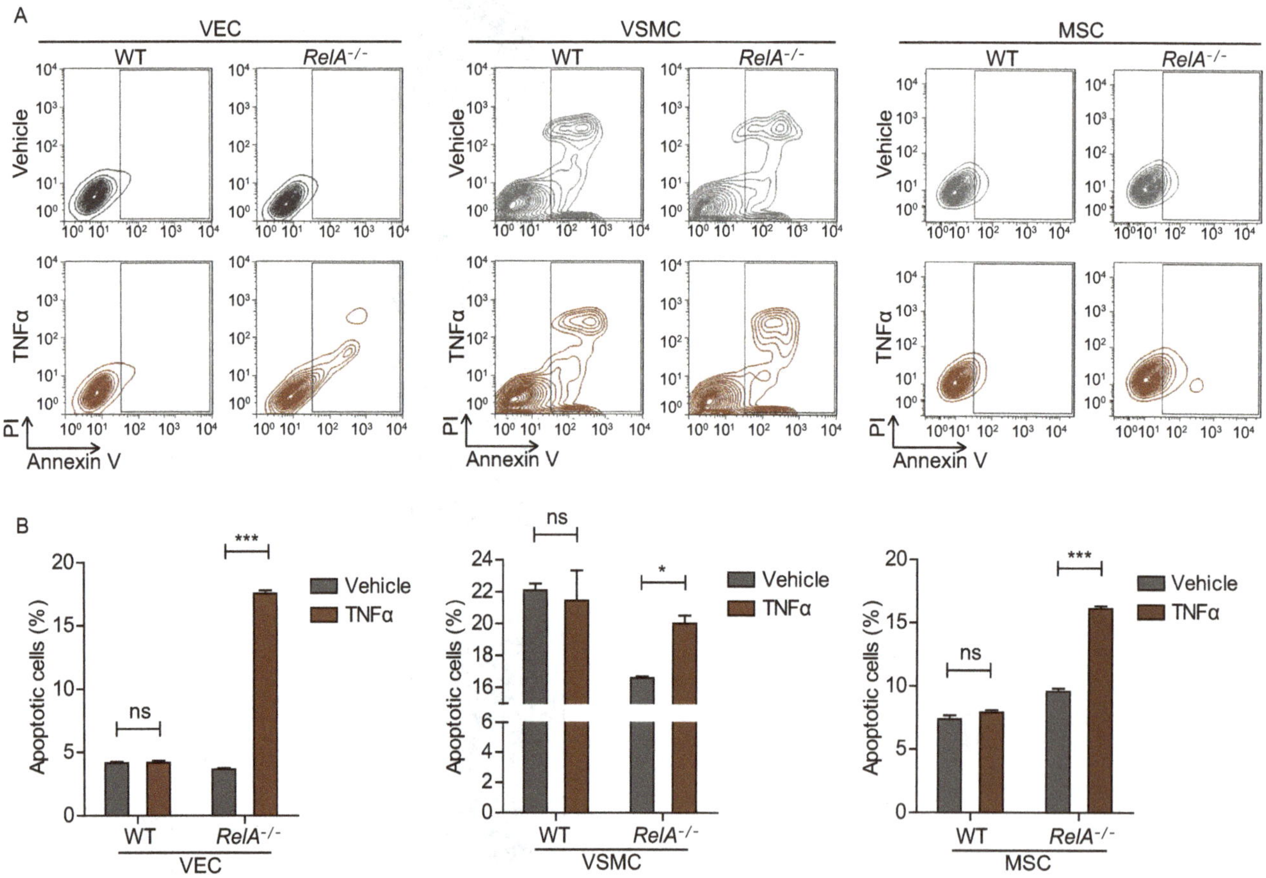

Figure 6. RelA deficiency promoted TNFα-induced apoptosis. (A) Flow cytometric analysis of apoptotic vascular cells under basal and 10 ng/mL TNFα-induced conditions. (B) Statistical analysis of apoptotic cells in WT and *RelA*$^{-/-}$ VECs, VSMCs, and MSCs upon 10 ng/mL TNFα treatment (n = 3). ns, not significant; * P < 0.05; *** P < 0.001.

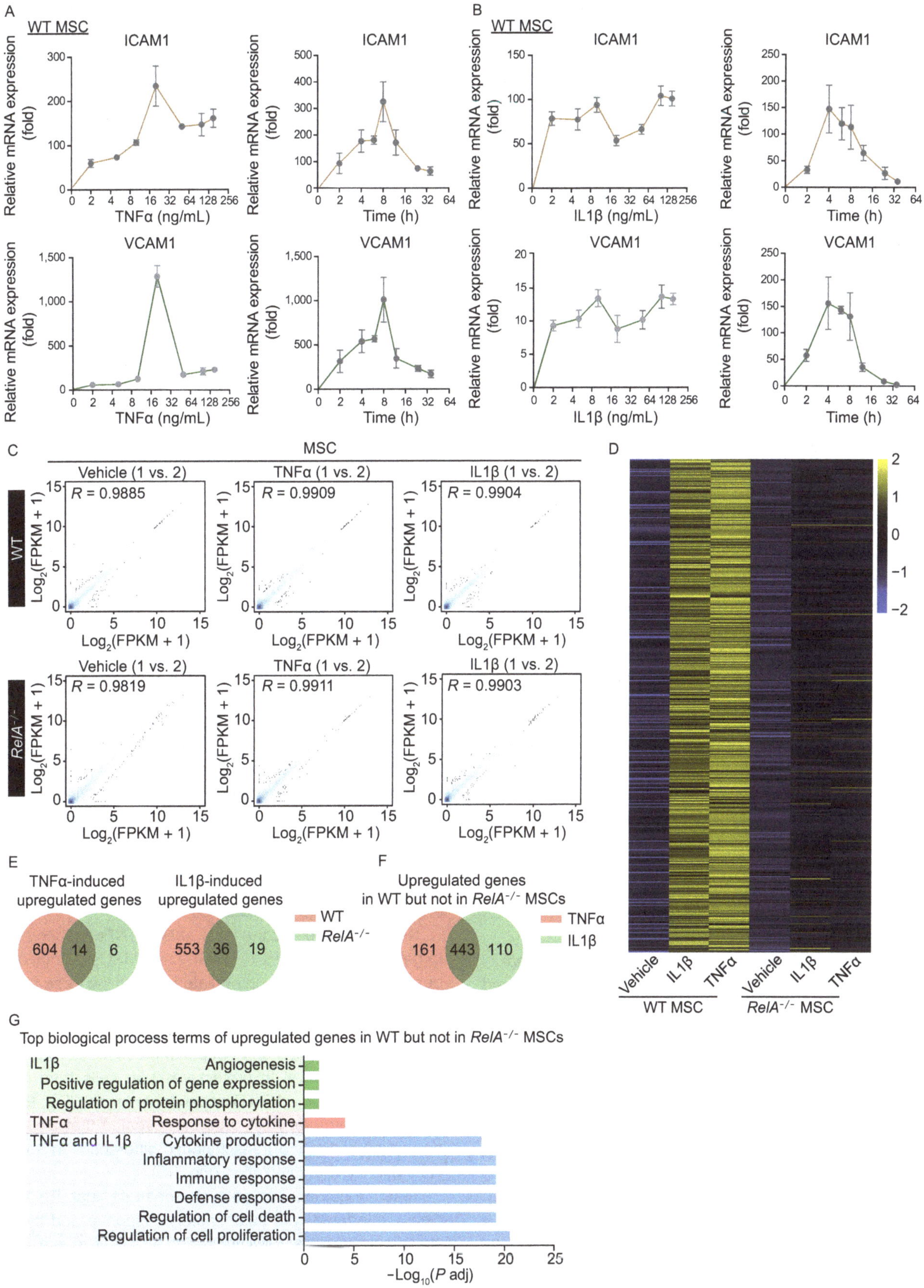
A
WT MSC
ICAM1
Relative mRNA expression (fold)
TNFα (ng/mL)
Time (h)
VCAM1
B
WT MSC
ICAM1
IL1β (ng/mL)
Time (h)
VCAM1
C
MSC
Vehicle (1 vs. 2)
TNFα (1 vs. 2)
IL1β (1 vs. 2)
WT
R = 0.9885
R = 0.9909
R = 0.9904
RelA-/-
R = 0.9819
R = 0.9911
R = 0.9903
Log2(FPKM + 1)
D
Vehicle
IL1β
TNFα
WT MSC
RelA-/- MSC
E
TNFα-induced upregulated genes
IL1β-induced upregulated genes
604
14
6
553
36
19
WT
RelA-/-
F
Upregulated genes in WT but not in RelA-/- MSCs
161
443
110
TNFα
IL1β
G
Top biological process terms of upregulated genes in WT but not in RelA-/- MSCs
IL1β
Angiogenesis
Positive regulation of gene expression
Regulation of protein phosphorylation
TNFα
Response to cytokine
TNFα and IL1β
Cytokine production
Inflammatory response
Immune response
Defense response
Regulation of cell death
Regulation of cell proliferation
-Log10(P adj)

◄ **Figure 7. MSCs exhibited similar transcriptional landscapes upon TNFα or IL1β stimulation.** (A) Transcriptional expression of *ICAM1* and *VCAM1* treated with TNFα at indicated concentrations for 1 h or indicated duration at 10 ng/mL via RT-qPCR detection (n = 4). (B) Transcriptional expression of *ICAM1* and *VCAM1* treated with IL1β at indicated concentrations for 1 h or indicated duration at 10 ng/mL via RT-qPCR detection (n = 4). (C) Scatter plot showing the correlation between RNA-seq replicates of WT and $RelA^{-/-}$ MSCs under basal, TNFα (10 ng/mL)- or IL1β (5 ng/mL)-treated conditions. Pearson correlation coefficient (R) is presented. (D) Heatmap showing the Z-score normalized expression levels (FPKM) of coordinately upregulated genes in WT but not in $RelA^{-/-}$ MSCs upon TNFα or IL1β treatment. (E) Venn diagrams showing the overlap of upregulated genes in WT and $RelA^{-/-}$ MSCs upon TNFα (left) or IL1β (right) treatment. (F) Venn diagram showing the overlap of upregulated genes in WT but not in $RelA^{-/-}$ MSCs upon TNFα or IL1β treatment. (G) Gene ontology (GO) enrichment analysis of upregulated genes presented in biological processes affected by the upregulated genes only in IL1β-treated MSCs (green bars), only in TNFα-treated MSCs (red bars), and in both IL1β- and TNFα-treated MSCs (blue bars).

stimulation in WT MSCs, and were abolished by RelA deficiency (Fig. 7D). A total of 618 upregulated genes induced by TNFα and 589 upregulated genes induced by IL1β were found in WT MSCs, while only 20 upregulated genes induced by TNFα and 55 upregulated genes induced by IL1β were found in $RelA^{-/-}$ MSCs (Fig. 7E). Most were shared in common (Fig. 7F) and contributed to cell proliferation, survival, defense response, immune response, and inflammatory response (Fig. 7G).

Due to the aforementioned similarity in gene expression patterns upon TNFα or IL1β induction, we used TNFα as the vascular inflammatory stimulus in the subsequent studies.

Transcriptomic analysis revealed a guardian role of RelA in maintaining blood vessel homeostasis

Genome-wide RNA-seq was performed in WT and $RelA^{-/-}$ VECs, VSMCs and MSCs under basal and TNFα-induced conditions (Fig. S1A and 8A). Under basal condition, RelA deficiency largely retained the transcriptional profile in vascular cells (Fig. 8B and 8C). Venn diagram analysis revealed only two commonly upregulated genes and three downregulated by RelA deficiency among VECs, VSMCs and MSCs (Fig. 8D), despite that RelA expressed substantially in all these cell types (Fig. S1B). These results suggest that RelA modulated vascular cells in a cell type-specific manner. The two commonly upregulated genes were transcription elongation factor A like 1 (*TCEAL1*) and peroxiredoxin 4 (*PRDX4*), and three downregulated genes were C-type lectin domain containing 11A (*CLEC11A*), DnaJ heat shock protein family member C15 (*DNAJC15*), and synaptotagmin 11 (*SYT11*) (Fig. 8E). The expression of *CLEC11A*, one of the downregulated genes and a growth factor that modulates stem cell proliferation and osteogenic development (Hiraoka et al., 2001; Yue et al., 2016), was re-confirmed by qPCR (Fig. 8F). Similarly to our observation with RelA deficiency, *CLEC11A* knockdown also impaired MSC proliferation (Fig. 8G, 8H and S1C), suggesting that RelA deficiency blocked MSC self-renewal at least in part through transcriptional silence of *CLEC11A*.

Further analysis revealed the specific aspects of biological process affected by RelA deficiency. In VECs, RelA deficiency repressed inflammatory response, cytokine production, and NF-κB signaling pathway, supporting a role of RelA in mediating vascular inflammatory response (Fig. 8I). In both VSMCs and MSCs, RelA deficiency led to aberrant extracellular matrix organization and skeletal development that were important to the maintenance of vascular structure and function. In addition, RelA deficiency affected blood vessel formation in VSMCs and promoted the ossification and inhibited the migration of MSCs (Fig. 8J and 8K).

TNFα-NF-κB axis modulated vascular gene expression in a cell type-specific manner

We next investigated how RelA regulated gene expression in TNFα-treated vascular cells. RelA deficiency dramatically abrogated TNFα-induced transcriptional changes in all types of vascular cells, especially in MSCs (Fig. 9A). Venn diagram analysis revealed that a total of 606 genes in WT VECs, 580 genes in WT VSMCs, and 604 genes in WT MSCs were upregulated by TNFα in a RelA-dependent manner (Fig. 9B). In VECs, TNFα induced a set of genes that modulate inflammatory response and cytokine signaling pathway (Fig. S1D). In VSMCs the upregulated genes contributed to defense response and cell proliferation (Fig. S1E). In MSCs the induced genes were associated with immune response and apoptotic process (Fig. S1F). All these effects were abolished in $RelA^{-/-}$ vascular cells.

Next, we compared the three sets of TNFα-induced RelA-target genes in WT VECs, VSMCs and MSCs. A total of 134 genes were commonly upregulated among the three types of vascular cells (Fig. 9C). These genes contributed to biological processes including immune response, cytokine signaling, cellular adhesion and apoptotic process, indicating that they may be critical to protect vascular cells from inflammatory damage in a NF-κB-dependent manner (Fig. 9D and S1G). Furthermore, PCA analysis revealed that RelA deficiency increased the transcriptomic differences in vascular cells upon TNFα stimulation (Fig. 9E).

Taken together, our data demonstrate that RelA deficiency interfered with inflammatory response and cytokine signaling in vascular cells, indicating a protective role of RelA in maintaining vascular homeostasis during vascular inflammation.

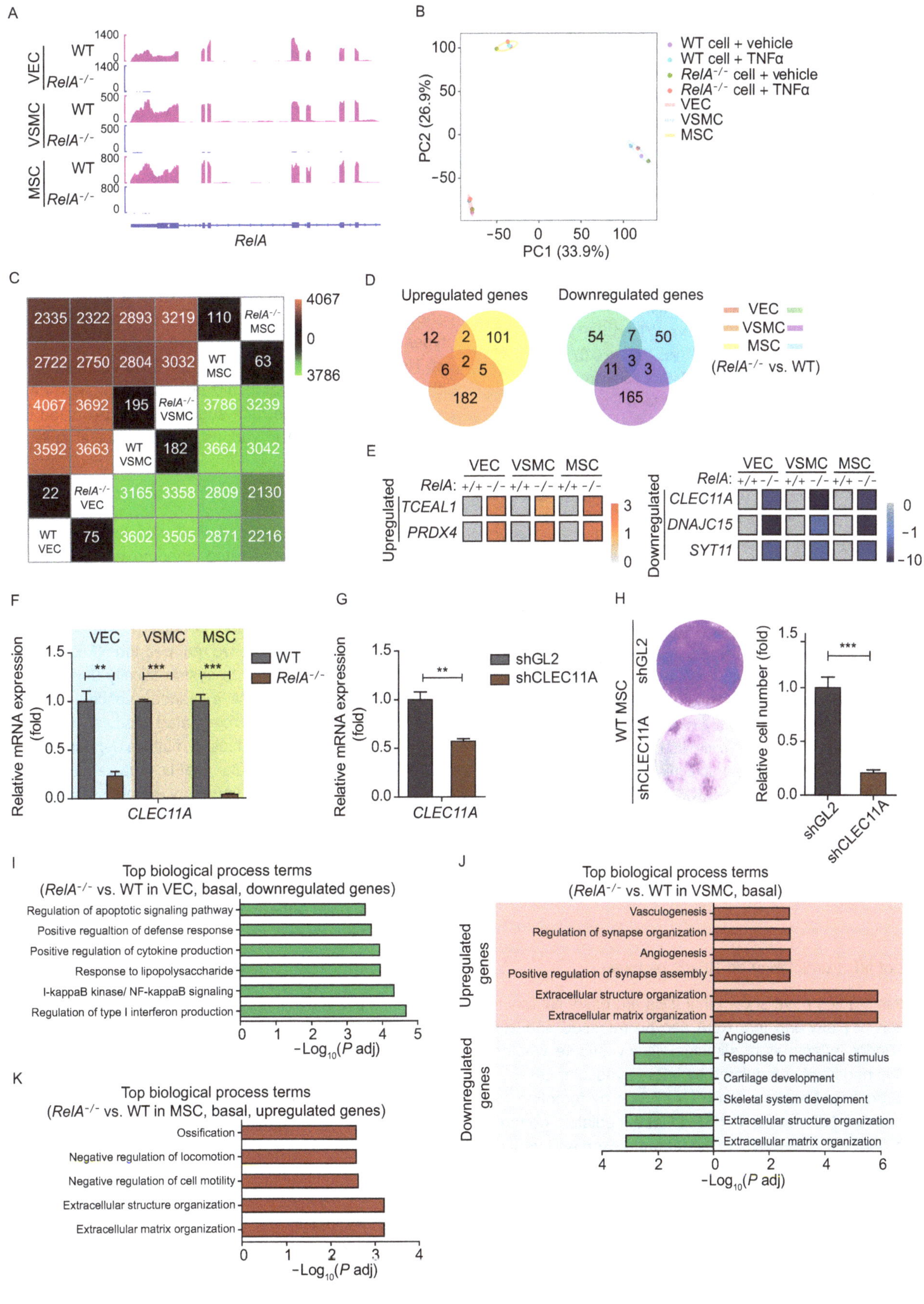
A
VEC
VSMC
MSC
WT
RelA-/-
RelA
B
WT cell + vehicle
WT cell + TNFα
RelA-/- cell + vehicle
RelA-/- cell + TNFα
VEC
VSMC
MSC
PC2 (26.9%)
PC1 (33.9%)
C
4067
0
3786
D
Upregulated genes
Downregulated genes
(RelA-/- vs. WT)
E
Upregulated
Downregulated
TCEAL1
PRDX4
CLEC11A
DNAJC15
SYT11
F
Relative mRNA expression (fold)
WT
RelA-/-
CLEC11A
G
shGL2
shCLEC11A
Relative mRNA expression (fold)
CLEC11A
H
WT MSC
Relative cell number (fold)
I
Top biological process terms
(RelA-/- vs. WT in VEC, basal, downregulated genes)
Regulation of apoptotic signaling pathway
Positive regualtion of defense response
Positive regulation of cytokine production
Response to lipopolysaccharide
I-kappaB kinase/ NF-kappaB signaling
Regulation of type I interferon production
-Log10(P adj)
J
Top biological process terms
(RelA-/- vs. WT in VSMC, basal)
Upregulated genes
Vasculogenesis
Regulation of synapse organization
Angiogenesis
Positive regulation of synapse assembly
Extracellular structure organization
Extracellular matrix organization
Downregulated genes
Angiogenesis
Response to mechanical stimulus
Cartilage development
Skeletal system development
Extracellular structure organization
Extracellular matrix organization
-Log10(P adj)
K
Top biological process terms
(RelA-/- vs. WT in MSC, basal, upregulated genes)
Ossification
Negative regulation of locomotion
Negative regulation of cell motility
Extracellular structure organization
Extracellular matrix organization
-Log10(P adj)

◄ **Figure 8. Transcriptomic analysis revealed RelA deficiency-induced vascular cell dysfunction under basal condition.** (A) Transcriptional signal of RelA in WT and *RelA*$^{-/-}$ VECs, VSMCs, and MSCs. Data were normalized by RPKM (reads per kilo bases per million mapped reads) at bin size of 10 bp. (B) Principle component analysis (PCA) of various vascular cells showing a cell type-specific transcriptional pattern. (C) Heatmap showing the number of differential expressed genes between WT and *RelA*$^{-/-}$ VECs, VSMCs and MSCs under basal condition. Green color represents downregulated genes; red color represents upregulated genes. (D) Venn diagrams showing the overlap of upregulated (left) and downregulated (right) genes in *RelA*$^{-/-}$ VECs, VSMCs and MSCs compared to WT. (E) Heatmap showing the transcriptional levels of upregulated genes (*RelA*$^{-/-}$ vs. WT > 1.5, P adj < 0.05) and downregulated genes (*RelA*$^{-/-}$ vs. WT < 0.67, P adj < 0.05) in *RelA*$^{-/-}$ VECs, VSMCs and MSCs compared to WT. All FPKMs of the genes were normalized to WT groups and the relative gene expression levels are presented via $Log_{1.5}$(*RelA*$^{-/-}$ / WT) in *RelA*$^{-/-}$ groups. (F) Transcriptional expression of *CLEC11A* in WT and *RelA*$^{-/-}$ VECs, VSMCs, and MSCs via RT-qPCR detection (n = 4). *18S* was used as a loading control. *** P < 0.001. (G) Transcriptional expression of *CLEC11A* in WT MSCs infected with shCLEC11A lentiviruses via RT-qPCR detection (n = 4). shGL2 lentiviruses were used as a negative control; *18S* was used as a loading control. ** P < 0.01. (H) Colony formation of WT MSCs infected with shCLEC11A (n = 3). *** P < 0.001. (I) GO enrichment analysis of differentially expressed genes in *RelA*$^{-/-}$ VECs compared to WT under basal condition. Enriched top GO biological process terms are presented with bars. Green bars represent downregulated genes. (J) GO enrichment analysis of differentially expressed genes in *RelA*$^{-/-}$ VSMCs compared to WT under basal condition. Red bars represent upregulated genes; green bars represent downregulated genes. (K) GO enrichment analysis of differentially expressed genes in *RelA*$^{-/-}$ MSCs compared to WT under basal condition. Red bars represent upregulated genes.

Effect of IκBα deficiency on NF-κB pathway

Previous studies have shown that IκBs are essential for the inhibition of RelA and that RelA can be released and activated under inflammatory stimulus. To investigate whether NF-κB is negatively modulated by IκBs in vascular cells, we generated *IκBα* knockout (*IκBα*$^{-/-}$) hESCs by targeting the first exon of *IκBα* using CRISPR/Cas9-mediated genome editing (Fig. 10A). Removal of the targeted exon was verified in *IκBα*$^{-/-}$ ESCs (Fig. 10B) and the resulting protein loss was confirmed by western blotting (Fig. 10C). To further investigate the role of IκBα in vascular cells, we differentiated WT and *IκBα*$^{-/-}$ ESCs to VECs, VSMCs and MSCs, respectively. Loss of IκBα expression was verified in diverse *IκBα*$^{-/-}$ cells (Fig. 10D).

To understand how IκBα regulates the homeostasis of different vascular cells, we performed genome-wide RNA-seq in WT and *IκBα*$^{-/-}$ VECs, VSMCs, and MSCs under basal and TNFα-induced conditions (Fig. S2A). IκBα deficiency resulted in 79, 78 and 99 genes upregulated, and 104, 97, 90 genes downregulated in untreated VECs, VSMCs, and MSCs, respectively (Fig. S2B). Venn diagram and GO enrichment analysis revealed that IκBα regulated gene expression in a cell type-specific manner (Fig. S2B–E). To further determine whether RelA modulates gene expression dependently of IκBα, we compared the upregulated genes in *IκBα*$^{-/-}$ vascular cells with the downregulated genes in *RelA*$^{-/-}$ vascular cells. Only a small fraction of genes were overlapped in each of the three vascular cell types (Fig. 10E–H), regardless of the absence or presence of TNFα. Consistently, we did not observe marked differences in the subcellular localization of RelA in IκBα-depleted cells (Fig. 10I). These data suggest that IκBα acted largely independent of RelA in various human vascular cells.

DISCUSSION

In this study, we generated RelA-deficient human ESCs using CRISPR/Cas9-mediated genome editing and differentiated them into different types of vascular cells to investigate how NF-κB/RelA regulates human vascular cells under basal and inflammatory conditions. Via genome-wide RNA sequencing analysis, we mapped the NF-κB-regulated transcriptomic landscapes to systematically understand the physiological roles of RelA in various vascular cells. Our findings suggest that RelA modulated vascular cells in a cell type-specific manner by affecting multiple aspects of cellular events including extracellular matrix organization, ossification, vasculogenesis, inflammatory response, proliferation and survival. In addition, our data indicate that IκBα regulated gene expression in vascular cells primarily in a RelA-independent manner (Fig. 11).

The correlation between NF-κB activation and vascular dysfunction have attracted great attention during the last decades. For instance, the nuclear translocation of RelA has frequently been observed in human atherosclerotic lesion vascular cells (Brand et al., 1996; Mallavia et al., 2013). In *ApoE*$^{-/-}$ mice the activation of NF-κB signaling pathway in endothelial cells and smooth muscle cells is accompanied with enhanced secretion of inflammatory factors and increased atherosclerotic lesions (Gareus et al., 2008; Mallavia et al., 2013). In addition, TNFα-induced NF-κB activation impairs cell proliferation and causes premature senescence in HUVECs (Khan et al., 2017); knockdown of RelA promotes cell proliferation and reduces apoptosis in high glucose-treated HUVECs (Chen et al., 2011a). Yet, the role of NF-κB in different human vascular cells has not been well studied via a side-by-side comparison.

Here, we took advantage of the various types of differentiated RelA-deficient vascular cells to reveal the

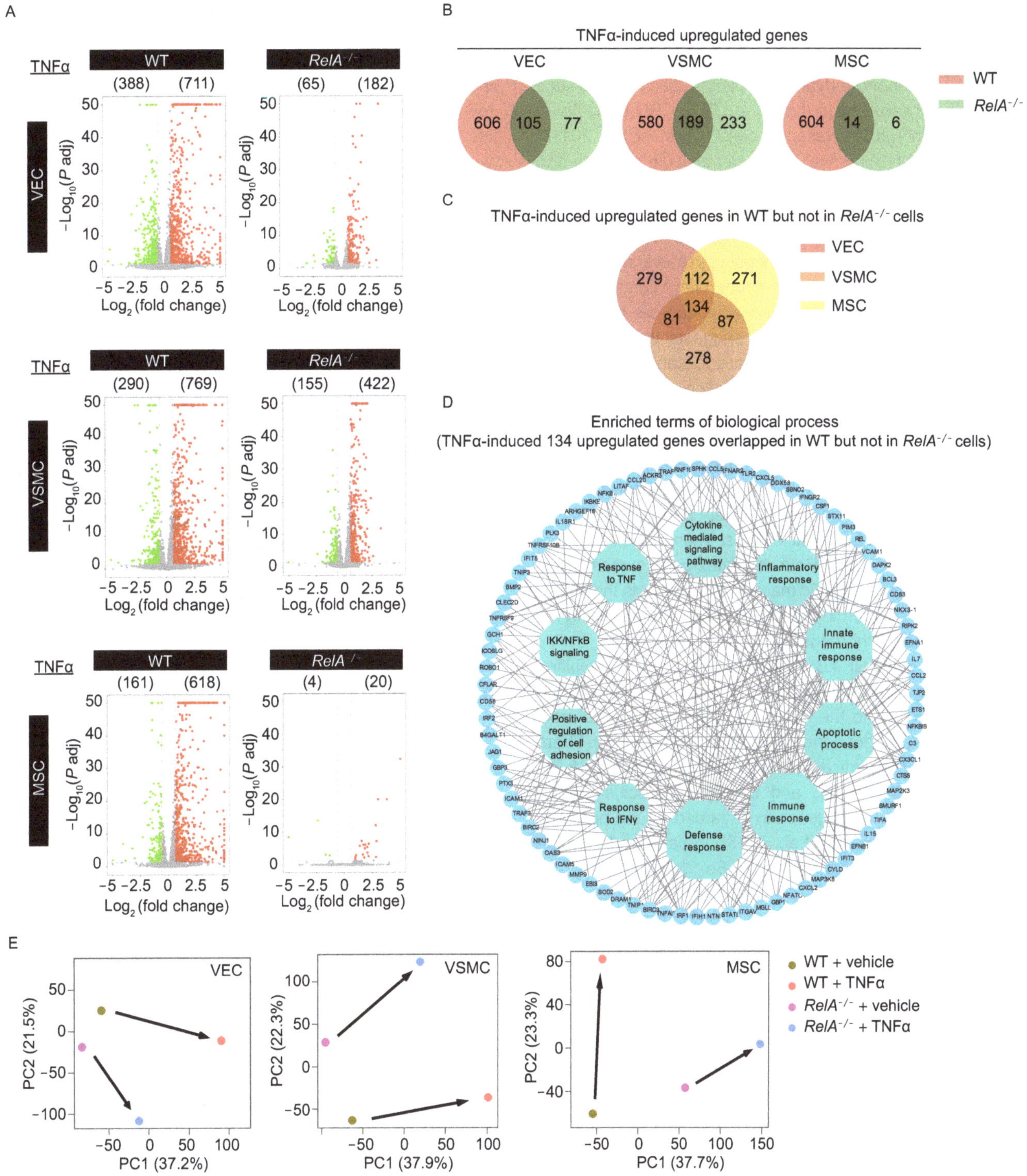

Figure 9. Transcriptomic analysis revealed RelA deficiency-induced vascular cell dysfunction upon TNFα treatment. (A) Volcano plots showing the differentially expressed genes between basal and TNFα-treated conditions in WT and *RelA*$^{-/-}$ VECs, VSMCs and MSCs. Green color indicates downregulated genes; red color indicates upregulated genes. The number listed in red or green indicates the number of differentially expressed genes. (B) Venn diagrams showing the overlap of upregulated genes in WT and *RelA*$^{-/-}$ VECs, VSMCs and MSCs upon TNFα induction. (C) Venn diagram showing the overlap of upregulated genes in WT but not in *RelA*$^{-/-}$ VECs, VSMCs, and MSCs upon TNFα induction. (D) Network diagram showing enriched GO biological process terms of upregulated genes in WT but not in *RelA*$^{-/-}$ VECs, VSMCs and MSCs upon TNFα treatment. The size of octagons represents the number of upregulated genes enriched in each term of biological processes. (E) Principal component analysis (PCA) showing a transcriptional pattern shift upon TNFα induction in WT and *RelA*$^{-/-}$ VECs, VSMCs and MSCs.

physiological functions of NF-κB in these cells, providing direct, unbiased characterization of the cell type-specific transcriptional regulation by NF-κB in vascular cells. Our platform is of superior advantages over traditional tools such as gene-engineered mice (Ijaz et al., 2016), RNAi-mediated knockdown in human vascular cells (Chen et al., 2011a; Chen et al., 2011b), as well as small-molecule agonists and antagonists of NF-κB (Jakkampudi et al., 2016; Janssen-Heininger et al., 2000).

Our data provided a comprehensive understanding on how NF-κB regulates the homeostasis of various human vascular cells under basal and inflammatory conditions. For instance, we observed enhanced adipogenesis and osteogenesis in *RelA*$^{-/-}$ MSCs, which suggests that RelA deficiency may cause atherosclerosis-associated abnormal lipid accumulation and vascular ossification. Our data also showed that the self-renewal and differentiation abilities of MSCs were directly regulated by RelA, implying that RelA inactivation might cause vessel dysfunction via compromising the regenerative ability of the adventitia-localized stem cells. Furthermore, we showed that RelA deficiency impaired vasculogenesis in VECs and VSMCs. Thus, our study offers detailed illustrations how RelA regulates gene expression in a vascular cell type-specific manner and supports the notion that RelA is implicated in the multi-perspective maintenance of human vascular homeostasis.

IκBs sequester NF-κB in the cytoplasm, thus inhibiting NF-κB activity. Degradation of IκBs releases NF-κB, leading to its translocation to the nucleus and the subsequent activation of NF-κB target genes (Simeonidis et al., 1999). There are at least eight dedicated IκB proteins: IκBα, IκBβ, IκBγ, IκBε, IκBz, IκBNS, Bcl-3 and Drosophila Cactus (Morris et al., 2016), with IκBα generally regarded as the predominant isoform (Fagerlund et al., 2015). In patients, heterozygous nonsense mutation in IκBα leads to NF-κB haploinsufficiency, resulting in ectodermal dysplasia and immune deficiency (Courtois et al., 2003). In this study, we genetically ablated IκBα in hESCs and obtained various vascular cells via directed differentiation to study the impact of IκBα deficiency on the functions of NF-κB signaling. Our results showed that IκBα acted largely independent of RelA signaling under basal condition as well as upon TNFα stimulation, suggesting that NF-κB is not predominantly regulated by IκBα, but perhaps by other IκB isoforms in vascular cells.

To date, inhibition of NF-κB during vascular inflammation has provided a promising therapeutic target for the treatment of various cardiovascular disorders. Various inhibitors that target NF-κB pathway have been developed to extend the choice for clinical applications (Tas et al., 2009). In our study, we have provided a platform that facilitates the in-depth mechanistic interpretation of how RelA regulates human vascular diseases and aging. The differential responses in various types of vascular cells to RelA deficiency sound a cautionary note for diseases being considered for anti-NF-κB treatment. Moderate suppression is likely to be beneficial to the maintenance of vascular homeostasis by alleviating the injury from vascular inflammation, whereas excessive suppression may lead to the abnormal behaviors for different layers of vascular cells. Lastly, given that the regulation of NF-κB is of cell type specificity, our study has demonstrated the importance of highly targeted therapy in the treatment of vascular inflammatory diseases via the inhibition of NF-κB activity.

MATERIALS AND METHODS

Cell culture

H9 human ESCs purchased from WiCell Research institute were cultured on mitomycin C-inactivated mouse embryonic fibroblasts (MEFs) or matrigel (BD Biosciences) (Liu et al., 2012). Human VECs were cultured in EGM-2 (Lonza) medium supplemented with 50 ng/mL VEGF (HumanZyme), 10 μmol/L SB431542 (Selleck) and 20 ng/mL FGF2 (Joint Protein Central, JPC). Human VSMCs were cultured in N2B27 medium (1:1 (*v/v*) of DMEM/F12 medium (Gibco) and Neurobasal medium (Gibco) supplemented with 2% B27 (Gibco), 1% N2 (Gibco), 55 μmol/L β-mercaptoethanol (Gibco) and 1% penicillin/streptomycin (Gibco)) supplemented with 10 ng/mL PDGF-BB (Peprotech). Human MSCs were cultured in 90% α-MEM medium (Gibco) supplemented with 10% fetal bovine serum (Ausbian, VS500T), 1% penicillin/streptomycin (Gibco), and 1 ng/mL FGF2 (JPC).

Generation of *RelA*$^{-/-}$ ESCs and *IκBα*$^{-/-}$ ESCs

CRISPR/Cas9 mediated gene targeting was performed as previously described (Ding et al., 2013; Wang et al., 2017; Wang et al., 2018). In brief, a donor plasmid containing homologous arms was constructed with a neomycin resistant cassette. sgRNAs targeting *RelA* (sequence: CTC GTC TGT AGT GCA CGC CGC GG) and *IκBα* (sequence: GCT CCT TCT TCA GCC CGT CG) were also constructed in the plasmid respectively. 5 × 10^6 ESCs were digested by TrypLE and were electroporated with 7 μg Cas9 plasmid, 7 μg *RelA* or *IκBα* targeted sgRNA plasmid and 7 μg donor plasmid. Cells were then cultured on DR4 MEF feeder. G418 (Gibco) was used for targeting screening for 14 days, and the drug resistant clones were transported to a 96-well plate for expansion and RelA or IκBα knockout verification. The targeted clones were verified by genomic PCR with primers listed as follows, *RelA* P1: CCA TCC GGG CTG TAG GCT CCG CAA AGC TC, P2: TCA TAG CCC GCC TCC TGT CCC CTC ATG CTG. *IκBα* P1: GCC CAG CCA TCA TTT CCA CTC TTG CGT TT, P2: AAG CAA CAA AAT GAG GGC TGA TCC TAC CAC. The neomycin-resistant cassette was later removed after verification according to the method previously described (Duan et al., 2015).

VEC generation and characterization

VECs were differentiated as previously described (Pan et al., 2016; Wu et al., 2018). In brief, human ESC clones were seeded in matrigel coated 6-well plates in mTeSR medium at day 0. ESCs were washed with IMDM medium (Gibco) and cultured in EGM-2 medium (Lonza) supplemented with 25 ng/mL BMP4 (R&D), 3 μmol/

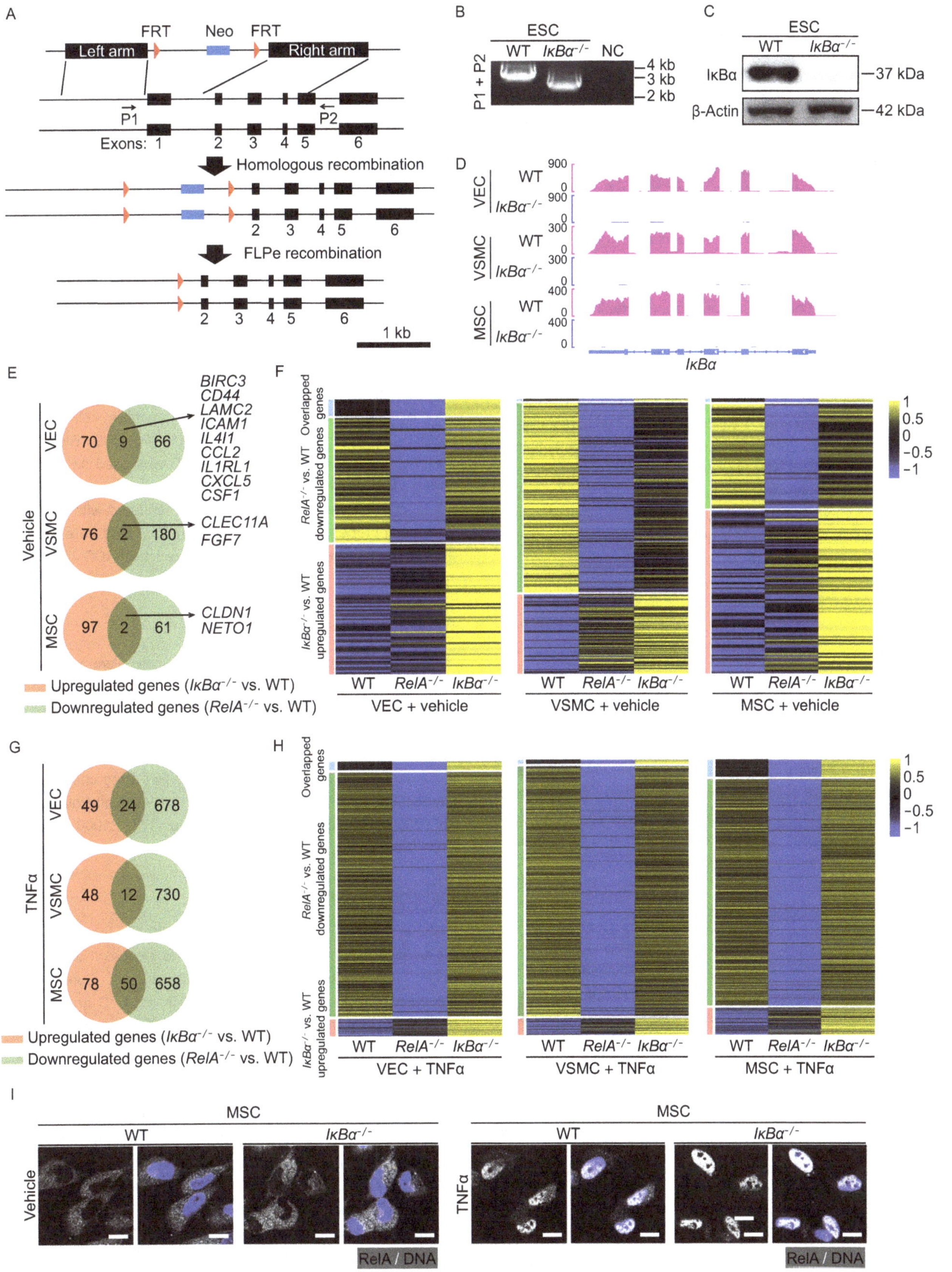
A
FRT
Neo
FRT
Left arm
Right arm
P1
P2
Exons: 1 2 3 4 5 6
Homologous recombination
2 3 4 5 6
FLPe recombination
2 3 4 5 6
1 kb
B
ESC
WT IκBα-/- NC
P1 + P2
4 kb
3 kb
2 kb
C
ESC
WT IκBα-/-
IκBα
37 kDa
β-Actin
42 kDa
D
VEC
VSMC
MSC
WT
IκBα-/-
900
300
400
0
IκBα
E
Vehicle
VEC
VSMC
MSC
70 9 66
76 2 180
97 2 61
BIRC3
CD44
LAMC2
ICAM1
IL4I1
CCL2
IL1RL1
CXCL5
CSF1
CLEC11A
FGF7
CLDN1
NETO1
Upregulated genes (IκBα-/- vs. WT)
Downregulated genes (RelA-/- vs. WT)
F
Overlapped genes
RelA-/- vs. WT downregulated genes
IκBα-/- vs. WT upregulated genes
WT RelA-/- IκBα-/-
VEC + vehicle
VSMC + vehicle
MSC + vehicle
1
0.5
0
-0.5
-1
G
TNFα
VEC
VSMC
MSC
49 24 678
48 12 730
78 50 658
Upregulated genes (IκBα-/- vs. WT)
Downregulated genes (RelA-/- vs. WT)
H
Overlapped genes
RelA-/- vs. WT downregulated genes
IκBα-/- vs. WT upregulated genes
WT RelA-/- IκBα-/-
VEC + TNFα
VSMC + TNFα
MSC + TNFα
1
0.5
0
-0.5
-1
I
MSC
WT
IκBα-/-
Vehicle
TNFα
RelA / DNA

◄ **Figure 10. Transcriptomic analysis revealed the effect of IκBα deficiency on RelA signaling.** (A) Schemic diagram of *IκBα* knockout strategy via CRISPR/Cas9 in human ESCs. A neomycin-resistant cassette (Neo) was included for positive selection. (B) Genomic PCR verification of the deletion of *IκBα* exon 1 in *IκBα*$^{-/-}$ ESCs. Water was used as a negative control (NC). (C) Western blot analysis showing IκBα protein levels in WT and *IκBα*$^{-/-}$ ESCs. β-Actin was used as a loading control. (D) Transcriptional signal of IκBα in WT and *IκBα*$^{-/-}$ in VECs, VSMCs and MSCs. Transcriptional signals were normalized by RPKM at bin size 10 bp. (E) Venn diagrams showing the overlap between upregulated genes in *IκBα*$^{-/-}$ vascular cells and downregulated genes in *RelA*$^{-/-}$ vascular cells compared to WT vascular cells under basal condition. (F) Heatmaps revealing the transcriptional patterns of genes upregulated only in *IκBα*$^{-/-}$ vascular cells (pink), downregulated only in *RelA*$^{-/-}$ vascular cells (green), and genes overlapped (blue) under basal condition. (G) Venn diagrams showing the overalp between upregulated genes in *IκBα*$^{-/-}$ vascular cells and downregulated genes in *RelA*$^{-/-}$ vascular cells compared to WT vascular cells upon TNFα treatment. (H) Heatmaps revealing the transcriptional patterns of genes upregulated only in *IκBα*$^{-/-}$ vascular cells (pink), downregulated only in *RelA*$^{-/-}$ vascular cells (green) and genes overlapped (blue) upon TNFα treatment. (I) Immunostaining of RelA in WT and *IκBα*$^{-/-}$ MSCs under basal and TNFα-treated conditions. DNA was labeled by Hoechst 33342. Scale bar, 10 μm.

L CHIR99021 (Selleck), 3 μmol/L IWP2 (Selleck) and 4 ng/mL FGF2 (JPC) for 3 days. At day 4, Cells were rinsed with IMDM medium and cultured in EGM-2 medium supplemented with 50 ng/mL VEGF (HumanZyme), 10 ng/mL IL6 (Peprotech) and 20 ng/mL FGF2 (JPC) for 3 days. Endothelial cells were sorted with CD201-PE (Biolegend, 351904, 1:300), CD34-FITC (BD biosciences, 555821, 1:100). IgG-FITC (BD biosciences, 555748) and IgG-PE (BD biosciences, 555749) were used as isotype controls.

VSMC generation and characterization

VSMCs were differentiated as previously reported (Patsch et al., 2015). Human ESC clones were digested by Accutase (Gibco) and 3 × 10^5 cells were seeded in matrigel coated 6-well plate and mTeSR medium containing 10 μmol/L Y27632 (Selleck) at day 0. Cells were cultured in the N2B27 medium with 25 ng/mL BMP4 (R&D), 8 μmol/L CHIR99021 (Selleck) for 3 days. At day 4, cells were cultured in N2B27 medium with 10 ng/mL PDGF-BB (Peprotech), 2 ng/mL Activin A (HumanZyme) and the medium was changed every day for 2 days. VSMCs were sorted with CD140b-APC (BD biosciences, 558821, 1:200), and IgG-APC (BD Biosciences, 555751) was used as an isotype control.

MSC generation and characterization

MSCs were derived from ESCs according to a protocol previously described (Zhang et al., 2015). Briefly, human ESC clones from embryoid bodies (EBs) were cultured in 90% α-MEM medium (Gibco) with 10% FBS (Ausbian, VS500T), 1% penicillin/streptomycin (Gibco), 5 ng/mL TGFβ (HumanZyme) and 10 ng/mL FGF2 (JPC) for 10 days until fibroblast-like cells emerged. The MSCs were sorted for CD73$^+$/CD90$^+$/CD105$^+$ cells by FACS. Antibodies for sorting and characterization were CD73-PE (BD Biosciences, 550257), CD90-FITC (BD Biosciences, 555595), CD105-APC (eBioscience, 555748). IgG-FITC, IgG-PE, and IgG-APC were used as isotype controls. The differentiation abilities of MSCs to adipocytes, osteoblasts and chondrocytes were assessed by oil red O, Von Kossa and Toluidine blue staining, respectively (Duan et al., 2015; Fang et al., 2018; Yang et al., 2017; Zhang et al., 2015).

Flow cytometry analysis

Cell apoptosis analysis was performed with Annexin V-EGFP Apoptosis Detection Kit (Vigrous Biotechnology). Cells were pretreated with 10 ng/mL TNFα (Peprotech) for 24 h before performing apoptosis assay. For ICAM1 expression analysis, cells were suspended in 10% FBS/PBS with ICAM1-PE (BD Biosciences, 560971, 1:100) for 30 min, and IgG-PE was used as an isotype control. The results were determined by FlowJo7.6.1 software. For Dil-Ac-LDL uptake analysis, VECs were cultured with Dil-Ac-LDL (Molecular probes, 1:400) for 6–8 h at 37 °C. VECs were harvested and washed twice by PBS and analyzed by FACS detected by FITC channel.

Western blotting

Protein was quantified by a BCA Kit after cells were lysed in RIPA buffer as previously described (Duan et al., 2015). 20 μg of protein lysates were subjected to 10% SDS-PAGE and electrotransferred to a 0.22 μm PVDF membrane (Millipore). The membrane was incubated in primary antibody overnight at 4 °C and HRP-conjugated secondary antibody followed by visualization using chemi-luminescence. The antibodies used were as follows: anti-RelA (Abcam, ab7970, 1:1000), anti-IκBα (Santa Cruz, sc-371, 1:1000) and anti-β-Actin (Santa Cruz, sc-69879, 1:4000).

Immunofluorescence

Cells were fixed in 4% formaldehyde for 30 min, permeabilized in 0.4% Triton X-100 in PBS for 30 min and incubated with blocking buffer (10% donkey serum in PBS) for 30 min. Samples were incubated with primary antibody overnight in 4 °C and with secondary antibody for 1 h at room temperature. Cell images were taken using confocal microscopy. Nuclear DNA was stained by Hoechst 33342 (Invitrogen, H3570). Antibodies used were as follows: anti-NANOG (Abcam, ab21624, 1:200), anti-SOX2 (Santa Cruz, sc-17320, 1:100), anti-OCT4 (Santa Cruz, sc-5279, 1:100), anti-TUJ1 (Sigma, T2200, 1:100), anti-αSMA (Sigma, A5228, 1:100), anti-FOXA2 (Cell Signaling Technology, 8186S, 1:100), anti-CD31-FITC (BD Biosciences, 555445, 1:50), anti-vWF (Dako, A0082, 1:200), Dil-Ac-LDL (Molecular probes, 1:400), anti-SM22 (Abcam, ab14106, 1:200), anti-Calponin (BD Biosciences, 2017-03, 1:200) and anti-Ki67 (Vector Labs, VP-RM04, 1:1,000) anti-RelA (Cell Signaling Technology, 8242, 1:200).

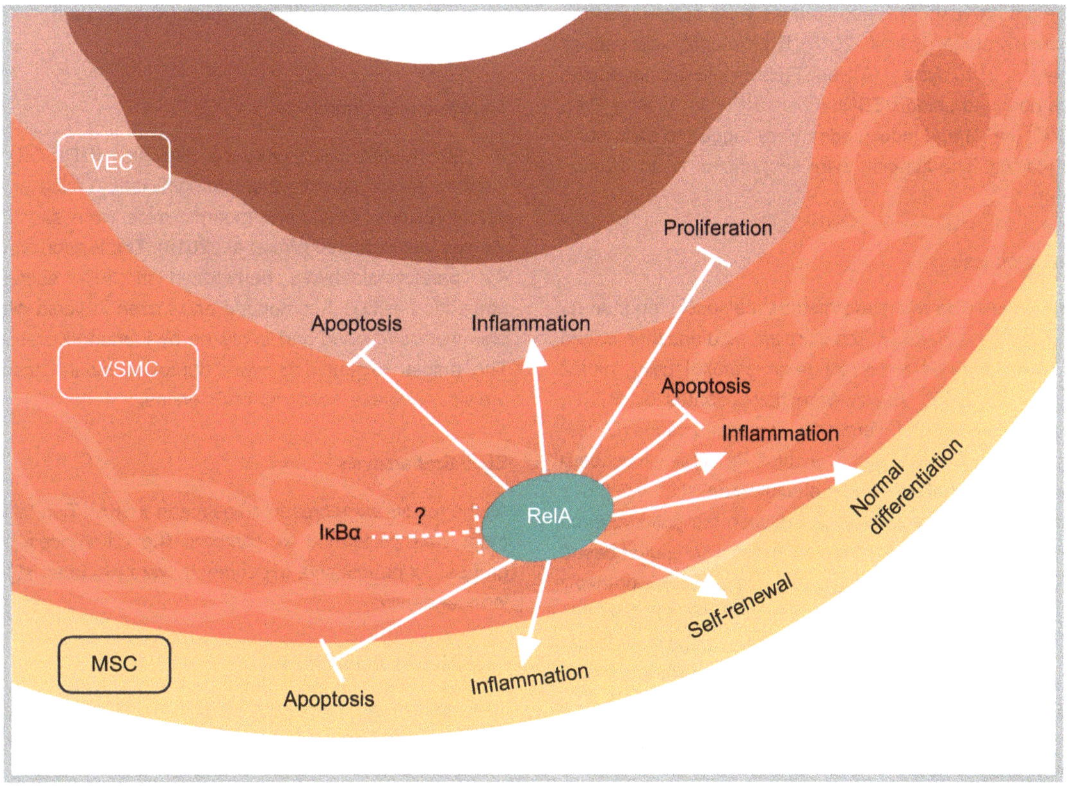

Figure 11. A proposed model showing a guardian role of RelA in maintaining human vascular cell homeostasis. (A) Under basal condition, NF-κB/RelA inhibits the proliferation of vascular endothelial cells (VECs) and promotes the self-renewal of mesenchymal stem cells (MSCs) that are the crucial cell type of the adventitia layer. NF-κB/RelA also maintains the multi-differentiation potentials of MSCs. Upon TNFα treatment, RelA mediates vascular inflammation and protects vascular cells from apoptosis. This model suggests a guardian role of RelA in maintaining human blood vessel homeostasis. Notably, IκBα appears to modulate vascular gene expression predominantly in a RelA-independent manner under basal and inflammatory conditions.

RNA extraction and RT-qPCR

Cells were harvested in TRIzol reagent (Invitrogen) and RNA was extracted according to the manufacturers' instruction. 2 μg RNA was used for cDNA synthesis using reverse transcription master Mix (Promega). RT-qPCR was performed in SYBR Green supermix (Thunderbird) on a CFX-384 RT-qPCR system. The relative expression of genes was normalized by *18S* rRNA transcript. All qPCR primers used were listed as follows: *18S* forward primer, GTA ACC CGT TGA ACC CCA TT, reverse primer, CCA TCC AAT CGG TAG TAG CG. *RelA* forward primer, GTG GGG ACT ACG ACC TGA ATG, reverse primer, GGG GCA CGA TTG TCA AAG ATG. *NANOG* forward primer, ACA ACT GGC CGA AGA ATA GCA, reverse primer, GGT TCC CAG TCG GGT TCA C. *OCT4* forward primer, GGG TTT TTG GGA TTA AGT TCT TCA, reverse primer, GCC CCC ACC CTT TGT GTT. *SOX2* forward primer, CAA AAA TGG CCA TGC AGG TT, reverse primer, AGT TGG GAT CGA ACA AAA GCT ATT. *FABP4* forward primer, ACT GGG CCA GGA ATT TGA CG, reverse primer, CTC GTG GAA GTG ACG CCT T. *PPARG* forward primer, ACC AAA GTG CAA TCA AAG TGG A, reverse primer, ATG AGG GAG TTG GAA GGC TCT. *LPL* forward primer, TCA TTC CCG GAG TAG CAG AGT, reverse primer, GGC CAC AAG TTT TGG CAC C. *BGLAP* forward primer, CAC TCC TCG CCC TAT TGG C, reverse primer, CCC TCC TGC TTG GAC ACA AAG. *SPP1* forward primer, CTC CAT TGA CTC GAA CGA CTC, reverse primer, CAG GTC TGC GAA ACT TCT TAG AT. *COL1A1* forward primer, GTG CGA TGA CGT GAT CTG TGA, reverse primer, CGG TGG TTT CTT GGT CGG T. *RUNX2* forward primer, CCG CCT CAG TGA TTT AGG GC, reverse primer, GGG TCT GTA ATC TGA CTC TGT CC. *ICAM1*, forward primer, ATG CCC AGA CAT CTG TGT CC, reverse primer, GGG GTC TCT ATG CCC AAC AA. *VCAM1* forward primer, TTT GAC AGG CTG GAG ATA GAC T, reverse primer, TCA ATG TGT AAT TTA GCT CGG CA. *MCP1* forward primer, CAG CCA GAT GCA ATC AAT GCC, reverse primer, TGG AAT CCT GAA CCC ACT TCT. *IL6* forward primer, ACT CAC CTC TTC AGA ACG AAT TG, reverse primer, CCA TCT TTG GAA GGT TCA GGT TG. *IL8* forward primer, ACT GAG AGT GAT TGA GAG TGG AC, reverse primer, AAC CCT CTG CAC CCA GTT TTC. *Selectin E* forward primer, CAG CAA AGG TAC ACA CAC CTG, reverse primer, CAG ACC CAC ACA TTG TTG ACT T. *CLEC11A* forward primer, GGG CCT CTA CCT CTT CGA AA, reverse primer, CAG TTC TCG AGC GTG CCA CC.

RNA-seq library construction

Vascular cells were cultured and treated with TNFα at 10 ng/mL for 4 h before harvest. Alternatively, MSCs were treated with IL1β at 5 ng/

mL for 4 h before harvest. RNA-seq libraries were constructed following a previous protocol (Li et al., 2016). RNA integrity was verified by the Bioanalyzer 2100 system (Agilent Technologies). Sequencing libraries were obtained using NEBNext® Ultra™ RNA Library Prep Kit for Illumina® (NEB) and index codes were tagged to each samples for sequencing. The libraries were sequenced on an Illumina Hiseq platform.

RNA-seq data processing

RNA-seq data processing was performed as previously reported (Li et al., 2016). Briefly, low quality reads were trimmed and then paired clean reads were mapped to hg19 human genome using hisat2 (v2.0.4) (Kim et al., 2015). Reads were then counted by HTSeq (v0.6.1) (Anders et al., 2015). Differentially expressed genes (DEGs) were calculated using DESeq2 with the cutoff Benjamini–Hochberg adjust P value 0.05 and absolute fold change more than 1.5 (Love et al., 2014). To evaluate the correlation between replicates of each sample, the Pearson correlation coefficient (R) was computed based on $\log_2$(FPKM + 1). Gene Ontology (GO) and KEGG pathway enrichment analysis was carried out by ClusterProfiler R package (Yu et al., 2012). The RNA-seq data have been deposited to the NCBI Gene Expression Omnibus (GEO) database with accession number GSE115311.

Matrigel tube formation assay

2×10^4 VECs were seeded in each well of a 24-well plate coated with matrigel (BD Biosciences) in triplicate and incubated for 6–8 h at 37 °C, and then stained by Calcein-AM (Invitrogen) (Wu et al., 2018). The tube formation was assessed by taking photomicrographs and measured by ImageJ2x 2.1.4.7 software.

Monocyte adhesion assay

2×10^5 VECs were seeded in each well of 12-well plate coated with collagen. The next day, VECs were treated with or without 10 ng/mL TNFα for 4 h. 2×10^6 monocytes were co-cultured with VECs for 1 h and then rinsed by PBS for 3 times carefully (Lee et al., 2015). The monocytes adhered on endothelium were analyzed by ImageJ2x 2.1.4.7 software.

Colony formation assay

VECs, VSMCs and MSCs were each seeded by 2,000 cells per well in 12-well plates (Pan et al., 2016), and cultured until clear cell colonies formed. Cells were fixed in 4% paraformaldehyde for 30 min and stained with crystal violet (Biohao Biotechnology). The images were analyzed by ImageJ software.

ShRNA plasmid construction and lentivirus packaging

ShRNA sequence (TGA GGA CAT CGT CAC TTA C) that targeting CLEC11A was cloned into pLVTHM vector. For lentivirus packaging, CLEC11A shRNA was co-transfected with psPAX2 (Addgene) and pMD2.G (Addgene) into HEK293T cells. Lentiviruses were harvested after 48 h and ultracentrifuged at 19,400 *g* for 2.5 h followed by resuspension in α-MEM medium (Kubben et al., 2016; Liu et al., 2011).

Teratoma experiment

3×10^6 human ESCs were injected into NOD-SCID (Non-obese diabetic severe combined immunodeficiency) mice under subcutaneous regions. After three months, mice were sacrificed and teratomas were excised (Wu et al., 2018). The teratomas were fixed in 4% paraformaldehyde, dehydrated in 30% sucrose solution, embedded in O.C.T compound and frozen in liquid nitrogen. Samples were sectioned and analyzed by immunofluorescent staining. The animal experiments were approved by the Institute of Biophysics, Chinese Academy of Sciences.

Statistical analysis

All the results were presented as mean ± SEM. Two-tailed Student's *t*-test was performed to compare the differences between two groups. All quantitative experiments were repeated at least 3 times independently.

ACKNOWLEDGEMENTS

This work was supported by the National Key Research and Development Program of China (2017YFA0103304), the Strategic Priority Research Program of the Chinese Academy of Sciences (XDA16010100), the National Key Research and Development Program of China (2015CB964800, 2017YFA0102802, 2014CB910503 and 2018YFA0107203), the National High Technology Research and Development Program of China (2015AA020307), the National Natural Science Foundation of China (NSFC: 31671429, 91749202, 91749123, 81625009, 81330008, 81371342, 81471414, 81422017, 81601233, 81671377, 31601109, 31601158, 81771515, 81701388), Program of Beijing Municipal Science and Technology Commission (Z151100003915072), Key Research Program of the Chinese Academy of Sciences (KJZDEW-TZ-L05), and the State Key Laboratory of Stem Cell and Reproductive Biology (2016SRLabKF13).

ABBREVIATIONS

EP, early passage; ESCs, embryonic stem cells; FACS, fluorescent-activated cell sorting; GO, gene ontology; ICAM1, intercellular adhesion molecule 1; IFNγ, interferon gamma; IκB, inhibitor of NF-κB; IKK, IκB kinase; LP, late passage; MCP1, monocyte chemoattractant protein 1; MSCs, mesenchymal stem cells; NOD-SCID, non-obese diabetic severe combined immunodeficiency; TNFα, tumor necrosis factor alpha; VCAM1, vascular cell adhesion molecule 1; VECs, vascular endothelial cells; VSMCs, vascular smooth muscle cells; WT, wildtype.

COMPLIANCE WITH ETHICS GUIDELINES

Ping Wang, Zunpeng Liu, Xiaoqian Zhang, Jingyi Li, Liang Sun, Zhenyu Ju, Jian Li, Piu Chan, Guang-Hui Liu, Weiqi Zhang, Moshi

Song and Jing Qu declare that they have no conflict of interest. All institutional and national guidelines for the care and use of laboratory animals were followed.

REFERENCES

Anders S, Pyl PT, Huber W (2015) HTSeq–a Python framework to work with high-throughput sequencing data. Bioinformatics 31:166–169

Baker RG, Hayden MS, Ghosh S (2011) NF-kappaB, inflammation, and metabolic disease. Cell Metab 13:11–22

Barnes PJ, Karin M (1997) Nuclear factor-kappaB: a pivotal transcription factor in chronic inflammatory diseases. N Engl J Med 336:1066–1071

Brand K, Page S, Rogler G, Bartsch A, Brandl R, Knuechel R, Page M, Kaltschmidt C, Baeuerle PA, Neumeier D (1996) Activated transcription factor nuclear factor-kappa B is present in the atherosclerotic lesion. J Clin Invest 97:1715–1722

Breitbach M, Kimura K, Luis TC, Fuegemann CJ, Woll PS, Hesse M, Facchini R, Rieck S, Jobin K, Reinhardt J et al (2018) In Vivo Labeling by CD73 Marks Multipotent Stromal Cells and Highlights Endothelial Heterogeneity in the Bone Marrow Niche. Cell Stem Cell 22(262–276):e267

Caplan AI, Correa D (2011) The MSC: an injury drugstore. Cell Stem Cell 9:11–15

Chen G, Chen Y, Chen H, Li L, Yao J, Jiang Q, Lin X, Wen J, Lin L (2011a) The effect of NF-kappaB pathway on proliferation and apoptosis of human umbilical vein endothelial cells induced by intermittent high glucose. Mol Cell Biochem 347:127–133

Chen G, Qiao Y, Yao J, Jiang Q, Lin X, Chen F, Lin F, Lin M, Lin L, Zhu P (2011b) Construction of NF-kappaB-targeting RNAi adenovirus vector and the effect of NF-kappaB pathway on proliferation and apoptosis of vascular endothelial cells. Mol Biol Rep 38:3089–3094

Chiba T, Kondo Y, Shinozaki S, Kaneko E, Ishigami A, Maruyama N, Umezawa K, Shimokado K (2006) A selective NFkappaB inhibitor, DHMEQ, reduced atherosclerosis in ApoE-deficient mice. J Atheroscler Thromb 13:308–313

Courtois G, Smahi A, Reichenbach J, Doffinger R, Cancrini C, Bonnet M, Puel A, Chable-Bessia C, Yamaoka S, Feinberg J et al (2003) A hypermorphic IkappaBalpha mutation is associated with autosomal dominant anhidrotic ectodermal dysplasia and T cell immunodeficiency. J Clin Invest 112:1108–1115

da Silva Meirelles L, Caplan AI, Nardi NB (2008) In search of the in vivo identity of mesenchymal stem cells. Stem Cells 26:2287–2299

Ding Q, Regan SN, Xia Y, Oostrom LA, Cowan CA, Musunuru K (2013) Enhanced efficiency of human pluripotent stem cell genome editing through replacing TALENs with CRISPRs. Cell Stem Cell 12:393–394

Duan S, Yuan G, Liu X, Ren R, Li J, Zhang W, Wu J, Xu X, Fu L, Li Y et al (2015) PTEN deficiency reprogrammes human neural stem cells towards a glioblastoma stem cell-like phenotype. Nat Commun 6:10068

Fagerlund R, Behar M, Fortmann KT, Lin YE, Vargas JD, Hoffmann A (2015) Anatomy of a negative feedback loop: the case of IkappaBalpha. J R Soc Interface 12:0262

Fang J, Yang J, Wu X, Zhang G, Li T, Wang X, Zhang H, Wang CC, Liu GH, Wang L (2018) Metformin alleviates human cellular aging by upregulating the endoplasmic reticulum glutathione peroxidase. Aging Cell 7:e12765

Galley HF, Webster NR (2004) Physiology of the endothelium. Br J Anaesth 93:105–113

Gareus R, Kotsaki E, Xanthoulea S, van der Made I, Gijbels MJ, Kardakaris R, Polykratis A, Kollias G, de Winther MP, Pasparakis M (2008) Endothelial cell-specific NF-kappaB inhibition protects mice from atherosclerosis. Cell Metab 8:372–383

Hajra L, Evans AI, Chen M, Hyduk SJ, Collins T, Cybulsky MI (2000) The NF-kappa B signal transduction pathway in aortic endothelial cells is primed for activation in regions predisposed to atherosclerotic lesion formation. Proc Natl Acad Sci USA 97:9052–9057

Hiraoka A, Yano Ki K, Kagami N, Takeshige K, Mio H, Anazawa H, Sugimoto S (2001) Stem cell growth factor: in situ hybridization analysis on the gene expression, molecular characterization and in vitro proliferative activity of a recombinant preparation on primitive hematopoietic progenitor cells. Hematol J 2:307–315

Ijaz T, Wakamiya M, Sun H, Recinos A III, Tilton RG, Brasier AR (2016) Generation and characterization of a novel transgenic mouse harboring conditional nuclear factor-kappa B/RelA knockout alleles. BMC Dev Biol 16:32

Jakkampudi A, Jangala R, Reddy BR, Mitnala S, Nageshwar Reddy D, Talukdar R (2016) NF-kappaB in acute pancreatitis: Mechanisms and therapeutic potential. Pancreatology 16:477–488

Janssen-Heininger YM, Poynter ME, Baeuerle PA (2000) Recent advances towards understanding redox mechanisms in the activation of nuclear factor kappaB. Free Radic Biol Med 28:1317–1327

Khan SY, Awad EM, Oszwald A, Mayr M, Yin X, Waltenberger B, Stuppner H, Lipovac M, Uhrin P, Breuss JM (2017) Premature senescence of endothelial cells upon chronic exposure to TNFα can be prevented by N-acetyl cysteine and plumericin. Sci Rep 7:39501

Kida Y, Kobayashi M, Suzuki T, Takeshita A, Okamatsu Y, Hanazawa S, Yasui T, Hasegawa K (2005) Interleukin-1 stimulates cytokines, prostaglandin E2 and matrix metalloproteinase-1 production via activation of MAPK/AP-1 and NF-kappaB in human gingival fibroblasts. Cytokine 29:159–168

Kim D, Langmead B, Salzberg SL (2015) HISAT: a fast spliced aligner with low memory requirements. Nat Methods 12:357–360

Kirton JP, Xu Q (2010) Endothelial precursors in vascular repair. Microvasc Res 79:193–199

Kubben N, Zhang W, Wang L, Voss TC, Yang J, Qu J, Liu GH, Misteli T (2016) Repression of the antioxidant NRF2 pathway in premature aging. Cell 165:1361–1374

Kucharczak J, Simmons MJ, Fan Y, Gelinas C (2003) To be, or not to be: NF-kappaB is the answer–role of Rel/NF-kappaB in the regulation of apoptosis. Oncogene 22:8961–8982

Lee TH, Sottile J, Chiang HY (2015) Collagen inhibitory peptide

R1R2 mediates vascular remodeling by decreasing inflammation and smooth muscle cell activation. PLoS ONE 10:e0117356

Li Y, Zhang W, Chang L, Han Y, Sun L, Gong X, Tang H, Liu Z, Deng H, Ye Y et al (2016) Vitamin C alleviates aging defects in a stem cell model for Werner syndrome. Protein Cell 7:478–488

Liu GH, Qu J, Shen X (2008) NF-kappaB/p65 antagonizes Nrf2-ARE pathway by depriving CBP from Nrf2 and facilitating recruitment of HDAC3 to MafK. Biochim Biophys Acta 1783:713–727

Liu GH, Barkho BZ, Ruiz S, Diep D, Qu J, Yang SL, Panopoulos AD, Suzuki K, Kurian L, Walsh C et al (2011) Recapitulation of premature ageing with iPSCs from Hutchinson-Gilford progeria syndrome. Nature 472:221–225

Liu GH, Qu J, Suzuki K, Nivet E, Li M, Montserrat N, Yi F, Xu X, Ruiz S, Zhang W et al (2012) Progressive degeneration of human neural stem cells caused by pathogenic LRRK2. Nature 491:603–607

Love MI, Huber W, Anders S (2014) Moderated estimation of fold change and dispersion for RNA-seq data with DESeq2. Genome Biol 15:550

Mallavia B, Recio C, Oguiza A, Ortiz-Munoz G, Lazaro I, Lopez-Parra V, Lopez-Franco O, Schindler S, Depping R, Egido J et al (2013) Peptide inhibitor of NF-kappaB translocation ameliorates experimental atherosclerosis. Am J Pathol 182:1910–1921

Morris O, Liu X, Domingues C, Runchel C, Chai A, Basith S, Tenev T, Chen H, Choi S, Pennetta G et al (2016) Signal integration by the IkappaB protein pickle shapes drosophila innate host defense. Cell Host Microbe 20:283–295

Nedeljkovic ZS, Gokce N, Loscalzo J (2003) Mechanisms of oxidative stress and vascular dysfunction. Postgrad Med J 79:195–199

Osborn L, Kunkel S, Nabel GJ (1989) Tumor necrosis factor alpha and interleukin 1 stimulate the human immunodeficiency virus enhancer by activation of the nuclear factor kappa B. Proc Natl Acad Sci USA 86:2336–2340

Pan H, Guan D, Liu X, Li J, Wang L, Wu J, Zhou J, Zhang W, Ren R, Zhang W et al (2016) SIRT6 safeguards human mesenchymal stem cells from oxidative stress by coactivating NRF2. Cell Res 26:190–205

Patsch C, Challet-Meylan L, Thoma EC, Urich E, Heckel T, O'Sullivan JF, Grainger SJ, Kapp FG, Sun L, Christensen K et al (2015) Generation of vascular endothelial and smooth muscle cells from human pluripotent stem cells. Nat Cell Biol 17:994–1003

Perkins ND (2007) Integrating cell-signalling pathways with NF-kappaB and IKK function. Nat Rev Mol Cell Biol 8:49–62

Perkins ND, Gilmore TD (2006) Good cop, bad cop: the different faces of NF-kappaB. Cell Death Diff 13:759–772

Rudijanto A (2007) The role of vascular smooth muscle cells on the pathogenesis of atherosclerosis. Acta Medica Indones 39:86–93

Salminen A, Huuskonen J, Ojala J, Kauppinen A, Kaarniranta K, Suuronen T (2008) Activation of innate immunity system during aging: NF-kB signaling is the molecular culprit of inflamm-aging. Ageing Res Rev 7:83–105

Schober A, Weber C (2005) Mechanisms of monocyte recruitment in vascular repair after injury. Antioxid Redox Signal 7:1249–1257

Simeonidis S, Stauber D, Chen G, Hendrickson WA, Thanos D (1999) Mechanisms by which IkappaB proteins control NF-kappaB activity. Proc Natl Acad Sci USA 96:49–54

Tas SW, Vervoordeldonk MJ, Tak PP (2009) Gene therapy targeting nuclear factor-kappaB: towards clinical application in inflammatory diseases and cancer. Curr Gene Therapy 9:160–170

Tilstra JS, Clauson CL, Niedernhofer LJ, Robbins PD (2011) NF-kappaB in Aging and Disease. Aging Dis 2:449–465

Uccelli A, Moretta L, Pistoia V (2008) Mesenchymal stem cells in health and disease. Nat Rev Immunol 8:726–736

Wang L, Yi F, Fu L, Yang J, Wang S, Wang Z, Suzuki K, Sun L, Xu X, Yu Y et al (2017) CRISPR/Cas9-mediated targeted gene correction in amyotrophic lateral sclerosis patient iPSCs. Protein Cell 8:365–378

Wang S, Hu B, Ding Z, Dang Y, Wu J, Li D, Liu X, Xiao B, Zhang W, Ren R et al (2018) ATF6 safeguards organelle homeostasis and cellular aging in human mesenchymal stem cells. Cell Discov 4:2

Wu Z, Zhang W, Song M, Wang W, Wei G, Li W, Lei J, Huang Y, Sang Y, Chan P et al (2018) Differential stem cell aging kinetics in Hutchinson-Gilford progeria syndrome and Werner syndrome. Protein Cell 9:333–350

Yang J, Li J, Suzuki K, Liu X, Wu J, Zhang W, Ren R, Zhang W, Chan P, Izpisua Belmonte JC et al (2017) Genetic enhancement in cultured human adult stem cells conferred by a single nucleotide recoding. Cell Res 27:1178–1181

Yu G, Wang LG, Han Y, He QY (2012) clusterProfiler: an R package for comparing biological themes among gene clusters. Omics 16:284–287

Yue R, Shen B,Morrison SJ (2016) Clec11a/osteolectin is an osteogenic growth factor that promotes the maintenance of the adult skeleton. eLife 5

Zhang W, Li J, Suzuki K, Qu J, Wang P, Zhou J, Liu X, Ren R, Xu X, Ocampo A et al (2015) Aging stem cells. A Werner syndrome stem cell model unveils heterochromatin alterations as a driver of human aging. Science 348:1160–1163

Zhang W, Song M, Qu J, Liu G (2018) Epigeneic modifications in cardiovascular aging and diseases. Circ Res (in press)

9

The SWI/SNF chromatin-remodeling factors BAF60a, b, and c in nutrient signaling and metabolic control

Ruo-Ran Wang[1], Ran Pan[1], Wenjing Zhang[1], Junfen Fu[2], Jiandie D. Lin[3], Zhuo-Xian Meng[1,3]✉

[1] Department of Pathology and Pathophysiology, Key Laboratory of Disease Proteomics of Zhejiang Province, School of Medicine, Chronic Disease Research Institute of School of Public Health, Zhejiang University, Hangzhou 310058, China
[2] Children's Hospital, Zhejiang University School of Medicine, Hangzhou 310003, China
[3] Life Sciences Institute and Department of Cell & Developmental Biology, University of Michigan Medical Center, Ann Arbor, MI 48109, USA
✉ Correspondence: zxmeng@zju.edu.cn (Z.-X. Meng)

ABSTRACT

Metabolic syndrome has become a global epidemic that adversely affects human health. Both genetic and environmental factors contribute to the pathogenesis of metabolic disorders; however, the mechanisms that integrate these cues to regulate metabolic physiology and the development of metabolic disorders remain incompletely defined. Emerging evidence suggests that SWI/SNF chromatin-remodeling complexes are critical for directing metabolic reprogramming and adaptation in response to nutritional and other physiological signals. The ATP-dependent SWI/SNF chromatin-remodeling complexes comprise up to 11 subunits, among which the BAF60 subunit serves as a key link between the core complexes and specific transcriptional factors. The BAF60 subunit has three members, BAF60a, b, and c. The distinct tissue distribution patterns and regulatory mechanisms of BAF60 proteins confer each isoform with specialized functions in different metabolic cell types. In this review, we summarize the emerging roles and mechanisms of BAF60 proteins in the regulation of nutrient sensing and energy metabolism under physiological and disease conditions.

KEYWORDS BAF60a, BAF60b, BAF60c, chromatin-remodeling, SWI/SNF, energy metabolism, nutrient sensing, glucose, lipid, skeletal muscle, liver

Ruo-Ran Wang and Ran Pan contributed equally to this work.

INTRODUCTION

Metabolic syndrome has become a global epidemic that markedly increases the risk for type 2 diabetes, cardiovascular diseases, and non-alcoholic fatty liver disease (NAFLD). Both genetic and environmental factors contribute to the pathogenesis of metabolic syndrome. In mammalian cells, the approximately 1.7 meters of genomic DNA are compacted more than 5,000 fold in order to squeeze into the nucleus, which is very important for the stability of chromatin structure. Chromatin is a complex of macromolecules found in eukaryotic cells. The basic structural unit of chromatin is nucleosome consisting of a segment of DNA wound in sequence around eight histone protein cores. Under certain circumstances such as DNA replication and transcription, the highly compacted chromatin structures have to be dissolved to allow the recruitment of replication or transcription complexes to local chromatin regions. At least three major epigenetic mechanisms regulate the landscape of chromatin and the accessibility of DNA to transcription factors and other proteins: DNA methylation, chromatin remodeling, and histone modification (Holliday, 2006; Cairns, 2009). Recent studies suggest that epigenetic processes play an important role in metabolic adaptation and the development of metabolic disorders through integrating the environmental signals (such as nutrient, hormone, exercise, and inflammation) to orchestrate the transcriptional networks involved in metabolic regulation (Keating and El-Osta, 2015).

The ATP-dependent chromatin-remodeling complexes can be divided into five main families: ISWI (imitation switch), NuRD/Mi2/CHD (chromodomain helicase DNA-binding),

INO80, SWR1, and SWI/SNF (switching defective/sucrose nonfermenting); the latter is the focus of this review. Recently, we and other groups have provided strong evidence suggesting that the SWI/SNF chromatin-remodeling factors, BAF60a, b, and c in particular, serve as key regulators in cell nutrient sensing, metabolic reprogramming, and the control of metabolic homeostasis in physiological and disease conditions (Lee et al., 2007; Li et al., 2008; Tao et al., 2011; Cotton et al., 2012; Meng et al., 2013; Wang et al., 2013; Meng et al., 2014; Meng et al., 2015; Lal et al., 2016; Meng et al., 2017). In this review, we focus on the roles and mechanisms of BAF60 proteins in the regulation of energy homeostasis and the development of metabolic disorders.

THE SWI/SNF COMPLEX

SWI/SNF was first discovered as a large multi-subunit complex in yeast, and named for switching defective/sucrose nonfermenting (Neigeborn and Carlson, 1984; Peterson and Herskowitz, 1992). In 1994, Brahma (Brm), the homolog of yeast transcriptional activator SWI2/SNF2, was identified in Drosophila (Elfring et al., 1994). In 1996, Wang et al. demonstrated that SWI/SNF complexes are present in multiple forms made up by 9–12 proteins in mammals (Wang et al., 1996). Since then, at least 11 subunits of SWI/SNF complex have been discovered, including 2 enzymatic ATPase subunits (Brm and Brg1) and 10 Brg1/Brm associated factors (BAFs), which are named according to their molecular weights (Sudarsanam and Winston, 2000; Wu et al., 2009). These SWI/SNF subunits, encoded by highly conserved genes, can be assembled into distinct chromatin-remodeling complexes in different cell types and developmental stages, thereby playing an essential role in tissue development (Ho and Crabtree, 2010). In addition, recent studies demonstrated that genes encoded for the SWI/SNF chromatin-remodeling complex are mutated in approximately 20% of all human tumors (Kadoch et al., 2013), and are considered to be critical epigenetic regulators of tumorigenesis due to their pleiotropic roles in the regulation of metabolism, cell cycle, and oncogenic pathway (Wiegand et al., 2010; Li et al., 2011; Varela et al., 2011; Wilson and Roberts, 2011; Shain et al., 2012; Shain and Pollack, 2013; Masliah-Planchon et al., 2015).

Mammalian SWI/SNF complexes are composed of one of two catalytic ATPase subunits (Brg1 or Brm) and 4–12 additional subunits known as BAFs. While BAF47, BAF170, and BAF155 form the core complex with Brg1 or Brm, other BAFs confer an extensive diversity of the complexes with specialized functions in different tissues through polymorphic assembly. During the past decade, most studies have focused on subunits in the core complex, revealing critical roles of the SWI/SNF in tissue development, stem cell renewal, and cancer pathogenesis. However, the physiological functions of other tissue specific subunits, which are responsible for the functional diversity of different SWI/SNF complexes, remain largely unknown (Hargreaves and Crabtree, 2011).

BAF60A, B, AND C

Among all the subunits, the BAF60 subunit is unique and important since it serves as a linker between the SWI/SNF core complex and transcription factors to regulate target gene expression (Surabhi et al., 1999; Hsiao et al., 2003; Debril et al., 2004; Flajollet et al., 2007; Takeuchi et al., 2007; Lamba et al., 2008; Oh et al., 2008). BAF60 subunit has three members, BAF60a, b, and c. The distinct tissue distribution patterns and regulatory mechanisms of BAF60s grant each isoform with specialized functions. We and other research groups have recently demonstrated that BAF60s display distinct expression profiles in metabolic tissues, and are exclusively required for the regulation of energy metabolism in physiological and disease states in a tissue specific manner. Here we summarize our current understanding of the important roles and underlying mechanisms of BAF60s in metabolic regulation and the pathogenesis of metabolic diseases.

BAF60A IN METABOLIC CONTROL AND DISEASES

BAF60a, encoded by *Smarcd1* gene, is highly expressed in liver, brain, and adipose tissues. Our previous studies demonstrated that BAF60a regulates hepatic fatty acid β-oxidation (FAO) through its interaction with PPARα and PPARγ coactivator-1α (PGC-1α) (Li et al., 2008) (Fig. 1). In addition, Tao et al. found that BAF60a is rhythmically expressed in mouse liver, and plays an important role in integrating peripheral circadian clock and energy metabolism in mammals (Tao et al., 2011) (Fig. 1). In a recent study, we demonstrated that hepatic BAF60a serves as a key component of a dietary sensitive pathway that conveys nutritional signals to transcriptional reprogramming to maintain whole body cholesterol homeostasis (Meng et al., 2015).

BAF60a and fatty liver disease

As one of the major metabolic organs, liver plays a key role in the control of systemic glucose and lipid homeostasis (Hardie and Ashford, 2014; Zhang et al., 2016). Liver regulates several major aspects of lipid metabolism, including *de novo* lipogenesis (DNL), FAO, as well as lipoprotein synthesis, uptake, and secretion. Dysregulation of hepatic lipid metabolism leads to NAFLD, which is a common and severe complication of obesity and type 2 diabetes (Kotronen and Yki-Jarvinen, 2008). NAFLD is the most prevalent cause of chronic liver injury and featured by an ectopic accumulation of triglycerides (TG), which comprises a disease spectrum ranging from simple fatty liver to non-alcoholic steatohepatitis (NASH), advanced fibrosis and cirrhosis. A better understanding of the molecular mechanisms mediating the

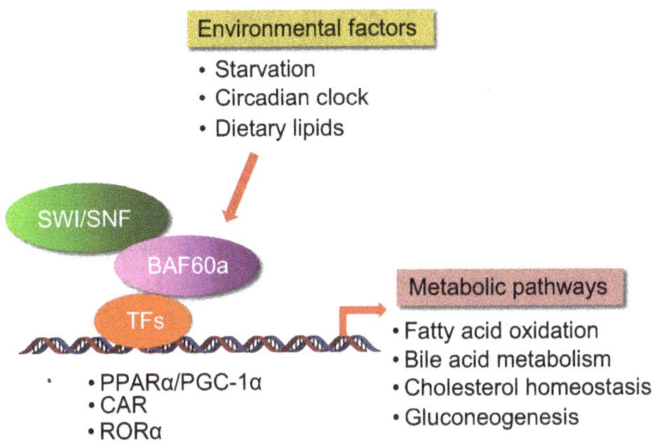

Figure 1. Metabolic regulation by BAF60a in liver. Hepatic BAF60a senses and integrates the environmental factors, such as starvation, circadian clock, and dietary fat consumption, to transcriptional reprogramming and metabolic adaptations, through its direct interaction with specific transcriptional factors/cofactors and SWI/SNF-mediated chromatin remodeling.

regulation of lipid metabolism in the liver is critical for the design of new therapeutic interventions for NAFLD.

PGC-1α is an inducible coactivator for nuclear hormone receptors and other transcriptional factors that regulates several major metabolic pathways including mitochondrial oxidative phosphorylation (OXPHOS) (Wu et al., 1999; Lehman et al., 2000) and hepatic gluconeogenesis (Herzig et al., 2001; Yoon et al., 2001). Through a genome-wide high-throughput coactivation screen, Li et al. identified BAF60a as an interaction partner of PGC-1α in the liver (Li et al., 2008). This study revealed that BAF60a is a key factor that bridges the SWI/SNF chromatin-remodeling complex and the fatty acid β-oxidation gene program in the liver. PGC-1α mediates the recruitment of BAF60a to PPARα binding sites, leading to transcriptional induction of genes involved in FAO. Adenoviral-mediated expression of BAF60a dose-dependently stimulates the transcription of genes involved in peroxisomal and mitochondrial fatty acid β-oxidation. Further *in vivo* studies demonstrated that BAF60a gain-of-function promotes FAO and ameliorates liver steatosis in diet-induced and genetically obese mice. In contrast, RNAi-mediated knockdown of BAF60a in the liver attenuates the induction of genes involved in FAO in response to starvation, leading to TG accumulation and liver steatosis. These results define BAF60a as a key regulator of hepatic FAO function and liver steatosis.

BAF60a couples circadian rhythms and energy metabolism

In mammals, many aspects of energy metabolism including glucose and lipid metabolism are coordinated by biological clocks residing in both brain and the peripheral tissues (Rutter et al., 2002; Wijnen and Young, 2006). By coordinating the metabolic and circadian transcriptional programs, PGC-1α serves as a molecular switch that synchronizes metabolic functions to the circadian clocks in mouse liver (Liu et al., 2007). As mentioned above, BAF60a is an interaction partner for PGC-1α in the regulation of hepatic lipid metabolism (Li et al., 2008). More interestingly, BAF60a is also rhythmically expressed in mouse liver, raising the possibility that BAF60a may also play a critical role in integrating the circadian signals and metabolic programs in the liver. Consistent with this hypothesis, Tao et al. demonstrated that adenoviral-mediated knockdown of BAF60a in liver significantly altered the rhythmic expression of clock genes including *Bmal1*, *Period 1* (*Per1*), *Per2*, *Rev-erbα*, and *Cryptochrome 1* (*Cry1*), and the genes involved in some major metabolic pathways such as gluconeogenesis, FAO, and mitochondrial respiration (Tao et al., 2011). Further mechanistic studies revealed that BAF60a induces *Bmal1* and *G6Pase* transcription by forming a transcriptional complex with retinoid-related orphan receptor alpha (RORα), allowing the recruitment of the SWI/SNF complex to activate local chromatin. These results suggest that BAF60a plays an important role in coupling circadian cues to metabolic gene expression programs and functions.

MiR-122, a hepatocyte-enriched microRNA previously linked to the regulation of cholesterol and lipid metabolism, also exhibits rhythmic expression pattern. It has been reported that depletion of miR-122 in the liver disrupts the mRNA expression of genes associated with cholesterol and lipid metabolism. Interestingly, PPARβ/δ and BAF60a were recently identified as miR-122 targets (Gatfield et al., 2009), suggesting an involvement of PPARβ/δ and BAF60a in coupling the circadian clocks to miR-122-mediated metabolic regulation.

BAF60a and whole body cholesterol homeostasis

Atherosclerosis is responsible for most of the cardiovascular mortality and morbidity in Western societies (Ross, 1993). Elevated plasma level of low-density lipoprotein (LDL)-cholesterol is an independent risk factor for atherosclerosis (Glass and Witztum, 2001; Steinberg, 2002). However, the currently available cholesterol lowering drugs such as statins and bile acid sequestrants could not reduce plasma cholesterol concentrations to the levels recommended by clinical guidelines in a large proportion of patients (Grundy et al., 2004; Waters et al., 2009), suggesting that other mechanisms may also play significant roles in the regulation and maintenance of cholesterol homeostasis.

Liver plays a central role in the control of systemic cholesterol homeostasis by regulating cholesterol biosynthesis, uptake and secretion in lipoproteins, and its catabolism and excretion as bile acids (Chiang, 2009; Goldstein and Brown, 2015). However, how the dietary signals are sensed and integrated into transcriptional reprogramming and metabolic functions to maintain systemic cholesterol homeostasis remains largely unknown. Recent ChIP-Seq studies revealed that, among the candidate genes identified by GWAS as potential contributors of dyslipidemia (Kathiresan et al., 2009), over 70% (32 out of 45) are bound by components of the SWI/SNF complex within 2 kb of the respective transcriptional start sites (Euskirchen et al., 2011; Dunham et al., 2012), suggesting that SWI/SNF chromatin remodeling factors may serve as core components of dietary sensing machinery that couples nutrient signals to metabolic reprogramming involved in lipid metabolism. To test this possibility, we analyzed the expression of SWI/SNF subunits in the liver from mice fed standard chow or western diet (WD), the latter contains a high amount of cholesterol and has been widely used to induce hypercholesterolemia. While the expression of core SWI/SNF subunits, including Brg1, BAF170, BAF155, and BAF53a, and the other two BAF60 proteins BAF60b and BAF60c remained similar between the two groups, mRNA and protein levels of BAF60a were significantly elevated following WD feeding (Meng et al., 2015). These data suggest that BAF60a may serve as a diet-sensitive component of the SWI/SNF complexes in the liver that links dietary cholesterol intake to whole body cholesterol homeostasis. Consistently, hepatocyte-specific deficiency of BAF60a protected the mice from diet-induced hypercholesterolemia and atherosclerosis, accompanied by reduced liver bile acid production and lower intestinal cholesterol absorption. Conversely, adenoviral-mediated overexpression of BAF60a elevates plasma cholesterol levels. Further gene expression analyses revealed that BAF60a induces the expression of CAR, which regulates bile acid detoxification by stimulating the expression of hepatic genes responsible for the modification, conjugation, and transportation of bile acids (Pascussi et al., 2008; Li and Chiang, 2013). Mechanistically, BAF60a physically interacts with CAR and forms a feedforward regulatory loop to stimulate the expression of CAR itself and its downstream target genes involved in bile acid metabolism. BAF60a is required for the recruitment of the SWI/SNF chromatin-remodeling complexes to the proximal promoter regions of CAR and its target genes to facilitate an epigenetic transition from repressive to active states. These studies define BAF60a as the central node of a dietary sensitive regulatory pathway that transmits nutritional signals to the gene expression program that governs bile acid metabolism and cholesterol homeostasis. Therapeutic intervention of this pathway may be used as a new way to treat diet-induced hypercholesterolemia and reduce the risk of atherosclerosis.

BAF60C IN METABOLIC CONTROL AND DISEASES

BAF60c, encoded by *Smarcd3* gene, is preferentially expressed in skeletal muscle, brain, and heart, and has been identified as an important regulator of heart and retina development (Lickert et al., 2004; Lamba et al., 2008). Our recent studies demonstrated that BAF60c is the core component of a novel regulatory pathway that drives glycolytic metabolism in skeletal muscle (Meng et al., 2013) (Fig. 2). Expression of BAF60c and its downstream target gene Deptor is downregulated in skeletal muscle from diabetic mice, linking obesity associated-inflammation to skeletal muscle metabolism and systemic glucose homeostasis (Meng et al., 2014) (Fig. 2). Transgenic rescue of BAF60c expression in skeletal muscle restores Deptor-mediated AKT activation and improves whole body glucose homeostasis in diet-induced and genetically obese mice. Furthermore, in a recent study, we discovered BAF60c-Deptor-AKT as the target of a glucose sensing pathway in skeletal muscle that acts in concert with insulin signaling pathway to maintain postprandial glucose homeostasis (Meng et al., 2017) (Fig. 2). In addition, it has also been reported that BAF60c plays an important role in insulin-induced lipogenesis in mouse liver (Wang et al., 2013) (Fig. 3).

BAF60c and glycolytic muscle metabolism

As one of the largest organs in normal human body, skeletal muscle is the major site of postprandial glucose disposal (Thiebaud et al., 1982; Ferrannini et al., 1988). As such, impaired skeletal muscle insulin action and glucose utilization are the primary defects in the development of type 2 diabetes (DeFronzo and Tripathy, 2009). Based on the expression of myosin heavy chain (MHC) isoforms, mitochondrial content and metabolic properties, skeletal muscle fibers have been classified as either fast-twitch fibers (type IIa, IIb, and IIx) or slow-twitch fibers (type I) (Berchtold et al., 2000). Type IIb and IIx fibers generate ATP mainly through glycolytic pathways, while type I and type IIa contain high concentrations of mitochondria and rely predominantly on oxidative metabolism. Previous studies have demonstrated that glycolytic capacity is increased while oxidative capacity is decreased in skeletal muscle in patients with type 2 diabetes (Simoneau and Kelley, 1997; Gaster et al., 2001; He et al., 2001; Mogensen et al., 2007). Impaired fuel oxidation and mitochondrial function have been proposed to play a causal role in skeletal muscle insulin resistance (Mootha et al., 2003; Mootha et al., 2004; Sparks et al., 2005). That is why so far most of the studies in skeletal muscle have focused on the regulation of oxidative muscle metabolism, leading to the identification of a number of key regulators for oxidative muscle metabolism, such as PGC-1α, AMPK, and SIRT1. Paradoxically, muscle-specific PGC-1α transgenic mice have increased mitochondrial content and slow-twitch myofibers, yet are not protected from high fat diet (HFD)-induced insulin resistance (Lin et al., 2002; Calvo et al., 2008). Notably, recent studies indicated that enhanced glycolytic capacity in individual myofiber strongly correlates with whole body insulin sensitivity (Nader and Esser, 2001; Oberbach et al., 2006; Izumiya et al., 2008), suggesting that activation of the glycolytic myofiber program may represent a compensatory mechanism in response to increasing glucose level in type 2 diabetes. However, very little is known about the regulatory mechanisms that control the metabolic properties of glycolytic muscle fibers.

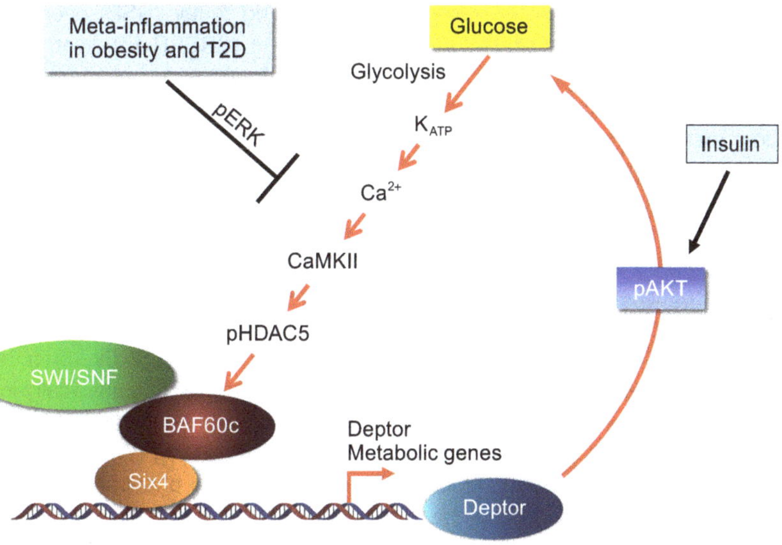

Figure 2. Metabolic regulation by BAF60c in skeletal muscle. BAF60c drives glycolytic muscle metabolism through Deptor-mediated AKT activation. Mechanistically, BAF60c forms a transcriptional complex with Six4 and recruits the SWI/SNF complex to the proximal Deptor promoter to induce Deptor gene expression. This pathway is important for muscle glucose metabolism and whole body glucose homeostasis in both physiological and diabetic states. In severe diabetic condition, expression of BAF60c and Deptor is downregulated due to activation of ERK pathway by meta-inflammation. Rescue of this pathway by either treatment with an ERK inhibitor or transgenic expression of BAF60c in skeletal muscle ameliorates insulin resistance and improves whole body glucose homeostasis in HFD-induced or genetically obese mice. Under physiological condition, BAF60c acts as a glucose sensor in skeletal muscle. Glucose triggers K_{ATP} channel-dependent Ca^{2+} response, which elicits the phosphorylation and nuclear exclusion of HDAC5 downstream of CaMKII activation, resulting in BAF60c and Deptor induction and insulin-independent AKT activation. This glucose sensing pathway works in concert with insulin signaling pathway to maintain postprandial glucose homeostasis.

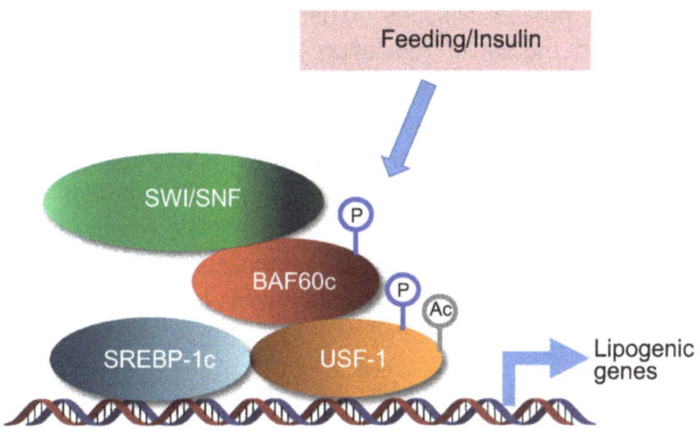

Figure 3. Regulation of insulin-induced lipogenesis by BAF60c in mouse liver. Upon insulin treatment, BAF60c is phosphorylated and translocated to the nucleus, which leads to its direct interaction with phosphorylated/ acetylated USF-1, allowing the recruitment of SWI/SNF chromatin-remodeling complex and transcriptional induction of lipogenic genes.

We recently identified BAF60c as a core component of a novel regulatory pathway that drives glycolytic metabolism in skeletal myofibers (Meng et al., 2013). BAF60c is preferentially expressed in fast/glycolytic muscle. We further demonstrated that BAF60c promotes the metabolic and contractile specification of fast-twitch muscle fibers, and is required for maintaining glycolytic metabolism in adult skeletal muscle in mice. At the mechanistic level, BAF60c forms a transcriptional complex with Six4 to stimulate the expression of Deptor, leading to Deptor-dependent AKT activation in a cell-autonomous manner. To our surprise, oxidative to glycolytic shift in skeletal muscle by BAF60c transgenic expression improves whole body glucose homeostasis in diet-induced obese mice. Our studies identified a new pathway that drives the specification of glycolytic muscle metabolism and provided evidence suggesting that enhanced glycolytic metabolism is a compensatory response to elevated blood glucose levels in type 2 diabetes, and that inadequate adaptation of glycolytic metabolism may accelerate the progression of this disorder. Consistently, we found that expression of BAF60c and Deptor was downregulated in skeletal muscle from severe diet-induced and genetically obese mice (Meng et al., 2013; Meng et al., 2014).

We next went on to dissect whether dysregulation of BAF60c-Deptor-AKT pathway in skeletal muscle contributes to the pathogenesis of type 2 diabetes and what are the upstream signalling cascades that are responsible for this dysregulation. Chronic low-grade inflammation, or meta-inflammation, has been implicated as a key risk factor for obesity-associated insulin resistance (Gregor and Hotamisligil, 2011; Osborn and Olefsky, 2012). However, the underlying mechanisms linking meta-inflammation to skeletal muscle insulin resistance and glucose intolerance remain unclear. We hypothesize that BAF60c-Deptor-AKT pathway may serve as a link between meta-inflammation and skeletal muscle insulin resistance. Indeed, we found that expression of BAF60c and Deptor is downregulated by proinflammatory cytokines, such as TNFα (Meng et al., 2014). Activation of ERK signalling cascade rather than other mitogen-activated protein kinases (MAPKs) or NF-κB pathways is responsible for this dysregulation. Importantly, transgenic rescue of this pathway by muscle specific BAF60c transgenic attenuates muscle insulin resistance and improves systemic glucose homeostasis in both HFD-induced or genetically obese mice. These studies identified BAF60c-Deptor-AKT pathway as a molecular link between meta-inflammation in skeletal muscle and systemic glucose homeostasis.

BAF60c serves as a glucose sensor in skeletal muscle

Having established the important roles of BAF60c-Deptor-AKT pathway in the specification of glycolytic muscle metabolism and in linking meta-inflammation in skeletal muscle to systemic glucose homeostasis, we went on to explore the role of this pathway under physiological conditions. Interestingly, we recently found that BAF60c functions as a glucose sensor in skeletal muscle and is important for the maintenance of postprandial glucose homeostasis (Meng et al., 2017). In cultured skeletal myocytes, glucose, but not other nutrients, such as fatty acids and amino acids, induces BAF60c and Deptor expression and stimulates AKT activation in a dose-dependent manner. Further mechanistic studies revealed that glucose triggers an ATP sensitive potassium channel (K_{ATP})-dependent Ca^{2+} response, which elicits the phosphorylation and nuclear exclusion of HDAC5 downstream of CaMKII activation, resulting in BAF60c and Deptor induction, and finally leading to insulin-independent AKT activation. This glucose sensing pathway works in concert with the insulin signaling pathway to maintain postprandial glucose homeostasis. Transgenic activation of this pathway in mice lowers postprandial blood glucose levels and improves glucose homeostasis. In contrast, inactivation of this pathway by muscle specific BAF60c deletion exhibits higher postprandial blood glucose levels and exacerbates obesity associated insulin resistance and glucose intolerance. In addition, we discovered that the BAF60c-mediated glucose sensing pathway functions as an extra-pancreatic target of sulfonylurea drugs to exert their full glucose lowering effects.

BAF60c in hepatic lipid metabolism

Hepatic transcriptional factors that play crucial roles in the regulation of lipogenic gene expression include the upstream stimulating factor (USF), sterol regulatory element-binding protein-1c (SREBP-1c), liver X receptor (LXR), and carbohydrate-responsive element-binding protein (ChREBP). SREBP-1c and USF belong to the helix-loop-helix-leucine zipper family. SREBP-1c binds as dimers to sterol regulatory elements (SREs) on the promoters of lipogenic genes (Yokoyama et al., 1993). USF acts as a molecular platform to recruit distinct transcription factors and cofactors in response to fasting/refeeding via its ability to strongly bind to E-box regulatory elements (Latasa et al., 2003). In response to insulin treatment, USF bound to the E-box recruits SREBP-1c to the nearby SREs. In a recent study, Wang et al. reported that BAF60c physically interacts with USF-1, and is important for the induction of lipogenic genes in response to insulin and feeding (Wang et al., 2013). Upon insulin treatment, USF-1 is phosphorylated by DNA-dependent protein kinase (DNA-PK) and acetylated by P300/CBP-associated factor (P/CAF). In the meantime, BAF60c is also phosphorylated, leading to its translocation from cytosol to nucleus. Direct interaction of phosphorylated BAF60c with phosphorylated/acetylated USF-1 allows the recruitment of SWI/SNF chromatin remodeling complex to activate the transcription of target genes involved in lipogenesis. These results indicate an important role of hepatic BAF60c in insulin-induced lipogenesis.

BAF60B IN METABOLIC CONTROL AND DISEASES

BAF60b, encoded by *Smarcd2* gene, is the third member of the BAF60 family and is highly expressed in immune cells instead of metabolic tissues. To date, very little is known about its role in metabolic regulation. Recent studies identified BAF60b as a critical regulator of granulocyte development and function (Michel and Kadoch, 2017; Priam et al., 2017; Witzel et al., 2017). Given the fact that metabolic disease is associated with chronic low-grade meta-inflammation, it is possible that BAF60b may indirectly regulate energy metabolism and the development of metabolic syndrome through modulating the immune system and meta-inflammation.

CONCLUSION

Taken together, accumulating evidence strongly suggests that the SWI/SNF chromatin-remodeling factors BAF60s are important regulators of energy metabolism and their dysregulation plays critical roles in the pathogenesis of metabolic syndrome. However, the current knowledge is limited to the roles of BAF60a and BAF60c in liver and skeletal muscle. Besides these two tissues, other tissues such as pancreas, adipose tissue, and brain are also critical for the development of metabolic syndrome. In addition, BAF60b may exert significant impacts on whole body energy homeostasis by regulating the development and function of the immune system and meta-inflammation. Further studies are needed to address these questions. A better understanding of the exact roles and the underlying mechanisms of BAF60 proteins in metabolic control may provide new insights into the pathogenesis of metabolic disorders such as type 2 diabetes and its associated complications, and form the basis for new therapeutic interventions for these diseases.

ACKNOWLEDGMENTS

This work was supported by the National Natural Science Foundation of China (Grant No. 81670740), the Thousand Young Talents Plan of China, and the National Key Research and Development Programme of China (No. 2016YFC1305303) to Z.X.M.; by National Natural Science Foundation of China (Grant Nos. 81570759 and 81270938), National Key Research and Development Programme of China (No. 2016YFC1305301), Zhejiang Provincial Key Science and Technology Project (No. 2014C03045-2), Key Disciplines of Medicine (Innovation discipline,11-CX24) to J.F.; and by NIH grant (No. DK112800) to J.D.L.

ABBREVIATIONS

BAFs, Brg1/Brm associated factors; Brm, Brahma; CAR, constitutive androstane receptor; CHD, chromodomain helicase DNA-binding; ChREBP, carbohydrate-responsive element-binding protein; Cry 1, cryptochrome 1; DNA-PK, DNA-dependent protein kinase; DNL, de novo lipogenesis; FAO, fatty acid β-oxidation; HFD, high fat diet; ISWI, imitation switch; LDL, low density lipoprotein; LXR, liver X receptor; MAPK, mitogen-activated protein kinase; NAFLD, non-alcoholic fatty liver disease; NASH, non-alcoholic steatohepatitis; OXPHOS, oxidative phosphorylation; PcG, polycomb group; Per 1, period 1; PGC-1α, PPARγ coactivator-1α; P/CAF, P300/CBP-associated factor; RORα, retinoid-related orphan receptor alpha; RORE, ROR response elements; SREBP-1c, sterol regulatory element-binding protein 1c; SREs, sterol regulatory elements; SWI/SNF, switching defective/sucrose nonferenting; TG, triglycerides; USF, upstream stimulating factor; WD, western diet

COMPLIANCE WITH ETHICS GUIDELINES

The authors declare that there is no conflict of interest regarding the publication of this paper. This article does not contain any studies with human or animal subjects performed by any of the authors.

REFERENCES

Berchtold MW, Brinkmeier H, Muntener M (2000) Calcium ion in skeletal muscle: its crucial role for muscle function, plasticity, and disease. Physiol Rev 80:1215–1265

Cairns BR (2009) The logic of chromatin architecture and remodelling at promoters. Nature 461:193–198

Calvo JA, Daniels TG, Wang X, Paul A, Lin J, Spiegelman BM, Stevenson SC, Rangwala SM (2008) Muscle-specific expression of PPARgamma coactivator-1alpha improves exercise performance and increases peak oxygen uptake. J Appl Physiol 104:1304–1312

Chiang JY (2009) Bile acids: regulation of synthesis. J Lipid Res 50:1955–1966

Cotton P, Soulard A, Wesolowski-Louvel M, Lemaire M (2012) The SWI/SNF KlSnf2 subunit controls the glucose signaling pathway to coordinate glycolysis and glucose transport in Kluyveromyces lactis. Eukaryot Cell 11:1382–1390

Debril MB, Gelman L, Fayard E, Annicotte JS, Rocchi S, Auwerx J (2004) Transcription factors and nuclear receptors interact with the SWI/SNF complex through the BAF60c subunit. J Biol Chem 279:16677–16686

DeFronzo RA, Tripathy D (2009) Skeletal muscle insulin resistance is the primary defect in type 2 diabetes. Diabetes Care 32(Suppl 2):S157–S163

Dunham I, Kundaje A, Aldred SF, Collins PJ, Davis CA, Doyle F, Epstein CB, Frietze S, Harrow J, Kaul R et al (2012) An integrated encyclopedia of DNA elements in the human genome. Nature 489:57–74

Elfring LK, Deuring R, McCallum CM, Peterson CL, Tamkun JW (1994) Identification and characterization of Drosophila relatives of the yeast transcriptional activator SNF2/SWI2. Mol Cell Biol 14:2225–2234

Euskirchen GM, Auerbach RK, Davidov E, Gianoulis TA, Zhong G, Rozowsky J, Bhardwaj N, Gerstein MB, Snyder M (2011) Diverse roles and interactions of the SWI/SNF chromatin remodeling complex revealed using global approaches. PLoS genetics 7: e1002008

Ferrannini E, Simonson DC, Katz LD, Reichard G Jr, Bevilacqua S, Barrett EJ, Olsson M, DeFronzo RA (1988) The disposal of an oral glucose load in patients with non-insulin-dependent diabetes. Metabolism 37:79–85

Flajollet S, Lefebvre B, Cudejko C, Staels B, Lefebvre P (2007) The core component of the mammalian SWI/SNF complex SMARCD3/BAF60c is a coactivator for the nuclear retinoic acid receptor. Mol Cell Endocrinol 270:23–32

Gaster M, Staehr P, Beck-Nielsen H, Schroder HD, Handberg A (2001) GLUT4 is reduced in slow muscle fibers of type 2 diabetic patients: is insulin resistance in type 2 diabetes a slow, type 1 fiber disease? Diabetes 50:1324–1329

Gatfield D, Le Martelot G, Vejnar CE, Gerlach D, Schaad O, Fleury-Olela F, Ruskeepaa AL, Oresic M, Esau CC, Zdobnov EM et al (2009) Integration of microRNA miR-122 in hepatic circadian gene expression. Genes Dev 23:1313–1326

Glass CK, Witztum JL (2001) Atherosclerosis the road ahead. Cell 104:503–516

Goldstein JL, Brown MS (2015) A century of cholesterol and coronaries: from plaques to genes to statins. Cell 161:161–172

Gregor MF, Hotamisligil GS (2011) Inflammatory mechanisms in obesity. Ann Rev Immunol 29:415–445

Grundy SM, Cleeman JI, Merz CN, Brewer HB Jr, Clark LT, Hunninghake DB, Pasternak RC, Smith SC Jr, Stone NJ (2004) Implications of recent clinical trials for the National Cholesterol Education Program Adult Treatment Panel III guidelines. Circulation 110:227–239

Hardie DG, Ashford ML (2014) AMPK: regulating energy balance at the cellular and whole body levels. Physiology 29:99–107

Hargreaves DC, Crabtree GR (2011) ATP-dependent chromatin

remodeling: genetics, genomics and mechanisms. Cell Res 21:396–420

He J, Watkins S, Kelley DE (2001) Skeletal muscle lipid content and oxidative enzyme activity in relation to muscle fiber type in type 2 diabetes and obesity. Diabetes 50:817–823

Herzig S, Long F, Jhala US, Hedrick S, Quinn R, Bauer A, Rudolph D, Schutz G, Yoon C, Puigserver P et al (2001) CREB regulates hepatic gluconeogenesis through the coactivator PGC-1. Nature 413:179–183

Ho L, Crabtree GR (2010) Chromatin remodelling during development. Nature 463:474–484

Holliday R (2006) Epigenetics: a historical overview. Epigenetics 1:76–80

Hsiao PW, Fryer CJ, Trotter KW, Wang W, Archer TK (2003) BAF60a mediates critical interactions between nuclear receptors and the BRG1 chromatin-remodeling complex for transactivation. Mol Cell Biol 23:6210–6220

Izumiya Y, Hopkins T, Morris C, Sato K, Zeng L, Viereck J, Hamilton JA, Ouchi N, LeBrasseur NK, Walsh K (2008) Fast/glycolytic muscle fiber growth reduces fat mass and improves metabolic parameters in obese mice. Cell Metab 7:159–172

Kadoch C, Hargreaves DC, Hodges C, Elias L, Ho L, Ranish J, Crabtree GR (2013) Proteomic and bioinformatic analysis of mammalian SWI/SNF complexes identifies extensive roles in human malignancy. Nat Genet 45:592–601

Kathiresan S, Willer CJ, Peloso GM, Demissie S, Musunuru K, Schadt EE, Kaplan L, Bennett D, Li Y, Tanaka T et al (2009) Common variants at 30 loci contribute to polygenic dyslipidemia. Nat Genet 41:56–65

Keating ST, El-Osta A (2015) Epigenetics and metabolism. Circ Res 116:715–736

Kotronen A, Yki-Jarvinen H (2008) Fatty liver: a novel component of the metabolic syndrome. Arterioscler Thromb Vasc Biol 28:27–38

Lal S, Alam MM, Hooda J, Shah A, Cao TM, Xuan Z, Zhang L (2016) The Swi3 protein plays a unique role in regulating respiration in eukaryotes. Biosci Rep 36:e00350

Lamba DA, Hayes S, Karl MO, Reh T (2008) Baf60c is a component of the neural progenitor-specific BAF complex in developing retina. Dev Dyn 237:3016–3023

Latasa MJ, Griffin MJ, Yang SM, Kang C, Sul HS (2003) Occupancy and function of the −150 sterol regulatory element and −65 E-box in nutritional regulation of the fatty acid synthase gene in living animals. Mol Cell Biol 23:5896–5907

Lee YS, Sohn DH, Han D, Lee HW, Seong RH, Kim JB (2007) Chromatin remodeling complex interacts with ADD1/SREBP1c to mediate insulin-dependent regulation of gene expression. Mol Cell Biol 27:438–452

Lehman JJ, Barger PM, Kovacs A, Saffitz JE, Medeiros DM, Kelly DP (2000) Peroxisome proliferator-activated receptor gamma coactivator-1 promotes cardiac mitochondrial biogenesis. J Clin Invest 106:847–856

Li M, Zhao H, Zhang X, Wood LD, Anders RA, Choti MA, Pawlik TM, Daniel HD, Kannangai R, Offerhaus GJ et al (2011) Inactivating mutations of the chromatin remodeling gene ARID2 in hepatocellular carcinoma. Nat Genet 43:828–829

Li S, Liu C, Li N, Hao T, Han T, Hill DE, Vidal M, Lin JD (2008) Genome-wide coactivation analysis of PGC-1 alpha identifies BAF60a as a regulator of hepatic lipid metabolism. Cell Metabolism 8:105–117

Li T, Chiang JY (2013) Nuclear receptors in bile acid metabolism. Drug Metab Rev 45:145–155

Lickert H, Takeuchi JK, Von Both I, Walls JR, McAuliffe F, Adamson SL, Henkelman RM, Wrana JL, Rossant J, Bruneau BG (2004) Baf60c is essential for function of BAF chromatin remodelling complexes in heart development. Nature 432:107–112

Lin J, Wu H, Tarr PT, Zhang CY, Wu Z, Boss O, Michael LF, Puigserver P, Isotani E, Olson EN et al (2002) Transcriptional co-activator PGC-1 alpha drives the formation of slow-twitch muscle fibres. Nature 418:797–801

Liu C, Li S, Liu T, Borjigin J, Lin JD (2007) Transcriptional coactivator PGC-1alpha integrates the mammalian clock and energy metabolism. Nature 447:477–481

Masliah-Planchon J, Bieche I, Guinebretiere JM, Bourdeaut F, Delattre O (2015) SWI/SNF chromatin remodeling and human malignancies. Annu Rev Pathol 24(10):145–171

Meng ZX, Gong J, Chen Z, Sun J, Xiao Y, Wang L, Li Y, Liu J, Xu XZS, Lin JD (2017) Glucose sensing by skeletal myocytes couples nutrient signaling to systemic homeostasis. Mol Cell 66(332–344):e334

Meng ZX, Li S, Wang L, Ko HJ, Lee Y, Jung DY, Okutsu M, Yan Z, Kim JK, Lin JD (2013) Baf60c drives glycolytic metabolism in the muscle and improves systemic glucose homeostasis through Deptor-mediated Akt activation. Nat Med 19:640–645

Meng ZX, Wang L, Chang L, Sun J, Bao J, Li Y, Chen YE, Lin JD (2015) A diet-sensitive BAF60a-mediated pathway links hepatic bile acid metabolism to cholesterol absorption and atherosclerosis. Cell Rep 13:1658–1669

Meng ZX, Wang L, Xiao Y, Lin JD (2014) The Baf60c/Deptor pathway links skeletal muscle inflammation to glucose homeostasis in obesity. Diabetes 63:1533–1545

Michel BC, Kadoch C (2017) A SMARCD2-containing mSWI/SNF complex is required for granulopoiesis. Nat Genet 49:655–657

Mogensen M, Sahlin K, Fernstrom M, Glintborg D, Vind BF, Beck-Nielsen H, Hojlund K (2007) Mitochondrial respiration is decreased in skeletal muscle of patients with type 2 diabetes. Diabetes 56:1592–1599

Mootha VK, Handschin C, Arlow D, Xie X, St Pierre J, Sihag S, Yang W, Altshuler D, Puigserver P, Patterson N et al (2004) Erralpha and Gabpa/b specify PGC-1alpha-dependent oxidative phosphorylation gene expression that is altered in diabetic muscle. Proc Natl Acad Sci U S A 101:6570–6575

Mootha VK, Lindgren CM, Eriksson KF, Subramanian A, Sihag S, Lehar J, Puigserver P, Carlsson E, Ridderstrale M, Laurila E et al (2003) PGC-1alpha-responsive genes involved in oxidative phosphorylation are coordinately downregulated in human diabetes. Nat Genet 34:267–273

Nader GA, Esser KA (2001) Intracellular signaling specificity in skeletal muscle in response to different modes of exercise. J Appl Physiol 90:1936–1942

Neigeborn L, Carlson M (1984) Genes affecting the regulation of SUC2 gene expression by glucose repression in Saccharomyces cerevisiae. Genetics 108:845–858

Oberbach A, Bossenz Y, Lehmann S, Niebauer J, Adams V, Paschke R, Schon MR, Bluher M, Punkt K (2006) Altered fiber distribution and fiber-specific glycolytic and oxidative enzyme activity in skeletal muscle of patients with type 2 diabetes. Diabetes Care 29:895–900

Oh J, Sohn DH, Ko M, Chung H, Jeon SH, Seong RH (2008) BAF60a interacts with p53 to recruit the SWI/SNF complex. J Biol Chem 283:11924–11934

Osborn O, Olefsky JM (2012) The cellular and signaling networks linking the immune system and metabolism in disease. Nat Med 18:363–374

Pascussi JM, Gerbal-Chaloin S, Duret C, Daujat-Chavanieu M, Vilarem MJ, Maurel P (2008) The tangle of nuclear receptors that controls xenobiotic metabolism and transport: crosstalk and consequences. Annu Rev Pharmacol Toxicol 48:1–32

Peterson CL, Herskowitz I (1992) Characterization of the yeast SWI1, SWI2, and SWI3 genes, which encode a global activator of transcription. Cell 68:573–583

Priam P, Krasteva V, Rousseau P, D'Angelo G, Gaboury L, Sauvageau G, Lessard JA (2017) SMARCD2 subunit of SWI/SNF chromatin-remodeling complexes mediates granulopoiesis through a CEBP varepsilon dependent mechanism. Nat Genet 49:753–764

Ross R (1993) The pathogenesis of atherosclerosis: a perspective for the 1990s. Nature 362:801–809

Rutter J, Reick M, McKnight SL (2002) Metabolism and the control of circadian rhythms. Annu Rev Biochem 71:307–331

Shain AH, Giacomini CP, Matsukuma K, Karikari CA, Bashyam MD, Hidalgo M, Maitra A, Pollack JR (2012) Convergent structural alterations define SWItch/Sucrose NonFermentable (SWI/SNF) chromatin remodeler as a central tumor suppressive complex in pancreatic cancer. Proc Natl Acad Sci USA 109:E252–E259

Shain AH, Pollack JR (2013) The spectrum of SWI/SNF mutations, ubiquitous in human cancers. PLoS ONE 8:e55119

Simoneau JA, Kelley DE (1997) Altered glycolytic and oxidative capacities of skeletal muscle contribute to insulin resistance in NIDDM. J Appl Physiol 83:166–171

Sparks LM, Xie H, Koza RA, Mynatt R, Hulver MW, Bray GA, Smith SR (2005) A high-fat diet coordinately downregulates genes required for mitochondrial oxidative phosphorylation in skeletal muscle. Diabetes 54:1926–1933

Steinberg D (2002) Atherogenesis in perspective: hypercholesterolemia and inflammation as partners in crime. Nature medicine 8:1211–1217

Sudarsanam P, Winston F (2000) The SWI/SNF family nucleosome-remodeling complexes and transcriptional control. Trends Genet 16:345–351

Surabhi RM, Daly LD, Cattini PA (1999) Evidence for evolutionary conservation of a physical linkage between the human BAF60b, a subunit of SWI/SNF complex, and thyroid hormone receptor interacting protein-1 genes on chromosome 17. Genome 42:545–549

Takeuchi JK, Lickert H, Bisgrove BW, Sun X, Yamamoto M, Chawengsaksophak K, Hamada H, Yost HJ, Rossant J, Bruneau BG (2007) Baf60c is a nuclear Notch signaling component required for the establishment of left-right asymmetry. Proc Natl Acad Sci U S A 104:846–851

Tao W, Chen S, Shi G, Guo J, Xu Y, Liu C (2011) SWItch/sucrose nonfermentable (SWI/SNF) complex subunit BAF60a integrates hepatic circadian clock and energy metabolism. Hepatology 54:1410–1420

Thiebaud D, Jacot E, DeFronzo RA, Maeder E, Jequier E, Felber JP (1982) The effect of graded doses of insulin on total glucose uptake, glucose oxidation, and glucose storage in man. Diabetes 31:957–963

Varela I, Tarpey P, Raine K, Huang D, Ong CK, Stephens P, Davies H, Jones D, Lin ML, Teague J et al (2011) Exome sequencing identifies frequent mutation of the SWI/SNF complex gene PBRM1 in renal carcinoma. Nature 469:539–542

Wang W, Xue Y, Zhou S, Kuo A, Cairns BR, Crabtree GR (1996) Diversity and specialization of mammalian SWI/SNF complexes. Genes Dev 10:2117–2130

Wang Y, Wong RH, Tang T, Hudak CS, Yang D, Duncan RE, Sul HS (2013) Phosphorylation and recruitment of BAF60c in chromatin remodeling for lipogenesis in response to insulin. Mol Cell 49:283–297

Waters DD, Brotons C, Chiang CW, Ferrieres J, Foody J, Jukema JW, Santos RD, Verdejo J, Messig M, McPherson R et al (2009) Lipid treatment assessment project 2: a multinational survey to evaluate the proportion of patients achieving low-density lipoprotein cholesterol goals. Circulation 120:28–34

Wiegand KC, Shah SP, Al-Agha OM, Zhao Y, Tse K, Zeng T, Senz J, McConechy MK, Anglesio MS, Kalloger SE et al (2010) ARID1A mutations in endometriosis-associated ovarian carcinomas. N Engl J Med 363:1532–1543

Wijnen H, Young MW (2006) Interplay of circadian clocks and metabolic rhythms. Annu Rev Genet 40:409–448

Wilson BG, Roberts CWM (2011) SWI/SNF nucleosome remodellers and cancer. Nat Rev Cancer 11:481–492

Witzel M, Petersheim D, Fan Y, Bahrami E, Racek T, Rohlfs M, Puchalka J, Mertes C, Gagneur J, Ziegenhain C et al (2017) Chromatin-remodeling factor SMARCD2 regulates transcriptional networks controlling differentiation of neutrophil granulocytes. Nat Genet 49:742–752

Wu JI, Lessard J, Crabtree GR (2009) Understanding the words of chromatin regulation. Cell 136:200–206

Wu Z, Puigserver P, Andersson U, Zhang C, Adelmant G, Mootha V, Troy A, Cinti S, Lowell B, Scarpulla RC et al (1999) Mechanisms controlling mitochondrial biogenesis and respiration through the thermogenic coactivator PGC-1. Cell 98:115–124

Yokoyama C, Wang X, Briggs MR, Admon A, Wu J, Hua X, Goldstein JL, Brown MS (1993) SREBP-1, a basic-helix-loop-helix-leucine zipper protein that controls transcription of the low density lipoprotein receptor gene. Cell 75:187–197

Yoon JC, Puigserver P, Chen G, Donovan J, Wu Z, Rhee J, Adelmant G, Stafford J, Kahn CR, Granner DK et al (2001) Control of hepatic gluconeogenesis through the transcriptional coactivator PGC-1. Nature 413:131–138

Zhang P, Li L, Bao Z, Huang F (2016) Role of BAF60a/BAF60c in chromatin remodeling and hepatic lipid metabolism. Nutr Metab 13:30

10

Targeted elimination of mutant mitochondrial DNA in MELAS-iPSCs by mitoTALENs

Yi Yang[1], Han Wu[2], Xiangjin Kang[1], Yanhui Liang[2], Ting Lan[2], Tianjie Li[3], Tao Tan[4], Jiangyun Peng[2], Quanjun Zhang[2], Geng An[1], Yali Liu[1], Qian Yu[1], Zhenglai Ma[1], Ying Lian[3], Boon Seng Soh[1,5,6], Qingfeng Chen[1,5], Ping Liu[3], Yaoyong Chen[1], Xiaofang Sun[1], Rong Li[3], Xiumei Zhen[3], Ping Liu[3], Yang Yu[3]✉, Xiaoping Li[2]✉, Yong Fan[1]✉

[1] Key Laboratory for Major Obstetric Diseases of Guangdong Province, Key Laboratory of Reproduction and Genetics of Guangdong Higher Education Institutes, The Third Affiliated Hospital of Guangzhou Medical University, Guangzhou 510150, China

[2] Key Laboratory of Regenerative Biology of the Chinese Academy of Sciences and Guangdong Provincial Key Laboratory of Stem Cells and Regenerative Medicine, South China Institute for Stem Cell Biology and Regenerative Medicine, Guangzhou Institutes of Biomedicine and Health, Chinese Academy of Sciences, Guangzhou 510530, China

[3] Center of Reproductive Medicine, Department of Obstetrics and Gynecology, Peking University Third Hospital, Beijing 100191, China

[4] Yunnan Key Laboratory of Primate Biomedical Research, Institute of Primate Translational Medicine, Kunming University of Science and Technology, Kunming 650500, China

[5] Disease Modeling and Therapeutics Laboratory, A*STAR Institute of Molecular and Cell Biology, 61 Biopolis Drive Proteos, Singapore 138673, Singapore

[6] Department of Biological Sciences, National University of Singapore, Singapore 117543, Singapore

✉ Correspondence: yuyang5012@hotmail.com (Y. Yu), li_xiaoping@gibh.ac.cn (X. Li), yongfan011@gzhmu.edu.cn (Y. Fan)

ABSTRACT

Mitochondrial diseases are maternally inherited heterogeneous disorders that are primarily caused by mitochondrial DNA (mtDNA) mutations. Depending on the ratio of mutant to wild-type mtDNA, known as heteroplasmy, mitochondrial defects can result in a wide spectrum of clinical manifestations. Mitochondria-targeted endonucleases provide an alternative avenue for treating mitochondrial disorders via targeted destruction of the mutant mtDNA and induction of heteroplasmic shifting. Here, we generated mitochondrial disease patient-specific induced pluripotent stem cells (MiPSCs) that harbored a high proportion of m.3243A>G mtDNA mutations and caused mitochondrial encephalomyopathy and stroke-like episodes (MELAS). We engineered mitochondrial-targeted transcription activator-like effector nucleases (mitoTALENs) and successfully eliminated the m.3243A>G mutation in MiPSCs. Off-target mutagenesis was not detected in the targeted MiPSC clones. Utilizing a dual fluorescence iPSC reporter cell line expressing a 3243G mutant mtDNA sequence in the nuclear genome, mitoTALENs displayed a significantly limited ability to target the nuclear genome compared with nuclear-localized TALENs. Moreover, genetically rescued MiPSCs displayed normal mitochondrial respiration and energy production. Moreover, neuronal progenitor cells differentiated from the rescued MiPSCs also demonstrated normal metabolic profiles. Furthermore, we successfully achieved reduction in the human m.3243A>G mtDNA mutation in porcine oocytes via injection of mitoTALEN mRNA. Our study shows the great potential for using mitoTALENs for specific targeting of mutant mtDNA both in iPSCs and mammalian oocytes, which not only provides a new avenue for studying mitochondrial biology and disease but also suggests a potential therapeutic approach for the treatment of mitochondrial disease, as well as the prevention of germline transmission of mutant mtDNA.

Yi Yang, Han Wu, and Xiangjin Kang have contributed equally to this work.

KEYWORDS mitochondria, iPSCs, TALEN, MELAS

INTRODUCTION

Mitochondria, the so-called "powerhouses" of cells, are double-membrane cellular organelles that are found in almost all eukaryotic cells. They play major roles in multiple cellular processes, including energy production, calcium homoeostasis, cellular signaling, and apoptosis (Chan, 2006). Unlike most other organelles, mitochondria contain their own circular genomic DNA (mtDNA) and reproduce independently of cells. mtDNA encodes a host of enzymes, as well as ribosomes for protein synthesis, which critically catalyze many metabolic steps of cellular respiration (Anderson et al., 1981). In contrast to nuclear DNA, mtDNA is exclusively transmitted through maternal inheritance (Taylor and Turnbull, 2005). In contrast to cells, mitochondria only possess a base-excision repair pathway to repair oxidative DNA damage; a double-stranded break (DSB) repair pathway has not been identified in mitochondria. Mutations in mtDNA occur at a higher rate than in nuclear DNA (Moretton et al., 2017), and mutations in mtDNA cause a group of maternally inherited genetic disorders termed mitochondrial diseases, which affect 1 in 5,000 live births and cause significant morbidity and mortality (Haas et al., 2007). In most mitochondrial diseases, mutant mtDNA co-exists with wild-type mtDNA, resulting in a situation of mtDNA heteroplasmy in which the residual wild-type mtDNA can partially compensate for the mutated mtDNA, averting a complete bioenergetic crisis (Wallace, 2013). However, when the percentage of mutant mtDNA exceeds a threshold in the range of 60%–95%, depending on the severity of the mutation, pathogenic mtDNA mutations can lead to various clinical manifestations (Russell and Turnbull, 2014).

To date, the only therapeutic option for preventing the transmission of mtDNA mutations is transferring embryos below the threshold of clinical expression based on preimplantation genetic diagnosis (PGD) (Brown et al., 2006). However, PGD can only partially reduce, not completely eliminate, the risk of transmitting mutant mtDNA. Moreover, embryo viability may be affected by the PGD procedure, which includes micromanipulation of the blastomeres (Reddy et al., 2015). Recently, mitochondrial replacement therapy (MRT) has emerged as a strategy for preventing the inheritance of mtDNA by using enucleated donor oocytes to replace defective maternal mitochondria (Wang et al., 2014; Wu et al., 2017). However, these techniques require combining genetic material from different individuals, which has raised medical, safety, and ethical concerns (Vogel, 2014). In addition, a recent study also showed that embryos reconstructed by MRT still show levels of heteroplasmy, which might impact the stability of the mtDNA genotype and compromise the efficacy of mitochondrial replacement (Yamada et al., 2016).

By contrast, the use of mitochondria-targeted nucleases has shown promise regarding their ability to target mutant mtDNA. Previous studies have demonstrated that mitochondria-targeted restriction endonucleases can be used to eliminate mutant mtDNA and to alter mtDNA heteroplasmy in living cells (Alexeyev et al., 2008) and mice (Reddy et al., 2015; Bacman et al., 2010). Moreover, customized endonucleases, such as transcription activator-like effector nucleases (TALENs) (Bacman et al., 2013) and zinc finger nucleases (ZFNs) (Minczuk et al., 2006), targeted to mitochondria have been utilized for the specific elimination of mitochondrial genomes carrying mutations responsible for mitochondrial diseases. In addition, these customized nucleases allow the development of specific nucleases against essentially any DNA sequence, overcoming the main limitation of restriction endonucleases, which typically have strict and limited DNA sequence recognition capacities (Bacman et al., 2013). However, these approaches have only been demonstrated in human tumor-derived cell lines or mouse models harboring mtDNA mutations. Characterizing the effects of mitochondria-targeted nucleases in patient-derived primary cells, especially pluripotent cells, is critical for our understanding of the practical application of mitochondria-targeted nucleases for the treatment of mitochondrial diseases and the prevention of germline transmission of mutant mtDNA.

To address this issue, we generated induced pluripotent stem cells (iPSCs) from a patient with mitochondrial encephalomyopathy and stroke-like episodes (MELAS) (Goto et al., 1990) caused by an m.3243A>G heteroplasmic mutation in $tRNA^{Leu}$ (MT-TL1). Multiple iPSC lines demonstrating high m.3243A>G heteroplasmy were generated. We then engineered transcription activator-like effector nucleases (TALENs) to specifically localize to the mitochondria and cleave pathogenic m.3243A>G mutant mtDNA. Transient expression of mitochondria-targeted TALENs (mitoTALENs) achieved a remarkable reduction in mutant mtDNA levels in patient-specific iPSCs. Importantly, genetically rescued patient-specific iPSCs and differentiated NPCs displayed normal metabolic functions, and no off-target effects on either mitochondrial or nuclear genomes were detected with the mitoTALENs. In addition, we also successfully achieved a reduction in the levels of human m.3243A>G mtDNA mutation in porcine oocytes using mitoTALENs. This study demonstrated the capacity of mitoTALENs to specifically target mtDNA mutations and induce metabolic heteroplasmic shifting in patient-derived iPSCs.

RESULTS

iPSCs derived from a patient with MELAS

Primary fibroblasts carrying 90% 3243A>G mutant mtDNA were derived from skin biopsy samples of a MELAS patient. MIPSCs were generated by reprogramming fibroblast cells under transgene-free, serum-free, and feeder-free conditions. The mutation rate of mtDNA 3243A>G in the patient-derived MiPSCs clones was measured by PCR restriction fragment length polymorphism (RFLP) (Zhang et al., 2015) and Sanger sequencing. As previously reported, the reprogramming of somatic cells from mitochondrial disease patients resulted in varying mutation rates in the iPSCs (Fig. S1A) (Folmes et al., 2013; Kodaira et al., 2015). Nevertheless, primary MiPSCs lines harboring 3243A>G heteroplasmy levels >80% were established (Fig. 1A). While the mutation rates continued to vary during long-term culture, most MiPSC lines maintained stable heteroplasmy levels (Fig. 1B). The observed fluctuations in heteroplasmy levels could be due to the different reprogramming states of the cell lines, i.e., highly pluripotent iPSCs predominantly produce energy by anaerobic glycolysis and carry a high proportion of mutations, even above the pathogenic threshold level. The MiPSC5 cell line was selected for the remainder of the study, as the heteroplasmy level was maintained stably in each passage.

To characterize the MELAS-iPSCs, we analyzed their pluripotency and differentiation capacity at passage 30. The expression levels of the pluripotent markers OCT4, SOX2, TRA1-60, SSEA4 and NANOG (Fig. S1B) were measured by immunofluorescence staining. We further assessed the differentiation ability of the reprogrammed MiPSCs based on the formation of teratomas *in vivo*. Histological examination revealed that the teratomas consisted of three embryonic germ layers, including intestine-like tissues (endoderm), cartilage (mesoderm), and neuron-like tissues (ectoderm) (Fig. S1C). Further analysis showed that the MiPSCs contained normal chromosomes (Fig. S1C).

Engineered mitoTALENs specifically target the mitochondrial genome

As the mitochondrial genome encodes only approximately 1% of the total mitochondrial proteome, and the remaining 99% of the mitochondrial proteins encoded by the nuclear genome, mitochondrial proteins synthesized in the cytosol require mitochondrial targeting signals (MTS) to facilitate their import into the mitochondria (Lionaki et al., 2016). Using MTS, gene therapies for mitochondrial disorders have been developed by delivering functional proteins and endonucleases to mitochondria with endogenous MTS (Farrar et al., 2013). To engineer specific mitochondrial-targeted TALENs (MitoTALENs), we first assessed the mitochondrial localization of exogenous proteins driven by different endogenous MTSs using fluorescence colocalization analysis. MTS derived from nuclear genes (APEX1 (Li et al., 2010), ATP5B (Reddy et al., 2015), COX8A (Bacman et al., 2013), COX10 (Smith et al., 2012), and SOD2 (Bacman et al., 2013)) were used to construct mitochondria-targeted EGFP (mito-EGFP) and TALEN-EGFP (mitoTALEN-EGFP) by fusing the MTS at the N-terminus (Figs. 2A, S2A, and S2B). The mito-EGFP and mitoTALEN-EGFP expression vectors were transiently transfected into human iPSCs and HEK293 cells by electroporation. EGFP and TALEN-EGFP with a SV40-nuclear

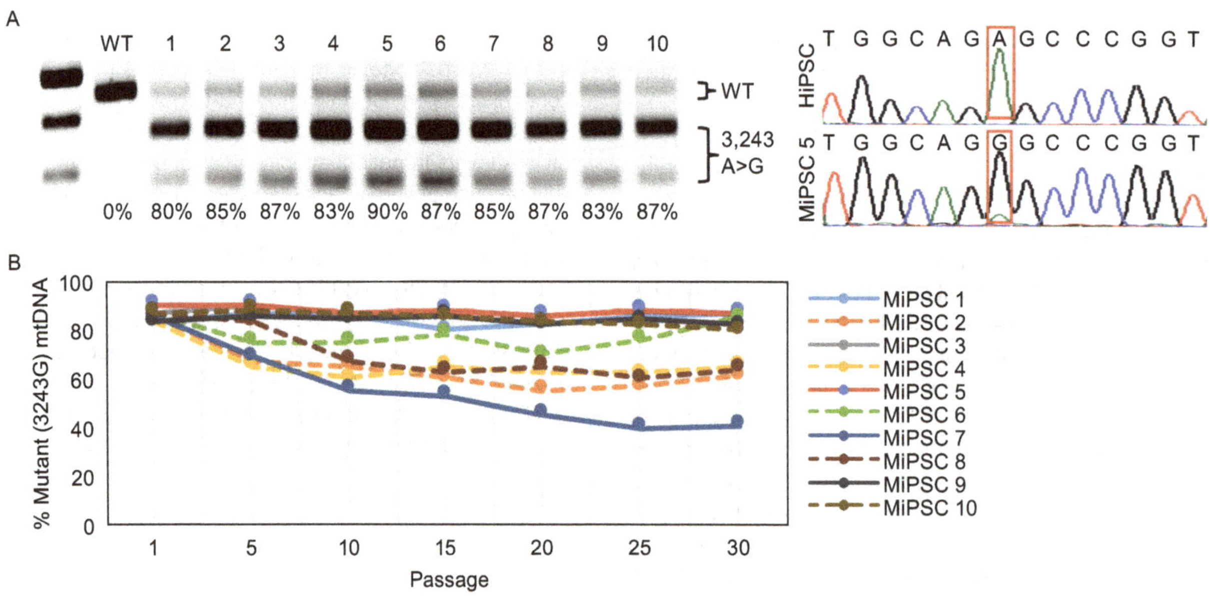

Figure 1. Derivation and characterization of MELAS patient-specific iPSCs harboring a mtDNA 3243A>G mutation. (A) MELAS patient-specific iPSC (MiPSC) lines were established that harbored high 3243A>G heteroplasmy levels (>80%). The mtDNA 3243A>G mutation rate was determined by RFLP analysis and Sanger sequencing. An iPSC line derived from healthy human fibroblasts (HiPSC) was used as the control. (B) The MiPSC lines maintained high 3243A>G heteroplasmy levels after long-term culture. The 3243A>G heteroplasmy levels were assessed in MiPSC lines every 5 passages from passage 1 to passage 30.

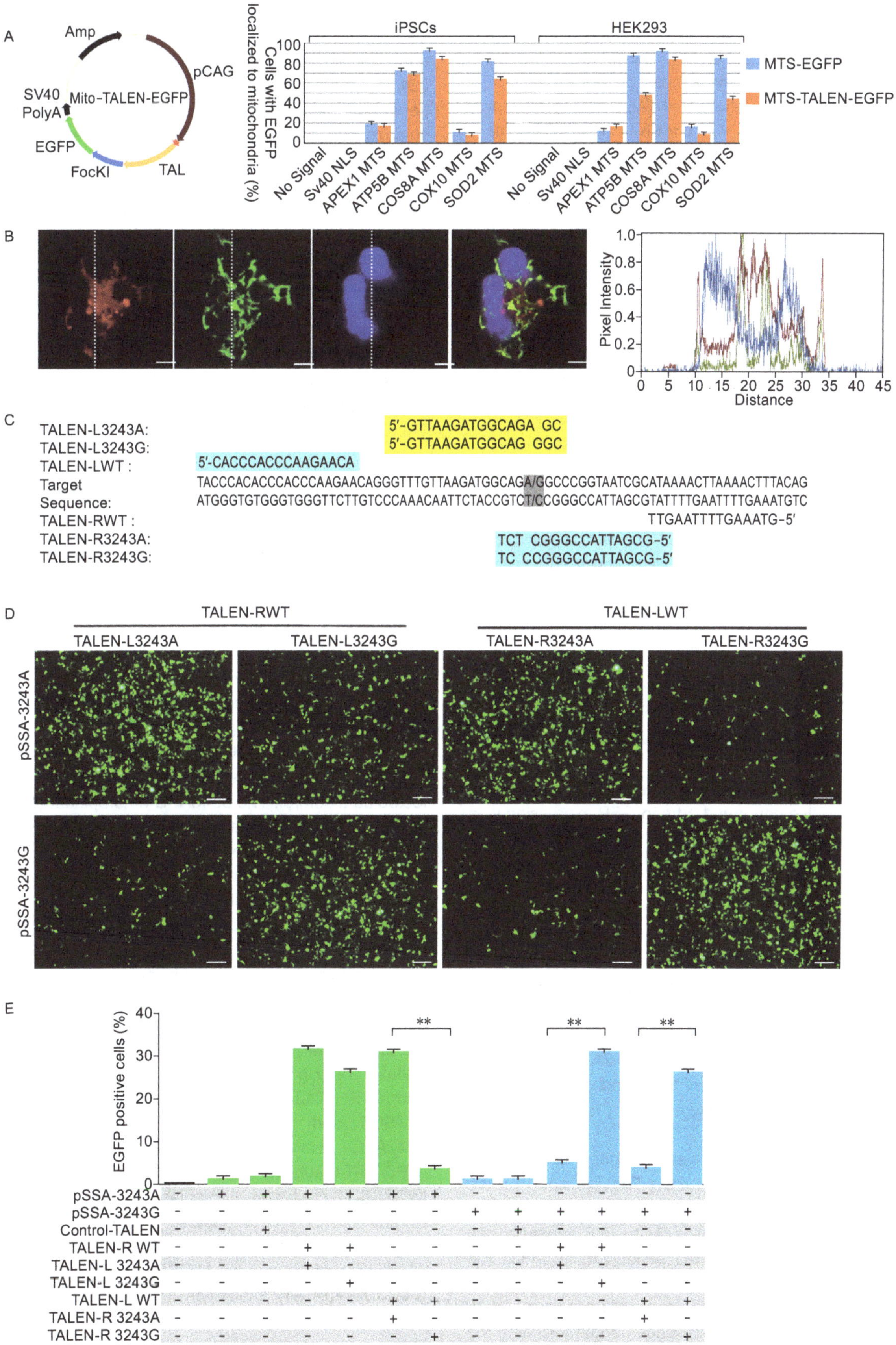
A
Amp
pCAG
SV40 PolyA
Mito-TALEN-EGFP
EGFP
FocKI
TAL
iPSCs
HEK293
Cells with EGFP localized to mitochondria (%)
MTS-EGFP
MTS-TALEN-EGFP
No Signal
Sv40 NLS
APEX1 MTS
ATP5B MTS
COS8A MTS
COX10 MTS
SOD2 MTS
B
Pixel Intensity
Distance
C
TALEN-L3243A: 5′-GTTAAGATGGCAGA GC
TALEN-L3243G: 5′-GTTAAGATGGCAG GGC
TALEN-LWT: 5′-CACCCACCCAAGAACA
Target Sequence:
TACCCACACCCACCCAAGAACAGGGTTTGTTAAGATGGCAGA/GGCCCGGTAATCGCATAAAACTTAAAACTTTACAG
ATGGGTGTGGGTGGGTTCTTGTCCCAAACAATTCTACCGTCT/CCGGGCCATTAGCGTATTTTGAATTTTGAAATGTC
TALEN-RWT: TTGAATTTTGAAATG-5′
TALEN-R3243A: TCT CGGGCCATTAGCG-5′
TALEN-R3243G: TC CCGGGCCATTAGCG-5′
D
TALEN-RWT
TALEN-LWT
TALEN-L3243A
TALEN-L3243G
TALEN-R3243A
TALEN-R3243G
pSSA-3243A
pSSA-3243G
E
EGFP positive cells (%)
pSSA-3243A
pSSA-3243G
Control-TALEN
TALEN-R WT
TALEN-L 3243A
TALEN-L 3243G
TALEN-L WT
TALEN-R 3243A
TALEN-R 3243G

◀ **Figure 2. Engineering of mitoTALENs that specifically target 3243G mutant mtDNA.** (A) The specificity of mitochondrial localization of EGFP and the TALEN-EGFP fusion proteins mediated by MTS derived from different nuclear genes. Left, schematic drawing of the engineered TALEN-EGFP polypeptide monomer containing an MTS in the N-terminus (MLS). The specificity and efficiency of MTS targeting to the mitochondria were analyzed based on the percentage of the EGFP and MitoTracker co-localizing cells in all of the transfected iPSCs and HEK293 cells, as shown in Fig. S1B and S1C. (B) Mitochondrial localization of mitoTALEN monomers mediated by COX8A-MTS in iPSCs 24 h after transfection. Mitochondria were visualized by MitoTracker Red. Scale bar = 10 μmol/L. Co-localization was visualized by the overlapping peaks of the relative fluorescence intensity (y-axis) on lines that passed through areas with marker signal. The position of the lines is indicated on the images, with lines running from top to bottom. (C) TALENs designed for targeting the mtDNA 3243 locus. (D and E) A single-strand annealing (SSA) assay to determine the specific targeting of the 3243G mtDNA mutation by TALENs. The mutated EGFP coding sequence was divided into two segments, which were separated by a stop codon and targeting sequence. Both segments contained an identical homology region. Once double-strand breaks were introduced into the target site by TALENs, the mutated EGFP coding sequence was repaired by annealing the two homologous sequences. Expression of EGFP was detected with a confocal microscope using appropriate filters after 48 h (D), (scale bars 50 μm), and the proportion of the EGFP-positive cells was measured by flow cytometry (E) ($n = 3$, error bars represent ±SEM; $^{**}P < 0.05$).

localization signal (NLS) and without any localization signal (controls) were also generated. Strong GFP expression was detected in all transfected cells 24 h post-transfection. The cells were fixed with 4% PFA and stained with a mitochondrial marker (Mitotracker) and a nuclear marker (Hoechst 33342). The specificity of the selected MTS targeting of mitochondria was determined by analyzing the co-localization of green fluorescence with Mitotracker and not Hoechst 33342 (Fig. S2B and S2C). Among the five selected MTS, three were derived from ATP5B, COX8A and SOD2 and showed higher specificity towards targeting the mitochondria in both HEK293 and in hiPS cells (Fig. 2A). COX8A-MTS was selected for construction of the mitochondrial-targeted TALENs, as it showed the highest capacity for mitochondrial localization (Fig. 2A and 2B).

Two pairs of mitoTALENs were designed to target the 3243G mtDNA mutation (Fig. 2C). The principle for the design of mitoTALENs is that each monomer for targeting the mitochondrial DNA contains 14.5–16.5 repeats: one monomer can bind the mutated sequence, in which the mutated site is adjacent to the 3′ end of the targeting sequence, and the other monomer is targeted to the wild-type sequence with a 14–17 bp spacer length, dictating the specific cleavage of the mutant mtDNA upon dimerization of the FokI nuclease. To evaluate the specificity and cleavage activity of the mitoTALENs, we generated a single-strand annealing (SSA) assay vector harboring either the mutant 3243G or the wild-type 3243A mtDNA sequence. We also generated mitoTALENs targeting the wild-type sequence as controls (Fig. 2C). After co-transfection of mitoTALEN pairs with the SSA assay vector into HEK293 cells, the percentage of restored EGFP-positive cells was determined via fluorescence microscopy and flow cytometry. We found that the mitoTALEN pairs TALEN-LWT/TALEN-R3243, which targeted to the antisense strand mtDNA 3,243 locus, showed significantly different cleavage activities on mutated and wild-type sequences. However, the mitoTALEN pairs TALEN-L3243/TALEN-RWT had similar efficiencies in the binding and cleavage of both mutant and wild-type target sequences (Fig. 2D and 2E). Therefore, the subsequent experiments were performed using the mitoTALEN pairs TALEN-LWT/TALEN-R324.

Specific targeting of mutant mtDNA in patient-derived iPSCs by mitoTALENs

Encouraged by the high specificity and cleavage activity of the mitoTALENs targeting the SSA assay vectors, we next tested the mitoTALENs TALEN-LWT + TALEN-R3243G in MiPSC5 sub-lines harboring approximately 90% mutant m.3243A>G mtDNA. The mitoTALENs, together with an EGFP-expression plasmid, were electroporated into the iPSCs. EGFP-positive cells were sorted by FACS 48 h after electroporation and propagated into sub-clones for heteroplasmy testing. The EGFP-expression plasmid and a pair of untargeted mitoTALENs (TALEN-control) were also transfected as a control. The 3243A>G heteroplasmy levels were assessed using RFLP after the sub-clones were expanded for one week. Compared with the controls, 3243A>G Mito-TALENs significantly reduced the percentage of mutant mtDNA in all of the sub-clones derived from the MELAS-iPSCs (Figs. 3A, 3B, and S3A). Interestingly, the mutant mtDNA was completely undetectable in two 3243G mitoTALEN-transfected MiPSC subclones (MiPSC5-T3 and MiPSC5-T7) by RFLP and Sanger sequencing. We then performed sequencing of the entire mtDNA genome using the Illumina MiSeq platform and quantitated the 3243A>G mutation ratio in MiPSC5 and the sub-clones (Fig. 3C). 92% and 89% 3243A>G heteroplasmy was detected in the MiPSC5 and TALEN-control cells, respectively, 27% 3243A>G heteroplasmy was detected in the targeted MiPSC5-T1 sub-clone, and the targeted MiPSC5-T3 and MiPSC5-T7 cells were homoplasmic for the wild-type allele. In addition to the pathogenic 3243A>G mutations, two heteroplasmic variants in the 16S rRNA gene and in MT-ATP6 were different in the MiPSC5 and targeted sub-clones (Fig. 3D, Table S1). In addition, in all of those clones, we also detected another 7 single nucleotide polymorphisms (SNPs) in the D-loop region, 2 in the 12S rRNA gene, 2 in the 16S

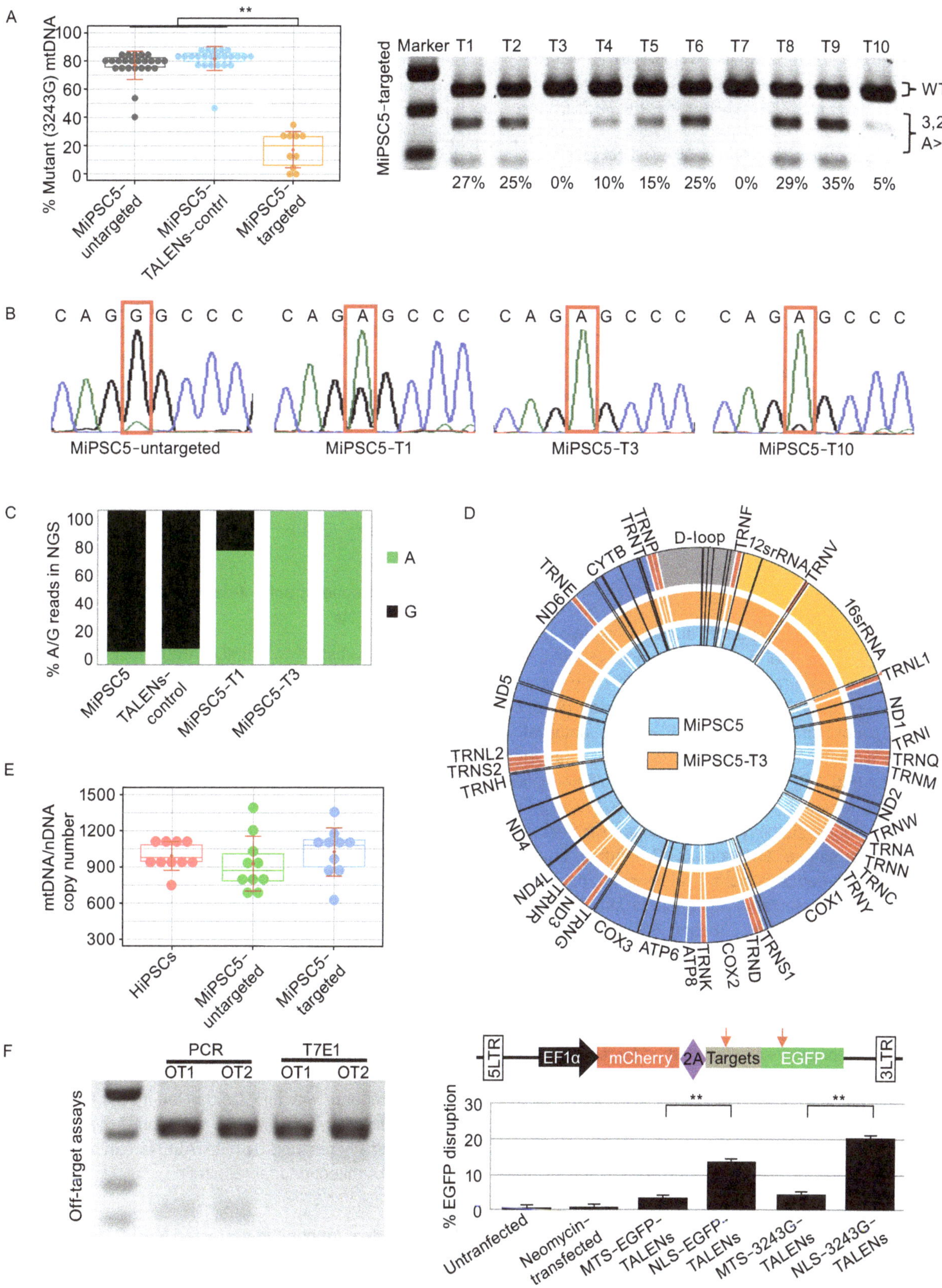
A
% Mutant (3243G) mtDNA
MiPSC5-untargeted
MiPSC5-TALENs-contrl
MiPSC5-targeted
MiPSC5-targeted
Marker T1 T2 T3 T4 T5 T6 T7 T8 T9 T10
WT
3,243 A>G
27% 25% 0% 10% 15% 25% 0% 29% 35% 5%
B
C A G G G C C C
C A G A G C C C
MiPSC5-untargeted
MiPSC5-T1
MiPSC5-T3
MiPSC5-T10
C
% A/G reads in NGS
MiPSC5
TALENs-control
MiPSC5-T1
MiPSC5-T3
A
G
D
D-loop
12srRNA
16srRNA
MiPSC5
MiPSC5-T3
E
mtDNA/nDNA copy number
HiPSCs
MiPSC5-untargeted
MiPSC5-targeted
F
PCR
T7E1
OT1 OT2 OT1 OT2
Off-target assays
5LTR
EF1α
mCherry
2A
Targets
EGFP
3LTR
% EGFP disruption
Untranfected
Neomycin-transfected
MTS-EGFP-TALENs
NLS-EGFP-TALENs
MTS-3243G-TALENs
NLS-3243G-TALENs

◄ **Figure 3. MitoTALENs specifically target and eliminate mutant mtDNA in MELAS-iPSCs.** (A) RFLP analysis and quantification of mtDNA 3243A>G heteroplasmy in MELAS-iPSCs and subclones targeted by mitoTALENs (MiPSC5-untargeted n = 24; MiPSC5-TALEN-control n = 24, MiPSC5-targeted n = 10, error bars represent ±SEM; $^{**}P < 0.05$). (B) Sanger sequencing to determine the mtDNA 3243A>G heteroplasmy in targeted MiPSC clones. Chromatographs show mtDNA genotyping at the 3,243 position (red box) in representative targeted MELAS-iPSC subclones. (C) Percentages of A and G reads at the mtDNA 3,243 position of the MELAS-iPSCs and targeted subclones were quantified using next-generation sequence analysis. (D) Single-nucleotide variations (SNVs) in iPSCs via exome sequencing. The black bars represent individual SNVs. Compared with untargeted b-thal iPSCs, iPSC-C2 had 12 SNVs, and the remaining corrected colonies had 21 SNVs. (E) mtDNA/nuclearDNA (nDNA) ratio in wild-type hiPSCs, MELAS-iPSCs and targeted clones (n = 10, error bars represent ±SEM). (F) T7E1 assays assessed mutagenesis at 2 predicted off-target sites; PCR products were used as a control. (G) Dual-fluorescence reporter-based assay assessing the nuclear targeting of mitoTALENs (n = 3, error bars represent ±SEM; $^{**}P < 0.05$).

rRNA gene, 1 in the tRNA-R gene, and 23 in protein genes (Fig. 3D, Table S1). Clinical symptoms associated with these variants have not been reported. The mtDNA copy number was determined by qPCR as previously reported (Rooney et al., 2015); compared with WT hiPSC and untargeted MiPSCs, no significant changes were detected in targeted MiPSC clones (Fig. 3E). The pluripotency and differentiation capacity of the targeted sub-clones were also assessed by immunofluorescence staining of pluripotency markers and by teratoma formation. Cytogenetic G-banding analysis revealed that targeted sub-clones retained normal diploid karyotypes with no detectable numerical or structural chromosomal abnormalities (Fig. S3B and S3C). We also performed fingerprinting by short tandem repeat analysis (STR) and confirmed that all of the MiPSC5 and targeted sub-clones were derived from the same patient (Fig. S3D).

To assess potential off-target effects on the nuclear genome, we first computationally predicted off-target sites in the human genome potentially created by the mitoTALENs. Only two potential off-target sites (Table S2) were found in the human genome with fewer than four mismatches/gaps in the spacer region. A T7 endonuclease I (T7EI) assay and Sanger sequencing were performed to detect the off-target cleavage of the mitoTALENs. No mutations were found in the potential off-target sites in any of the targeted sub-clones (Fig. 3F). Additionally, we did not find a target site for the mtDNA 3243G mitoTALENs in NMTRL-TAA3-1, the nuclear-encoded mitochondrial transfer RNA-Leu gene.

To further assess the off-target effects of mitoTALENs on nuclear genomes, we generated a dual-fluorescence reporter lentivirus vector, pEF1α-mCherry-EGFP, which harbored the mtDNA 3243G mitoTALEN targeting sequence between mCherry and the EGFP cassette (Fig. 3G). Approximately 27 h after iPSCs were infected with the dual-fluorescence reporter lentivirus, dual-fluorescence-positive cells were sorted by fluorescence-activated cell sorting (FACS) and then used to determine the off-target effects of mitoTALENs on nuclear genomes (Fig. S3E). We also generated TALENs with a nuclear localization signal (NLS) as controls, which target the 3243G sequence and the EGFP gene. To enrich the transfected cells, MitoTALENs or NLS-TALENs were co-electroporated with a *Puro* expression plasmid into the dual-fluorescence reporter cells. After selection with puromycin (0.5 μg/mL) for 2 days, FACS was performed to analyze the expression levels of the dual fluorescence markers, which showed that NLS-TALENs were highly efficient in targeting nuclear sequences and disrupted the expression of EGFP in 13%–20% of the transfected cells. In contrast, MitoTALENs targeted to the same sequence demonstrated a limited targeting ability for nuclear sequences, with only 3%–6% of the transfected cells shown to be $mCherry^{+}/EGFP^{-}$ (Figs. 3F and S3E).

Metabolic rescue in patient-derived iPSCs by mitoTALENs

The A to G substitution at mtDNA nucleotide position 3,243 causes 80% of mitochondrial encephalomyopathy, lactic acidosis, and stroke-like episodes (MELAS), which affects many of the body's systems, particularly the nervous system and the muscles (Goto et al., 1990). The 3243A>G mtDNA mutation disturbs the function of tRNA leucine 1 (UUA/G) and impairs the ability of mitochondria to make proteins, use oxygen, and produce energy. To evaluate the mitochondrial function of MiPSCs and to determine the genetic rescue of the sub-clones by mitoTALENs, oxygen consumption rates (OCRs) were determined using XF24 extracellular flux analyzers (Seahorse Biosciences), which indicated the mitochondrial respiration and energy production capacities. Compounds (oligomycin, FCCP, and a mix of rotenone and antimycin A) were serially injected to measure ATP production, maximal respiration, and non-mitochondrial respiration, respectively (Fig. 4A). MiPSCs harboring high 3243A>G heteroplasmy levels demonstrated significantly reduced OCRs compared with hiPSCs derived from a healthy person (Fig. 4A and 4B), while MiPSC sub-clones (MiPSC5-T3 and T7) genetically rescued by mitoTALENs exhibited functional recovery of mitochondrial respiration.

Mitochondrial function in corrected MiPSC-derived NPC

The distinguishing clinical feature of MELAS syndrome is the recurrence of stroke-like episodes. To further evaluate whether genetic targeting of mutant mtDNA could restore the mitochondrial function of neural progenitor cells (NPCs),

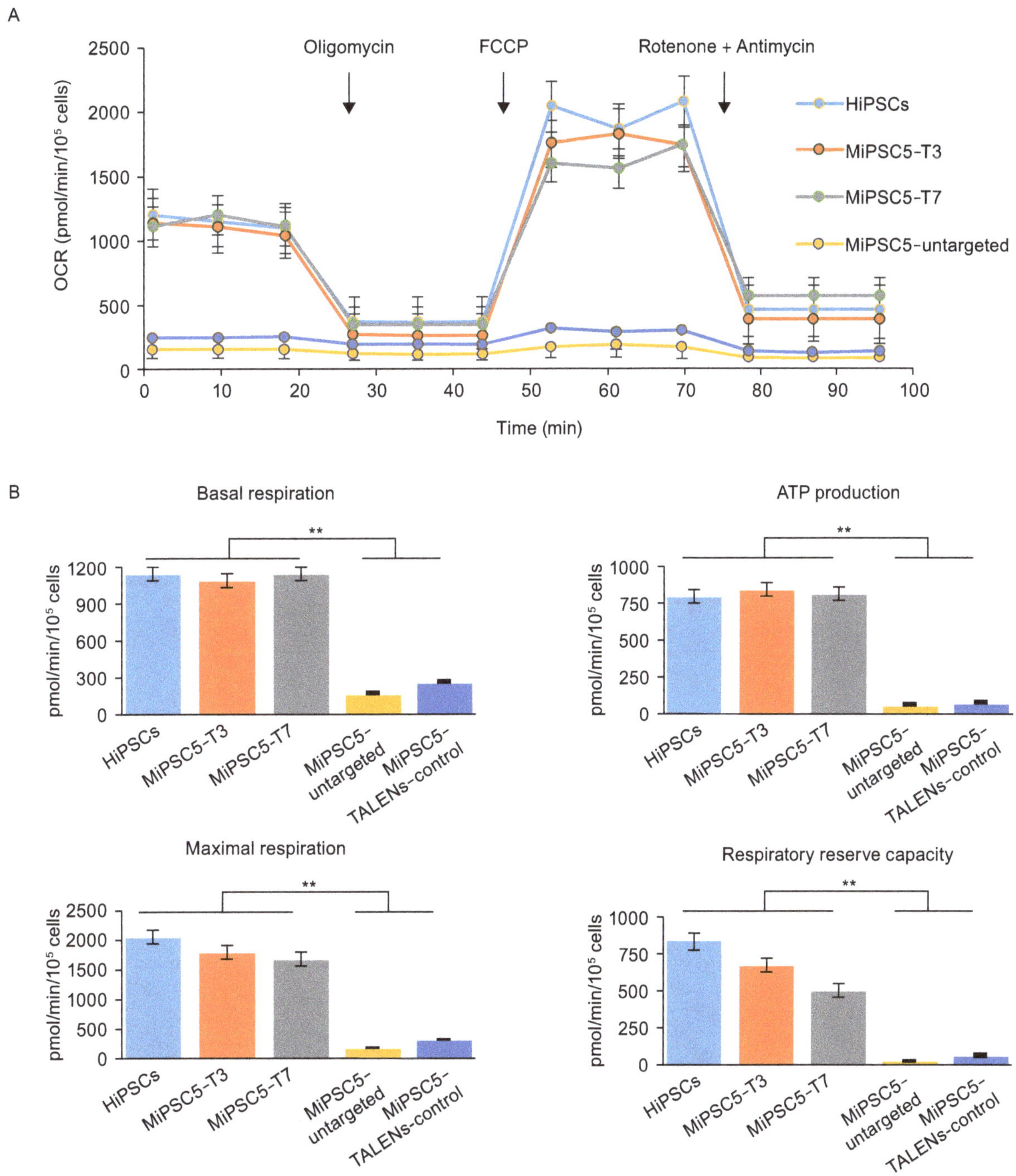

Figure 4. Mitochondrial respiratory function of MELAS-iPSCs and targeted subclones. (A) Mitochondrial function based on *in vitro* oxygen capacity in response to 0.5 μg/mL oligomycin, 1 μmol/L 4-(trifluoromethoxy) phenylhydrazone (FCCP), 0.5 μmol/L rotenone and 1 μmol/L antimycin. (B) Quantitative analysis of basal oxygen consumption, ATP production, maximal respiration and proton leak of iPSCs (n = 3, error bars represent ±SEM; $^{**}P < 0.05$).

NPCs were differentiated from WT hiPSC, MiPSCs and the targeted sub-clone MiPSC5-T3. Immunofluorescence staining was performed to confirm neural differentiation of iPSCs with the neural progenitor markers Nestin and Sox2 (Fig. 5A). The mutation rate of mtDNA 3243A>G in the NPCs was detected by RFLP (Fig. 5B) and showed no apparent differences with the iPSCs. Next, mitochondrial functions of the NPCs were investigated using XF24 extracellular flux

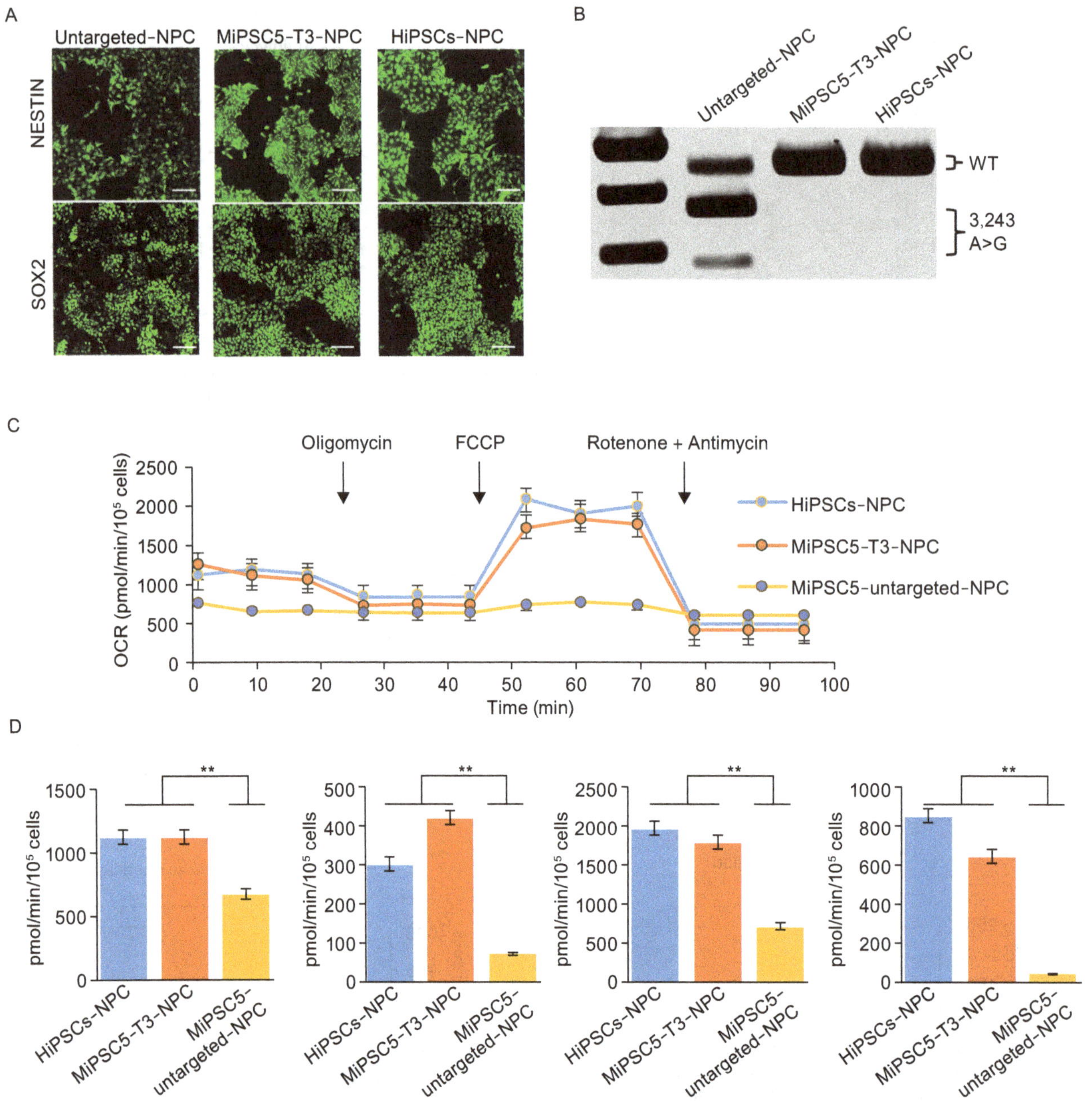

Figure 5. Mitochondrial respiratory function of neural progenitor cells differentiated from MELAS-iPSCs. (A) Immunofluorescence analysis of neural progenitor markers in MELAS-iPSC-derived NPCs. Scale bar, 100 mm. (B) RFLP analysis and quantification of m.3243A>G heteroplasmy in MELAS-iPSC-derived NPCs. (C) OCR of the NPCs differentiated from wild-type hiPSCs, MELAS-iPSCs and targeted subclones. (D) Quantitative analysis of basal oxygen consumption, ATP production, maximal respiration and proton leak of NPCs (n = 3, error bars represent ±SEM; $^{**}P < 0.05$).

analyzers (Fig. 5C and 5D). OCR data indicated that mitochondrial function of the iPSC-derived NPCs was successfully restored after gene correction.

Specific reduction of human mutant mtDNA in porcine oocytes

To evaluate the specific elimination of mutant mtDNA in patient-derived iPSCs and the potential use of mitoTALENs to prevent the germline transmission of mutant mtDNA, we

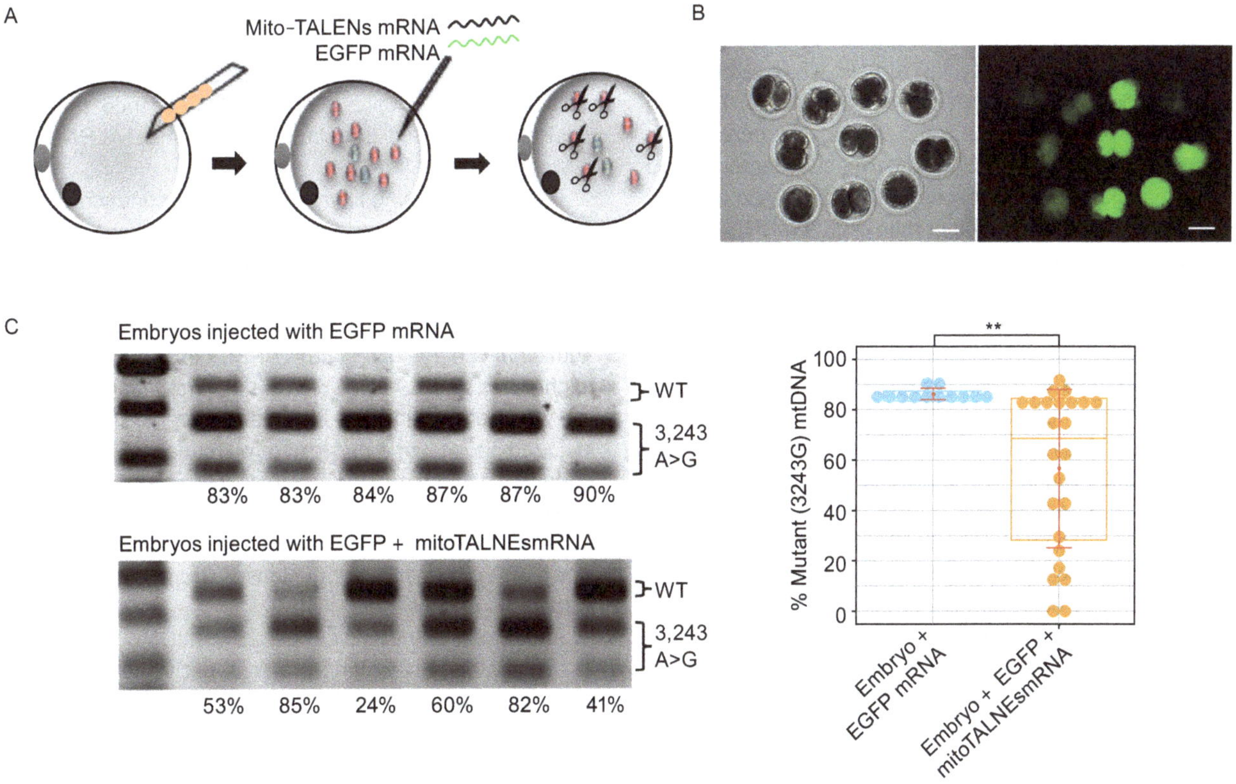

Figure 6. Specific targeting of human mutant mtDNA in porcine oocytes using MitoTALENs. (A) Construction of porcine oocytes carrying human m.3243G>A mutations by injection of the cytoplasm of MiPSCs into porcine MII oocytes, followed by injection of EGFP and mitoTALENs mRNA targeting the 3243G mutant mtDNA. (B) Expression of EGFP in artificial porcine oocytes 48 h after injection of mRNA. (C) RFLP analysis and quantification of m.3243A>G heteroplasmy in individual oocytes 3 days after mRNA injection (EGFP n = 12; EGFP + TALENs n = 24, error bars represent ±SEM; **P < 0.05).

analyzed the ability of mitoTALENs to reduce mutant mtDNA levels in mammalian oocytes. A previous study showed that mitochondria-targeted nucleases could prevent germline transmission of mutant mtDNA in mice (Bacman et al., 2010). Several other mitoTALENs have also successfully reduced human m.14459G>A and m.9176T>C mutant mtDNA in murine oocytes (Reddy et al., 2015). Using porcine oocytes, which have a similar developmental timeline as human oocytes (Griffin et al., 2006), we generated artificial porcine oocytes carrying the human m.3423A>G mtDNA mutation by direct injection of the cytoplasm of MiPSCs into porcine MII oocytes (Fig. 6A). *In vitro* transcribed mitoTALENs mRNA was then injected into the oocytes harboring human m.3423A>G mtDNA. To monitor gene expression, EGFP mRNA was co-injected into the oocytes. The expression of EGFP was assessed by fluorescence microscopy after 48 h (Fig. 6B), after which RFLP analysis was performed to detect the levels of 3243A>G heteroplasmy. Compared with the control (where only EGFP mRNA was injected), the injection of mitoTALEN mRNA significantly reduced the human 3243A>G mutant mtDNA (Figs. 6C and S4). Collectively, these results demonstrated the potential for custom-designed mitoTALENs to specifically eliminate disease-relevant mtDNA mutations responsible for human mitochondrial diseases.

DISCUSSION

Mitochondrial diseases refer to a group of clinically heterogeneous disorders caused by mitochondrial dysfunction. Because mitochondria are not only regulated by the nuclear genome but also by their own mtDNA, mutations within either set of DNA may cause mitochondrial disorders. Unlike the nuclear genome, mutations in mtDNA occur at a higher rate (Ingman et al., 2000). However, there is no proven technology that allows the generation of targeted mtDNA mutations to study the basic characteristics of mitochondrial biogenesis. This has prompted keen interest in the development of relevant cellular and animal models and in exploring innovative therapeutic strategies to modulate the mitochondrial deficiencies observed in these diseases. In this study, we have focused on genetically correcting mitochondrial disease patient iPSCs by targeting the disruption of mutant mtDNA using mitochondrial-targeted TALENs.

MiPSCs harboring a high ratio of the pathogenic mtDNA mutation m.3243A>G were generated from a MELAS patient. We first engineered mitoTALENs, which exhibited a highly specific mutant mtDNA targeting capability. Through the transient expression of mitoTALENs, we achieved a remarkable reduction in mutant mtDNA and demonstrated the functional metabolic rescue of MiPSCs and derived NPCs. Moreover, the injection of mitoTALEN mRNA led to a significant reduction in human mutant mtDNA in reconstructed porcine oocytes.

Induced pluripotent stem cells (iPSCs) (Takahashi et al., 2007) derived from patients with genetic disorders hold enormous promise for basic research and drug screening. Specifically, mitochondrial disease-specific iPSCs carrying various heteroplasmic mtDNA mutations have been reported, which has opened new avenues for understanding the definitive genotype-phenotype relationship of affected tissues and organs in various types of mitochondrial diseases triggered by mtDNA mutations (Hatakeyama and Goto, 2016). iPSC lines harboring the m.3243A>G mtDNA mutation were generated from patient fibroblasts with various levels of heteroplasmy. In addition, isogenic iPSCs carrying exclusively wild-type or mutant mtDNA were reported and were generated though the spontaneous segregation of heteroplasmic mtDNA during reprogramming or proliferation of patient fibroblast (Ma et al., 2015). Although we did not obtain these iPSCs in this experiment, this might be because the fibroblasts derived from the patient carried a ratio of m.3243A>G mtDNA mutation above the pathogenic threshold level, which limited the proliferative ability of the cells, resulting in the failure of segregation of the heteroplasmic mtDNA. As reported, we also found mtDNA mutations that did not affect the reprogramming of the fibroblasts or the stability and pluripotency of iPSCs (Hamalainen et al., 2013). This might have been because the iPSCs predominantly produced energy via anaerobic glycolysis (Ma et al., 2015).

Due to the lack of a DSB repair pathway, DSBs caused by various effectors will lead to degradation of damaged mtDNA (Moretton et al., 2017). Engineered mitochondria-targeted nucleases, including restriction endonucleases and customized endonucleases, have been used to specifically eliminate mutant mtDNA in human tumor-derived cell lines and murine models. Compared with restriction endonucleases and ZFN, TALENs can be more flexibly designed and constructed using a simple "one-to-one" correspondence between single DNA base pairs in a target site and two-amino-acid sequences in one TAL effector repeat (Boch and Bonas, 2010). Although CRISPR/Cas9 technology has been widely used in targeting nuclear DNA, only one recent study showed that mtDNA editing is possible using CRISPR/Cas9 (Jo et al., 2015). Additionally, unlike other endonucleases, the specific DNA targeting of Cas9 was guided by an engineered guide-RNA by means of base-pair complementarity. Thus, reengineered guide-RNA would require the development of mitochondrial localization- and single-base-pair-distinguishing capabilities. While our current study and those previously reported have demonstrated that engineered mitoTALENs can specifically target single-base pair mutant mtDNA, we have utilized a dual fluorescence reporter to illustrate the limited targeting ability of mitoTALENs to the nuclear genome.

Interestingly, using Illumina MiSeq to sequence complete mtDNA, two heteroplasmic variants were found in rescued MiPSC sub-clones that were different from the untargeted MiPSCs. As the current study suggested that the mtDNA mutation frequency is significantly increased in iPSCs, the additional variants of rescued MiPSCs observed in this study might have developed during the *in vitro* culturing and editing. Another possibility is that the nontargeted MiPSCs also included those variants at a rare frequency, but their frequency accumulated in the mito-TALEN-induced mtDNA heteroplasmy shifts. Regardless, this implied that comprehensive assessment of variants in mtDNA is necessary when using engineered nucleases to genetically correct mitochondrial diseases.

In contrast to the nuclear genome, mtDNA is segregated in a non-Mendelian manner and is exclusively transmitted through maternal inheritance (Ingman et al., 2000). Therefore, MRT has been developed as a therapeutic approach to prevent germline transmission of mutant mtDNA and has been approved in the UK as well as declared ethically permissible in the US. Despite MRT showing great potential, there are serious ethical concerns surrounding this approach, and more studies are required to show the effects of MRT in human embryonic development and reproduction (Yamada et al., 2016). Our study, together with a previous study, highlights the significant potential of mitoTALENs for the specific elimination or correction of disease-relevant mtDNA mutations that are responsible for mitochondrial diseases in humans. Furthermore, the approach used to reconstruct porcine oocytes via injection of the cytoplasm of iPSCs combined with mitoTALEN mRNA serves as a proof-of-concept for the successful prevention of transmission of mutant mtDNA as well as therapeutic compensation for the reduced mitochondrial copies.

MATERIALS AND METHODS

Animals and ethics statement

The patients in this study provided written informed consent prior to donating fibroblasts for stem cell generation. The experiments regarding animal research were approved by the Institutional Review Board at The Third Affiliated Hospital of Guangzhou Medical University. The experiments using human cells and mice were approved by the ethics committee of The Third Affiliated Hospital of Guangzhou Medical University, and all animal care and experiments were performed in accordance with the institutional ethical guidelines for animal experiments.

MELAS patient iPSC generation

A skin biopsy was obtained from a MELAS patient. Human fibroblasts were cultured in fibroblast medium (DMEM) supplemented with 10% fetal bovine serum (FBS) (HyClone), 1 mmol/L glutamine (Gibco), 1% non-essential amino acids (NEAA) (Gibco), and 100 IU/mL penicillin/streptomycin (Gibco). To generate iPSCs, fibroblast cells were transfected with a Sendai virus reprogramming kit (Life Technologies, A16517). The transfected cells were then plated onto Matrigel-coated culture dishes according to the manufacturer's instructions. To remove the Sendai viruses, iPSCs were incubated at 38.5°C for 4 days as instructed. All iPSCs were cultured on Matrigel-coated tissue culture dishes (ES-qualified, BD Biosciences) with mTeSR1 (STEMCELL Technologies) at 37°C and 5% CO_2 in a 100% humidified atmosphere incubator. The culture medium was changed daily until the cells were ready for passage or harvest. The cells were passaged every 3–4 days using Accutase (Stemcell Technologies).

MitoTALEN construction and activity determination

TALENs targeting the mtDNA 3423 locus were constructed through Golden Gate TALEN Assembly. TALEN expression vectors were pCAG-T7-TALEN (Sangamo)-Destination with heterodimeric (ELD, KKR) domains obtained from Addgene (Plasmids #37184 and #40131).

A single strand annealing (SSA) assay was performed to detect the activity of TALENs as previously reported. TALEN-targeted sequences, both mtDNA 3243A and 3243G, were amplified from MiPSC and HiPSC genomes using the following primers with additional *Bam*HI and *Hin*dIII enzyme sites: forward, 5′gcgcggatccgagaaataaggcctacttca3′; reverse, 5′gcccaagcttatgccattgcgattagaatg 3′. The PCR fragment digested with *Bam*HI and *Hin*dIII was inserted into a pSSA-EGFP vector, and the correct clones were confirmed by Sanger sequencing. For SSA assays, 0.5 μg TALEN expression plasmid and SSA reporter were co-electroporated into 5.0 × 10^5 HEK293 cells using a Neon transfection system according to the manufacturer's instructions (Life Technologies). The expression of the EGFP was observed under a fluorescence microscope using appropriate filters after 48 h, and the proportion of the EGFP-positive cells was measured by flow cytometry.

To construct the mitoTALENs, the MTS coding sequences were first synthesized and cloned into the pEF1α-puro vector by replacement of the *puro* cassette, designated pEF1α-MTS. The sequences used for encoding MTS are listed below:

APE1-MTS:
ATGCACTCTCTGTTACCTGCATTGTGTGACAGCAAGATCCGTTCCAAGGCCCTCGGCAGTGATCACTGTCCTATCACCCTATACCTAGCACTG
ATP5B-MTS:
ATGTTGGGGTTTGTGGGTCGGGTGGCCGCTGCTCCGGCCTCCGGGGCCTTGCGGAGACTCACCCCTTCAGCGTCGCTGCCCCCAGCTCAGCTCTTACTGCGGGCCGCTCCGACGGCGGTCCATCCTGTCAGGGACTATGCGGCGCAAACATCTCCTTCG
COX8A–MTS:
ATGTCCGTCCTGACGCCGCTGCTGCTGCGGGGCTTGACACGCTCGGCCCGGCGGCTCCCAGTGCCGCGCGCCAAGATCCATTCGTTG
COX10-MTS:
ATGGCCGCATCTCCGCACACTCTCTCCTCACGCCTCCTGACAGGTTGCGTAGGAGGCTCTGTCTGGTATCTTGAAGTCGACGCG
SOD2-MTS:
ATGTTGAGCCGGGCAGTGTGCGGCACCAGCAGGCAGCTGGCTCCGGTTTTGGGGTATCTGGGCTCCAGGCAGAAGCACAGCCTCCCCGACCGCGTCGACCGC

The mitoTALENs targeted to the 3243G mtDNA were constructed by subcloning the TALEN monomers from pCAG-T7-TALENs into pEF1α-MTS and then removing the nuclear localization signal. We generated the mitoEGFP vectors by cloning an EGFP cassette without the start codon ATG into the pEF1α-MTS vectors. Meanwhile, mitoTALEN-EGFP was constructed by one-step subcloning of the TALEN-LWT monomer and EGFP into pEF1α-MTS using a Gibson Assembly kit from New England Biolabs.

Mitochondrial genome targeting by mitoTALENs

Approximately 1 × 10^6 MiPSCs were electroporated using a Neon transfection system (Life Technologies) at 1,150 V, 30 ms, and 1 pulse in 100 μL of buffer B containing 4 μg of each TALEN monomer and 2 μg pEGFP-N1(Clontech) as a selection maker. The cells were then recovered in mTeSR1 medium supplemented with 10 mmol/L Rho-associated kinase (ROCK) inhibitor Y-27632 (10 mmol/L, Sigma) after electroporation. At 48 h post-transfection, the GFP-positive cells were collected via FACS and plated in 6-well plates. After 10–14 days of culture, single colonies were picked and cultured in 96-well plates for further expansion and identification.

Heteroplasmy determination by RFLP

Total DNA from cells was extracted using the TIANamp genomic DNA kit (Tiangen). For RFLP analysis, the mtDNA 3243 locus was amplified using the following primers: forward, 5′cctcggagcagaacccaacct3′; reverse, 5′cgaagggttgtagtagcccgt3′, which produced a PCR product of 634 bp. In the presence of the A3243G mutation, the PCR product was digested by ApaI into two fragments of 424 and 210 bp. The digested PCR products were separated on a 1% agarose gel and stained with ethidium bromide. ImageJ software (https://imagej.nih.gov/ij/) was used to analyze the signal intensity of the bands.

Neural progenitor cell differentiation and culture

A STEMdiffTM Neural System (Stemcell technologies) was used to differentiate neural progenitor cells from iPSCs. iPSCs were harvested by treatment with 2 mg/mL dispase (Invitrogen) and then washed twice with DPBS. The cells were resuspended in STEMdiffTM Neural Induction Medium with 10 μmol/L Y-27632 and plated onto Matrigel-coated culture dishes at an approximate density of 2 × 10^5 cells/cm^2. A daily full medium change was performed with warm STEMdiffTM Neural Induction Medium. Cells were passaged three times upon reaching 70% to 80% confluence using Accutase. Then, NPCs were maintained on Matrigel-coated dishes in STEMdiffTM Neural Progenitor Medium.

Oxygen consumption detection

To assess the mitochondrial respiration and energy production of the MiPSCs, oxygen consumption rates (OCR) were measured using XF24 extracellular flux analyzers (Seahorse Biosciences). Briefly, 5×10^4 iPSCs were plated into the wells of a Matrigel-coated XF24 cell culture microplate and incubated with 10 μmol/L Y-27632 for 24 h to ensure attachment. Before the assay, the cells were equilibrated in unbuffered XF assay medium supplemented with 25 mmol/L glucose, 1 mmol/L sodium pyruvate, 2 mmol/L Glutamax, 1× nonessential amino acids and 1% FBS in a non-CO_2 incubator for 1 h. The mitochondrial processes were interrogated by sequential injection of oligomycin (0.5 mg/mL), carbonyl cyanide 4-(trifluoromethoxy) phenylhydrazone (FCCP, 1 mmol/L) and rotenone (0.5 mmol/L)/antimycin A (1 mmol/L). The results were normalized to the cell number and analyzed using Seahorse XF24 software.

Whole mtDNA sequencing

Mitochondrial whole-genome sequencing was performed using new next-generation sequencing (NGS) technology, VariantProTM Capture Technology (VPCT), designed by LC Sciences. Briefly, the sequence libraries were prepared by multiplex PCR using VariantPro primer pools, which contained 110 primer pairs in two PCR tubes with an average amplicon length of 200 bp. The amplicons were purified with Agen-court AMPure XP beads (Beckman Coulter Genomics, High Wycombe, UK) mixed in equimolar concentrations, and mitochondrial genome sequencing was performed on an Illumina HiSeq 4000. Globally, approximately 89.12% of the reads mapped to the reference sequence. The average counts per amplicon were over 1000. Mitochondrial heteroplasmy rates in each sample were calculated as the percentage of the number of alternative reads to that of the total reads. The clinical significance of the variants was then analyzed with MitoMaster (http:// www.mitomap.org/MITOMASTER/WebHome).

Immunofluorescence staining and fluorescence analysis

For pluripotent analysis, iPSCs were fixed with 3% paraformaldehyde for 15 min, permeabilized with 0.1% Triton-X and blocked using Super-block. Immunofluorescence (IF) staining was performed using primary antibodies (all at 1:200 dilutions) to detect OCT4 (Abcam), SOX2 (Abcam), TRA-1-60 (Abcam), and SSEA-4 (Abcam). The nuclei were stained with DAPI at a final concentration of 0.01 mg/mL for 10 min.

For the mitochondrial localization analysis, mitoEGFP- and mitoTALEN-EGFP-transfected cells were incubated in the presence of 200 nmol/L MitoTracker Red CMXRos and 0.5 μg/mL Hoechst (Invitrogen) at 37°C for 30 min and fixed with 1% paraformaldehyde in PBS for 10 min. Confocal image acquisition was performed using a Zeiss LSM 710 laser-scanning microscope (Carl Zeiss Jena). Line profiles were performed on images using the RGB Profiles Tool in ImageJ.

Reconstruction of porcine oocytes and mRNA injection

Porcine ovaries were obtained from a local slaughterhouse. Oocytes were aspirated from 3 to 6 mm follicles and cultured in TCM-199 supplemented with 0.1% polyvinyl alcohol, d-glucose (3.05 mmol/L), sodium pyruvate (0.91 mmol/L), penicillin (75 μg/mL), streptomycin (50 μg/mL), epidermal growth factor (10 ng/mL), cysteine (0.57 mmol/L), follicle-stimulating hormone (0.5 μg/mL), and luteinizing hormone (0.5 μg/mL) at 38.5°C at 5% CO_2 for 42 to 44 h. Oocytes with an extruded first polar body, a round shape and an intact cytoplasm were selected and maintained in the manipulation medium for subsequent experiments. Oocyte activation was performed in medium containing 0.3 mol/L mannitol, 1.0 mmol/L $CaCl_2 \cdot 2H_2O$, 1.0 mmol/L $MgCl_2 \cdot 6H_2O$ and 0.5 mmol/L HEPES, using two 2-DC pulses with a voltage of 1.2 kV/cm for 30 μs on a BTX Electro Cell Manipulator 2001 (BTX, San Diego, CA, USA). Oocytes were then cultured in PZM-3 (with 3 mg/mL BSA) supplemented with 7.5 μg/mL cytochalasin B for 4 h. The cytoplasm of the MiPSCs was extracted by gentle aspiration of cells in and out of the injection pipette. Cytoplasm from 4–5 cells was injected inside the ooplasm using the micropipette. After 3–4 h of recovery in PZM-3 (with 3 mg/mL BSA), reconstructed oocytes were injected with mitoTALENs (200 ng/μL) and EGFP (50 ng/μL) mRNA and then further cultured in PZM-3 (with 3 mg/mL BSA) at 38.5°C in a humidified atmosphere at 5% CO_2 for 3 days before analysis. The mRNA was produced using a mMESSAGE mMACHINE SP6 ULTRA kit (Life Technologies) according to the manufacturer's instructions using linearized and gel-purified (QIAGEN) plasmid templates.

Teratoma formation and karyotype analysis

Approximately 1–2 × 10^6 iPSCs from a confluent 10-cm plate were harvested by digestion with 2 mg/mL dispase, resuspended in Matrigel, and injected into the inguinal grooves of 6-week-old male SCID mice. Eight weeks later, the resulting tumors were removed, fixed for 4–8 h in 4% paraformaldehyde, and embedded in paraffin. After staining with hematoxylin and eosin (H&E), the sections were examined using a light microscope to determine the presence of derivatives from the three germ layers.

For the chromosome analysis, iPSCs were incubated in culture medium with 0.25 mg/mL colcemid (Gibco, Invitrogen) for 4 h, harvested, and incubated in 0.4% sodium citrate and 0.4% chloratum Kaliumat (1:1, *v/v*) at 37°C for 5 min. The cells were then fixed three times in a methanol:acetic acid solution (3:1, *v/v*). Subsequently, after Giemsa staining, at least 20 cells were examined in each group for the chromosome analysis.

Short tandem repeat analysis (STR)

For STR analysis, the genomic DNA was extracted from MiPSCs, targeted MiPSCs and the patient's fibroblasts. The extracted DNA was amplified for 15 different genetic loci using a Promega PowerPlex 16 system kit (Promega). Capillary electrophoresis was performed on an automated ABI 3100 genetic analyzer (Applied Biosystems).

Statistical analysis

All statistical analyses were performed using SPSS 19.0 software to detect significant differences in the measured variables among the groups. A value of $P < 0.05$ was considered to indicate a statistically significant difference.

ACKNOWLEDGMENTS

This work was supported in part by the "Reproductive health and major birth defects prevention and control research" Key Special Fund (No. 2016YFC1000601), the National Natural Science Foundation of China (Grant Nos. 31371521, 81370766, 81401254, 81570101, 81671121, 31601187, 81521002), the Guangdong Province Science and Technology Project (2014TQ01R683, 2017A020214005, 2016A020216023, 2015A030310119, 2016B030229008), the Bureau of Science and Technology of Guangzhou Municipality (201505011111498), the "Reproductive health and major birth defects prevention and control research" Key Special Fund (Nos. 2016YFC1000201 and 2016YFC1000302), the Ministry of Science and Technology of China Grants (973 program; 2014CB943203), and the Beijing Nova Program (xxjh2015011).

ABBREVIATIONS

MELAS, mitochondrial encephalomyopathy and stroke-like episodes; mitoTALENs, mitochondrial-targeted transcription activator-like effector nucleases; MiPSCs, mitochondrial disease patient-specific induced pluripotent stem cells; MRT, mitochondrial replacement therapy; NLS, nuclear localization signal; NPCs, neural progenitor cells; SSA, single strand annealing.

COMPLIANCE WITH ETHICS GUIDELINES

For human subjects used in the present study, all procedures followed were in accordance with the ethical standards of the responsible committee on human experimentation (institutional and national) and with the Helsinki Declaration of 1975, as revised in 2000 (5). Informed consent was obtained from all patients for being included in the study. For mouse subjects used in the present study, all institutional and national guidelines for the care and use of laboratory animals were followed.

Yi Yang, Han Wu, Xiangjin Kang, Yanhui Liang, Ting Lan, Tianjie Li, Tao Tan, Jiangyun Peng, Quanjun Zhang, Geng An, Yali Liu, Qian Yu, Zhenglai Ma, Ying Lian, Boon Seng Soh, Qingfeng Chen, Ping Liu, Yaoyong Chen, Xiaofang Sun, Rong Li, Xiumei Zhen, Ping Liu, Yang Yu, Xiaoping Li and Yong Fan declare that they have no conflicts of interest.

REFERENCES

Alexeyev MF, Venediktova N, Pastukh V, Shokolenko I, Bonilla G, Wilson GL (2008) Selective elimination of mutant mitochondrial genomes as therapeutic strategy for the treatment of NARP and MILS syndromes. Gene Ther 15(7):516–523

Anderson S, Bankier AT, Barrell BG, de Bruijn MH, Coulson AR, Drouin J, Eperon IC, Nierlich DP, Roe BA, Sanger F et al (1981) Sequence and organization of the human mitochondrial genome. Nature 290(5806):457–465

Bacman SR, Williams SL, Garcia S, Moraes CT (2010) Organ-specific shifts in mtDNA heteroplasmy following systemic delivery of a mitochondria-targeted restriction endonuclease. Gene Ther 17(6):713–720

Bacman SR, Williams SL, Pinto M, Peralta S, Moraes CT (2013) Specific elimination of mutant mitochondrial genomes in patient-derived cells by mitoTALENs. Nat Med 19(9):1111–1113

Boch J, Bonas U (2010) Xanthomonas AvrBs3 family-type III effectors: discovery and function. Annu Rev Phytopathol 48:419–436

Brown DT, Herbert M, Lamb VK, Chinnery PF, Taylor RW, Lightowlers RN, Craven L, Cree L, Gardner JL, Turnbull DM (2006) Transmission of mitochondrial DNA disorders: possibilities for the future. Lancet 368(9529):87–89

Chan DC (2006) Mitochondria: dynamic organelles in disease, aging, and development. Cell 125(7):1241–1252

Farrar GJ, Chadderton N, Kenna PF, Millington-Ward S (2013) Mitochondrial disorders: aetiologies, models systems, and candidate therapies. Trends Genet 29(8):488–497

Folmes CD, Martinez-Fernandez A, Perales-Clemente E, Li X, McDonald A, Oglesbee D, Hrstka SC, Perez-Terzic C, Terzic A, Nelson TJ (2013) Disease-causing mitochondrial heteroplasmy segregated within induced pluripotent stem cell clones derived from a patient with MELAS. Stem Cells 31(7):1298–1308

Goto Y, Nonaka I, Horai S (1990) A mutation in the tRNA(Leu)(UUR) gene associated with the MELAS subgroup of mitochondrial encephalomyopathies. Nature 348(6302):651–653

Griffin J, Emery BR, Huang I, Peterson CM, Carrell DT (2006) Comparative analysis of follicle morphology and oocyte diameter in four mammalian species (mouse, hamster, pig, and human). J Exp Clin Assist Reprod 3:2

Haas RH, Parikh S, Falk MJ, Saneto RP, Wolf NI, Darin N, Cohen BH (2007) Mitochondrial disease: a practical approach for primary care physicians. Pediatrics 120(6):1326–1333

Hamalainen RH, Manninen T, Koivumaki H, Kislin M, Otonkoski T, Suomalainen A (2013) Tissue- and cell-type-specific manifestations of heteroplasmic mtDNA 3243A>G mutation in human induced pluripotent stem cell-derived disease model. Proc Natl Acad Sci USA 110(38):E3622–E3630

Hatakeyama H, Goto Y (2016) Concise review: heteroplasmic mitochondrial DNA mutations and mitochondrial diseases: toward iPSC-Based disease modeling, drug discovery, and regenerative therapeutics. Stem Cells 34(4):801–808

Ingman M, Kaessmann H, Paabo S, Gyllensten U (2000) Mitochondrial genome variation and the origin of modern humans. Nature 408(6813):708–713

Jo A, Ham S, Lee GH, Lee YI, Kim S, Lee YS, Shin JH, Lee Y (2015) Efficient Mitochondrial Genome Editing by CRISPR/Cas9. Biomed Res Int 2015:305716

Kodaira M, Hatakeyama H, Yuasa S, Seki T, Egashira T, Tohyama S, Kuroda Y, Tanaka A, Okata S, Hashimoto H et al (2015) Impaired respiratory function in MELAS-induced pluripotent stem cells with high heteroplasmy levels. FEBS Open Bio 5:219–225

Li M, Zhong Z, Zhu J, Xiang D, Dai N, Cao X, Qing Y, Yang Z, Xie J, Li Z et al (2010) Identification and characterization of mitochondrial targeting sequence of human apurinic/apyrimidinic endonuclease 1. J Biol Chem 285(20):14871–14881

Lionaki E, Gkikas I, Tavernarakis N (2016) Differential protein distribution between the nucleus and mitochondria: implications in aging. Front Genet 7:162

Ma H, Folmes CD, Wu J, Morey R, Mora-Castilla S, Ocampo A, Ma L, Poulton J, Wang X, Ahmed R et al (2015) Metabolic rescue in pluripotent cells from patients with mtDNA disease. Nature 524 (7564):234–238

Minczuk M, Papworth MA, Kolasinska P, Murphy MP, Klug A (2006) Sequence-specific modification of mitochondrial DNA using a chimeric zinc finger methylase. Proc Natl Acad Sci USA 103 (52):19689–19694

Moretton A, Morel F, Macao B, Lachaume P, Ishak L, Lefebvre M, Garreau-Balandier I, Vernet P, Falkenberg M, Farge G (2017) Selective mitochondrial DNA degradation following double-strand breaks. PLoS ONE 12(4):e0176795

Reddy P, Ocampo A, Suzuki K, Luo J, Bacman SR, Williams SL, Sugawara A, Okamura D, Tsunekawa Y, Wu J et al (2015) Selective elimination of mitochondrial mutations in the germline by genome editing. Cell 161(3):459–469

Rooney JP, Ryde IT, Sanders LH, Howlett EH, Colton MD, Germ KE, Mayer GD, Greenamyre JT, Meyer JN (2015) PCR based determination of mitochondrial DNA copy number in multiple species. Methods Mol Biol 1241:23–38

Russell O, Turnbull D (2014) Mitochondrial DNA disease-molecular insights and potential routes to a cure. Exp Cell Res 325(1):38–43

Smith AJ, Bainbridge JW, Ali RR (2012) Gene supplementation therapy for recessive forms of inherited retinal dystrophies. Gene Ther 19(2):154–161

Takahashi K, Tanabe K, Ohnuki M, Narita M, Ichisaka T, Tomoda K, Yamanaka S (2007) Induction of pluripotent stem cells from adult human fibroblasts by defined factors. Cell 131(5):861–872

Taylor RW, Turnbull DM (2005) Mitochondrial DNA mutations in human disease. Nat Rev Genet 6(5):389–402

Vogel G (2014) Assisted reproduction. FDA considers trials of 'three-parent embryos'. Science 343(6173):827–828

Wallace DC (2013) A mitochondrial bioenergetic etiology of disease. J Clin Invest 123(4):1405–1412

Wang T, Sha H, Ji D, Zhang HL, Chen D, Cao Y, Zhu J (2014) Polar body genome transfer for preventing the transmission of inherited mitochondrial diseases. Cell 157(7):1591–1604

Wu K, Chen T, Huang S, Zhong C, Yan J, Zhang X, Li J, Gao Y, Zhao H, Chen ZJ (2017) Mitochondrial replacement by pre-pronuclear transfer in human embryos. Cell Res 27(6):834–837

Yamada M, Emmanuele V, Sanchez-Quintero MJ, Sun B, Lallos G, Paull D, Zimmer M, Pagett S, Prosser RW, Sauer MV et al (2016) Genetic drift can compromise mitochondrial replacement by nuclear transfer in human oocytes. Cell Stem Cell 18(6):749–754

Zhang J, Guo J, Fang W, Jun Q, Shi K (2015) Clinical features of MELAS and its relation with A3243G gene point mutation. Int J Clin Exp Pathol 8(10):13411–13415

11

An inducible CRISPR-ON system for controllable gene activation in human pluripotent stem cells

Jianying Guo[1], Dacheng Ma[2], Rujin Huang[1], Jia Ming[1], Min Ye[1], Kehkooi Kee[1], Zhen Xie[2], Jie Na[1]✉

[1] Department of Basic Medical Sciences, School of Medicine, Center for Stem Cell Biology, Tsinghua University, Beijing 100084, China

[2] MOE Key Laboratory of Bioinformatics and Bioinformatics Division, Center for Synthetic and System Biology, TNLIST/ Department of Automation, Tsinghua University, Beijing 100084, China

✉ Correspondence: jie.na@tsinghua.edu.cn (J. Na)

ABSTRACT

Human pluripotent stem cells (hPSCs) are an important system to study early human development, model human diseases, and develop cell replacement therapies. However, genetic manipulation of hPSCs is challenging and a method to simultaneously activate multiple genomic sites in a controllable manner is sorely needed. Here, we constructed a CRISPR-ON system to efficiently upregulate endogenous genes in hPSCs. A doxycycline (Dox) inducible dCas9-VP64-p65-Rta (*dCas9-VPR*) transcription activator and a reverse Tet transactivator (rtTA) expression cassette were knocked into the two alleles of the *AAVS1* locus to generate an iVPR hESC line. We showed that the *dCas9-VPR* level could be precisely and reversibly controlled by the addition and withdrawal of Dox. Upon transfection of multiplexed gRNA plasmid targeting the *NANOG* promoter and Dox induction, we were able to control *NANOG* gene expression from its endogenous locus. Interestingly, an elevated *NANOG* level promoted naïve pluripotent gene expression, enhanced cell survival and clonogenicity, and enabled hESCs to integrate with the inner cell mass (ICM) of mouse blastocysts *in vitro*. Thus, iVPR cells provide a convenient platform for gene function studies as well as high-throughput screens in hPSCs.

KEYWORDS CRISPR, transcription activation, human pluripotent stem cells, NANOG, pluripotency

INTRODUCTION

Human pluripotent stem cells (hPSCs), including human embryonic stem cells (hESCs) and human induced pluripotent stem cells (hiPSCs), are capable of self-renewal indefinitely and have the potential to differentiate into all cell types in the human body. Therefore this system offers a useful platform to study early human embryogenesis and a potential cell source for regenerative medicine. Moreover, functional cells derived from hESCs can be used to model human diseases in the context of drug toxicity tests and new drug development. These applications rely on methods to precisely control gene expression. However, because of difficulties in culture and transfection, targeted regulation of gene expression in hPSCs remains a technically challenging task. A method for efficient, rapid, and controllable gene activation is sorely needed.

Recently, the clustered regularly interspaced short palindromic repeat (CRISPR)/Cas9 system emerged as a powerful and versatile tool for genome editing (Wiedenheft et al., 2012). CRISPR was initially discovered as the adaptive immune system of bacteria and archaea (Wiedenheft et al., 2012). In response to viral and plasmid infection, bacteria and archaea could cut and degrade the foreign DNA recognized by a matching spacer RNA with the help of the Cas9 enzyme (Wiedenheft et al., 2012). CRISPR was rapidly transformed to a genome editing tool, and it has been shown to work in a wide range of systems, from plants to human cells, since the Cas9 nuclease can be directed easily to virtually anywhere in the genome using a short guide RNA and cutting the target DNA (Hsu et al., 2014). In pluripotent stem cells, the CRISPR system has been used to perform highly efficient gene knock-out and knock-in studies (Hsu et al., 2014). In addition to genome editing, a nuclease inactivated Cas9 (dCas9) was developed (Gilbert et al., 2014). By fusing dCas9 with transcription activators and repressors, such as VP64, and KRAB (Balboa et al., 2015; Gilbert et al., 2014; Mandegar et al., 2016; Genga et al., 2016), or with epigenetic modifiers, such as the catalytic domain of acetyltransferase p300 (Hilton et al., 2015) and Tet (ten eleven translocation) dioxygenase (Xu et al., 2016), one can use the CRISPR system to activate or inhibit gene expression or modify the histone and DNA methylation status at the desired locus.

Because of its potential applications in regenerative medicine, random insertion of foreign DNA into the genome of hPSCs should be avoided, since this may cause harmful mutations. The Adeno-Associated Virus Integration Site 1 (*AAVS1*) locus resides in the first intron of the *PPP1R12C* gene and has been used as a safe harbor for transgene integration (Smith et al., 2008; Hockemeyer et al., 2009; Lombardo et al., 2011; Qian et al., 2014; Zhu et al., 2014; Genga et al., 2016). Here we generated an iVPR hESC line by knocking-in the inducible dCas9-VPR system into the two alleles of the *AAVS1* locus. Detailed characterization of the iVPR hESC demonstrated that dCas9-VPR protein could be induced by Dox within 12 h and disappear after Dox withdrawal. An inducible *NANOG* overexpression line (iNANOG) was established based on the iVPR system. We found a significant increase in NANOG protein after Dox induction. INANOG cells upregulated naïve pluripotency genes and were able to grow for a significant length of time in a naïve state medium containing ERK and GSK3 inhibitors and human LIF. The iVPR system can be a valuable system to control gene expression from endogenous loci and serve as platform for genome wide screens to identify new genes that can regulate stem cell self-renewal and differentiation.

RESULTS

DCas9-VPR mediated robust ectopic and endogenous gene activation in human cell lines

To construct a robust and tunable gene activation system in hPSCs, we first compared the activation efficiency of dCas9-VPR (Chavez et al., 2015) with dCas9-VP64 (Kearns et al., 2014) and the Doxycycline (Dox) inducible Tet-On transactivator (rtTA) (Fig. 1A). We constructed plasmids to express gRNA targeting the TetO sequence (gTetO), and tested the ability of dCas9-VPR + gTetO or dCas9-VP64 + gTetO to activate the synthetic TRE promoter driving enhanced blue fluorescent protein expression (TRE-BFP) in 293FT cells (Fig. 1A). The Tet transactivator (rtTA) was used as positive control (Fig. 1B). DCas9-VPR strongly activated BFP fluorescence, 43.1% of cells were BFP positive, while in the rtTA + Dox and dCas9-VP64 groups, only 28.2% and 5.8% of cells activated BFP, respectively (Fig. 1C and 1D). Moreover, dCas9-VPR resulted in the strongest mean BFP fluorescence intensity, indicating that it is the strongest activator among the three (Fig. 1D).

We next tested the dCas9-VPR function in hESCs. DCas9-VPR, gTetO, and TRE-BFP plasmids were co-transfected into H9 hESCs. In another group, rtTA and TRE-BFP plasmids were co-transfected. FACS analysis showed that nearly 17% of cells in the dCas9-VPR group turned on BFP, while 24.7% of cells in the rtTA group were BFP positive after Dox induction, and only 0.6% of cells exhibited BFP fluorescence without Dox (Fig. 1E). Interestingly, the dCas9-VPR group showed the strongest mean fluorescence intensity (Fig. 1F). This is consistent with our result based on 293FT cells and proves that dCas9-VPR is a robust transcription activator, even compared with rtTA. We also tested the activation effect of dCas9-VPR in mouse embryonic stem cells (mESCs) and mouse embryonic fibroblasts (MEFs) and obtained similar results (Fig. S1A and S1B).

We then tested the efficiency of dCas9-VPR to activate normally silenced pluripotency genes in human cells. Two gRNAs targeting the -254 and -144 positions upstream of the transcription start site (TSS) of the pluripotency gene *NANOG* were selected (Fig. 2A). A GFP-2A-Puromycin resistant gene expression cassette was placed after the gRNA cassette both to monitor the transfection efficiency and for selection (Fig. 2A). *NANOG* cannot be activated by gNANOG alone or by dCas9-VPR together with the control gTetO. However, introducing gNANOG and dCas9-VPR together could elevate the *NANOG* transcript level by up to 150-fold in 293FT cells, indicating that it has a robust gene activation function (Fig. 2C).

Next, we tested whether the dCas9-VPR system could simultaneously activate multiple genes in human cells, we designed 2 different gRNAs per gene promoter for *HOXA10, SNAIL1, MESP1, GATA5* and *HOXA9*. First we tested the activation efficiency of these gRNAs towards their target genes when transfected separately in 293FT cells (Fig. 2D). Q-PCR analysis showed all of the five pairs of gRNAs can activate their target gene upon co-transfection with dCas9-VPR (Fig. 2D). We next pooled gRNA pairs of two genes (2× gRNAs: *MESP1, GATA5*), three genes (3× gRNAs: *HOXA10, SNAIL1, HOXA9*) or five genes (5× gRNAs: *HOXA10, SNAIL1, MESP1, GATA5* and *HOXA9*) to test the co-activation efficiency. Upon co-transfection with dCas9-VPR, different combination of gRNAs upregulated their target genes together (Fig. 2E), indicating that dCas9-VPR system could be a useful tool for multiplexed endogenous gene activation.

To validate the utility of the dCas9-VPR system in hESCs, we transfected H9 hESCs with either dCas9-VPR and gNANOG or with rtTA and *NANOG* coding DNA sequence (CDS) joined to H2B-mCherry through a 2A peptide driven by a TRE promoter. As shown in Fig. 3A, for the dCas9-VPR group, increased NANOG protein expression (in

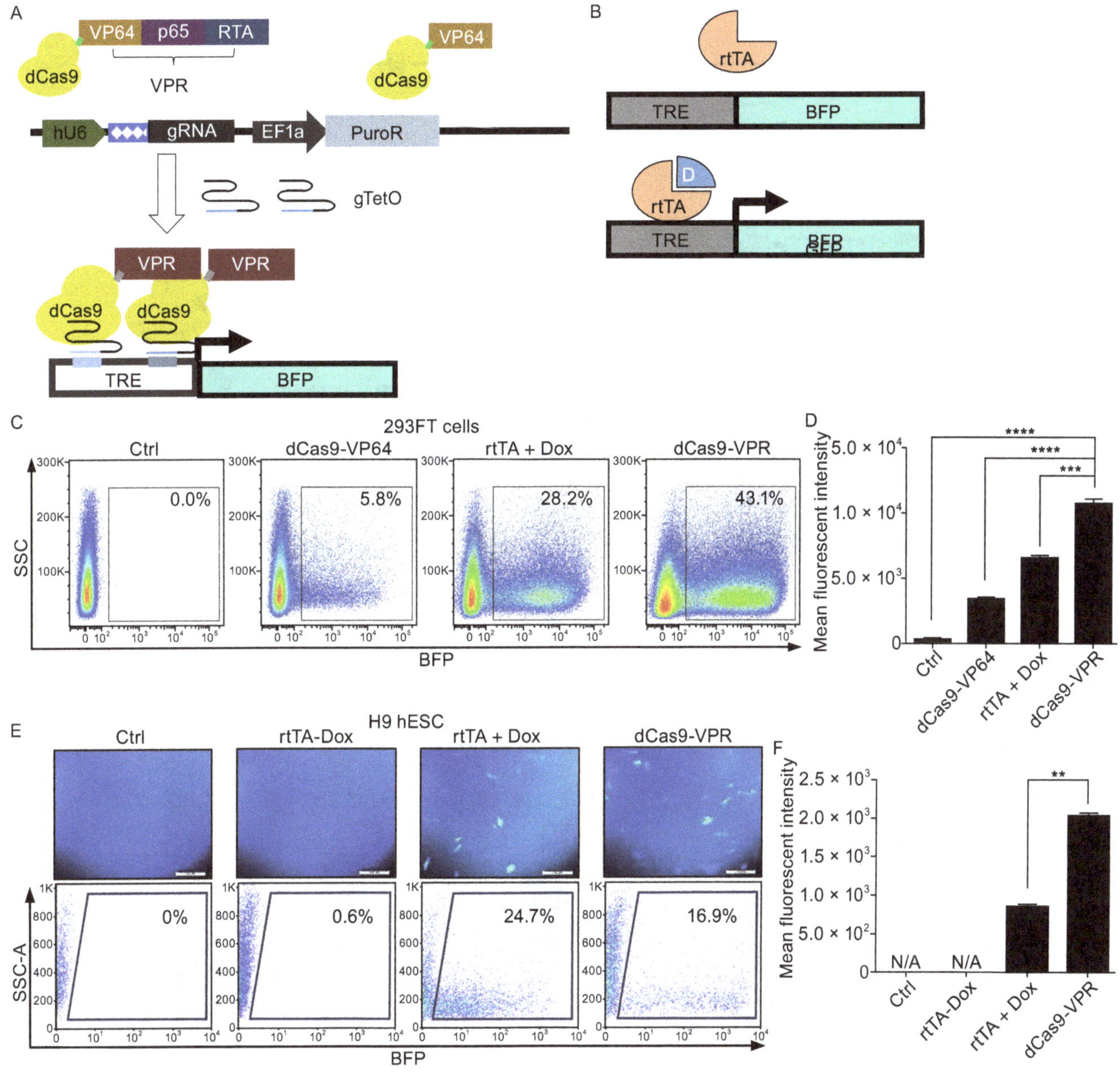

Figure 1. The dCas9-VPR system leads to robust transcription activation in human cell lines. (A) Schematic diagram of the gRNA guided dCas9-VPR gene activation system that consists of two parts: one plasmid contains dCas9-VPR driven by a CAG promoter; another plasmid contains gRNA targeting the promoter of the gene of interest driven by the human U6 promoter, in this case gTetO, and a PuroR selection cassette driven by an EF1α promoter. Upon co-transfection of the two plasmids, dCas9-VPR can activate the BFP transcription downstream of the TRE promoter. (B) Tet-On system: rtTA protein can bind to the TRE promoter and drive expression of the down-stream BFP gene in the presence of Dox. (C) 293FT cells were transfected with the reporter plasmid containing BFP driven by the TRE promoter. They were either co-transfected with dCas9-VPR or dCas9-VP64 and gTetO plasmids, or with the CAG-rtTA plasmid. Dox was added immediately after transfection. Cells were harvested 2 days after transfection and the fluorescence was analyzed using flow cytometry. (D) Bar graph quantification of mean fluorescent intensity analyzed using the FlowJo software v7.6.1. ****P* < 0.001, *****P* < 0.0001, *n* = 3. (E) H9 hESCs were electroporated with either rtTA or dCas9-VPR + gTetO plasmids together with the TRE-BFP plasmid. Dox was added immediately after electroporation. Cells were harvested 3 days after electroporation and analyzed using flow cytometry. (F) Bar graph quantification of the mean fluorescent intensity analyzed using the FlowJo software v7.6.1. N/A, not applicable. ***P* < 0.01, *n* = 2.

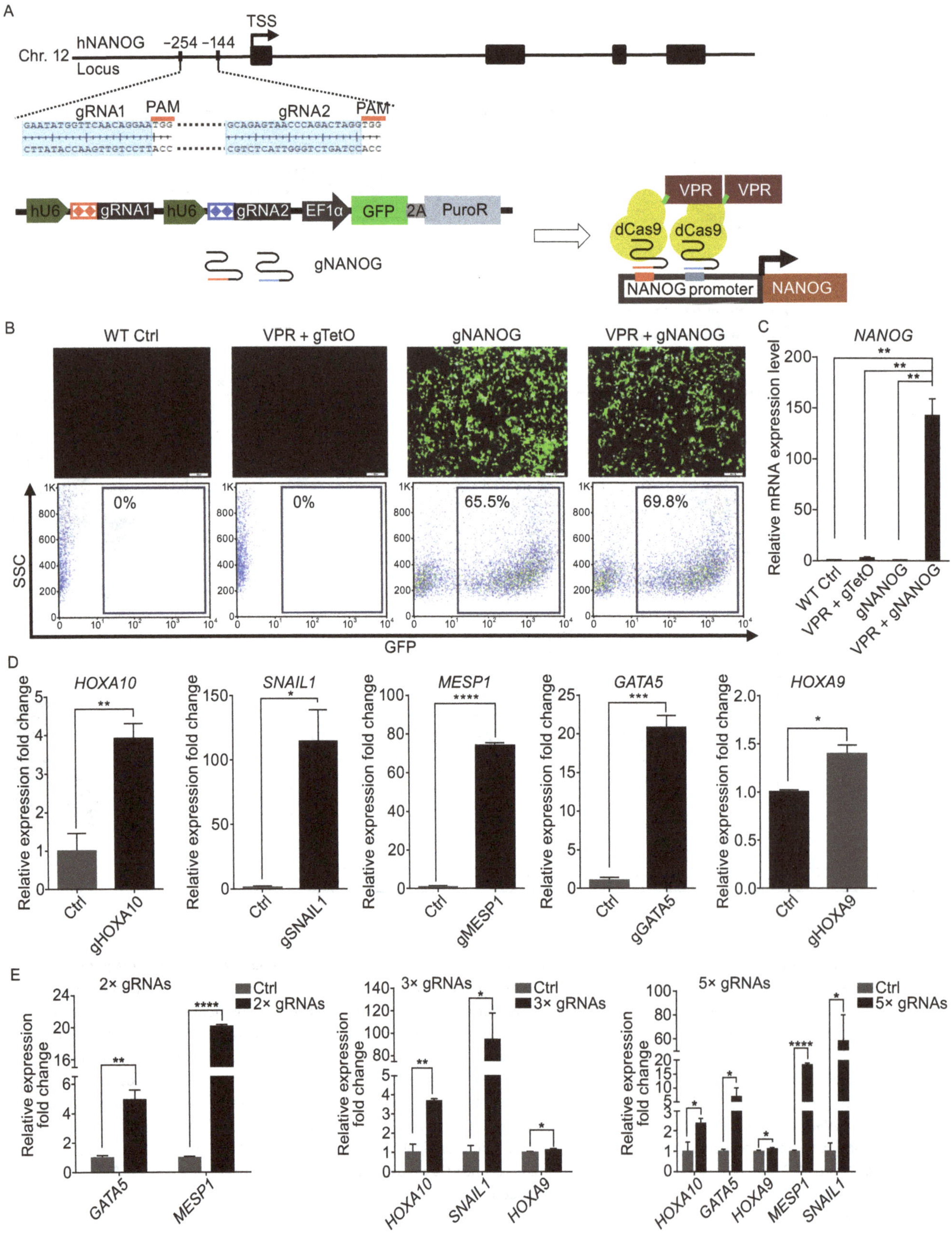
A
TSS
Chr. 12
hNANOG
Locus
-254 -144
gRNA1
PAM
gRNA2
PAM
GAATATGGTTCAACAGGAATGG
GCAGAGTAACCCAGACTAGGTGG
CTTATACCAAGTTGTCCTTACC
CGTCTCATTGGGTCTGATCCACC
hU6
gRNA1
hU6
gRNA2
EF1α
GFP
2A
PuroR
gNANOG
VPR
VPR
dCas9
dCas9
NANOG promoter
NANOG
B
WT Ctrl
VPR + gTetO
gNANOG
VPR + gNANOG
0%
0%
65.5%
69.8%
SSC
GFP
C
NANOG
Relative mRNA expression level
WT Ctrl
VPR + gTetO
gNANOG
VPR + gNANOG
D
HOXA10
SNAIL1
MESP1
GATA5
HOXA9
Relative expression fold change
Ctrl
gHOXA10
gSNAIL1
gMESP1
gGATA5
gHOXA9
E
2× gRNAs
3× gRNAs
5× gRNAs
Ctrl
2× gRNAs
3× gRNAs
5× gRNAs
Relative expression fold change
GATA5
MESP1
HOXA10
SNAIL1
HOXA9

◀ **Figure 2. DCas9-VPR can be used to activate single or multiple genes in 293FT cells.** (A) *NANOG* gRNA targeting sites were located at -254 bp and -144 bp upstream of the *NANOG* transcription starting site (TSS); protospacer-adjacent motif (PAM) sequences in red; black boxes indicate exons. (B) DCas9-VPR and gNANOG plasmids were co-transfected into 293FT cells. DCas9-VPR and gTetO plasmids were used as control. Top panels, fluorescence images of transfected cells; gNANOG plasmid transfected cells showed strong GFP fluorescence. Bottom panel, flow cytometry analysis of GFP^+ cells in each group. (C) Q-PCR analysis of *NANOG* expression 2 days after transfection; the dCas9-VPR system showed nearly 150-fold up-regulation of *NANOG* mRNA. Relative gene expression values were normalized against *GAPDH*. Error bars represent SEM. $^{**}P < 0.01$, $n = 3$. (D) Activation of endogenous genes by dCas9-VPR. DCas9-VPR was co-transfected with gRNA pairs targeting *HOXA10*, *SNAIL1*, *MESP1*, *GATA5* or *HOXA9*, respectively. Cells were harvested 2 days after transfection and subjected to Q-PCR analysis. All tested genes showed significant upregulation compared to the control group. All expression levels were normalized against *GAPDH*. Error bars represent SEM. $^{*}P < 0.05$, $^{**}P < 0.01$, $^{***}P < 0.001$, $^{****}P < 0.0001$, $n = 3$. (E) Simultaneously activation of multiple endogenous genes in 293FT cells. DCas9-VPR was co-transfected with 2× gRNAs (gMESP1, gGATA5), 3× gRNAs (gHOXA10, gSNAIL1, gHOXA9) or 5× gRNAs (gHOXA10, gSNAIL1, gMESP1, gGATA5 and gHOXA9). Cells were harvested 2 days after transfection. Q-PCR analysis confirmed co-upregulation of multiple genes targeted by pooled gRNAs. All expression levels normalized against *GAPDH*. Error bars represent SEM. $^{*}P < 0.05$, $^{**}P < 0.01$, $^{****}P < 0.0001$, $n = 3$.

red) can be detected in colonies with GFP fluorescence. Upon Dox induction, stronger NANOG was also visible in Tet-On system transfected cells and co-localized with the H2B-mCherry (Fig. 3A). Quantitative PCR (Q-PCR) and Western blot confirmed the elevated NANOG level induced by either dCas9-VPR + gNANOG or *NANOG* CDS. The transcript level of another pluripotency marker gene, *OCT4*, was increased synergistically (Fig. 3B). Western blot analysis confirmed the upregulation of NANOG and OCT4 proteins in transiently transfected H9 cells (Fig. 3C). We generated a transgenic hESC line constitutively expressing dCas9-VPR and observed no cytotoxicity, decrease in pluripotency gene expression, or change in cell morphology for long-term cultures (Fig. 3D and 3E). This suggests that the dCas9-VPR system is suitable for gene activation studies in hPSCs.

Generation of an inducible idCas9-VPR hESC knock-in line

To achieve efficient, tunable, and reversible gene activation while avoiding compromising the genome integrity of hPSCs, we engineered an iVPR system by inserting the CAG promoter driving the rtTA expression cassette and the TRE promoter driving the dCas9-VPR cassette into the two alleles of the *AAVS1* locus on chromosome 19. H9 hESCs were co-transfected with two donor plasmids containing dCas9-VPR and M2rtTA, as well as a pair of Cas9 nickase plasmids with *AAVS1* targeting gRNAs to induce DNA double-strand break (DSB) and homology recombination (HR) (Fig. S2A). After puromycin and neomycin double selection for 2 weeks, we picked and expanded 17 clones. Upon addition of Dox, all the clones showed clear induction of dCas9-VPR protein expression (Fig. S2B). Genomic DNA PCR was performed to select correct targeted clones and rule out random insertions (Fig. S2C). Clone 2, 6 and 8 had targeted insertion at both *AAVS1* alleles and without any random insertion (Fig. S2C). They were chosen for further analysis. Southern blot confirmed that in all three clones, both alleles of *AAVS1* contained the correct insertion (Fig. 4A and 4B). Q-PCR analysis showed that in hESCs, without Dox treatment, little *dCas9-VPR* transcript could be detected, while after Dox addition, strong *dCas9-VPR* expression was induced (Fig. 4C). Karyotype analysis showed that all three clones had normal 46XX karyotype (Fig. S2D). IVPR clone 2 was chosen for further study. Without Dox, we could not detect any dCas9-VPR protein in iVPR cells. The dCas9-VPR protein appeared after 12 h of Dox addition and reached a plateau at 24 h (Fig. 4D). While 6 h after Dox withdrawal, the dCas9-VPR protein decreased, by 12 h, it decreased to a low level and could not be detected anymore after 24 h (Fig. 4D). The induction of dCas9-VPR from the *AAVS1* locus was not affected by differentiation. We induced mesoderm differentiation by culturing cells in an RPMI medium supplemented with albumin, ascorbic acid, transferrin, selenite, BMP4 (5 ng/mL) and CHIR99021 (2 μmol/L) as described by Burridge et al. (2015). Q-PCR analysis showed that after 3 days of differentiation, pluripotency marker genes *OCT4* and *SOX2* were significantly downregulated, while *dCas9-VPR* was highly expressed as long as Dox was present, regardless whether cells were in hESC culture medium E8 or in the differentiation medium (Fig. 4E). Genes related to mesoderm differentiation and epithelial to mesenchymal transition, such as *SNAIL*, were strongly upregulated by BMP4 and CHIR99021, confirming that hESCs had taken a mesoderm fate (Fig. 4E). These results suggest that the iVPR hESC line can be used for efficient and reversible gene activation.

Upregulation of *NANOG* by dCas9-VPR promoted naïve state of pluripotency

The iVPR system provided a unique platform to investigate gene functions through activation from the endogenous locus. *NANOG* is a key regulator of pluripotency. We generated iNANOG hESCs by transfecting the PiggyBac based gNANOG plasmid described earlier into iVPR clone 2, 6, and 8, followed by FACS selection of GFP^+ cells. Q-PCR analysis showed that after 2 days of Dox treatment, only

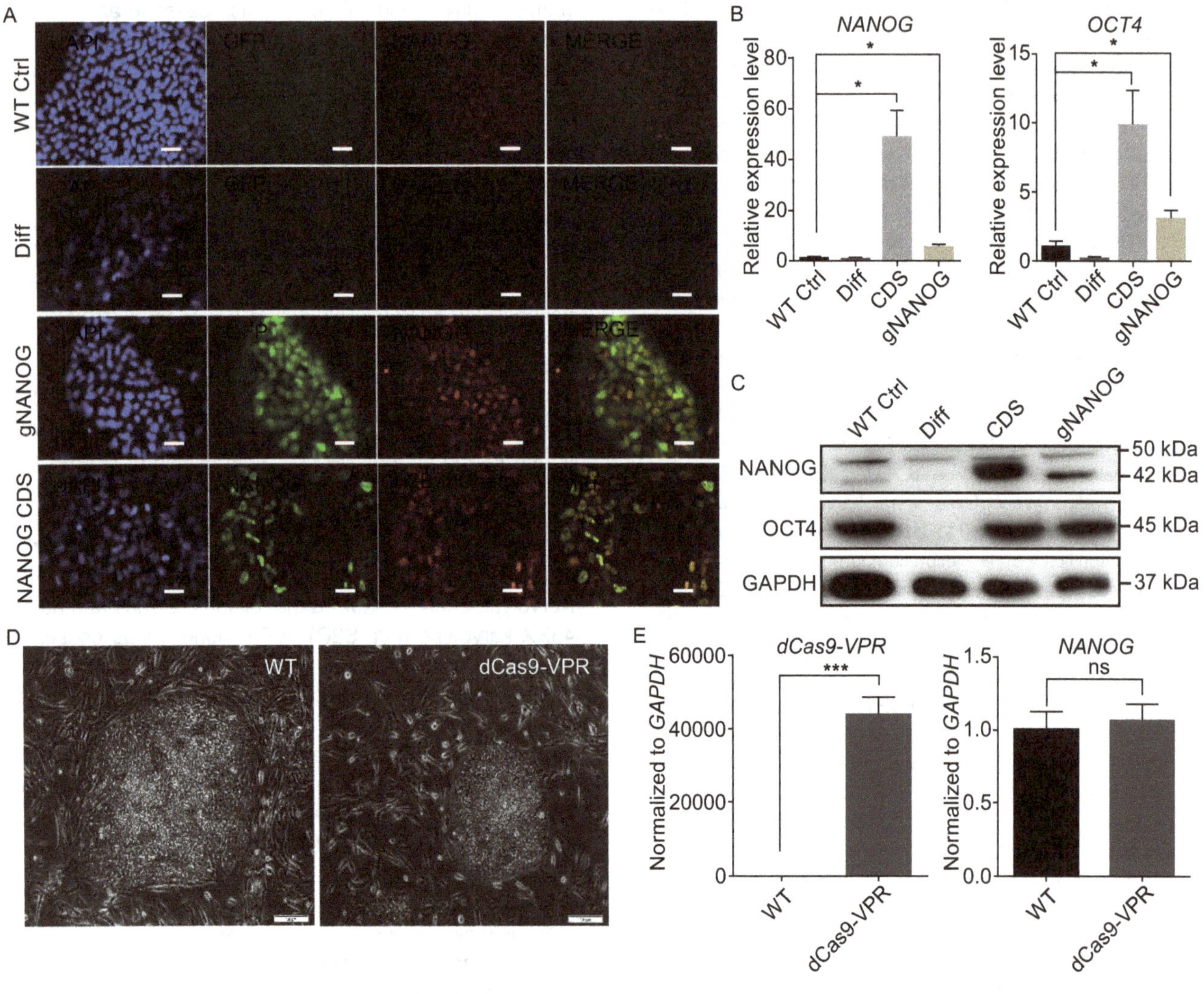

Figure 3. Activation of endogenous *NANOG* gene in hESCs by dCas9-VPR. (A) Immunostaining showing upregulation of NANOG protein by the dCas9-VPR system. Cells were fixed 5 days after transfection. WT Ctrl, untransfected H9 cells; Diff, differentiated H9 cells induced by 10 μmol/L retinoic acid (RA); gNANOG, H9 cells co-transfected with dCas9-VPR and gNANOG plasmids; *NANOG* CDS, cells co-transfected with CAG-rtTA and TRE driving NANOG-2A-H2B-mCherry. Dox were added immediately after electroporation. All plasmids were based on the PiggyBac system and co-transfected with a plasmid containing HyperPB transposase driven by a CAG promoter. Scale bar, 20 μm. (B) Q-PCR analysis of *NANOG* and *OCT4* expression in H9 cells 5 days after transfection. All expression levels normalized against *GAPDH*. Error bars represent SEM. *$P < 0.05$, $n = 3$. (C) Western blot analysis of NANOG and OCT4 protein expression in H9 hESCs. Cells were harvested 5 days after transfection without selection. (D) DCas9-VPR constitutive expressing H9 cells showed similar clone morphology after long-term culture. Scale bar, 100 μm. (E) Q-PCR result showing dCas9-VPR constitutive expressing H9 cells and wild-type H9 cells expressed similar amount of *NANOG*. All expression levels normalized against *GAPDH*. Error bars represent SEM. ns. $P > 0.05$, ***$P < 0.001$, $n = 3$.

iNANOG cells showed a significant increase (about 18 folds) in the *NANOG* mRNA level, while iVPR cells, with or without Dox, or iNANOG cells without Dox did not show any change in *NANOG* expression, indicating that the iNANOG system is tightly regulated (Fig. 5A). We also tested the time window of *NANOG* down-regulation after Dox withdrawal. *NANOG* mRNA was unchanged during the first 12 h and decreased after 24 h. It approached the background level after 48 h (Fig. 5B). We next examined the change in NANOG protein level after Dox addition and withdrawal. Western blot revealed that dCas9-VPR protein became detectable 12 h after Dox induction and reached a significant level after 24 h (Fig. 5C, dCas9, long exposure; LE). Accordingly, NANOG protein showed an obvious increase after 24 h and maintained at high level as long as dCas9-VPR was present (Fig. 5C, NANOG, LE). On the other hand, 6 h after Dox

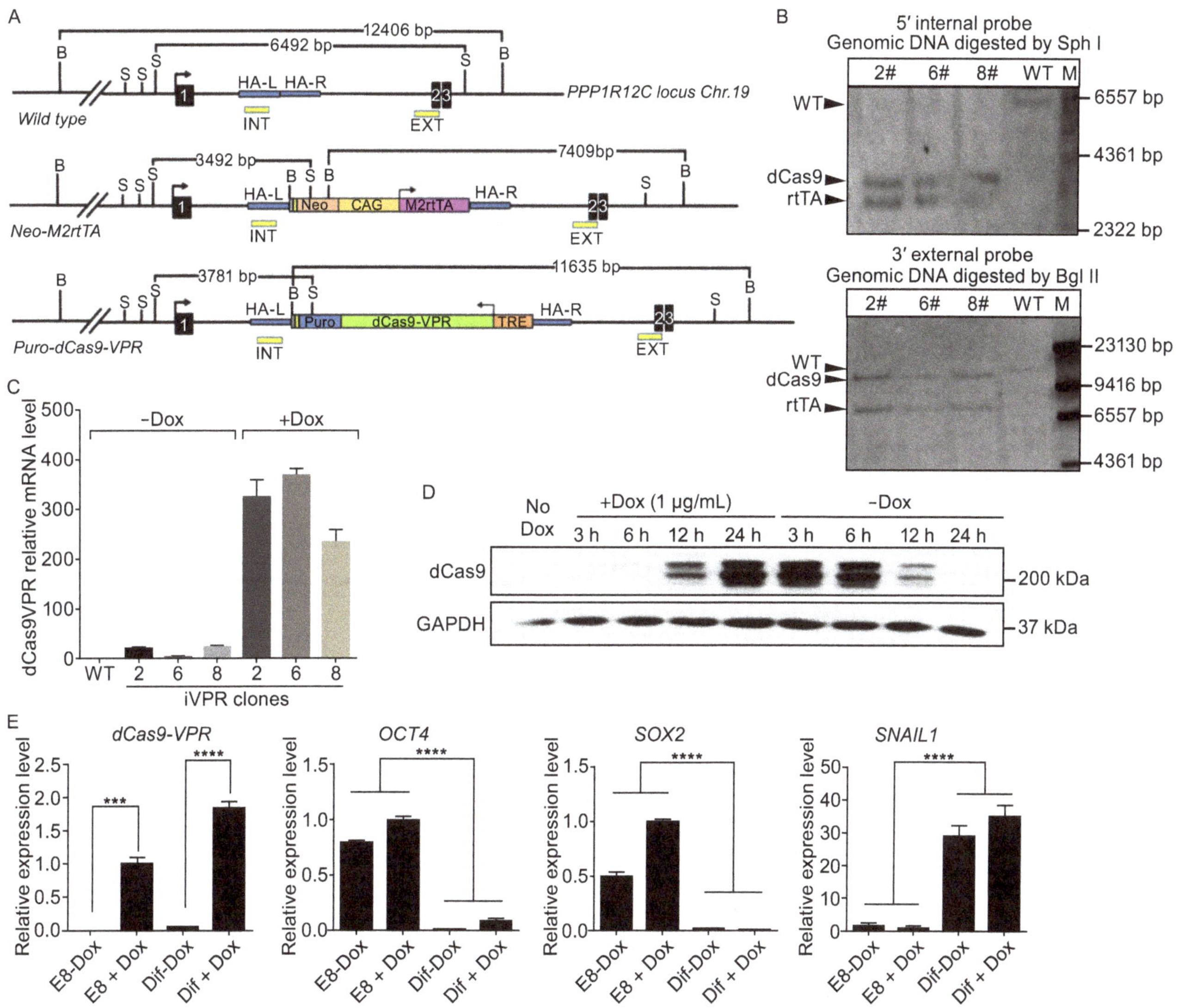

Figure 4. Generation of the iVPR hESC line. (A) Schematic view of wild type, targeted *AAVS1* locus, and positions of Southern blot probes. B (Bgl II site), S (Sph I site), EXT (external probe), INT (internal probe). The sizes of the expected bands are indicated at the top. Blue lines indicate homology to the *PPP1R12C* intron. HA-L and HA-R, left and right homology arms. (B) Southern blot confirmed the correct targeted *AAVS1* locus in the iVPR clone 2#, 6#, 8#. M, marker. (C) Q-PCR analysis of dCas9-VPR transcript levels with or without Dox treatment. Expression levels were normalized against *GAPDH*. Error bar represents SEM. (D) Western blot of dCas9-VPR protein level upon Dox addition and after Dox withdrawal in idCas9-VPR clone 2. The time points are indicated at the top. (E) Q-PCR showing that the induction of *dCas9-VPR* was not affected by differentiation. Cells were induced to undergo mesoderm differentiation for 3 days in the presence or absence of Dox. Gene expression levels were all normalized against *GAPDH*. Error bar indicates SEM. ***P < 0.001, ****P < 0.0001, n = 3.

removal, the dCas9-VPR protein decreased significantly (Fig. 5D, dCas9, short exposure; SE). The decline of the dCas9-VPR protein was most apparent during the first 24 h. After 4 days without Dox, dCas9-VPR protein became almost undetectable (Fig. 5D). Similarly, the NANOG protein level dropped to the background level after 4 days of Dox withdrawal (Fig. 5D). Q-PCR analysis showed that after Dox induction, iNANOG significantly upregulated naïve state related genes such as *OCT4*, *PRDM14*, *GDF3*, and *LEF-TYB*, while the early differentiation genes such as *AFP* was significantly downregulated (Fig. 5E). *XIST*, a long non-coding RNA involved in X chromosome inactivation were also downregulated after *NANOG* induction (Fig. 5F). The expression of SSEA3, a more rigorous pluripotency cell

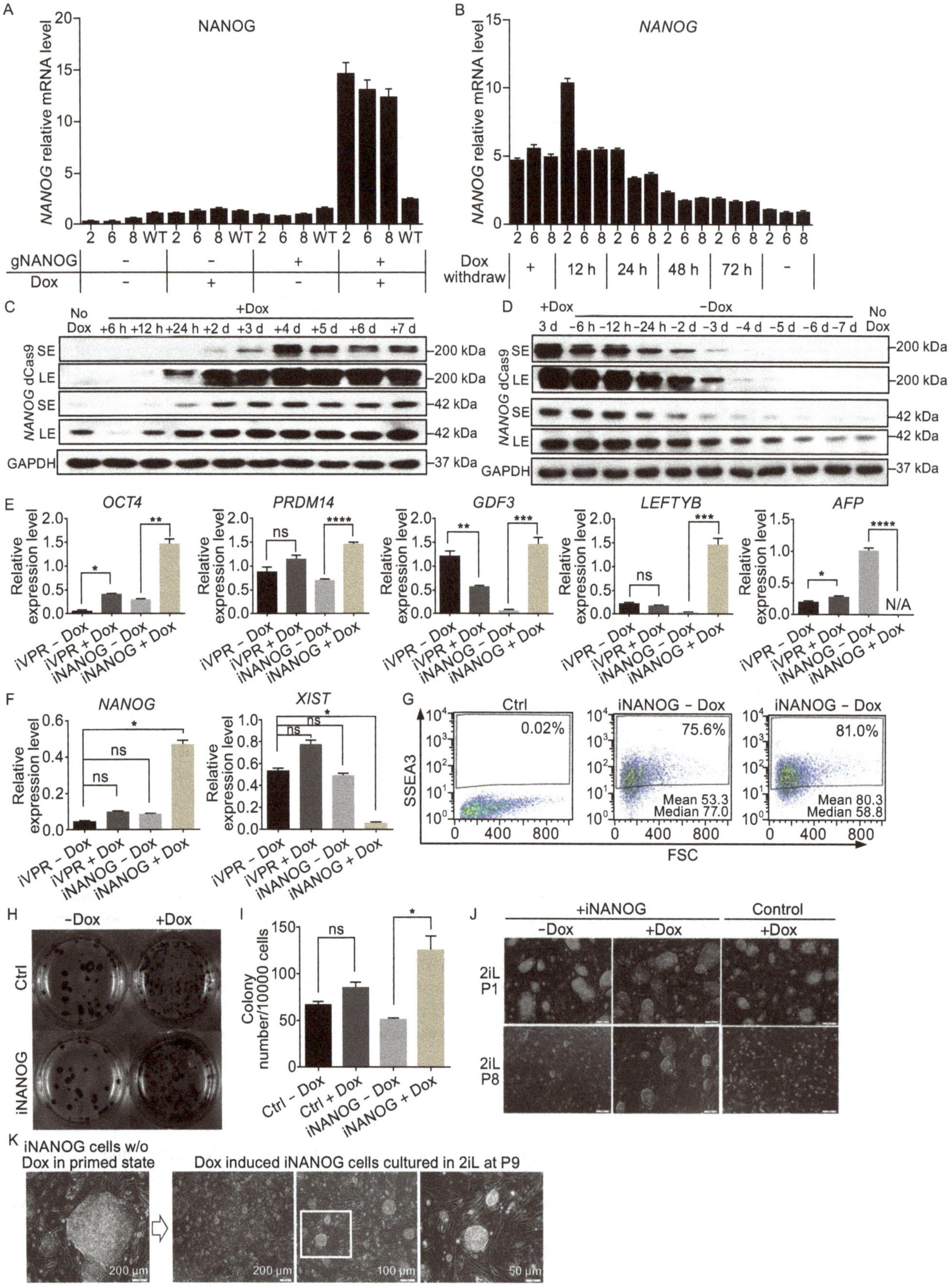
A
NANOG
NANOG relative mRNA level
gNANOG
Dox
B
NANOG
NANOG relative mRNA level
Dox withdraw
12 h
24 h
48 h
72 h
C
No Dox
+Dox
+6 h +12 h +24 h +2 d +3 d +4 d +5 d +6 d +7 d
NANOG dCas9
SE
LE
GAPDH
200 kDa
42 kDa
37 kDa
D
+Dox
3 d
−Dox
−6 h −12 h −24 h −2 d −3 d −4 d −5 d −6 d −7 d
No Dox
E
OCT4
PRDM14
GDF3
LEFTYB
AFP
Relative expression level
N/A
iVPR − Dox
iVPR + Dox
iNANOG − Dox
iNANOG + Dox
F
NANOG
XIST
G
Ctrl
iNANOG − Dox
0.02%
75.6%
81.0%
Mean 53.3
Median 77.0
Mean 80.3
Median 58.8
SSEA3
FSC
H
−Dox
+Dox
Ctrl
iNANOG
I
Colony number/10000 cells
Ctrl − Dox
Ctrl + Dox
iNANOG − Dox
iNANOG + Dox
J
+iNANOG
Control
−Dox
+Dox
2iL P1
2iL P8
K
iNANOG cells w/o Dox in primed state
Dox induced iNANOG cells cultured in 2iL at P9
200 µm
100 µm
50 µm

◀ **Figure 5. Upregulation of *NANOG* by dCas9-VPR promoted clonogenicity and the naïve state of pluripotency.** (A) Q-PCR analysis of *NANOG* upregulation in iNANOG cells. IVPR clones 2, 6, and 8 were electroporated with gRNA expression plasmid targeting the *NANOG* promoter, as shown in Fig. 2A. GFP positive cells were purified by FACS and maintained as iNANOG cells. They were treated with or without Dox (1 μg/mL) for 2 days. *NANOG* expression level was normalized against *GAPDH*. Error bar represents SEM. (B) Q-PCR analysis of *NANOG* down-regulation in iNANOG cells. Dox was added for 2 days, then removed. Cells were harvested at different time points, as indicated. *NANOG* expression was normalized against *GAPDH*. Error bar represents SEM. (C) Western blot showing increased NANOG protein expression in iNANOG cells at different time points after Dox treatment. SE, short exposure; LE, long exposure; d, day; h, h. (D) Western blot showing NANOG protein expression decrease in iNANOG cells at different time points after Dox withdrawal. SE, short exposure; LE, long exposure; d, day; h, h. (E) Q-PCR analysis showing upregulation of pluripotency gene *OCT4, PRDM14, GDF3,* and *LEFTYB,* and down-regulation of differentiation gene *AFP.* Expression level all normalized against *GAPDH*. Error bar represents SEM. N/A, not applicable. ns. $P > 0.05$, $^{*}P < 0.05$, $^{**}P < 0.01$, $^{***}P < 0.001$, $^{****}P < 0.0001$, $n = 3$. (F) Q-PCR analysis showing downregulation of *XIST* after *NANOG* induction. Expression level normalized against *GAPDH*. Error bar represents SEM. ns. $P > 0.05$, $^{*}P < 0.05$, $n = 3$. (G) Flow cytometry analysis showing increased SSEA3 expression after *NANOG* induction. Data analyzed using the FlowJo software v7.6.1. (H) Clonogenicity assay of iNANOG cells. Alkaline phosphatase assay (dark blue) was used to visualize undifferentiated colonies. (I) Bar graph quantification of the clonogenicity assay. ns. $P > 0.05$, $^{*}P < 0.05$, $n = 3$. (J) Morphology of iNANOG cells cultured in the 2iL medium. *NANOG* overexpression (iNANOG + Dox) promoted long-term cell growth in the 2iL medium. Representative images of passages 1 and 8 (P1 and P8) are shown. Scale bar, 100 μm. (K) Morphology of primed state iNANOG cells (without Dox) and Dox induced iNANOG cells (passage 9, P9) in the 2iL medium.

surface marker, was increased and became more homogeneous after *NANOG* elevation (Figs. 5G and S3). In addition to elevated expression of pluripotency genes, iNANOG cells also showed enhanced survival and proliferation abilities. Clonogenicity assay showed that after Dox induction, twice as many clones formed from dissociated iNANOG single cells (Fig. 5H and 5I). Finally, we tested whether *NANOG* upregulation by iVPR may facilitate hESCs to enter the naïve state of pluripotency. IVPR cells and iNANOG cells were cultured in 2iL medium which supplemented with ERK inhibitor PD0325901, GSK3 inhibitor CHIR99021, human LIF, and bFGF proteins with or without Dox addition (Silva et al., 2009; Takashima et al., 2014). Upon changing to the 2iL medium, hESCs colonies changed into a domed-shaped morphology and became more compact (Fig. 5J and 5K). INANOG cells without induction can only survive for no more than three passages in the 2iL medium (Fig. 5J). Interestingly, Dox induced iNANOG cells can grow in the 2iL medium for longer than 9 passages with single cell dissociation and a 1:15 passage ratio (Fig. 5J and 5K). In contrast to iNANOG cells, Dox treated iVPR cells could not survive in 2iL conditions (Fig. 5J). Thus, upregulation of *NANOG* from its endogenous locus significantly improved single cell clonogenicity and permitted hESCs to grow in a naïve state culture environment.

Upregulation of *NANOG* enabled hESCs to integrate with mouse ICM *in vitro*

Entering the pluripotent ICM lineage is considered a more stringent test for naïve state ESCs (Gafni et al., 2013; Takashima et al., 2014). We next used *in vitro* human-mouse blastocyst chimera assay to assess the functionality of iNANOG cells (Fig. 6A). To exclude the influence of Dox treatment only, wild type hESCs stably carrying gNANOG (WTSG) were used as the control. For this series of experiments, we also added Forskolin (a cAMP agonist) into the 2iL medium, since it had been shown to promote hPSCs to enter the naïve state (Hanna et al., 2010; Ware et al., 2014; Duggal et al., 2015). We refer to this medium as 2iL/FK. iNANOG cells showed further enhanced proliferation in the 2iL/FK medium and were able to form large, dome-shaped colonies (Fig. 6B), while cells without *NANOG* overexpression could only form small colonies (Fig. S4A). E3.5 blastocysts were collected from ICR mice for hESC injection. iNANOG cells and WTSG cells cultured with or without Dox, in either the E8 or 2iL/FK medium, were dissociated into single cells. 10–15 single cells were injected into the blastocoel cavity and cultured in a 1:1 mixed KSOM:2iL/FK medium for 24 h (Fig. 6A). Because cells without *NANOG* overexpression only formed small colonies on feeder in the 2iL/FK medium, we could not obtain sufficient pure hESCs for blastocyst injection. Therefore, this group was omitted from this series of experiments. Since all cells used for injection contained GFP transgene expressed from the gNANOG plasmid, the location of human cells in the mouse blasocysts could be followed directly under the fluorescence microscope. 4–6 h after injection, most blastocysts contained GFP positive human cells (Fig. 6B and 6C). After 24 h of culture, many embryos still contained hESCs (Fig. 6B). We used time-lapse imaging to monitor the activity of hESCs in mouse blastocysts over time (Supplementary movie S1). Interestingly, endogenous *NANOG* overexpression strongly enhanced the survival of hESCs in mouse blastocysts. 12 h after injection, 2iL/FK cultured Dox induced gNANOG cells could be found in approximately 82% of blastocysts, while E8 cultured Dox induced gNANOG cells were alive in 73% of blastocysts (Figs. 6C and S4B). In contrast, without Dox induction, E8 cultured INANOG cells could only be seen in 49% of injected blastocysts (Figs. 6C and S4B). We next

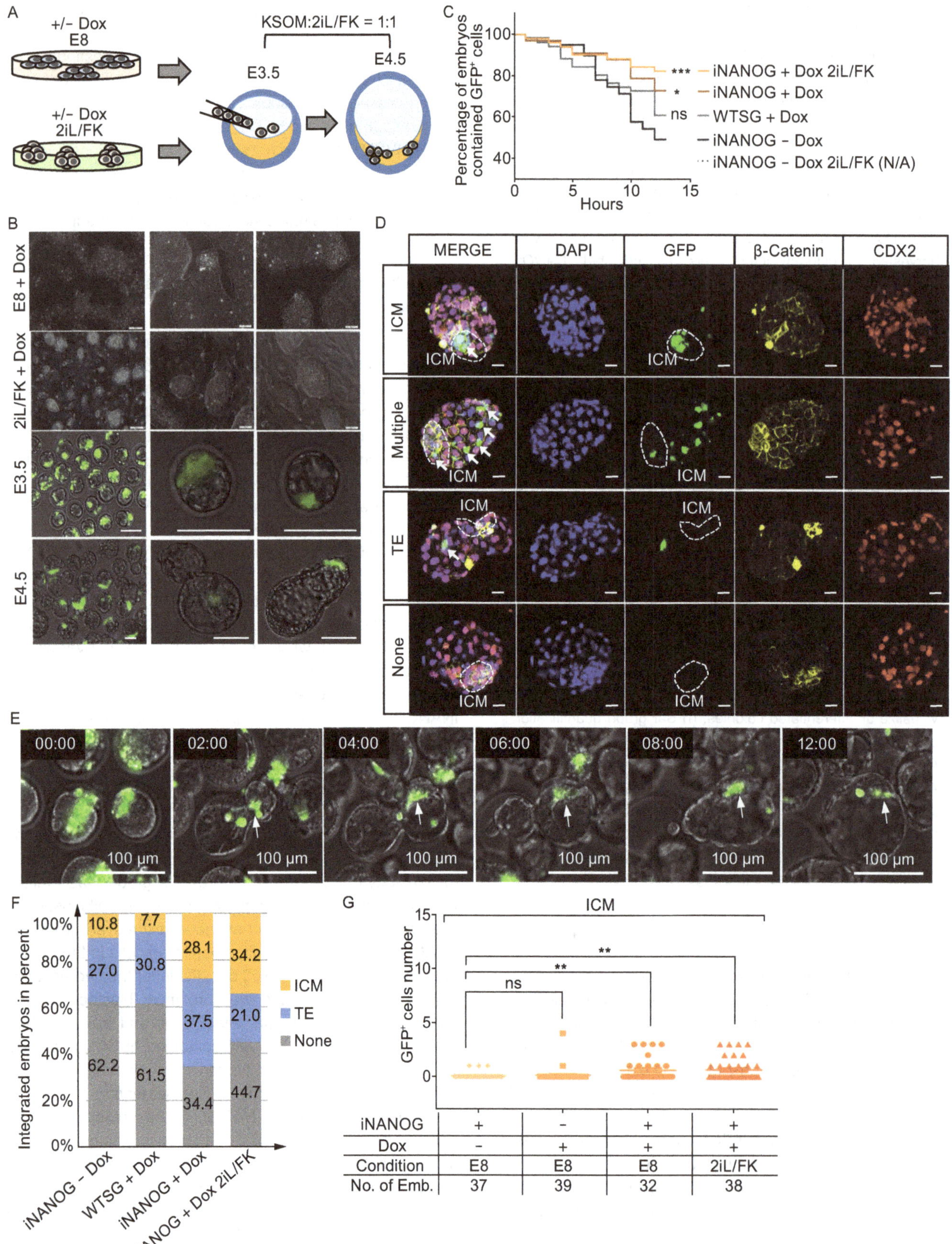
A
+/- Dox
E8
+/- Dox
2iL/FK
KSOM:2iL/FK = 1:1
E3.5
E4.5
C
Percentage of embryos contained GFP+ cells
Hours
*** iNANOG + Dox 2iL/FK
* iNANOG + Dox
ns WTSG + Dox
iNANOG - Dox
iNANOG - Dox 2iL/FK (N/A)
B
E8 + Dox
2iL/FK + Dox
E3.5
E4.5
D
MERGE
DAPI
GFP
β-Catenin
CDX2
ICM
Multiple
TE
None
E
00:00
02:00
04:00
06:00
08:00
12:00
100 μm
F
Integrated embryos in percent
ICM
TE
None
iNANOG - Dox
WTSG + Dox
iNANOG + Dox
iNANOG + Dox 2iL/FK
G
GFP+ cells number
ICM
ns
**
iNANOG
Dox
Condition
No. of Emb.
E8
2iL/FK
37
39
32
38

◀ **Figure 6. Upregulation of *NANOG* by idCas9-VPR promoted hESC survival and ICM integration in mouse blastocysts *in vitro*.** (A) Cartoon showing *in vitro* hESC-mouse blastocyst chimera formation assay. iNANOG cells were cultured in E8 or 2il/FK medium with or without Dox, then injected into E3.5 mouse blastocysts and cultured to E4.5 in KSOM: 2il/FK = 1:1 medium *in vitro*. (B) Morphology of iNANOG cells in culture and chimeric embryos. Top 2 rows, cells cultured in E8 or 2iL/FK on feeders; scale bar, 100 μm. Bottom 2 rows, E3.5 and E4.5 mouse blastocysts with iNANOG cells (GFP); scale bar, 100 μm. (C) Survival curve of hESC in mouse blastocysts over time. WTSG, wild type hESCs expressing *NANOG* gRNA. The *P* value was calculated using the Log-rank (Mantel-Cox) test. ns. $P > 0.05$, *$P < 0.05$, ***$P < 0.001$. Detailed information is provided in Fig. S4B. (D) Confocal images of E4.5 chimeric embryos. β-Catenin, yellow; CDX2, red; DNA, blue; iNANOG cells, green. The ICM region is highlighted by a dashed circle. Scale bar, 20 μm. (E) Selected frame from time-lapse movie of iNANOG-mouse blastocyst chimera. Arrows indicating iNANOG cells moved with mouse inner cell mass cells during blastocyst hatching. Scale bar, 100 μm. (F) The proportion of blastocysts with hESC integration for E4.5 embryos. Blastocysts with GFP cells in both ICM and TE were counted as ICM. The percentage of embryos with ICM/TE/None integration was labeled in the colored bar. (G) Dot graph showing the number of iNANOG cells in the ICM region of E4.5 embryos. Cells, culture condition before injection, and number of embryos were as listed. Error bars represent SEM. ns. $P > 0.05$, **$P < 0.01$.

analyzed the locations of the transplanted hESCs. Injected embryos were fixed after 24 h of culture, stained with CDX2 (a trophectoderm marker) and β-Catenin, and observed with a confocal microscope. Different integration patterns were shown: hPSCs integrated into the ICM region (ICM), in both the ICM and the trophectoderm (Multiple), only in the trophectoderm (TE), and disappeared (None) (Fig. 6D). We also performed live imaging to monitor the behavior of iNANOG cells in the mouse blastocyst. Interetingly, 2iL/FK cultured iNANOG cells tend to migrate with mouse inner cell mass cells as blastocyst hatching from the zona pellucida (Fig. 6E). *NANOG* overexpression significantly improved the percentage of cells remaining in blastocysts, and the 2iL/FK culture further increased the ICM integration proportion (Figs. 6F, 6G, and S4C). On average, two 2iL/FK or E8 cultured iNANOG cells could be found in the ICM region 24 h after injection, while without *NANOG* overexpression, hardly any GFP cells were seen in the ICM (Fig. 6G). Thus, upregulation of *NANOG* from its endogenous locus greatly enhanced cell survival and their subsequent ICM integration in hPSC-mouse blastocyst chimeras.

DISCUSSION

In this study, we generated an inducible CRISPR-ON hESC line by targeting the *AAVS1* locus. Based on both our results and those of Chavez et al. (2015), dCas9-VPR appeared to be a stronger activator than VP64 to induce gene expression from both ectopic and endogenous promoters. It even led to a higher level of reporter gene activation compared with Tet-ON rtTA, where VP64 was fused with Tet protein directly bound to the TRE elements. This is likely due to the combined effects of VP64, NF-κB transcactivating subunit p65, and the viral transcription factor Rta, which together can recruit a multitude of endogenous factors to achieve dramatically enhanced transcriptional activation. Other dCas9 based transcription activators have been generated. For example, Balboa et al. found increased activation ability with more VP16 fusing together. Using the longest version of dCas9-VP192 combined with inducible systems, they sucessfully facilitated human cell reprogramming and differentiation (Balboa et al., 2015). Konermann et al. engineered a structure-guided CRISPR synergistic activation mediator system (SAM), where they engineered gRNA2.0 by replacing the tetraloop and stem loop 2 of the original gRNA with a minimal hairpin aptamer that specifically binds to MS2 bacteriophage coat proteins (Konermann et al., 2014). By co-expression of dCas9-VP64, gRNA2.0, and MS2 fused with p65 and the activation domain of the human heat-shock factor 1 (HSF1), highly effective gene activation can be achieved (Konermann et al., 2014). Tanenbaum et al. constructed a SunTag system: dCas9 was joined with 10 copies of GCN4 peptide (SunTag), while VP64 was fused with scFv-GCN4 (the single-chain variable fragment (scFv) antibody of GCN4) (Tanenbaum et al., 2014). When co-expressed in the cell, SunTag was bound by scFv-GCN4, and multiple copies of VP64 resulted activation of the target gene (Tanenbaum et al., 2014). Compared with the systems discussed above, which required introducing tandem repeat large cassette or the co-expression of two components in addition to the gRNA, dCas9-VPR is a simple and effective option.

In our study, we chose to insert the iVPR system into the *AAVS1* locus, since it has been used as a 'safe habor' for transgene insertion in human stem cell systems (Dekelver et al., 2010). For example, Genga et al. constructed a GFP labled H1 hESC line by knocking-in a CAG-GFP into the *AAVS1* locus. Besides, an inducible dCas9-KRAB gene inhibition system was also introduced into the GFP-H1 cells. By infecting sgRNA targeting the exdogenous CAG promoter, they successfully realized CRISPR based inhibition of exdogenous gene in hESCs (Genga et al., 2016). González et al. inserted the Dox inducible Cas9 system into the 2 alleles of the *AAVS1* locus of HUES8 hESCs (González et al., 2014; Zhu et al., 2014; Zhu et al., 2015). The resulting iCRISPR hESC line enabled selection-free gene knock-out and the generation of lineage-specific knock-in reporters. This demonstrated that when Cas9 was expressed in a controllable manner from a suitable locus, the resulting cell line can be a powerful platform for genome editing in normally hard to transfect human stem cells (Zhu et al., 2015). Similarly, using the iVPR line, we found that the efficiency to generate an iNANOG line was much improved. Upon Dox addition and withdrawal, *NANOG* transcripts and proteins could be up- and down-regulated in a highly repeatable manner, which greatly facilitated downstream experiments. Recently, Ordovás et al. reported *AAVS1*-locus mediated transgene inhibition in hESCs, and that inhibition may due to different cassettes inserted into the locus (Ordovás et al., 2015). We tested the iVPR expression in both undifferentiated hESCs and after induction of mesoderm differentiation. The level of *dCas9-VPR* transcripts was even higher upon Dox treatment after mesoderm induction (Fig. 4E). The iVPR and iNANOG cells have been maintained for more than 6 months, and we did not observe any reduction in the level of *dCas9-VPR* or *NANOG* induced by Dox. Thus, results of us and other groups suggested that, in most cases, *AAVS1* locus integration is a reliable approach to generate transgenic hPSCs.

NANOG is a master transcription factor for pluripotency in both human and mouse ESCs (Mitsui et al., 2003; Boyer et al., 2005; Chambers et al., 2007). During somatic cell reprogramming to pluripotent stem cells, ectopic expression of *NANOG* helped to speed up reprogramming and restrict partially reprogrammed cells to the ground state (Hanna et al., 2009; Silva et al., 2009). Different from mESCs, conventional cultured hESCs are in a primed state, similar to the epiblast stem cells in mice (Brons et al., 2007; Tesar et al., 2007). Recently, multiple groups reported methods to obtain naïve state hPSCs that resemble ground-state mESCs (Gafni et al., 2013; Duggal et al., 2015; Takashima et al., 2014; Theunissen et al., 2014). Takashima et al. showed that ectopic expression of *NANOG* and *KLF2* could reset the self-renewal requirements of hPSCs so that they can be grown in a medium containing ERK1/2 inhibitor PD0325901 and GSK3 inhibitor CHIR99021, and adopt a domed-shaped morphology similar to that of mESCs (Takashima et al., 2014). Here we increased the expression of endogenous *NANOG* by targeting a strong transcription activator, dCas9-VPR, to its promoter. As expected, we observed upregulation of naïve state genes such as *GDF3*, *PRDM14*, and *LEFTYB* and downregulation of early differentiation gene *AFP*. Interestingly, these iNANOG cells showed a significantly improved survival ability and clonogenicity when cultured in the primed state, and they could grow in 2i plus LIF conditions for more than nine passages. The improved survival and self-renewal of iNANOG cells was not due to the effect of Dox treatment as described by Chang et al. (2014), because Dox treated iNANOG cells showed significantly higher clonogenicity over Dox treated iVPR cells (Fig. 5H and 5I). The enhanced survival ability seemed to have a significant influence on whether hPSCs can integrate with the ICM of mouse blastocysts during *in vitro* culture. We found that even when iNANOG cells were in the primed state, after injection into mouse blastocysts, more cells remained inside the blastocysts and some of the cells were able to integrate with mouse ICM cells (Fig. 6F and 6G). Culturing iNANOG cells in 2iL/FK naïve state medium (Duggal et al., 2015) further improved the ICM integration rate (Fig. 6F and 6G). INANOG cells displayed highly dynamic interactions with mouse ICM cells, as observed in time-lapse movies (Supplementary movie S1). They migrated with mouse ICM cells as blastocysts hatched from zona pellucida. However, despite enhanced survival ability of iNANOG cells, many injected cells died over time. After 24 h, more than 30% of injected blastocysts lost all iNANOG cells and more than 60% of blastocysts lost the injected hESCs if *NANOG* was not overexpressed (Fig. 6F). This was partially caused by poor survival of hESCs in the IVC-1 and -2 media designed to culture peri-implantation mouse and human embryos (Bedzhov and Zernicka-Goetz, 2014; Deglincerti et al., 2016; Shahbazi et al., 2016) (Fig. S4D). Thus, to achieve better naïve hPSC and mouse ICM integration, a culture medium suitable for both mouse blastocysts and hPSCs may be needed. The effect of *NANOG* overexpression on cell survival and self-renewal is also in accordance with the observation that chromosome 12, where the *NANOG* gene is located, is the most frequently gained chromosome in culture adapted hPSCs (Baker et al., 2007) and during hiPSC generation (Taapken et al., 2011). Moreover, *NANOG* was reported to be upregulated by a number of factors such as *STAT3*, Hedgehog signaling, hypoxia, etc., in human cancers, and repression or ablation of *NANOG* inhibited tumor initiation (Gong et al., 2015). Thus, iNANOG hESCs, where the endogenous *NANOG* can be activated by dCas9-VPR in a controllable manner, may also be a good system to study the process of hPSC adaptation and cancerous transformation.

In summary, the iVPR hESC line generated and characterized in this study offered a convenient, stable, and highly controllable platform for gene activation studies. It can also be used to investigate the function of regulatory elements in the genome such as super enhancers as well as for genome wide screens using established human gRNA libraries.

MATERIALS AND METHODS

HESC culture

H9 hESCs (WiCell Institute) were maintained on inactivated mouse embryonic fibroblast (MEF) cells in standard hESC medium consisting of KO-DMEM (Invitrogen) supplemented with 1× Nonessential Amino Acids (NEAA) (Invitrogen), 0.1 mmol/L 2-mercaptoethanol (Sigma-Aldrich), 1 mmol/L GlutaMAX (Invitrogen), 20% Knock-out serum-replacement (KOSR) (Invitrogen) and 4 ng/mL bFGF (Peprotech). Cells were cultured at 37°C in a humidified atmosphere with 5% CO_2 in air. They were passaged with 1 mg/mL collagenase IV (Invitrogen) and seeded onto MEFs. For feeder-free culture, hESCs were grown for more than three passages on Matrigel (growth factor reduced, BD Biosciences) in the absence of feeders in E8 medium (Invitrogen).

Plasmid construction

DCas9-VPR was constructed by fusing the nuclease deficient Cas9 (dCas9) with transcription activator VP64, p65, and Rta in tandem as described by Chavez et al. (Maeder et al., 2013). For constitutive expression, dCas9-VPR was placed behind a CAG promoter in a PiggyBac vector also containing a PGK promoter driving a hygromycin resistance gene. For inducible expression from the *AAVS1* locus, dCas9-VPR was placed behind a TRE promoter in the *AAVS1* homologous recombineering donor plasmid, as shown in Fig. S2A. DCas9-VP64 was constructed by fusing dCas9 with VP64. Tet-On system was obtained from Clontech (http://www.clontech.com). PiggyBac plasmids were generous gift from the Sanger institute, Cambridge, UK (http://www.sanger.ac.uk). The multiple *NANOG* gRNA expression plasmid was constructed by SynGene (http://syngen.tech) as depicted in Fig. 2A.

Naïve state culture condition for hPSCs

For naïve state conversion, cells cultured in standard hESC medium on MEFs were dissociated to single cells using 0.05% trypsin/EDTA solution (Invitrogen), replated on MEFs, and cultured overnight in standard hESC medium supplement with 10 μmol/L Rho Kinase (ROCK)-inhibitor Y-27632 (Calbiochem). The next day, the standard medium was changed to the 2iL or 2iL/FK (for injection) medium, which consisted of KO-DMEM (Invitrogen), 20% KOSR, 1×NEAA, 0.1 mmol/L 2-mercaptoethanol, 1 mmol/L GlutaMAX, 12 ng/mL bFGF, 10 ng/mL human recombinant LIF (Peprotech), 1 μmol/L ERK1/2 inhibitor PD0325901 (Peprotech), 3 μmol/L GSK3 inhibitor CHIR99021 (Peprotech), 10 μmol/L Forskolin (Peprotech), and 50 μg/mL ascorbic acid (Sigma). HESCs changed to a dome-shaped morphology within 4–6 days after culturing in the 2iL or 2iL/FK medium and were passaged every 4 days as single cells using 0.05% trypsin/EDTA.

Cardiac mesoderm differentiation from hESCs

For cardiac mesoderm differentiation, hESCs maintained on Matrigel (growth factor reduced, BD Biosciences) in E8 were dissociated into single cells with Accutase (Invitrogen), then seeded onto Matrigel-coated tissue culture dishes at a density of 5×10^4 cells/cm^2 and cultured in E8 for 3 days. Then the medium was switched to the RPMI1640 medium supplemented with Albumin, Ascorbic acid, transferrin, selenite, 5 ng/mL BMP4 (R&D Systems), and CHIR99021 to induce cardiac mesoderm formation.

Quantitative PCR

Total RNA was extracted with TRIZOL (Invitrogen). 1 μg RNA of each sample was used for reverse transcription with Superscript III (Invitrogen). Q-PCR reactions were performed using GoTaq qPCR Master Mix (Promega) in a CFX96 Real-Time System (Bio-Rad). The relative expression level of each gene was normalized against the Ct (Critical Threshold) value of the house-keeping gene *GAPDH* using the Bio-Rad CFX Manager program. Primer sequences are listed in table S2.

Antibodies, immunostaining, Western blot, and FACS analysis

For immunostaining, cells were fixed in 4% paraformaldehyde (PFA) in PBS, permeabilized in 0.5% Triton X-100 (Sigma), blocked in 5% normal goat serum (Origene) and incubated with primary antibodies against NANOG (1:200), SSEA3 (1:200) in 4°C overnight and detected by DyLight 488- or 549-conjugated secondary antibodies (Thermo). Nuclei were stained with DAPI (Sigma). A Nikon Ti-U fluorescence microscope was used for image acquisition. For western blot, cells were lysed in a RIPA buffer (Applygen, http://applygen.com.cn) with Protease Inhibitor Cocktail (Roche). Total proteins were separated on a 12% SDS/PAGE gel, transferred to nitrocellulose membrane (Whatman). The membrane was blocked with 5% non-fat dry milk in TBST and then incubated with primary antibodies against Cas9 (Genetex, 1:1000), GAPDH (CWBio, 1:1000), OCT4 (Santa Cruz, 1:1000) and NANOG (Cell Signaling Technology, 1:1000). After washing, the membrane was incubated with anti-mouse or anti-rabbit peroxidase-conjugated secondary antibodies (ZSGB-Bio http://www.zsbio.com/). Bands recognized by antibodies were revealed by ECL reagent (Pierce). For FACS analysis, cells were first dissociated with 0.05% Trypsin in 0.2% EDTA and PBS. FACS was performed on a Fortessa flow cytometer (Becton Dickinson).

Mouse blastocyst injection and *in vitro* culture

The animal facility of Tsinghua University has been accredited by the AAALAC (Association for Assessment and Accreditation of Laboratory Animal Care International) and all animal protocols used in this study were approved by the IACUC (Institutional Animal Care and Use Committee) of Laboratory Animal Research Center of Tsinghua University. Mouse morula were collected from ICR females 2.5 days post-coitus and cultured in KSOM medium (95 mmol/L NaCl, 2.5 mmol/L KCl, 0.35 mmol/L KH_2PO_4, 0.2 mmol/L $MgSO_4 \cdot 7H_2O$, 0.2 mmol/L glucose, 10 mmol/L sodium lactate, 25 mmol/L $NaHCO_3$, 0.2 mmol/L sodium pyruvate, 1.71 mmol/L $CaCl \cdot 2H_2O$, 0.01 mmol/L EDTA, 1 mmol/L L-glutamine, 0.1 mmol/L EAA, 0.1 mmol/L NEAA, 4 mg/mL BSA) at 37°C, 5% CO_2 for 24 h to get blastocysts (Hogan et al., 1986). HESCs were briefly treated with Accutase for single cell and injected (~10–15 cells for each embryo) on a Nikon microscope fitted with piezo-driven Eppendorf NK2 micromanipulator, CellTram air and CellTram Vario. After injection, embryos containing hESCs were cultured in medium supplemented with naïve culture medium:KOSM (1:1) (Chen et al., 2015) in 37°C, 5% CO_2 incubator. After injected embryos reformed blastocoel, the chimera embryos were live cell imaged using Leica microscope fitted with a live cell imaging system and fixed after 24–36 h post-injection for staining and confocal imaging. For embryo immunostaining, zona pellucida-free injected embryos were fixed with 3.5% paraformaldehyde, permeabilized in 0.5% Triton X-100 (Sigma) and blocked with 5% BSA and then incubated with primary antibodies against CDX2 (BioGenex), β-Catenin (1:50, Abcam) and detected by DyLight 549- or 633- conjugated secondary antibodies (Thermo). Nuclei were stained with DAPI (Sigma). A Nikon-A1 fluorescence microscope was used for image acquisition.

Statistical analysis

Data are presented as mean ± standard error of the mean (SEM). Statistical significance was determined by Student's *t*-test (two-tail) for two groups or one-way Analysis of Variance (ANOVA) for multiple groups using Graphpad software. $P < 0.05$ was considered significant.

ACKNOWLEDGMENTS

This work was supported by the National Basic Research Program (973 Program) (No. 2012CB966701), the National Natural Science Foundation of China (Grant No. 31171381 to J.N.), and core facilities of the Tsinghua-Peking University Center for Life Sciences. TNLIST Interdisciplinary research foundation grant 042003171 (to Z.X. and N.J.). We thank Dr. Danwei Huangfu for the *AAVS1* homologous recombineering donor plasmids and Dr. Xiaohua Shen lab for assistance in Southern blot.

AUTHOR CONTRIBUTIONS

J.G.: concept and design, collection and/or assembly of data, data analysis and interpretation, manuscript writing; D.M., R.H., M.Y., J. M.: collection and/or assembly of data; K.K. provided essential reagents, technical and scientific advice to the experiments and manuscript; Z.X.: concept and design; J.N.: concept and design, manuscript writing, and final approval of the manuscript.

COMPLIANCE WITH ETHICAL GUIDELINES

Jianying Guo, Dacheng Ma, Rujin Huang, Jia Ming, Min Ye, Kehkooi Kee, Zhen Xie, and Jie Na declare no conflict of interests.

This article does not contain any studies with human subjects performed by the any of the authors. All institutional and national guidelines for the care and use of laboratory animals were followed.

REFERENCES

Baker DEC, Harrison NJ, Maltby E, Smith K, Moore HD, Shaw PJ, Heath PR, Holden H, Andrews PW (2007) Adaptation to culture of human embryonic stem cells and oncogenesis in vivo. Nat Biotechnol 25:207–215

Balboa D, Weltner J, Eurola S, Trokovic R, Wartiovaara K, Otonkoski T (2015) Conditionally stabilized dCas9 activator for controlling gene expression in human cell reprogramming and differentiation. Stem Cell Rep 5:448–459

Bedzhov I, Zernicka-Goetz M (2014) Self-organizing properties of mouse pluripotent cells initiate morphogenesis upon implantation. Cell 156:1032–1044

Boyer LA, Lee TI, Cole MF, Johnstone SE, Levine SS, Zucker JP, Guenther MG, Kumar RM, Murray HL, Jenner RG (2005) Core transcriptional regulatory circuitry in human embryonic stem cells. Cell 122:947–956

Brons IGM, Smithers LE, Trotter MWB, Rugg-Gunn P, Sun B, de Sousa Lopes SMC, Howlett SK, Clarkson A, Ahrlund-Richter L, Pedersen RA (2007) Derivation of pluripotent epiblast stem cells from mammalian embryos. Nature 448:191–195

Burridge PW, Holmström A, and Wu JC (2015). Chemically defined culture and cardiomyocyte differentiation of human pluripotent stem cells. Curr Protoc Hum Genet: 21.23. 21–21.23. 15

Chambers I, Silva J, Colby D, Nichols J, Nijmeijer B, Robertson M, Vrana J, Jones K, Grotewold L, Smith A (2007) Nanog safeguards pluripotency and mediates germline development. Nature 450:1230–1234

Chang M-Y, Rhee Y-H, Yi S-H, Lee S-J, Kim R-K, Kim H, Park C-H, Lee S-H (2014) Doxycycline enhances survival and self-renewal of human pluripotent stem cells. Stem Cell Rep 3:353–364

Chavez A, Scheiman J, Vora S, Pruitt BW, Tuttle M, Iyer EPR, Lin S, Kiani S, Guzman CD, Wiegand DJ (2015) Highly efficient Cas9-mediated transcriptional programming. Nat Methods 12:326–328

Chen Y, Niu Y, Li Y, Ai Z, Kang Y, Shi H, Xiang Z, Yang Z, Tan T, Si W (2015) Generation of cynomolgus monkey chimeric fetuses using embryonic stem cells. Cell Stem Cell 17:116–124

Deglincerti A, Croft GF, Pietila LN, Zernicka-Goetz M, Siggia ED, Brivanlou AH (2016) Self-organization of the in vitro attached human embryo. Nature 533:251–254

DeKelver RC, Choi VM, Moehle EA, Paschon DE, Hockemeyer D, Meijsing SH, Sancak Y, Cui X, Steine EJ, Miller JC (2010) Functional genomics, proteomics, and regulatory DNA analysis in isogenic settings using zinc finger nuclease-driven transgenesis into a safe harbor locus in the human genome. Genome Res 20:1133–1142

Duggal G, Warrier S, Ghimire S, Broekaert D, Van der Jeught M, Lierman S, Deroo T, Peelman L, Van Soom A, Cornelissen R (2015) Alternative routes to induce naive pluripotency in human embryonic stem cells. Stem Cells 33:2686–2698

Gafni O, Weinberger L, Mansour AA, Manor YS, Chomsky E, Ben-Yosef D, Kalma Y, Viukov S, Maza I, Zviran A (2013) Derivation of novel human ground state naive pluripotent stem cells. Nature 504:282–286

Genga RM, Kearns NA, Maehr R (2016) Controlling transcription in human pluripotent stem cells using CRISPR-effectors. Methods 101:36–42

Gilbert LA, Horlbeck MA, Adamson B, Villalta JE, Chen Y, Whitehead EH, Guimaraes C, Panning B, Ploegh HL, Bassik MC (2014) Genome-scale CRISPR-mediated control of gene repression and activation. Cell 159:647–661

Gong S, Li Q, Jeter CR, Fan Q, Tang DG, Liu B (2015) Regulation of NANOG in cancer cells. Mol Carcinog 54:679–687

González F, Zhu Z, Shi Z-D, Lelli K, Verma N, Li QV, Huangfu D (2014) An iCRISPR platform for rapid, multiplexable, and inducible genome editing in human pluripotent stem cells. Cell Stem Cell 15:215–226

Hanna J, Saha K, Pando B, Van Zon J, Lengner CJ, Creyghton MP, van Oudenaarden A, Jaenisch R (2009) Direct cell reprogramming is a stochastic process amenable to acceleration. Nature 462:595–601

Hanna J, Cheng AW, Saha K, Kim J, Lengner CJ, Soldner F, Cassady JP, Muffat J, Carey BW, Jaenisch R (2010) Human embryonic stem cells with biological and epigenetic characteristics similar to those of mouse ESCs. Proc Natl Acad Sci 107:9222–9227

Hilton IB, D'Ippolito AM, Vockley CM, Thakore PI, Crawford GE, Reddy TE, Gersbach CA (2015) Epigenome editing by a CRISPR-Cas9-based acetyltransferase activates genes from promoters and enhancers. Nat Biotechnol 33:510–517

Hockemeyer D, Soldner F, Beard C, Gao Q, Mitalipova M, DeKelver RC, Katibah GE, Amora R, Boydston EA, Zeitler B (2009) Efficient targeting of expressed and silent genes in human ESCs and iPSCs using zinc-finger nucleases. Nat Biotechnol 27:851–857

Hogan B, Costantini F, Lacy E (1986) Manipulating the mouse

embryo: a laboratory manual, vol 34. Cold spring harbor laboratory, Cold Spring Harbor, NY

Hsu PD, Lander ES, Zhang F (2014) Development and applications of CRISPR-Cas9 for genome engineering. Cell 157:1262–1278

Kearns NA, Genga RMJ, Enuameh MS, Garber M, Wolfe SA, Maehr R (2014) Cas9 effector-mediated regulation of transcription and differentiation in human pluripotent stem cells. Development 141:219–223

Konermann S, Brigham MD, Trevino AE, Joung J, Abudayyeh OO, Barcena C, Hsu PD, Habib N, Gootenberg JS, Nishimasu H (2014) Genome-scale transcriptional activation by an engineered CRISPR-Cas9 complex. Nature 517:583–588

Lombardo A, Cesana D, Genovese P, Di Stefano B, Provasi E, Colombo DF, Neri M, Magnani Z, Cantore A, Riso PL (2011) Site-specific integration and tailoring of cassette design for sustainable gene transfer. Nat Methods 8:861–869

Maeder ML, Linder SJ, Cascio VM, Fu Y, Ho QH, Joung JK (2013) CRISPR RNA-guided activation of endogenous human genes. Nat Methods 10:977–979

Mandegar MA, Huebsch N, Frolov EB, Shin E, Truong A, Olvera MP, Chan AH, Miyaoka Y, Holmes K, Spencer CI (2016) CRISPR interference efficiently induces specific and reversible gene silencing in human iPSCs. Cell Stem Cell 18:541–553

Mitsui K, Tokuzawa Y, Itoh H, Segawa K, Murakami M, Takahashi K, Maruyama M, Maeda M, Yamanaka S (2003) The homeoprotein Nanog is required for maintenance of pluripotency in mouse epiblast and ES cells. Cell 113:631–642

Ordovás L, Boon R, Pistoni M, Chen Y, Wolfs E, Guo W, Sambathkumar R, Bobis-Wozowicz S, Helsen N, Vanhove J (2015) Efficient recombinase-mediated cassette exchange in hPSCs to study the hepatocyte lineage reveals AAVS1 locus-mediated transgene inhibition. Stem cell Rep 5:918–931

Qian K, Huang CL, Chen H, Blackbourn LW, Chen Y, Cao J, Yao L, Sauvey C, Du Z, Zhang SC (2014) A simple and efficient system for regulating gene expression in human pluripotent stem cells and derivatives. Stem Cells 32:1230–1238

Shahbazi MN, Jedrusik A, Vuoristo S, Recher G, Hupalowska A, Bolton V, Fogarty NME, Campbell A, Devito LG, Ilic D (2016) Self-organization of the human embryo in the absence of maternal tissues. Nature cell Biol 18:700–708

Silva J, Nichols J, Theunissen TW, Guo G, van Oosten AL, Barrandon O, Wray J, Yamanaka S, Chambers I, Smith A (2009) Nanog is the gateway to the pluripotent ground state. Cell 138:722–737

Smith JR, Maguire S, Davis LA, Alexander M, Yang F, Chandran S, Pedersen RA (2008) Robust, persistent transgene expression in human embryonic stem cells is achieved with AAVS1-targeted integration. Stem Cells 26:496–504

Taapken SM, Nisler BS, Newton MA, Sampsell-Barron TL, Leonhard KA, McIntire EM, Montgomery KD (2011) Karyotypic abnormalities in human induced pluripotent stem cells and embryonic stem cells. Nat Biotechnol 29:313–314

Takashima Y, Guo G, Loos R, Nichols J, Ficz G, Krueger F, Oxley D, Santos F, Clarke J, Mansfield W (2014) Resetting transcription factor control circuitry toward ground-state pluripotency in human. Cell 158:1254–1269

Tanenbaum ME, Gilbert LA, Qi LS, Weissman JS, Vale RD (2014) A protein-tagging system for signal amplification in gene expression and fluorescence imaging. Cell 159:635–646

Tesar PJ, Chenoweth JG, Brook FA, Davies TJ, Evans EP, Mack DL, Gardner RL, McKay RDG (2007) New cell lines from mouse epiblast share defining features with human embryonic stem cells. Nature 448:196–199

Theunissen TW, Powell BE, Wang H, Mitalipova M, Faddah DA, Reddy J, Fan ZP, Maetzel D, Ganz K, Shi L (2014) Systematic identification of culture conditions for induction and maintenance of naive human pluripotency. Cell Stem Cell 15:471–487

Ware CB, Nelson AM, Mecham B, Hesson J, Zhou W, Jonlin EC, Jimenez-Caliani AJ, Deng X, Cavanaugh C, Cook S (2014) Derivation of naive human embryonic stem cells. Proc Natl Acad Sci 111:4484–4489

Wiedenheft B, Sternberg SH, Doudna JA (2012) RNA-guided genetic silencing systems in bacteria and archaea. Nature 482:331–338

Xu X, Tao Y, Gao X, Zhang L, Li X, Zou W, Ruan K, Wang F, G-l Xu, Hu R (2016) A CRISPR-based approach for targeted DNA demethylation. Cell Discov 2:16009

Zhu Z, González F, Huangfu D (2014) The iCRISPR platform for rapid genome editing in human pluripotent stem cells. Methods Enzymol 546:215

Zhu Z, Verma N, González F, Shi Z-D, Huangfu D (2015) A CRISPR/Cas-mediated selection-free knockin strategy in human embryonic stem cells. Stem Cell Rep 4:1103–1111

12

Biogenesis and regulation of the *let-7* miRNAs and their functional implications

Hosuk Lee[1], Sungwook Han[1], Chang Seob Kwon[2], Daeyoup Lee[1]✉

[1] Department of Biological Sciences, Korea Advanced Institute of Science and Technology, Daejeon 305-701, Korea
[2] Department of Chemistry and Biology, Korea Science Academy of KAIST, Busan 614-822, Korea
✉ Correspondence: daeyoup@kaist.ac.kr (D. Lee)

ABSTRACT

The *let-7* miRNA was one of the first miRNAs discovered in the nematode, *Caenorhabditis elegans*, and its biological functions show a high level of evolutionary conservation from the nematode to the human. Unlike in *C. elegans*, higher animals have multiple isoforms of *let-7* miRNAs; these isoforms share a consensus sequence called the 'seed sequence' and these isoforms are categorized into *let-7* miRNA family. The expression of *let-7* family is required for developmental timing and tumor suppressor function, but must be suppressed for the self-renewal of stem cells. Therefore, *let-7* miRNA biogenesis must be carefully controlled. To generate a *let-7* miRNA, a primary transcript is produced by RNA polymerase II and then subsequently processed by Drosha/DGCR8, TUTase, and Dicer. Because dysregulation of *let-7* processing is deleterious, biogenesis of *let-7* is tightly regulated by cellular factors, such as the RNA binding proteins, LIN28A/B and DIS3L2. In this review, we discuss the biological functions and biogenesis of *let-7* miRNAs, focusing on the molecular mechanisms of regulation of *let-7* biogenesis in vertebrates, such as the mouse and the human.

KEYWORDS miRNA processing, miRNA biogenesis, let-7 family, TUTase, LIN28A/B

INTRODUCTION

MicroRNAs (miRNAs) are short (~22-nucleotide-long) non-coding RNAs found in diverse eukaryotes from plants to animals. They inhibit gene expression largely in a post-transcriptional manner, by recognizing a specific complementary sequence usually located in the 3′ UTR of a target mRNA. The binding of a miRNA to this complementary sequence decreases translation of the target mRNA via several mechanisms, including mRNA degradation, inhibition of translational initiation and elongation (Eulalio et al., 2008; Filipowicz et al., 2008; Ameres and Zamore, 2013; Ha and Kim, 2014).

Let-7 (*lethal-7*) was one of the first miRNAs to be discovered. It was originally identified as a regulator of developmental timing in the nematode, *C. elegans*, and was therefore regarded as a heterochronic gene (Reinhart et al., 2000). The *let-7* miRNA is evolutionarily conserved across various animal species, including flies and mammals, but it is not found in plants (Pasquinelli et al., 2000; Hertel et al., 2012). The nematode and fruit fly have a single isoform, whereas higher animals have multiple *let-7* isoforms. In the human, for instance, the *let-7* family is composed of nine mature *let-7* miRNAs encoded by 12 different genomic loci, some of which are clustered together (Ruby et al., 2006; Roush and Slack, 2008).

As *let-7* expression gradually increases during development, and this miRNA plays important roles in many biological processes, it could be expected that the biogenesis of *let-7* should be tightly regulated (Pasquinelli et al., 2000; Sempere et al., 2002; Thomson et al., 2006; Liu et al., 2007). Indeed, studies have shown that LIN28A/B blocks *let-7* biogenesis in several different ways to maintain self-renewal and pluripotency in stem cells (Heo et al., 2008; Newman et al., 2008; Rybak et al., 2008; Viswanathan et al., 2008; Heo et al., 2009; Piskounova et al., 2011; Kim et al., 2014). In addition, TUTase has been shown to be involved in degrading the *let-7* precursor (*pre-let-7*) to block the generation of mature *let-7* in the cytoplasm (Hagan et al., 2009; Heo et al., 2009; Thornton et al., 2012).

Hosuk Lee and Sungwook Han have contributed equally to this work.

In this review, we briefly summarize the current state of knowledge regarding the *let-7* miRNA family and its biological functions, focusing on *let-7* biogenesis in higher animals. In addition, we discuss recent progress in better understanding the regulatory mechanisms that act upon *let-7*.

GENERAL FEATURES OF THE *let-7* FAMILY

The discovery of *let-7* in *C. elegans*

Experiments using forward genetics originally identified *let-7* (*lethal-7*) as a heterochronic gene in *C. elegans* (Reinhart et al., 2000). Heterochronic genes act sequentially to regulate cell fates in a stage-specific manner during the different larval transitions in *C. elegans* (Moss, 2007). For instance, *miR-48*, *miR-84*, and *miR-241* regulate the second larval (L2) to third larval (L3) transition, while *let-7* regulates the fourth larval (L4) to adult transition (Fig. 1) (Reinhart et al., 2000; Abbott et al., 2005). During the development of *C. elegans*, hypodermal seam cells undergo asymmetric division in a manner similar to that seen in stem cells. As a result, one daughter cell undergoes differentiation, while the other undergoes self-renewal at each larval stage. At the final transition (the L4-to-adult transition), all of the daughter cells stop proliferation and undergo differentiation. After this terminal differentiation, the seam cells form alae. In contrast, seam cells harboring the *let-7* mutation fail to finish the L4-to-adult transition and instead exhibit extra cell division without

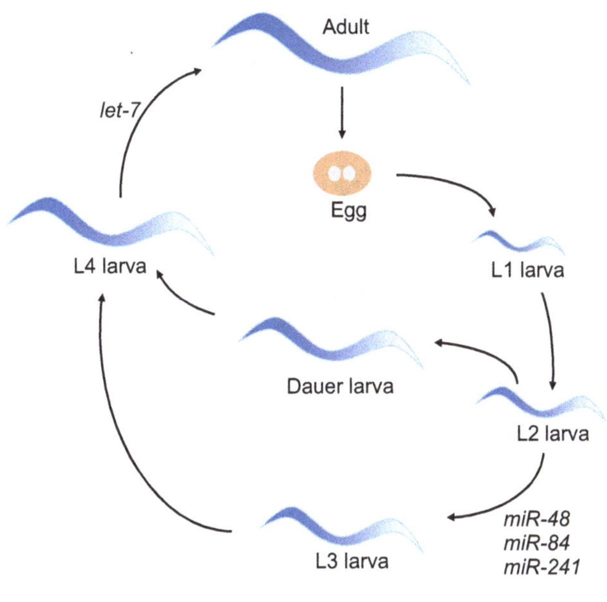

Figure 1. Life cycle of the nematode, *Caenorhabditis elegans*. Schematic diagrams of the *C. elegans* life cycle. Eggs laid by adult *C. elegans* go through four developmental stages: L1, L2, L3, and L4 larva. If the environment is harsh, L2 larva can go through the Dauer larva stage instead of the L3 larva stage. During the life cycle of *C. elegans*, *miR-48*, *miR-84*, and *miR-241* regulate the L2-to-L3 transition, whereas *let-7* regulates the L4-to-adult transition.

proper formation of the adult alae (Reinhart et al., 2000). As a result, the majority of *let-7* mutants die due to bursting of the vulva, earning this mutation its name: *lethal-7*. The expression pattern of *let-7* is consistent with its mutant phenotype, as its expression is first detected at the L3 stage and peaks at the L4 stage (Reinhart et al., 2000; Esquela-Kerscher et al., 2005). In addition, precocious expression of *let-7* at the L2 stage yielded an early adult-like phenotype at the L4 stage (Hayes and Ruvkun, 2006). These studies collectively support the notion that *let-7* is a key regulator of proper developmental timing in *C. elegans*.

Characteristics of the *let-7* family

Let-7 miRNAs are found in various animal species, including the human. This conservation suggests that *let-7* may act as a regulator of gene expression across diverse animal species (Pasquinelli et al., 2000; Hertel et al., 2012). Using computational analyses, such as BLAST (Basic Local Alignment Search Tool), researchers have discovered a total of 28,645 miRNAs from 223 species that have been recorded in miRBase release 21.0 (http://www.mirbase.org). This substantial total includes 401 *let-7* sequences from various organisms. According to miRBase, *Caenorhabditis elegans* (nematode), *Drosophila melanogaster* (fly), *Xenopus tropicalis* (frog), *Danio rerio* (zebra fish), *Gallus gallus* (chicken), *Canis familiaris* (dog), *Mus musculus* (mouse) and *Homo sapiens* (human) all express a version of *let-7* (*let-7a*) that possesses the exact consensus sequence of 'UGAGGUAGUAGGUUGUAUAGUU' (Fig. 2A). Most of *let-7* sequences include the 'seed sequence'. This highly preserved sequence that spans nucleotides 2 through 8 in some miRNAs (Ruby et al., 2006), and is an essential component required for target recognition by the RNA-induced silencing complex (RISC) (Brennecke et al., 2005; Grimson et al., 2007; Hibio et al., 2012). This conserved feature of the *let-7* miRNAs suggests that their targets and functions may be similar across diverse animal species.

Although the *let-7* sequence is well conserved from the nematode to the human, several differences distinguish the closely related *let-7* family members of various animal species (Roush and Slack, 2008). For one, whereas the nematode and the fly have only one *let-7* miRNA, higher animals (e.g., fishes and mammals) have diverse *let-7* family members including *let-7a*, *-7b*, *-7c*, *-7d*, *-7e*, *-7f*, *-7g*, *-7h*, *-7i*, *-7j*, *-7k* (see below for a discussion of this nomenclature) and *miR-98* (Table 1) (Lagos-Quintana et al., 2001; Lau et al., 2001; Chen et al., 2005; Landgraf et al., 2007). Higher animals have generally similar sets of *let-7* family members, although slight differences may be observed (for example, *let-7h* exists in the zebrafish but not in the human). Notably, each *let-7* family member is often present in multiple copies across the genomes of higher animals (Table 1). To distinguish between the various isoforms, a letter and/or number are placed after the term '*let-7*'. Sequence differences are indicated by letters (e.g., *let-7a* and *-7b*), while different

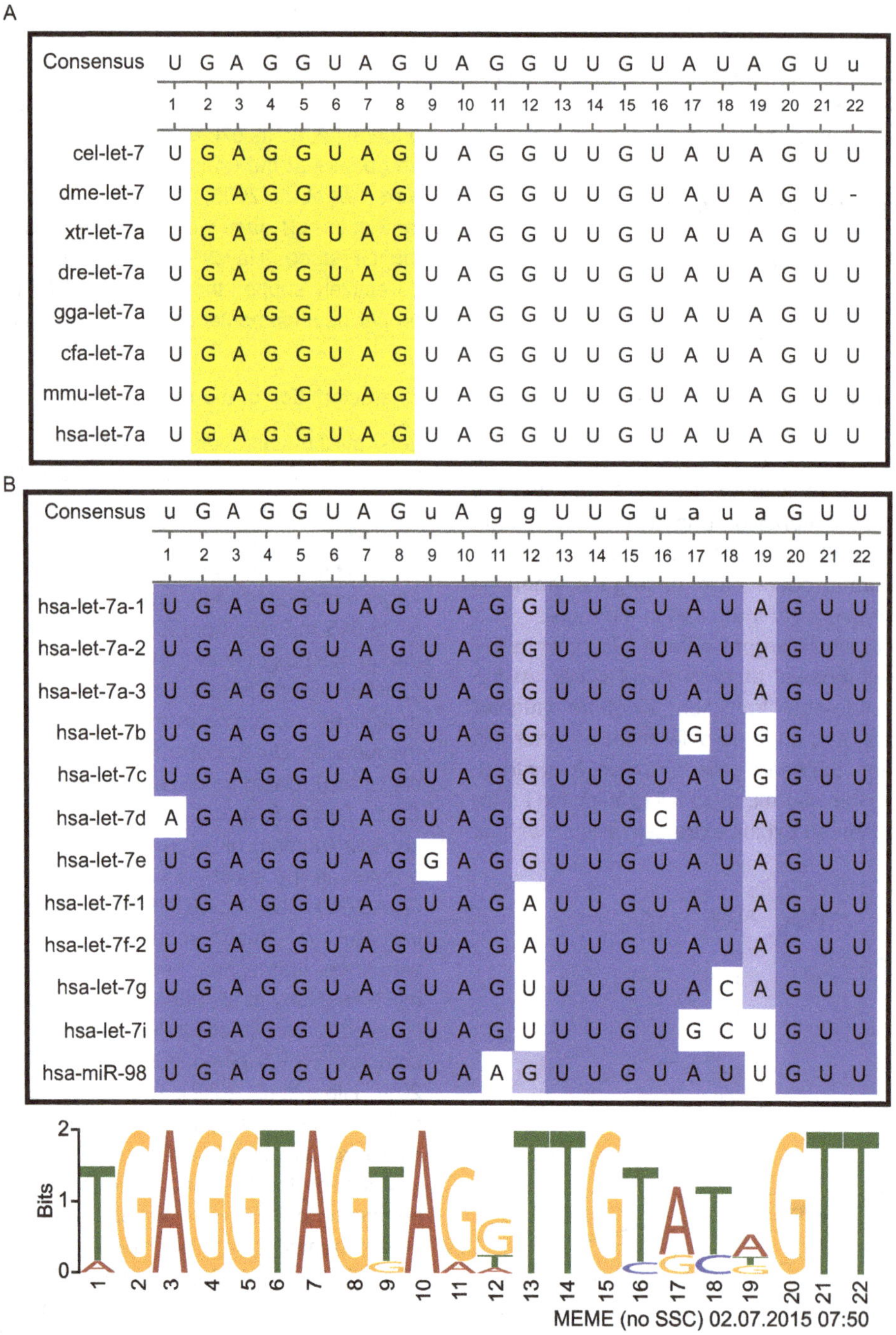

Figure 2. Sequence comparison of *let-7* family members across diverse animal species. (A) *C. elegans* (cel), *D. melanogaster* (dme), *X. tropicalis* (xtr), *D. rerio* (dre), *G. gallus* (gga), *C. familiaris* (cfa), *M. musculus* (mmu), and *H. sapiens* (hsa) all possess the consensus mature *let-7* (*let-7a*) sequence of 'UGAGGUAGUAGGUUGUAUAGUU'. The seed sequence is indicated as a yellow box. Consensus mature sequences are placed at the top of the box, where only perfectly aligned sequences are capitalized. (B) Sequence alignment of the mature forms of human *let-7* family members (upper panel). Dark blue box represents percentage identity over 70%, whereas light blue box indicates percentage of over 50%. Consensus mature sequences are placed at the top of the box, where only perfectly aligned sequences are capitalized. Consensus sequences of the mature human *let-7* family members, as assessed by MEME (http://meme-suite.org, bottom panel).

genomic loci expressing the same sequence are indicated by numbers. As an example of the latter, the precursors (also known as the stem-loop sequence in miRBase) of human *let-7a-1*, *let-7a-2*, and *let-7a-3* are encoded on chromosomes 9, 11, and 12, respectively, but all produce the same *let-7a* miRNA (Fig. 2B and Table 1). Thus, the numbers of

Table 1. Characterization of *let-7* family across different species

Mature let-7	*C. elegans* nematode	*D. melanogaster* fly	*D. rerio* zebrafish	*M. musculus* mouse	*H. sapiens* human
let-7a	*let-7*	*let-7*	*let-7a-1, 2, 3, 4, 5, 6*	*let-7a-1, 2*	*let-7a-1, 2, 3*
let-7b			*let-7b*	*let-7b*	*let-7b*
let-7c			*let-7c-1, 2*	*let-7c-1, 2*	*let-7c*
let-7d			*let-7d-1, 2*	*let-7d*	*let-7d*
let-7e			*let-7e*	*let-7e*	*let-7e*
let-7f			*let-7f*	*let-7f-1, 2*	*let-7f-1, 2*
let-7g			*let-7g-1, 2*	*let-7g*	*let-7g*
let-7h			*let-7h*		
let-7i			*let-7i*	*let-7i*	*let-7i*
let-7j			*let-7j*		
let-7k					
miR-98				*miR-98*	*miR-98*

Mature *let-7* family members (*let-7a*, *-7b*, *-7c*, *-7d*, *-7e*, *-7f*, *-7g*, *-7h*, *-7i*, *-7j*, *-7k*, and *miR-98*) and their corresponding precursors in *Caenorhabditis elegans*, *Drosophila melanogaster*, *Danio rerio*, *Mus musculus*, and *Homo sapiens* are presented.

Table 2. Genomic location and conserved clusters of *let-7* family in human and fly

let-7 Family	Genome context	Clusters
H. sapiens		
hsa-let-7a-2	chr11: 122146522-122146593 –	Cluster1-a (*let-7a-2*, *miR-100*, *miR-125b-1*)
hsa-let-7c	chr21: 16539828-16539911 +	Cluster1-b (*let-7c*, *miR-99a*, *miR-125b-2*)
hsa-let-7e	chr19: 51692786-51692864 +	Cluster1-c (*let-7e*, *miR-99b*, *miR-125a*)
hsa-let-7a-1	chr9: 94175957-94176036 +	Cluster2 (*let-7a-1*, *-7d*, *-7f-1*)
hsa-let-7d	chr9: 94178834-94178920 +	
hsa-let-7f-1	chr9: 94176347-94176433 +	
hsa-let-7a-3	chr22: 46112749-46112822 +	Cluster3 (*let-7a-3*, *-7b*)
hsa-let-7b	chr22: 46113686-46113768 +	
hsa-let-7f-2	chrX: 53557192-53557274 –	Cluster4 (*let-7f-2*, *miR-98*)
hsa-miR-98	chrX: 53556223-53556341 –	
hsa-let-7g	chr3: 52268278-52268361 –	
hsa-let-7i	chr12: 62603686-62603769 +	
D. melanogaster		
dme-let-7a-2	2L: 18472034-18472111 +	Cluster1 (*let-7*, *miR-100*, *miR-125*)

Precursors of human and fly *let-7* family can be located individually (*let-7g*, *-7i*) or as clusters (cluster 1 to 4). Genomic location and four clusters of these precursors are described.

precursor sequences encoded in the genome of a given species may differ from the number of mature miRNAs expressed in that species. In the human, for example, 12 distinct loci encode nine mature *let-7* miRNAs (Fig. 2B and Table 2).

In animal genomes, the *let-7* family members can be encoded individually or as clusters with other family members and/or unrelated miRNAs. Comparison of *let-7* family members in *D. melanogaster* and higher animals has revealed that such sequences tend to show similar genomic positions, suggesting that they form well-preserved clusters (Lagos-Quintana et al., 2001; Bashirullah et al., 2003; Sempere et al., 2003). In the human, *let-7g* and *let-7i* are located individually on chromosomes 3 and 12, respectively, while the other *let-7* family members are distributed among four clusters (clusters 1 to 4) (Table 2). Cluster 1, which contains three miRNAs, including *let-7a*, *miR-100*, and *miR-125*, is also conserved in *D. melanogaster* (Table 2).

Cluster 1 can be further sub-classified into three clusters (cluster 1-a, 1-b, and 1-c) by its location and components. Interestingly, cluster 1-a and cluster 1-b are involved in hematopoietic stem and progenitor cell (HSPC) homeostasis by regulating the balance between TGFβ and Wnt signaling (Emmrich et al., 2014), whereas cluster 1-c is highly expressed in HSPC and confers hematopoietic phenotypes (Gerrits et al., 2012). However, *miR-125a* is responsible for most of these properties in cluster 1-c and the transcription of miRNAs in cluster 1-a (*let-7a-2*, *miR-100*, and *miR-125b-1*) are loosely related (Sempere et al., 2004; Gerrits et al., 2012). Cluster 2 contains *let-7a*, *-7d*, and *-7f-1*, whereas cluster 3 is composed of *let-7a-3* and *-7b*. Lastly, cluster 4 is consisted of *let-7f-2* and *miR-98* (Table 2). Vertebrate-specific genomic duplication events are thought to be responsible for the formation of these clusters (Hertel et al., 2012), which may confer proper regulation and correct biogenesis of the involved miRNAs.

Biological roles of *let-7* family members

The high degree of conservation among *let-7* miRNAs across different animal species suggests that they may play important (and potentially similar) roles in the biological processes of various organisms (Pasquinelli et al., 2000; Hertel et al., 2012). Indeed, recent studies have shown that *let-7* family members generally promote differentiation during development and function as tumor suppressors in various cancers (Reinhart et al., 2000; Takamizawa et al., 2004; Grosshans et al., 2005; Johnson et al., 2005; Yu et al., 2007; Caygill and Johnston, 2008; Kumar et al., 2008).

In *C. elegans*, *let-7* controls the crucial developmental timing of the last larval transition (L4-to-adult) via regulation of transcription factors (*daf-12*, *pha-4*, *die1*, and *lss4*) in different tissues (Fig. 1) (Reinhart et al., 2000; Grosshans et al., 2005). *let-7* has also been shown to function as a heterochronic gene in *D. melanogaster* (Caygill and Johnston, 2008; Sokol et al., 2008), wherein *let-7* mutants show abnormal (delayed) cell cycle exit in the wing (Caygill and Johnston, 2008) and an irregular maturation of neuromuscular junctions in the adult abdominal muscles that results in immaturity of the neuromusculature and defects in adult fertility, motility, and flight (Sokol et al., 2008). Consistent with this mutant phenotype, *let-7* expression in *D. melanogaster* gradually increases during the third larval instar stage and peaks in the pupa (Pasquinelli et al., 2000; Bashirullah et al., 2003). Thus, the *let-7* miRNAs of *C. elegans* and *D. melanogaster* both act as essential regulators for proper development at the larva-to-adult transition. In chicken and mice, *let-7* is involved in limb development (Mansfield et al., 2004; Lancman et al., 2005; Schulman et al., 2005).

In mammals, *let-7* expression is high during embryogenesis and brain development (Thomson et al., 2004; Schulman et al., 2005; Thomson et al., 2006; Wulczyn et al., 2007) and remains high in adult tissues (Sempere et al., 2004; Thomson et al., 2004). Moreover, *let-7* is known to regulate hematopoietic stem cell fate along with *miR-99a/100*, *miR-125b-1/2*, and LIN28B (Copley et al., 2013; Lee et al., 2013b; Emmrich et al., 2014). Cluster1-a (*let-7a-2*, *miR-100*, *miR-125b-1*) and Cluster1-b (*let-7c*, *miR-99a*, *miR-125b-2*) are involved in HSPC (hematopoietic stem and progenitor cell) homeostasis such as self-renewal, proliferation, quiescence, and differentiation by blocking TGFβ pathway and amplifying Wnt signaling (Emmrich et al., 2014), whereas LIN28B represses *let-7* to inhibit erythroid development and maintain stemness (Copley et al., 2013; Lee et al., 2013b). However, the exact role of *let-7* family members in mammalian development has not yet been fully elucidated (Lancman et al., 2005; Schulman et al., 2005; Wulczyn et al., 2007), in large part because it is technically difficult to knock out multiple *let-7* family members in the same individual. Moreover, these multiple *let-7* family members are likely to have functionally redundant roles.

With respect to the function of *let-7* as tumor suppressor, the targets of *C. elegans let-7* were initially predicted using computational analysis, and the 3′ UTR of *let-60* [also known as an ortholog of the RAS (human Rat sarcoma) oncogene] was identified as having the highest identified sequence complementarity to *let-7* (Johnson et al., 2005). Subsequently, *let-7* was shown to interact with *let-60* and RAS in *C. elegans* and human cancers, respectively (Johnson et al., 2005). Moreover, up-regulation of RAS was found to require down-regulation of *let-7* in lung cancer and non-small cell lung cancer (NSCLC) (Takamizawa et al., 2004; Johnson et al., 2005; Kumar et al., 2008), and *let-7g* was shown to block tumorigenesis by suppressing RAS in NSCLC (Kumar et al., 2008). In addition to the role of *let-7* in modulating the RAS oncogene, multiple *let-7* members were found to be down-regulated in human cancers and cancer stem cells, strengthening the notion that *let-7* may also function as a tumor suppressor (Takamizawa et al., 2004; Shell et al., 2007; Yu et al., 2007; Dahiya et al., 2008; O'Hara et al., 2009). Several other lines of evidence strongly suggest that *let-7* functions as tumor suppressor in general. For example, *let-7* family members have been shown to repress cell cycle regulators (e.g., cyclin A, cyclin D1, cyclin D3, and CDK4) and block cell cycle progression and anchorage-independent growth in cancer cells (Johnson et al., 2007; Schultz et al., 2008). Additionally, *let-7a* reportedly inhibits MYC-induced cell growth in Burkitt lymphoma cells by blocking MYC expression (Sampson et al., 2007). Moreover, HuR, RNA-binding protein, binds and represses MYC mRNA by recruiting the *let-7*/RISC complex to 3′ UTR region of MYC (Ma et al., 1996; Kim et al., 2009). In addition, recruitment of HuR and *let-7* to the transcript of MYC is interdependent (Kim et al., 2009; Gunzburg et al., 2015). Interestingly, MYC can also negatively regulate *let-7* family members such as *let-7a*, *-7d*, and *-7g* by binding to their promoters, thus, forming a negative-feedback loop (Chang et al., 2008; Wang et al., 2011).

The involvement of *let-7* miRNA in stem cell regulation also provided a clue as to how *let-7* may function as a tumor suppressor. *let-7* was shown to regulate the expression of high-mobility group AT-hook 2 (HMGA2), which is an early embryonic oncofetal gene that is overexpressed in stem cells and contributes to their self-renewal (Yu et al., 2007; Nishino et al., 2008). Thus, one of the mechanisms of maintaining undifferentiated state in stem cells is upregulation of HMGA2 by maintaining the low level of *let-7* miRNA. During differentiation, increased expression of *let-7* down-regulates HMGA2 by interacting with its 3′ UTR (Yu et al., 2007; Boyerinas et al., 2008; Nishino et al., 2008). The inverse relationship between the expression levels of *let-7* and HMGA2 was further supported by recent studies demonstrating that ectopic *let-7* expression can inhibit cell growth and mammosphere formation by down-regulating RAS and HMGA2 in mouse breast cancers (Sempere et al., 2007; Yu et al., 2007). Together, these lines of evidence strongly suggest that the *let-7* family members act as crucial tumor suppressors that inhibit diverse oncogenes.

In summary, two major biological roles have been elucidated for the *let-7* miRNA: as an essential regulator of terminal differentiation, and as a fundamental tumor suppressor. It thus seems that *let-7* should be expressed at specific stages of terminal differentiation, but down-regulated in stem cells being maintained in their undifferentiated state.

PATHWAYS OF MIRNA BIOGENESIS

Canonical miRNA biogenesis pathway

The canonical miRNA biogenesis pathway is dependent on two microprocessors: Drosha and Dicer (Fig. 3). RNA polymerase II produces a primary miRNA transcript with a 5′ cap and a 3′ poly(A) tail from the encoding genomic locus (Bracht et al., 2004). Internal base-pairing within the primary miRNA (pri-miRNA) forms a characteristic hairpin stem-loop structure with a stem of ~33 bp in length. The pri-miRNA is subsequently processed by a microprocessor complex composed of the RNase III enzyme, Drosha, and the double-stranded RNA binding protein, DiGeorge syndrome critical region 8 (DGCR8; also known as Pasha), which cleaves the stem-loop structure into a 60–70-nt-long pre-miRNA that has a two-nt-long 3′ overhang (Lee et al., 2003; Denli et al., 2004; Gregory et al., 2004; Landthaler et al., 2004). The Drosha/DGCR8 microprocessor is a heterotrimeric complex consisting of one Drosha and two DGCR8 proteins. Following its processing by this Drosha/DGCR8 complex, the pre-miRNA is exported from the nucleus to the cytoplasm by the Ran-GTP-dependent transporter, exportin 5 (EXP5). When the pre-miRNA/EXP5/Ran-GTP complex is exported to the cytoplasm through the nuclear pore complex, GTP is hydrolyzed and the pre-miRNA subsequently dissociates (Yi et al., 2003; Bohnsack et al., 2004; Lund et al., 2004).

Following its transport into the cytoplasm, the pre-miRNA is further processed by Dicer into an RNA duplex of ~22 bp (Bernstein et al., 2001; Grishok et al., 2001; Hutvagner et al., 2001; Ketting et al., 2001; Knight and Bass, 2001). Dicer cleaves the pre-miRNA at a fixed length away from the base of the stem-loop, removing the loop to produce the 22-bp RNA duplex (Zhang et al., 2002; Zhang et al., 2004; Vermeulen et al., 2005; Macrae et al., 2006; MacRae et al., 2007; Park et al., 2011). Dicer may act together with trans-activation response RNA-binding protein (TRBP) or protein activator of PKR (PACT; also known as PRKRA) in mammals (Lee et al., 2006; Lee et al., 2013a). These cofactors are dsRNA-binding proteins that have differential preferences for siRNA and miRNA. TRBP recruits Argonaute (AGO); however, the exact role of TRBP and PACT in miRNA biogenesis have not yet been fully elucidated.

One strand of the small dsRNA processed by Dicer, called a guide strand, is loaded onto an AGO protein to form RISC, which recognizes a target sequence that is usually embedded within the 3′ UTR region of a target mRNA in the P-body (Gregory et al., 2005; Liu et al., 2005; Eulalio et al., 2007). The *Drosophila* expresses two AGO proteins: AGO1, which preferentially associates with miRNAs, and AGO2, which binds to siRNAs (Okamura et al., 2004). The human has four AGO proteins; all of them have affinities for both siRNAs and miRNAs, and there does not appear to be any sorting mechanism to distinguish between siRNAs and miRNAs (Liu et al., 2004; Meister et al., 2004; Azuma-Mukai et al., 2008; Su et al., 2009; Dueck et al., 2012). RISC-incorporated mature miRNAs can block gene expression via a post-transcriptional mechanism, such as by inhibiting translation or facilitating mRNA degradation (Eulalio et al., 2008; Filipowicz et al., 2008).

Although *let-7* maturation generally follows the canonical miRNA biogenesis pathway, some family members require an additional step. Three members of the *let-7* family (*pre-let-7a-2*, -7c, and -7*e*) carry the typical two-nucleotide 3′ overhang in their precursors (group I pre-miRNAs), while the rest possess one-nucleotide 3′ overhang (group II pre-miRNAs) (Heo et al., 2012). The group II *pri-let-7* precursors have a bulged adenosine (*pri-let-7d*) or uridine (all other members of the group) next to the processing site (Heo et al., 2012). Drosha may fail to recognize this uridine/adenosine bulge, resulting in the generation of a one-nucleotide 3′ overhang. Due to this structural difference, an additional step is required to ensure efficient Dicer activity during biogenesis (Heo et al., 2012). In this step, terminal uridylyl transferases (TUT2/PAPD4/GLD2, TUT4/ZCCHC11, and TUT7/ZCCHC6) specifically mono-uridylate the 3′ end of the group II *pre-let-7*s, yielding the two-nucleotide 3′ overhang preferred by Dicer (Heo et al., 2012).

The non-canonical miRNA biogenesis pathway

Although *let-7* family is generated through canonical miRNA biogenesis pathway, it would be helpful to understand the *let-7* biogenesis when comparing with the non-canonical miRNA biogenesis. The non-canonical miRNA pathways are well summarized in recent reviews (Ameres and Zamore, 2013; Ha and Kim, 2014). While the canonical miRNA

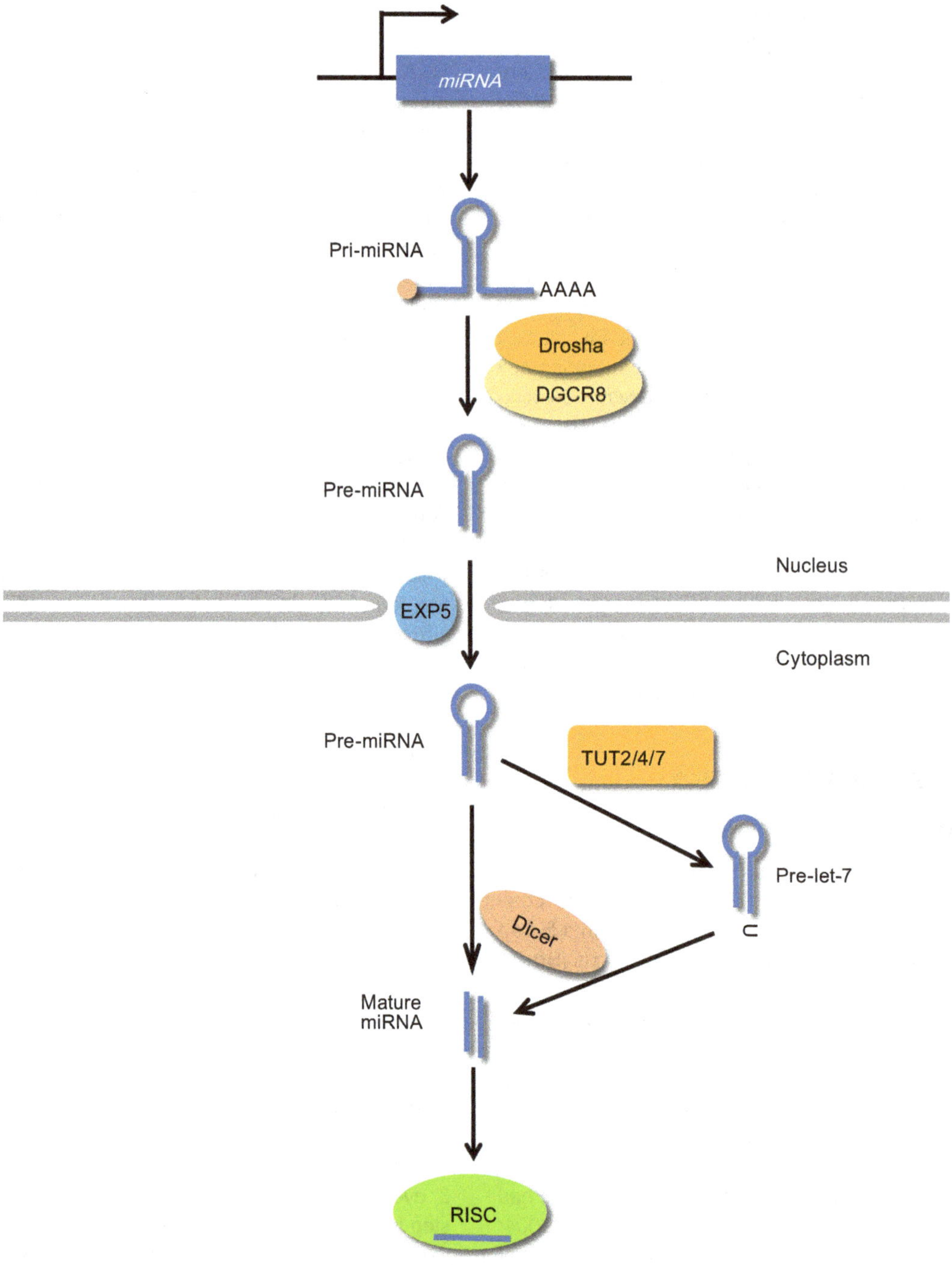

Figure 3. Canonical pathway of miRNA biogenesis. Schematic diagram of the canonical miRNA biogenesis process. A primary miRNA transcript produced by RNA polymerase II is processed by the Drosha microprocessor in the nucleus. The generated pre-miRNA is transported to the cytoplasm in an EXP5-Ran-GTP-dependent manner and further processed by the Dicer microprocessor to generate a mature miRNA. *Pre-let-7* is mono-uridylated at the 3′ end by LIN28A and TUTases prior to Dicer-mediated processing. The mature miRNA is loaded onto RISC to inhibit the translation of a target mRNA.

biogenesis pathway depends on Drosha and Dicer, a small subset of miRNAs is processed independent of Drosha or Dicer. The biogenesis of mirtron, which was the first non-canonical biogenesis pathway to be discovered, is a Drosha-independent pathway (Berezikov et al., 2007; Okamura et al., 2007; Ruby et al., 2007). Mirtrons are encoded in an intronic region, such that the precursor is generated through an mRNA splicing mechanism that does not require Drosha. After splicing, lariat debranching and refolding converts the lariat to a pre-miRNA-like structure that is then subjected to Dicer cleavage (Okamura et al., 2007; Ruby et al., 2007). Mirtrons that contain additional sequences at their 5′ or 3′

ends are further trimmed by an exonuclease (Flynt et al., 2010). After trimming, the mirtrons can be processed by Dicer in a manner similar to that seen in the canonical miRNA pathway.

A Dicer-independent biogenesis pathway was also recently reported for a miRNA, as the maturation of *miR-451* was shown to require Drosha but not Dicer (Cheloufi et al., 2010; Cifuentes et al., 2010; Yang et al., 2010). Drosha-dependently processed *pre-miR-451* has a stem of only ~18 bp, which is too short for Dicer-mediated cleavage. Instead, *pre-miR-451* is directly loaded onto RISC, where AGO2-dependent cleavage generates *ac-pre-miR-451* (AGO-cleaved *pre-miR-451*) (Cheloufi et al., 2010; Cifuentes et al., 2010; Yang et al., 2010). Thereafter, the poly(A)-specific ribonuclease (PARN) further trims the 3′ end of *ac-pre-miR-451* to generate mature versions of *miR-451* harboring divergent 3′ ends (Yoda et al., 2013).

REGULATION OF *let-7* BIOGENESIS

Dysregulation of *let-7* family members leads to abnormal physiological processes. The *let-7* mutant is lethal in the nematode (Reinhart et al., 2000), and decreased *let-7* expression or genomic deletion has been detected in several human cancer types (Takamizawa et al., 2004; Dahiya et al., 2008; O'Hara et al., 2009). In addition, while the mature *let-7* miRNA is not detected, *pri-let-7* exists in some cell types including mESCs (Suh et al., 2004; Thomson et al., 2006; Wulczyn et al., 2007). The observation that *let-7* expression gradually increases during development suggests that *let-7* biogenesis may be tightly regulated by additional factors (Pasquinelli et al., 2000; Sempere et al., 2002; Thomson et al., 2006; Liu et al., 2007). To date, several transcriptional and post-transcriptional mechanisms have been proposed as regulators of *let-7* biogenesis.

Transcriptional regulation of *let-7*

C. elegans harbors a feedback circuit between *let-7* and the nuclear hormone receptor, DAF-12, in that DAF-12 is a target of *let-7*, but also regulates the transcription of *let-7* in a ligand-dependent manner. In an unfavorable environment, ligand-unbound DAF-12 suppresses *let-7* expression with its co-repressor, DIN-1. When environmental conditions favor developmental progression, however, ligand-bound DAF-12 activates the transcription of *let-7*. This feedback loop may regulate cellular fate and developmental arrest (Bethke et al., 2009; Hammell et al., 2009). Interestingly, a similar feedback loop has also been demonstrated in mammals: MYC is a target of *let-7*, but it can also repress the transcription of *let-7* during MYC-mediated tumorigenesis by directly binding to the promoter and upstream region of the *let-7a-1/let-7f-1/let-7d* cluster (Chang et al., 2008; Wang et al., 2011). Consistent with this idea of a negative feedback loop, shRNA-mediated suppression of endogenous MYC was found to up-regulate *let-7* (Wang et al., 2011), whereas *let-7* expression was shown to suppress MYC expression in a Burkitt lymphoma cell line (Sampson et al., 2007). Based on this, it seems reasonable to speculate that other transcription factors may also participate in the transcriptional regulation of *let-7* family members.

Even though *let-7* is ubiquitously expressed in adult mammalian tissues (Sempere et al., 2004), expression of individual *let-7* family members is also context-dependent. For example, *let-7i* is relatively enriched in thyroid compared to the other tissues (Lee et al., 2008). In addition, a subset of *let-7* family member would be expressed in specific tissues, cell lines, and cancers (Boyerinas et al., 2010; Chiu et al., 2014). This context-dependent expression of *let-7* family members would be tightly related with the expression of LIN28A/B as well as transcription factors (Thornton and Gregory, 2012). Despite *let-7* is one of the first discovered miRNAs, the details on transcriptional regulation of *let-7* family, especially individual members of *let-7* family, are not clearly understood. For this reason, mechanistic studies of transcriptional regulation should be further determined.

Oligo-uridylation by TUTases is a marker for *pre-let-7* degradation

It has been reported that *let-7* is also post-transcriptionally regulated by additional factors. As discussed above, TUTase is essential for the processing of the group II *pre-let-7* miRNAs, which have a unique 3′ overhang (Fig. 3) (Heo et al., 2012). Interestingly, the TUTases play a second role in the degradation of *pre-let-7* through their terminal uridylation activity (Fig. 4) (Heo et al., 2008; Hagan et al., 2009; Heo et al., 2009; Thornton et al., 2012). When LIN28A is over-expressed in HEK293T cells, the 3′-terminal oligo-uridylation of *pre-let-7* yields a uridine tail of ~14 nt (Heo et al., 2008). This oligo-uridylated *pre-let-7* resists Dicer cleavage and is instead susceptible to degradation. TUT4 and TUT7 were recently shown to oligo-uridylate *pre-let-7* in embryonic stem cells and cancer cells (Hagan et al., 2009; Heo et al., 2009; Thornton et al., 2012). The machinery responsible for degrading oligo-uridylated *pre-let-7* was recently identified as the catalytic subunit of the cytoplasmic exosome, DIS3L2 (Chang et al., 2013; Malecki et al., 2013; Ustianenko et al., 2013). The activity of DIS3L2 is stimulated when the uridine tail is at least 10 nt long, and it shows maximal activity against tails of 14 nt or longer. X-ray crystallography has shown that the three RNA binding domains of DIS3L2 form an open funnel that facilitates uridine-specific interactions with the first 12 uridines of the *pre-let-7* tail. This structural feature forms the basis for the substrate specificity of DIS3L2 (Faehnle et al., 2014).

LIN28A/B negatively regulates *let-7* biogenesis

As noted above, LIN28A is required for the oligo-uridylation of *pre-let-7* by TUTases (Heo et al., 2008; Hagan et al., 2009; Heo et al., 2009; Piskounova et al., 2011; Thornton et al.,

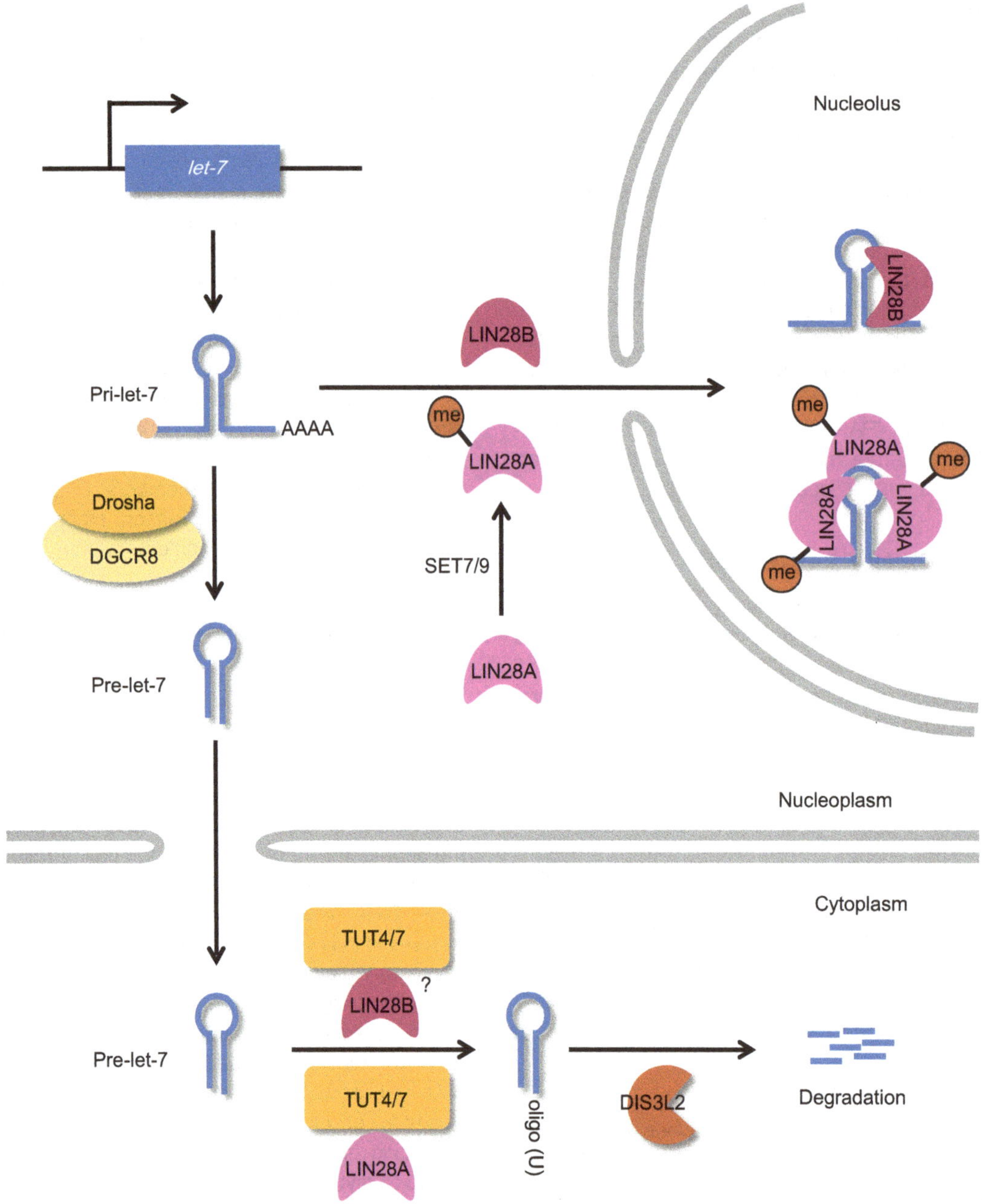

Figure 4. Regulation of *let-7* biogenesis by LIN28A/B. LIN28A and LIN28B inhibit the biogenesis of *let-7* via both TUTase-dependent and -independent pathways. LIN28A helps TUTases to oligo-uridylate *pre-let-7*. Methylated LIN28A binds to *pri-let-7* in the nucleus and sequesters it into the nucleolus to prevent Drosha-mediated processing. LIN28B blocks the biogenesis of the *let-7* miRNA via TUTase-independent pathways. The detailed relationship between LIN28B and TUTases needs to be further understood.

2012). LIN28, which was originally identified as a heterochronic gene in *C. elegans*, is evolutionarily conserved in animals. Mammals have two paralogs of LIN28, LIN28A (also known as LIN28) and LIN28B, which can bind to both *pri-* and *pre-let-7* to block the activities of Drosha and Dicer (Fig. 4) (Heo et al., 2008; Newman et al., 2008; Rybak et al., 2008; Viswanathan et al., 2008). LIN28A and LIN28B each have two RNA-binding domains, a cold-shock domain and a zinc finger motif (Moss and Tang, 2003). Through its RNA-binding activity, LIN28A associates with the bulging GGAG motif in the terminal loop of *pre-let-7* and recruits TUT4/7 (Nam et al., 2011). The terminal loop of *pre-let-7* has three independent binding sites for LIN28A, which can be multiply assembled in a stepwise fashion (Desjardins et al., 2014). This multimerization of LIN28A is likely to be required for the efficient blockade of Dicer-dependent *pre-let-7* processing. LIN28A reportedly competes with Dicer for *pre-let-7* and blocks processing of the precursor (Rybak et al., 2008); in the absence of LIN28A, *pre-let-7* is mono-uridylated by TUT2/4/7 and further processed by Dicer to generate the mature *let-7* (Heo et al., 2012). Thus, LIN28A blocks the Dicer activity in the cytoplasm, which is a TUTase-dependent pathway.

Interestingly, LIN28A also blocks Drosha-mediated processing in the nucleus (Newman et al., 2008; Viswanathan et al., 2008). Purified LIN28A inhibits *pri-let-7* processing *in vitro* and its ectopic expression selectively blocks *pri-let-7* processing *in vivo* (Newman et al., 2008; Viswanathan et al., 2008). In addition, *pri-let-7* processing is rescued by knockdown of LIN28A in mouse embryonal carcinoma (Viswanathan et al., 2008). Thus, although it is not yet clear whether LIN28A directly inhibits Drosha activity, it appears to negatively regulate *let-7* biogenesis in the nucleus as well as in the cytoplasm. LIN28A is mainly localized in the cytoplasm, but it can enter the nucleus and shows affinity for both *pri-* and *pre-let-7* (Heo et al., 2008; Newman et al., 2008; Rybak et al., 2008; Viswanathan et al., 2008). These lines of evidence suggest that LIN28A might participate in multiple steps of *let-7* biogenesis, including both Dicer- and Drosha-mediated processing.

LIN28B has also been shown to inhibit *let-7* biogenesis (Fig. 4), but the similar functions of LIN28A and LIN28B are achieved through very different action mechanisms (Piskounova et al., 2011). LIN28B was originally reported to have no affinity for TUTases, and the expressions of LIN28A and LIN28B appear to be mutually exclusive (Piskounova et al., 2011). In addition, LIN28B has a NoLS (nucleolar-localization sequence), and thus could be localized in the nucleolus. LIN28B appears to directly bind to *pri-let-7* in the nucleus and sequester it to the nucleolus, which lacks Drosha, thereby suppressing *let-7* maturation via a TUTase-independent pathway. Interestingly, however, a recent study showed that LIN28B interacts with DIS3L2 in the cytoplasm of LIN28B-expressing cancer cell lines, indicating that it also participates in the TUTase-dependent pathway (Suzuki et al., 2015). In this context, the level of *pre-let-7* appears to influence the subcellular localization of LIN28B (Suzuki et al., 2015).

Post-translational modification changes the action mode of LIN28A

It was recently shown that LIN28A can prevent the biogenesis of *let-7* independent of TUT4/7 in hESCs, in a manner similar to that seen for LIN28B (Fig. 4) (Kim et al., 2014). The histone H3K4 methyltransferase, SET7/9, can monomethylate LIN28A at lysine 135, which is near a sequence that is homologous to the NoLS of LIN28B (Kim et al., 2014). This sequence might be required for the nuclear (and especially nucleolar) localization of methylated LIN28A, which is its nuclear form. Electrophoretic mobility shift assays (EMSAs) have shown that the nuclear form of LIN28A binds to *pri-let-7* in a stepwise manner similar to its multimerization with *pre-let-7* (Desjardins et al., 2014; Kim et al., 2014). In addition, methylated LIN28A has a higher binding affinity for *pri-let-7* compared to cytoplasmic unmethylated LIN28A, whereas the affinity for *pre-let-7* does not differ between the two (Kim et al., 2014). Thus, it appears that LIN28A may regulate *pri-let-7* processing in a TUTase-independent fashion in the nucleus as well as a TUTase-dependent pathway in the cytoplasm. Moreover, the SET7/9-mediated post-translational modification (methylation) appears to act as a switch that changes the action mode of LIN28A in the inhibition of *let-7* biogenesis.

SUMMARY AND PERSPECTIVES

In this review, we provide an overview of the features and biological roles of the *let-7* family members in higher eukaryotes. As *let-7* is induced during development and represses the expression of pluripotency factors, its biogenesis must be precisely regulated. In general, the *let-7* miRNA is generated through the canonical miRNA biogenesis pathway, which involves Drosha- and Dicer-dependent processing and is supported by TUTases. In the presence of LIN28A/B, TUTases instead inhibit *pre-let-7* processing by oligo-uridylation via LIN28A/B-mediated targeting. LIN28A/B proteins also regulate *let-7* biogenesis via TUTase-independent pathways. In the case of LIN28A, methylation seems to act as a switch, changing both its subcellular localization and its action mechanism. Although the expressions of LIN28A and LIN28B are mutually exclusive and these proteins play somewhat different inhibitory roles in *let-7* biogenesis, recent results suggest that they might share the consensus of their molecular mechanism. Indeed, compensatory redundancy between LIN28A and LIN28B has been observed (Wilbert et al., 2012).

At present, the detailed molecular mechanisms underlying *let-7* miRNA biogenesis are not fully understood. For instance, we do not yet know what happens to *pri-let-7* following its sequestration into the nucleolus by methylated LIN28A or LIN28B. The details of the relationship between DIS3L2-related cytoplasmic exosomes and *let-7* biogenesis are also unknown. Indeed, DIS3, other catalytic subunit of cytoplasmic exosome, also indirectly regulates the expression of *let-7* through degradation of LIN28B mRNAs in several mammalian cancer cell lines (Segalla et al., 2015). Emerging evidence suggests that the activities of the regulatory machineries are likely to be fine-tuned by post-translational modifications. In fact, the deacetylation of DGCR8 by HDAC1 was shown to increase the affinity for pri-miRNAs (Wada et al., 2012). Further studies examining the molecular mechanisms of *let-7* biogenesis and its regulation by nuclear/nucleolar and cytoplasmic factors should provide new insights into the biological roles of the *let-7* family members. Ultimately, detailed mechanistic studies for *let-7* biogenesis and its regulation involved in the developmental timing, cell division and differentiation in animals should be elucidated.

ACKNOWLEDGEMENTS

This work was supported by grants from the Stem Cell Research Program (2012M 3A9B 4027953) and the KAIST Future Systems Healthcare Project funded by the Ministry of Science, ICT and Future Planning.

ABBREVIATIONS

EMSAs, electrophoretic mobility shift assays; HMGA2, high-mobility group AT-hook 2; HSPC, hematopoietic stem and progenitor cell; *Let-7*, *lethal-7*; miRNAs, microRNAs; NSCLC, non-small cell lung cancer; PARN, poly(A)-specific ribonuclease; *pre-let-7*, *let-7* precursor; RISC, RNA-induced silencing complex.

COMPLIANCE WITH ETHICS GUIDELINES

Hosuk Lee, Sungwook Han, Chang Seob Kwon and Daeyoup Lee declare that they have no conflict of interest. This article does not contain any studies with human or animal subjects performed by the any of the authors.

REFERENCES

Abbott AL, Alvarez-Saavedra E, Miska EA, Lau NC, Bartel DP, Horvitz HR, Ambros V (2005) The let-7 MicroRNA family members mir-48, mir-84, and mir-241 function together to regulate developmental timing in *Caenorhabditis elegans*. Dev Cell 9:403–414

Ameres SL, Zamore PD (2013) Diversifying microRNA sequence and function. Nat Rev Mol Cell Biol 14:475–488

Azuma-Mukai A, Oguri H, Mituyama T, Qian ZR, Asai K, Siomi H, Siomi MC (2008) Characterization of endogenous human Argonautes and their miRNA partners in RNA silencing. Proc Natl Acad Sci USA 105:7964–7969

Bashirullah A, Pasquinelli AE, Kiger AA, Perrimon N, Ruvkun G, Thummel CS (2003) Coordinate regulation of small temporal RNAs at the onset of *Drosophila* metamorphosis. Dev Biol 259:1–8

Berezikov E, Chung WJ, Willis J, Cuppen E, Lai EC (2007) Mammalian mirtron genes. Mol Cell 28:328–336

Bernstein E, Caudy AA, Hammond SM, Hannon GJ (2001) Role for a bidentate ribonuclease in the initiation step of RNA interference. Nature 409:363–366

Bethke A, Fielenbach N, Wang Z, Mangelsdorf DJ, Antebi A (2009) Nuclear hormone receptor regulation of microRNAs controls developmental progression. Science 324:95–98

Bohnsack MT, Czaplinski K, Gorlich D (2004) Exportin 5 is a RanGTP-dependent dsRNA-binding protein that mediates nuclear export of pre-miRNAs. RNA 10:185–191

Boyerinas B, Park SM, Shomron N, Hedegaard MM, Vinther J, Andersen JS, Feig C, Xu J, Burge CB, Peter ME (2008) Identification of let-7-regulated oncofetal genes. Cancer Res 68:2587–2591

Boyerinas B, Park SM, Hau A, Murmann AE, Peter ME (2010) The role of let-7 in cell differentiation and cancer. Endocr Relat Cancer 17:F19–F36

Bracht J, Hunter S, Eachus R, Weeks P, Pasquinelli AE (2004) Trans-splicing and polyadenylation of let-7 microRNA primary transcripts. RNA 10:1586–1594

Brennecke J, Stark A, Russell RB, Cohen SM (2005) Principles of microRNA-target recognition. PLoS Biol 3:e85

Caygill EE, Johnston LA (2008) Temporal regulation of metamorphic processes in *Drosophila* by the let-7 and miR-125 heterochronic microRNAs. Curr Biol 18:943–950

Chang TC, Yu D, Lee YS, Wentzel EA, Arking DE, West KM, Dang CV, Thomas-Tikhonenko A, Mendell JT (2008) Widespread microRNA repression by Myc contributes to tumorigenesis. Nat Genet 40:43–50

Chang HM, Triboulet R, Thornton JE, Gregory RI (2013) A role for the Perlman syndrome exonuclease Dis3l2 in the Lin28-let-7 pathway. Nature 497:244–248

Cheloufi S, Dos Santos CO, Chong MM, Hannon GJ (2010) A dicer-independent miRNA biogenesis pathway that requires Ago catalysis. Nature 465:584–589

Chen PY, Manninga H, Slanchev K, Chien M, Russo JJ, Ju J, Sheridan R, John B, Marks DS, Gaidatzis D et al (2005) The developmental miRNA profiles of zebrafish as determined by small RNA cloning. Genes Dev 19:1288–1293

Chiu SC, Chung HY, Cho DY, Chan TM, Liu MC, Huang HM, Li TY, Lin JY, Chou PC, Fu RH et al (2014) Therapeutic potential of microRNA let-7: tumor suppression or impeding normal stemness. Cell Transplant 23:459–469

Cifuentes D, Xue H, Taylor DW, Patnode H, Mishima Y, Cheloufi S, Ma E, Mane S, Hannon GJ, Lawson ND et al (2010) A novel miRNA processing pathway independent of Dicer requires Argonaute2 catalytic activity. Science 328:1694–1698

Copley MR, Babovic S, Benz C, Knapp DJ, Beer PA, Kent DG, Wohrer S, Treloar DQ, Day C, Rowe K et al (2013) The Lin28b-let-7-Hmga2 axis determines the higher self-renewal potential of fetal haematopoietic stem cells. Nat Cell Biol 15:916–925

Dahiya N, Sherman-Baust CA, Wang TL, Davidson B, Shih le M, Zhang Y, Wood W 3rd, Becker KG, Morin PJ (2008) MicroRNA expression and identification of putative miRNA targets in ovarian cancer. PLoS One 3:e2436

Denli AM, Tops BB, Plasterk RH, Ketting RF, Hannon GJ (2004) Processing of primary microRNAs by the microprocessor complex. Nature 432:231–235

Desjardins A, Bouvette J, Legault P (2014) Stepwise assembly of multiple Lin28 proteins on the terminal loop of let-7 miRNA precursors. Nucleic Acids Res 42:4615–4628

Dueck A, Ziegler C, Eichner A, Berezikov E, Meister G (2012) microRNAs associated with the different human Argonaute proteins. Nucleic Acids Res 40:9850–9862

Emmrich S, Rasche M, Schoning J, Reimer C, Keihani S, Maroz A, Xie Y, Li Z, Schambach A, Reinhardt D et al (2014) miR-99a/100~125b tricistrons regulate hematopoietic stem and progenitor cell homeostasis by shifting the balance between TGFbeta and Wnt signaling. Genes Dev 28:858–874

Esquela-Kerscher A, Johnson SM, Bai L, Saito K, Partridge J, Reinert KL, Slack FJ (2005) Post-embryonic expression of *C. elegans* microRNAs belonging to the lin-4 and let-7 families in the hypodermis and the reproductive system. Dev Dyn 234:868–877

Eulalio A, Behm-Ansmant I, Schweizer D, Izaurralde E (2007) P-body formation is a consequence, not the cause, of RNA-mediated gene silencing. Mol Cell Biol 27:3970–3981

Eulalio A, Huntzinger E, Izaurralde E (2008) Getting to the root of miRNA-mediated gene silencing. Cell 132:9–14

Faehnle CR, Walleshauser J, Joshua-Tor L (2014) Mechanism of Dis3l2 substrate recognition in the Lin28-let-7 pathway. Nature 514:252–256

Filipowicz W, Bhattacharyya SN, Sonenberg N (2008) Mechanisms of post-transcriptional regulation by microRNAs: are the answers in sight? Nat Rev Genet 9:102–114

Flynt AS, Greimann JC, Chung WJ, Lima CD, Lai EC (2010) MicroRNA biogenesis via splicing and exosome-mediated trimming in *Drosophila*. Mol Cell 38:900–907

Gerrits A, Walasek MA, Olthof S, Weersing E, Ritsema M, Zwart E, van Os R, Bystrykh LV, de Haan G (2012) Genetic screen identifies microRNA cluster 99b/let-7e/125a as a regulator of primitive hematopoietic cells. Blood 119:377–387

Gregory RI, Yan KP, Amuthan G, Chendrimada T, Doratotaj B, Cooch N, Shiekhattar R (2004) The microprocessor complex mediates the genesis of microRNAs. Nature 432:235–240

Gregory RI, Chendrimada TP, Cooch N, Shiekhattar R (2005) Human RISC couples microRNA biogenesis and posttranscriptional gene silencing. Cell 123:631–640

Grimson A, Farh KK, Johnston WK, Garrett-Engele P, Lim LP, Bartel DP (2007) MicroRNA targeting specificity in mammals: determinants beyond seed pairing. Mol Cell 27:91–105

Grishok A, Pasquinelli AE, Conte D, Li N, Parrish S, Ha I, Baillie DL, Fire A, Ruvkun G, Mello CC (2001) Genes and mechanisms related to RNA interference regulate expression of the small temporal RNAs that control *C. elegans* developmental timing. Cell 106:23–34

Grosshans H, Johnson T, Reinert KL, Gerstein M, Slack FJ (2005) The temporal patterning microRNA let-7 regulates several transcription factors at the larval to adult transition in *C. elegans*. Dev Cell 8:321–330

Gunzburg MJ, Sivakumaran A, Pendini NR, Yoon JH, Gorospe M, Wilce MC, Wilce JA (2015) Cooperative interplay of let-7 mimic and HuR with MYC RNA. Cell Cycle. doi:10.1080/15384101.2015.1069930

Ha M, Kim VN (2014) Regulation of microRNA biogenesis. Nat Rev Mol Cell Biol 15:509–524

Hagan JP, Piskounova E, Gregory RI (2009) Lin28 recruits the TUTase Zcchc11 to inhibit let-7 maturation in mouse embryonic stem cells. Nat Struct Mol Biol 16:1021–1025

Hammell CM, Karp X, Ambros V (2009) A feedback circuit involving let-7-family miRNAs and DAF-12 integrates environmental signals and developmental timing in *Caenorhabditis elegans*. Proc Natl Acad Sci USA 106:18668–18673

Hayes GD, Ruvkun G (2006) Misexpression of the *Caenorhabditis elegans* miRNA let-7 is sufficient to drive developmental programs. Cold Spring Harb Symp Quant Biol 71:21–27

Heo I, Joo C, Cho J, Ha M, Han J, Kim VN (2008) Lin28 mediates the terminal uridylation of let-7 precursor MicroRNA. Mol Cell 32:276–284

Heo I, Joo C, Kim YK, Ha M, Yoon MJ, Cho J, Yeom KH, Han J, Kim VN (2009) TUT4 in concert with Lin28 suppresses microRNA biogenesis through pre-microRNA uridylation. Cell 138:696–708

Heo I, Ha M, Lim J, Yoon MJ, Park JE, Kwon SC, Chang H, Kim VN (2012) Mono-uridylation of pre-microRNA as a key step in the biogenesis of group II let-7 microRNAs. Cell 151:521–532

Hertel J, Bartschat S, Wintsche A, Otto C, Stadler PF (2012) Evolution of the let-7 microRNA family. RNA Biol 9:231–241

Hibio N, Hino K, Shimizu E, Nagata Y, Ui-Tei K (2012) Stability of miRNA 5'terminal and seed regions is correlated with experimentally observed miRNA-mediated silencing efficacy. Sci Rep 2:996

Hutvagner G, McLachlan J, Pasquinelli AE, Balint E, Tuschl T, Zamore PD (2001) A cellular function for the RNA-interference enzyme Dicer in the maturation of the let-7 small temporal RNA. Science 293:834–838

Johnson SM, Grosshans H, Shingara J, Byrom M, Jarvis R, Cheng A, Labourier E, Reinert KL, Brown D, Slack FJ (2005) RAS is regulated by the let-7 microRNA family. Cell 120:635–647

Johnson CD, Esquela-Kerscher A, Stefani G, Byrom M, Kelnar K, Ovcharenko D, Wilson M, Wang X, Shelton J, Shingara J et al (2007) The let-7 microRNA represses cell proliferation pathways in human cells. Cancer Res 67:7713–7722

Ketting RF, Fischer SE, Bernstein E, Sijen T, Hannon GJ, Plasterk RH (2001) Dicer functions in RNA interference and in synthesis of small RNA involved in developmental timing in *C. elegans*. Genes Dev 15:2654–2659

Kim HH, Kuwano Y, Srikantan S, Lee EK, Martindale JL, Gorospe M (2009) HuR recruits let-7/RISC to repress c-Myc expression. Genes Dev 23:1743–1748

Kim SK, Lee H, Han K, Kim SC, Choi Y, Park SW, Bak G, Lee Y, Choi JK, Kim TK et al (2014) SET7/9 methylation of the pluripotency factor LIN28A is a nucleolar localization mechanism that blocks let-7 biogenesis in human ESCs. Cell Stem Cell 15:735–749

Knight SW, Bass BL (2001) A role for the RNase III enzyme DCR-1 in RNA interference and germ line development in *Caenorhabditis elegans*. Science 293:2269–2271

Kumar MS, Erkeland SJ, Pester RE, Chen CY, Ebert MS, Sharp PA, Jacks T (2008) Suppression of non-small cell lung tumor development by the let-7 microRNA family. Proc Natl Acad Sci USA 105:3903–3908

Lagos-Quintana M, Rauhut R, Lendeckel W, Tuschl T (2001) Identification of novel genes coding for small expressed RNAs. Science 294:853–858

Lancman JJ, Caruccio NC, Harfe BD, Pasquinelli AE, Schageman JJ, Pertsemlidis A, Fallon JF (2005) Analysis of the regulation of

lin-41 during chick and mouse limb development. Dev Dyn 234:948–960
Landgraf P, Rusu M, Sheridan R, Sewer A, Iovino N, Aravin A, Pfeffer S, Rice A, Kamphorst AO, Landthaler M et al (2007) A mammalian microRNA expression atlas based on small RNA library sequencing. Cell 129:1401–1414
Landthaler M, Yalcin A, Tuschl T (2004) The human DiGeorge syndrome critical region gene 8 and Its D. melanogaster homolog are required for miRNA biogenesis. Curr Biol 14:2162–2167
Lau NC, Lim LP, Weinstein EG, Bartel DP (2001) An abundant class of tiny RNAs with probable regulatory roles in *Caenorhabditis elegans*. Science 294:858–862
Lee Y, Ahn C, Han J, Choi H, Kim J, Yim J, Lee J, Provost P, Radmark O, Kim S et al (2003) The nuclear RNase III Drosha initiates microRNA processing. Nature 425:415–419
Lee Y, Hur I, Park SY, Kim YK, Suh MR, Kim VN (2006) The role of PACT in the RNA silencing pathway. EMBO J 25:522–532
Lee EJ, Baek M, Gusev Y, Brackett DJ, Nuovo GJ, Schmittgen TD (2008) Systematic evaluation of microRNA processing patterns in tissues, cell lines, and tumors. RNA 14:35–42
Lee HY, Zhou K, Smith AM, Noland CL, Doudna JA (2013a) Differential roles of human Dicer-binding proteins TRBP and PACT in small RNA processing. Nucleic Acids Res 41:6568–6576
Lee YT, de Vasconcellos JF, Yuan J, Byrnes C, Noh SJ, Meier ER, Kim KS, Rabel A, Kaushal M, Muljo SA et al (2013b) LIN28B-mediated expression of fetal hemoglobin and production of fetal-like erythrocytes from adult human erythroblasts ex vivo. Blood 122:1034–1041
Liu J, Carmell MA, Rivas FV, Marsden CG, Thomson JM, Song JJ, Hammond SM, Joshua-Tor L, Hannon GJ (2004) Argonaute2 is the catalytic engine of mammalian RNAi. Science 305:1437–1441
Liu J, Valencia-Sanchez MA, Hannon GJ, Parker R (2005) MicroRNA-dependent localization of targeted mRNAs to mammalian P-bodies. Nat Cell Biol 7:719–723
Liu S, Xia Q, Zhao P, Cheng T, Hong K, Xiang Z (2007) Characterization and expression patterns of let-7 microRNA in the silkworm (*Bombyx mori*). BMC Dev Biol 7:88
Lund E, Guttinger S, Calado A, Dahlberg JE, Kutay U (2004) Nuclear export of microRNA precursors. Science 303:95–98
Ma WJ, Cheng S, Campbell C, Wright A, Furneaux H (1996) Cloning and characterization of HuR, a ubiquitously expressed Elav-like protein. J Biol Chem 271:8144–8151
Macrae IJ, Zhou K, Li F, Repic A, Brooks AN, Cande WZ, Adams PD, Doudna JA (2006) Structural basis for double-stranded RNA processing by Dicer. Science 311:195–198
MacRae IJ, Zhou K, Doudna JA (2007) Structural determinants of RNA recognition and cleavage by Dicer. Nat Struct Mol Biol 14:934–940
Malecki M, Viegas SC, Carneiro T, Golik P, Dressaire C, Ferreira MG, Arraiano CM (2013) The exoribonuclease Dis3L2 defines a novel eukaryotic RNA degradation pathway. EMBO J 32:1842–1854
Mansfield JH, Harfe BD, Nissen R, Obenauer J, Srineel J, Chaudhuri A, Farzan-Kashani R, Zuker M, Pasquinelli AE, Ruvkun G et al (2004) MicroRNA-responsive 'sensor' transgenes uncover Hox-like and other developmentally regulated patterns of vertebrate microRNA expression. Nat Genet 36:1079–1083
Meister G, Landthaler M, Patkaniowska A, Dorsett Y, Teng G, Tuschl T (2004) Human Argonaute2 mediates RNA cleavage targeted by miRNAs and siRNAs. Mol Cell 15:185–197
Moss EG (2007) Heterochronic genes and the nature of developmental time. Curr Biol 17:R425–R434
Moss EG, Tang L (2003) Conservation of the heterochronic regulator Lin-28, its developmental expression and microRNA complementary sites. Dev Biol 258:432–442
Nam Y, Chen C, Gregory RI, Chou JJ, Sliz P (2011) Molecular basis for interaction of let-7 microRNAs with Lin28. Cell 147:1080–1091
Newman MA, Thomson JM, Hammond SM (2008) Lin-28 interaction with the Let-7 precursor loop mediates regulated microRNA processing. RNA 14:1539–1549
Nishino J, Kim I, Chada K, Morrison SJ (2008) Hmga2 promotes neural stem cell self-renewal in young but not old mice by reducing p16 Ink4a and p19 Arf expression. Cell 135:227–239
O'Hara AJ, Wang L, Dezube BJ, Harrington WJ Jr, Damania B, Dittmer DP (2009) Tumor suppressor microRNAs are underrepresented in primary effusion lymphoma and Kaposi sarcoma. Blood 113:5938–5941
Okamura K, Ishizuka A, Siomi H, Siomi MC (2004) Distinct roles for Argonaute proteins in small RNA-directed RNA cleavage pathways. Genes Dev 18:1655–1666
Okamura K, Hagen JW, Duan H, Tyler DM, Lai EC (2007) The mirtron pathway generates microRNA-class regulatory RNAs in *Drosophila*. Cell 130:89–100
Park JE, Heo I, Tian Y, Simanshu DK, Chang H, Jee D, Patel DJ, Kim VN (2011) Dicer recognizes the 5' end of RNA for efficient and accurate processing. Nature 475:201–205
Pasquinelli AE, Reinhart BJ, Slack F, Martindale MQ, Kuroda MI, Maller B, Hayward DC, Ball EE, Degnan B, Muller P et al (2000) Conservation of the sequence and temporal expression of let-7 heterochronic regulatory RNA. Nature 408:86–89
Piskounova E, Polytarchou C, Thornton JE, LaPierre RJ, Pothoulakis C, Hagan JP, Iliopoulos D, Gregory RI (2011) Lin28A and Lin28B inhibit let-7 microRNA biogenesis by distinct mechanisms. Cell 147:1066–1079
Reinhart BJ, Slack FJ, Basson M, Pasquinelli AE, Bettinger JC, Rougvie AE, Horvitz HR, Ruvkun G (2000) The 21-nucleotide let-7 RNA regulates developmental timing in *Caenorhabditis elegans*. Nature 403:901–906
Roush S, Slack FJ (2008) The let-7 family of microRNAs. Trends Cell Biol 18:505–516
Ruby JG, Jan C, Player C, Axtell MJ, Lee W, Nusbaum C, Ge H, Bartel DP (2006) Large-scale sequencing reveals 21U-RNAs and additional microRNAs and endogenous siRNAs in *C. elegans*. Cell 127:1193–1207
Ruby JG, Jan CH, Bartel DP (2007) Intronic microRNA precursors that bypass Drosha processing. Nature 448:83–86
Rybak A, Fuchs H, Smirnova L, Brandt C, Pohl EE, Nitsch R, Wulczyn FG (2008) A feedback loop comprising lin-28 and let-7 controls pre-let-7 maturation during neural stem-cell commitment. Nat Cell Biol 10:987–993
Sampson VB, Rong NH, Han J, Yang Q, Aris V, Soteropoulos P, Petrelli NJ, Dunn SP, Krueger LJ (2007) MicroRNA let-7a down-

regulates MYC and reverts MYC-induced growth in Burkitt lymphoma cells. Cancer Res 67:9762–9770

Schulman BR, Esquela-Kerscher A, Slack FJ (2005) Reciprocal expression of lin-41 and the microRNAs let-7 and mir-125 during mouse embryogenesis. Dev Dyn 234:1046–1054

Schultz J, Lorenz P, Gross G, Ibrahim S, Kunz M (2008) MicroRNA let-7b targets important cell cycle molecules in malignant melanoma cells and interferes with anchorage-independent growth. Cell Res 18:549–557

Segalla S, Pivetti S, Todoerti K, Chudzik MA, Giuliani EC, Lazzaro F, Volta V, Lazarevic D, Musco G, Muzi-Falconi M et al (2015) The ribonuclease DIS3 promotes let-7 miRNA maturation by degrading the pluripotency factor LIN28B mRNA. Nucleic Acids Res 43:5182–5193

Sempere LF, Dubrovsky EB, Dubrovskaya VA, Berger EM, Ambros V (2002) The expression of the let-7 small regulatory RNA is controlled by ecdysone during metamorphosis in *Drosophila melanogaster*. Dev Biol 244:170–179

Sempere LF, Sokol NS, Dubrovsky EB, Berger EM, Ambros V (2003) Temporal regulation of microRNA expression in *Drosophila melanogaster* mediated by hormonal signals and broad-Complex gene activity. Dev Biol 259:9–18

Sempere LF, Freemantle S, Pitha-Rowe I, Moss E, Dmitrovsky E, Ambros V (2004) Expression profiling of mammalian microRNAs uncovers a subset of brain-expressed microRNAs with possible roles in murine and human neuronal differentiation. Genome Biol 5:R13

Sempere LF, Christensen M, Silahtaroglu A, Bak M, Heath CV, Schwartz G, Wells W, Kauppinen S, Cole CN (2007) Altered MicroRNA expression confined to specific epithelial cell subpopulations in breast cancer. Cancer Res 67:11612–11620

Shell S, Park SM, Radjabi AR, Schickel R, Kistner EO, Jewell DA, Feig C, Lengyel E, Peter ME (2007) Let-7 expression defines two differentiation stages of cancer. Proc Natl Acad Sci USA 104:11400–11405

Sokol NS, Xu P, Jan YN, Ambros V (2008) *Drosophila* let-7 microRNA is required for remodeling of the neuromusculature during metamorphosis. Genes Dev 22:1591–1596

Su H, Trombly MI, Chen J, Wang X (2009) Essential and overlapping functions for mammalian Argonautes in microRNA silencing. Genes Dev 23:304–317

Suh MR, Lee Y, Kim JY, Kim SK, Moon SH, Lee JY, Cha KY, Chung HM, Yoon HS, Moon SY et al (2004) Human embryonic stem cells express a unique set of microRNAs. Dev Biol 270:488–498

Suzuki HI, Katsura A, Miyazono K (2015) A role of uridylation pathway for blockade of let-7 microRNA biogenesis by Lin28B. Cancer Sci. doi:10.1111/cas.12721

Takamizawa J, Konishi H, Yanagisawa K, Tomida S, Osada H, Endoh H, Harano T, Yatabe Y, Nagino M, Nimura Y et al (2004) Reduced expression of the let-7 microRNAs in human lung cancers in association with shortened postoperative survival. Cancer Res 64:3753–3756

Thomson JM, Parker J, Perou CM, Hammond SM (2004) A custom microarray platform for analysis of microRNA gene expression. Nat Methods 1:47–53

Thomson JM, Newman M, Parker JS, Morin-Kensicki EM, Wright T, Hammond SM (2006) Extensive post-transcriptional regulation of microRNAs and its implications for cancer. Genes Dev 20:2202–2207

Thornton JE, Gregory RI (2012) How does Lin28 let-7 control development and disease? Trends Cell Biol 22:474–482

Thornton JE, Chang HM, Piskounova E, Gregory RI (2012) Lin28-mediated control of let-7 microRNA expression by alternative TUTases Zcchc11 (TUT4) and Zcchc6 (TUT7). RNA 18:1875–1885

Ustianenko D, Hrossova D, Potesil D, Chalupnikova K, Hrazdilova K, Pachernik J, Cetkovska K, Uldrijan S, Zdrahal Z, Vanacova S (2013) Mammalian DIS3L2 exoribonuclease targets the uridylated precursors of let-7 miRNAs. RNA 19:1632–1638

Vermeulen A, Behlen L, Reynolds A, Wolfson A, Marshall WS, Karpilow J, Khvorova A (2005) The contributions of dsRNA structure to Dicer specificity and efficiency. RNA 11:674–682

Viswanathan SR, Daley GQ, Gregory RI (2008) Selective blockade of microRNA processing by Lin28. Science 320:97–100

Wada T, Kikuchi J, Furukawa Y (2012) Histone deacetylase 1 enhances microRNA processing via deacetylation of DGCR8. EMBO Rep 13:142–149

Wang Z, Lin S, Li JJ, Xu Z, Yao H, Zhu X, Xie D, Shen Z, Sze J, Li K et al (2011) MYC protein inhibits transcription of the microRNA cluster MC-let-7a-1~let-7d via noncanonical E-box. J Biol Chem 286:39703–39714

Wilbert ML, Huelga SC, Kapeli K, Stark TJ, Liang TY, Chen SX, Yan BY, Nathanson JL, Hutt KR, Lovci MT et al (2012) LIN28 binds messenger RNAs at GGAGA motifs and regulates splicing factor abundance. Mol Cell 48:195–206

Wulczyn FG, Smirnova L, Rybak A, Brandt C, Kwidzinski E, Ninnemann O, Strehle M, Seiler A, Schumacher S, Nitsch R (2007) Post-transcriptional regulation of the let-7 microRNA during neural cell specification. FASEB J 21:415–426

Yang JS, Maurin T, Robine N, Rasmussen KD, Jeffrey KL, Chandwani R, Papapetrou EP, Sadelain M, O'Carroll D, Lai EC (2010) Conserved vertebrate mir-451 provides a platform for Dicer-independent, Ago2-mediated microRNA biogenesis. Proc Natl Acad Sci USA 107:15163–15168

Yi R, Qin Y, Macara IG, Cullen BR (2003) Exportin-5 mediates the nuclear export of pre-microRNAs and short hairpin RNAs. Genes Dev 17:3011–3016

Yoda M, Cifuentes D, Izumi N, Sakaguchi Y, Suzuki T, Giraldez AJ, Tomari Y (2013) Poly(A)-specific ribonuclease mediates 3'-end trimming of Argonaute2-cleaved precursor microRNAs. Cell Rep 5:715–726

Yu F, Yao H, Zhu P, Zhang X, Pan Q, Gong C, Huang Y, Hu X, Su F, Lieberman J et al (2007) let-7 regulates self renewal and tumorigenicity of breast cancer cells. Cell 131:1109–1123

Zhang H, Kolb FA, Brondani V, Billy E, Filipowicz W (2002) Human Dicer preferentially cleaves dsRNAs at their termini without a requirement for ATP. EMBO J 21:5875–5885

Zhang H, Kolb FA, Jaskiewicz L, Westhof E, Filipowicz W (2004) Single processing center models for human Dicer and bacterial RNase III. Cell 118:57–68

13

Effective gene editing by high-fidelity base editor 2 in mouse zygotes

Puping Liang[1,2,3✉], Hongwei Sun[1], Ying Sun[1], Xiya Zhang[1], Xiaowei Xie[1], Jinran Zhang[1], Zhen Zhang[1,4], Yuxi Chen[1], Chenhui Ding[3], Yuanyan Xiong[1], Wenbin Ma[1], Dan Liu[5], Junjiu Huang[1,2✉], Zhou Songyang[1,2,3,5✉]

[1] Key Laboratory of Gene Engineering of the Ministry of Education, Guangzhou Key Laboratory of Healthy Aging Research and State Key Laboratory of Biocontrol, SYSU-BCM Joint Research Center, School of Life Sciences, Sun Yat-sen University, Guangzhou 510275, China

[2] State Key Laboratory of Ophthalmology, Zhongshan Ophthalmic Center, Sun Yat-sen University, Guangzhou 510275, China

[3] Key Laboratory of Reproductive Medicine of Guangdong Province, School of Life Sciences and the First Affiliated Hospital, Sun Yat-sen University, Guangzhou 510275, China

[4] Guangzhou Magigen Biotechnology Co.Ltd, Guangzhou 510320, China

[5] Verna and Marrs Mclean Department of Biochemistry and Molecular Biology, Baylor College of Medicine, One Baylor Plaza, Houston, TX 77030, USA

✉ Correspondence: liangpp5@mail.sysu.edu.cn (P. Liang), hjunjiu@mail.sysu.edu.cn (J. Huang), songyanz@mail.sysu.edu.cn (Z. Songyang)

ABSTRACT

Targeted point mutagenesis through homologous recombination has been widely used in genetic studies and holds considerable promise for repairing disease-causing mutations in patients. However, problems such as mosaicism and low mutagenesis efficiency continue to pose challenges to clinical application of such approaches. Recently, a base editor (BE) system built on cytidine (C) deaminase and CRISPR/Cas9 technology was developed as an alternative method for targeted point mutagenesis in plant, yeast, and human cells. Base editors convert C in the deamination window to thymidine (T) efficiently, however, it remains unclear whether targeted base editing in mouse embryos is feasible. In this report, we generated a modified high-fidelity version of base editor 2 (HF2-BE2), and investigated its base editing efficacy in mouse embryos. We found that HF2-BE2 could convert C to T efficiently, with up to 100% biallelic mutation efficiency in mouse embryos. Unlike BE3, HF2-BE2 could convert C to T on both the target and non-target strand, expanding the editing scope of base editors. Surprisingly, we found HF2-BE2 could also deaminate C that was proximal to the gRNA-binding region. Taken together, our work demonstrates the feasibility of generating point mutations in mouse by base editing, and underscores the need to carefully optimize base editing systems in order to eliminate proximal-site deamination.

KEYWORDS base editor, high-fidelity, mouse embryos, proximal-site deamination, whole-genome sequencing

INTRODUCTION

The human genome project has revealed unprecedented genetic diversity in human, manifested predominantly as single nucleotide variations (SNVs). Probing the physiological significance of these SNVs is both essential and challenging. Researchers have traditionally relied on homologous recombination (HR) to generate SNVs, a process that is inefficient (usually $<10^{-5}$), labor-intensive, and often ineffective in non-dividing primary cells (Capecchi, 2005; Porteus and Carroll, 2005; Thomas and Capecchi, 1987). Molecular scissors such as ZFN, TALEN, and CRISPR/Cas9 can promote HR at target sites, however, the

Puping Liang and Hongwei Sun have contributed equally to this work.

more efficient non-homologous end joining (NHEJ) pathway invariably outcompetes HR in these cases (Cho et al., 2013; Cong et al., 2013; Gaj et al., 2013; Jiang et al., 2013; Jinek et al., 2013; Kim and Kim, 2014; Komor et al., 2017; Porteus, 2006; Suzuki et al., 2016; Tesson et al., 2011; Yang et al., 2013). A programmable cytidine deaminase built on the CRISPR/Cas9 platform has recently been developed to more efficiently edit target bases (Komor et al., 2016; Ma et al., 2016; Nishida et al., 2016). This base editor (BE) has an effector that fuses cytidine deaminase (rAPOBEC1) with Cas9 and the uracil DNA glycosylase inhibitor (UGI), enabling targeted cytidine (C) to uridine (U) conversion in the desired DNA sequence (Komor et al., 2016; Ma et al., 2016; Nishida et al., 2016). Following DNA replication, this conversion will lead to C-to-T (or G-to-A) substitution. BE-directed base editing at single-base resolution has been successfully carried out in plant, yeast, and human cells (Komor et al., 2016; Li et al., 2016; Lu and Zhu, 2016; Nishida et al., 2016), and shown to be >100-fold more efficient than HR at generating point mutations (Komor et al., 2016; Ma et al., 2016; Nishida et al., 2016). It has also been found that base editors could efficiently deaminate cytidines within a deamination window, typically several nucleotides long (positions 4–8) in the gRNA-binding region (Komor et al., 2016). Previous studies using CRISPR-based genome editing methods to generate point mutations in mice could not achieve 100% efficiency and resulted in mosaicism (Inui et al., 2014; Wu et al., 2013). Whether BE-mediated genome editing proves more efficient and reliable remains to be studied.

Of the different base editors, base editor 3 (BE3, rAPOBEC1-nCas9-UGI) uses the Cas9 nickase (nCas9, D10A), whereas base editor 2 (BE2, rAPOBEC1-dCas9-UGI) utilizes the nuclease activity dead Cas9 mutant (dCas9, D10A/H840A). In cells, BE2 appeared to have lower base editing efficiency than BE3, although still more efficient than HR. Unlike nCas9, dCas9 does not cleave DNA, which should help reduce off-target indels and increase the specificity of BE2. Since the gRNA/Cas9 units within base editors are responsible for their targeting, improving Cas9 specificity, such as using high-fidelity Cas9 variants, should improve the specificity of base editors and further reduce off-targets. The Cas9 high-fidelity 1 variant (Cas9-HF1), which contains four point mutations (N497A/R661A/Q695A/Q926A), is thought to have less binding energy with DNA than wild type Cas9. The mutations presumably disrupt hydrogen bonding with the phosphate backbone of the complementary DNA strand, thereby decreasing Cas9 binding with mismatched sequences and increasing its overall specificity (Anders et al., 2014; Kleinstiver et al., 2016; Nishimasu et al., 2014). Cas9 high-fidelity 2 (Cas9-HF2), which contains one additional mutation (D1135E) compared to Cas9-HF1 and exhibits altered PAM preference (from NGG/A to NGG only), has been proven highly specific based on genome-wide sequencing and targeted deep sequencing analyses (Kleinstiver et al., 2016; Kleinstiver et al., 2015). We have generated a high-fidelity variant of base editor 2 (HF2-BE2) by introducing the five point mutations into dCas9 (Kleinstiver et al., 2016). Here, we report our findings on using HF2-BE2 to edit target genes in mouse zygotes.

Figure 1. HF2-BE2 edits target bases efficiently in human cells. (A) Schematic representation of HF2-BE2. Point mutations that inactivate Cas9 nuclease activity are shown in red, and mutations that enhance its fidelity are shown in black. (B) The HF2-BE2 expression vector was respectively co-transfected into 293T cells with gRNA expression vectors targeting HEK293 site 3 and RNF2 (Komor et al., 2016). Genomic DNA was extracted from the edited cell populations for PCR amplification of the target sites. Sequencing chromatographs of the PCR amplicons are shown. WT, untransfected wild-type control cells. Edited, base-edited cells. Red arrowheads, successfully edited base. (C) The PCR amplicons from (B) were subcloned into pGEM-T vectors and sequenced. The number of clones for each sequence pattern is indicated. Underlined, gRNA target regions. Green, PAM sequence. Red, point mutations. ▶

We found that both HF2-BE2 and BE2 could convert target C to T efficiently in mouse embryos, where the editing efficiency of HF2-BE2 appeared higher than that of BE2. We found biallelic mutant embryos and pups, indicating 100% efficiency in base conversion. Moreover, we found that both HF2-BE2 and BE2 could deaminate cytidines on non-target strand as well as target strand. Surprisingly, both HF2-BE2 and BE2 could deaminate C proximal to the deamination window, which we termed proximal-site deamination. Taken together, our data highlight the potential of base editors in generating point mutations in mouse, and underscore the need to optimize base editors in order to avoid proximal-site deamination.

RESULTS

The high-fidelity version of BE2 (HF2-BE2) mediates efficient editing in mouse embryos

We first examined the ability of HF2-BE2 (rAPOBEC1-XTEN-dCas9-HF2-UGI) (Fig. 1A) to edit two previously published base editor target sites in human cells (HEK293 site 3 and RNF2) (Komor et al., 2016), by co-transfecting HF2-BE2 with the respective gRNAs into 293T cells. Genomic DNA analysis clearly indicated the presence of thymidine peaks in the target region (Fig. 1B). PCR amplicons of the target sequences from both untransfected wild-type (WT) control cells and edited cells were then subcloned, and 15 single bacterial clones from each pool were sequenced (Fig. 1C). As expected, both wildtype and deaminated alleles were found in the edited cells, indicating successful base editing by HF2-BE2.

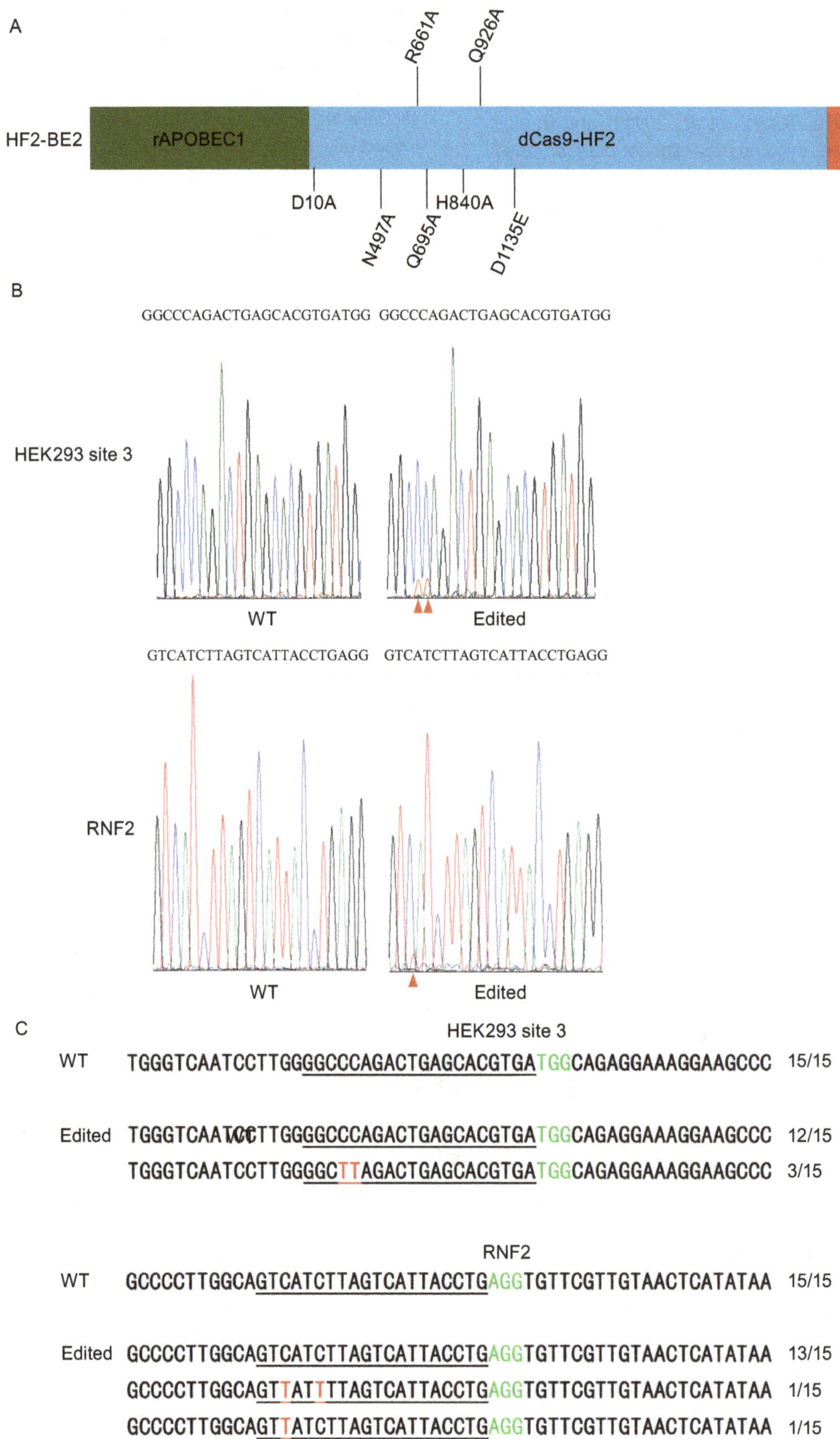
A
HF2-BE2
rAPOBEC1
dCas9-HF2
UGI
R661A
Q926A
D10A
N497A
Q695A
H840A
D1135E
B
GGCCCAGACTGAGCACGTGATGG GGCCCAGACTGAGCACGTGATGG
HEK293 site 3
WT
Edited
GTCATCTTAGTCATTACCTGAGG GTCATCTTAGTCATTACCTGAGG
RNF2
WT
Edited
C
HEK293 site 3
WT TGGGTCAATCCTTGGGGCCCAGACTGAGCACGTGATGGCAGAGGAAAGGAAGCCC 15/15
Edited TGGGTCAATCCTTGGGGCCCAGACTGAGCACGTGATGGCAGAGGAAAGGAAGCCC 12/15
TGGGTCAATCCTTGGGGCTTAGACTGAGCACGTGATGGCAGAGGAAAGGAAGCCC 3/15
RNF2
WT GCCCCTTGGCAGTCATCTTAGTCATTACCTGAGGTGTTCGTTGTAACTCATATAA 15/15
Edited GCCCCTTGGCAGTCATCTTAGTCATTACCTGAGGTGTTCGTTGTAACTCATATAA 13/15
GCCCCTTGGCAGTTATTTTAGTCATTACCTGAGGTGTTCGTTGTAACTCATATAA 1/15
GCCCCTTGGCAGTTATCTTAGTCATTACCTGAGGTGTTCGTTGTAACTCATATAA 1/15

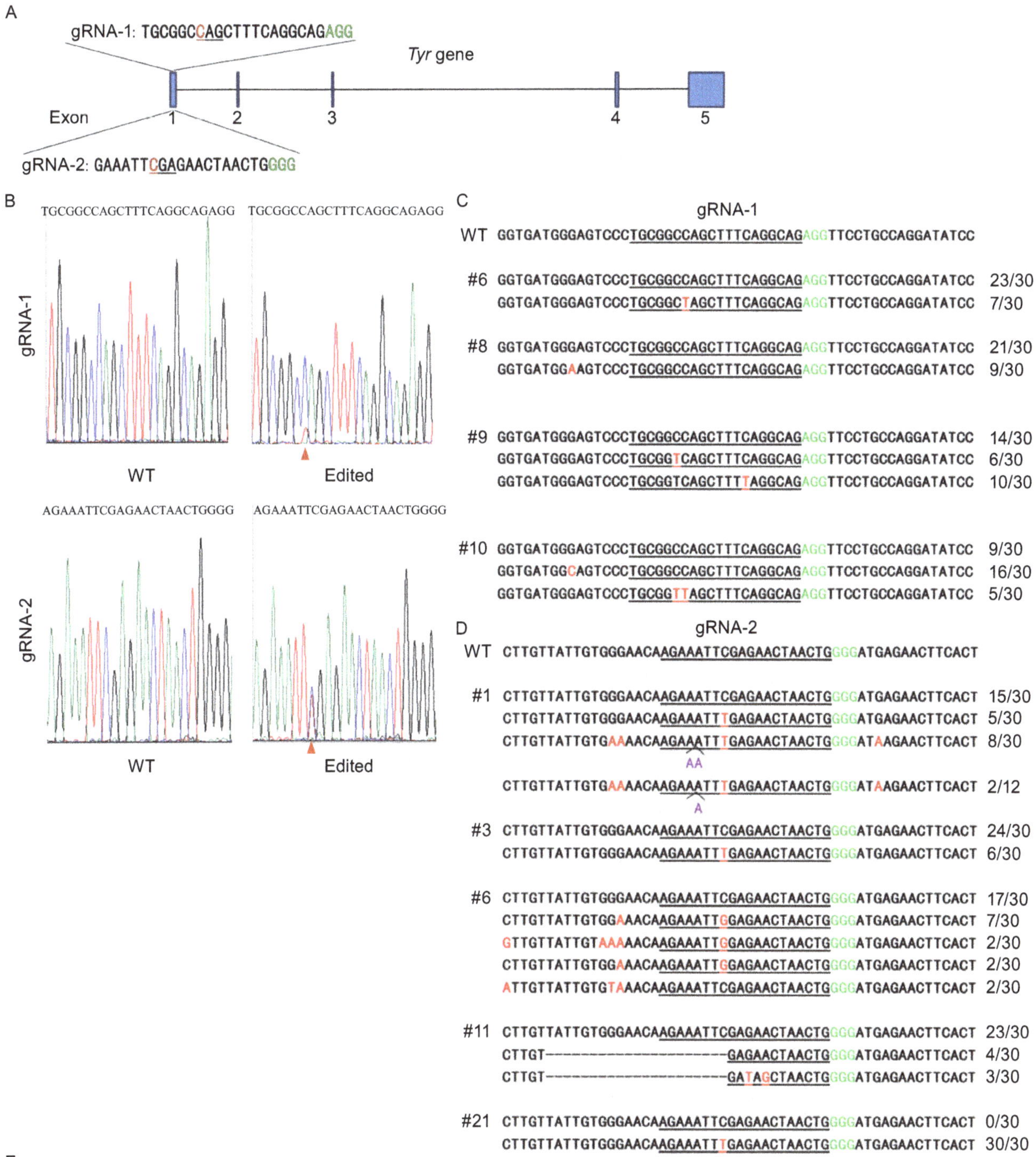

E

gRNA	Injected embryo	Survived embryo (%)	PCR amplified	Mutant (%)	Biallelic mutant
gRNA-1	75	55 (73.3)	43	5 (11.6)	0
gRNA-2	70	55 (78.6)	44	22 (50)	1 (2.3)[a]

◄ **Figure 2. HF2-BE2 mediates efficient base editing of *Tyr* in mouse embryos.** (A) Schematic representation of the *Tyr* locus and gRNA target sites. gRNA target sequences are also shown. The codon to be modified is underlined, with the nucleotide to be deaminated in red. The protospacer adjacent motif (PAM) is in green. (B) The two gRNAs were respectively co-injected into 1-cell zygotes with HF2-BE2 mRNA, and the embryos were analyzed 48 h later. Representative sequencing chromatographs of the PCR amplicons of target sites are shown here. WT, wild-type embryo. Edited, embryos edited by HF2-BE2 with the successfully edited base indicated by red arrowheads. (C) PCR amplicons of gRNA-1 target site from the genomic DNA of selected embryos were subcloned into pGEM-T vectors and sequenced. The number of clones for each sequence pattern is indicated. Underlined, gRNA target regions. Green, PAM sequence. Red, point mutations. Purple, insertions. Dash, deletions. (D) PCR amplicons of gRNA-2 target site from the genomic DNA of selected embryos were subcloned into pGEM-T vectors and sequenced. (E) Summary of base editing by HF2-BE2 in mouse embryos. a, this biallelic mutant embryo is homozygous.

We next investigated HF2-BE2-mediated base editing in mouse embryos, by generating two gRNAs targeting exon 1 of the *Tyr* gene (Fig. 2A) and respectively co-injecting them into the cytoplasm of 1-cell zygotes with HF2-BE2 mRNA. The injected embryos were harvested after 48 h for genomic DNA extraction and genotyping by Sanger sequencing (Fig. 2B) and subcloning/sequencing analysis (Fig. 2C and 2D). Both gRNAs were able to direct efficient C–T conversion on the non-target strand in the target region (Fig. 2C and 2D), and to a lesser extent, C–G/A conversion (Figs. 2C, 2D, and S1). Unexpectedly, we found C–T conversion on the target strand and deamination at cytidines proximal to gRNA binding sites (which we termed proximal-site deamination), even at cytidines 38 bps upstream or 3 bps downstream of the gRNA target site (Figs. 2C, 2D, and S1). In addition, we found indels in two embryos edited by gRNA-2 (#1 and #11) (Fig. 2D). Given the absence of nuclease activity in dCas9, these findings suggest that cytidine deamination alone can result in indels. Of the examined embryos, 11.6% and 50% respectively were edited by gRNA-1 and gRNA-2 (Fig. 2E). One gRNA-2 edited embryo was a homozygous mutant (#21), indicating 100% base editing efficiency (Figs. 2D, 2E, and S2). Genomic DNA from this homozygous mutant embryo was further examined by whole-genome sequencing, which found no off-target deamination, suggesting that HF2-BE2 was able to bind specifically the target site in mouse embryos (Table S1).

One-step generation of base-edited mouse by HF2-BE2

We then proceeded to generate base-edited mice. To rule out possible embryonic toxicity and better determine base editing efficacy, HF2-BE2 mRNA was injected alone or together with gRNA-1 or gRNA-2. Nuclease-free water was also included as a control. The injected embryos were then transplanted into pseudopregnant mice. The rate of pups obtained after transplantation appeared similar between different groups (Table 1), indicating low toxicity of HF2-BE2. Genotyping revealed that 2 out of 11 (18.2%) pups from gRNA-1 group and 7 out of 11 (63.6%) pups from gRNA-2 group were mutants (Table 1, Figs. 3A–C, S3, and S4). Of the gRNA-2 group, we obtained 3 (27.3%) biallelic mutant founder mice (P3, P6, P11), in line with the mouse embryo data (Fig. 2E and Table 1). Furthermore, as was observed in mouse embryos (Fig. 2C and 2D), C-to-T conversion occurred on both target and non-target strands in founder mice (Fig. 3C). Additionally, we also found proximal-site deamination in 4 pups (Fig. 3C), one of which lies 42 bps downstream of the PAM sequence on the target strand (Fig. S5).

Successful C-to-T conversion by HF2-BE2 is expected to yield a premature stop codon in the gRNA target regions, leading to albinism in C57B/6 J mice. Two black pups from gRNA-1 group (P8 and P9) showed ~50% base editing efficiency, suggesting that they were heterozygous mutants.

Table 1. Summary of base editing by HF2-BE2 in founder mice

Group	Survived/Injected embryos (%)	Pups/ Transferred (%)	Albino pups (%)	Mosaic pups (%)	Mutant black pups (%)	Mutants (%)	Biallelic mutants (%)
gRNA-1 + HF2-BE2 mRNA*	120/162 (74.1)	13/120 (10.8)	0	0	2 (18.2)*	2 (18.2)	0
gRNA-2 + HF2-BE2 mRNA	106/145 (73.1)	11/106 (10.4)	1 (9.1)	4 (36.4)	2 (18.2)	7 (63.6)	3 (27.3)
HF2-BE2 mRNA#	108/142 (76.1)	14/108 (13.0)	0	0	0	0	0
H_2O	103/146 (70.5)	9/103 (8.7)	0	0	0	0	0

*, # Pups were cannibalized by the mother (2 for * and 1 for #).

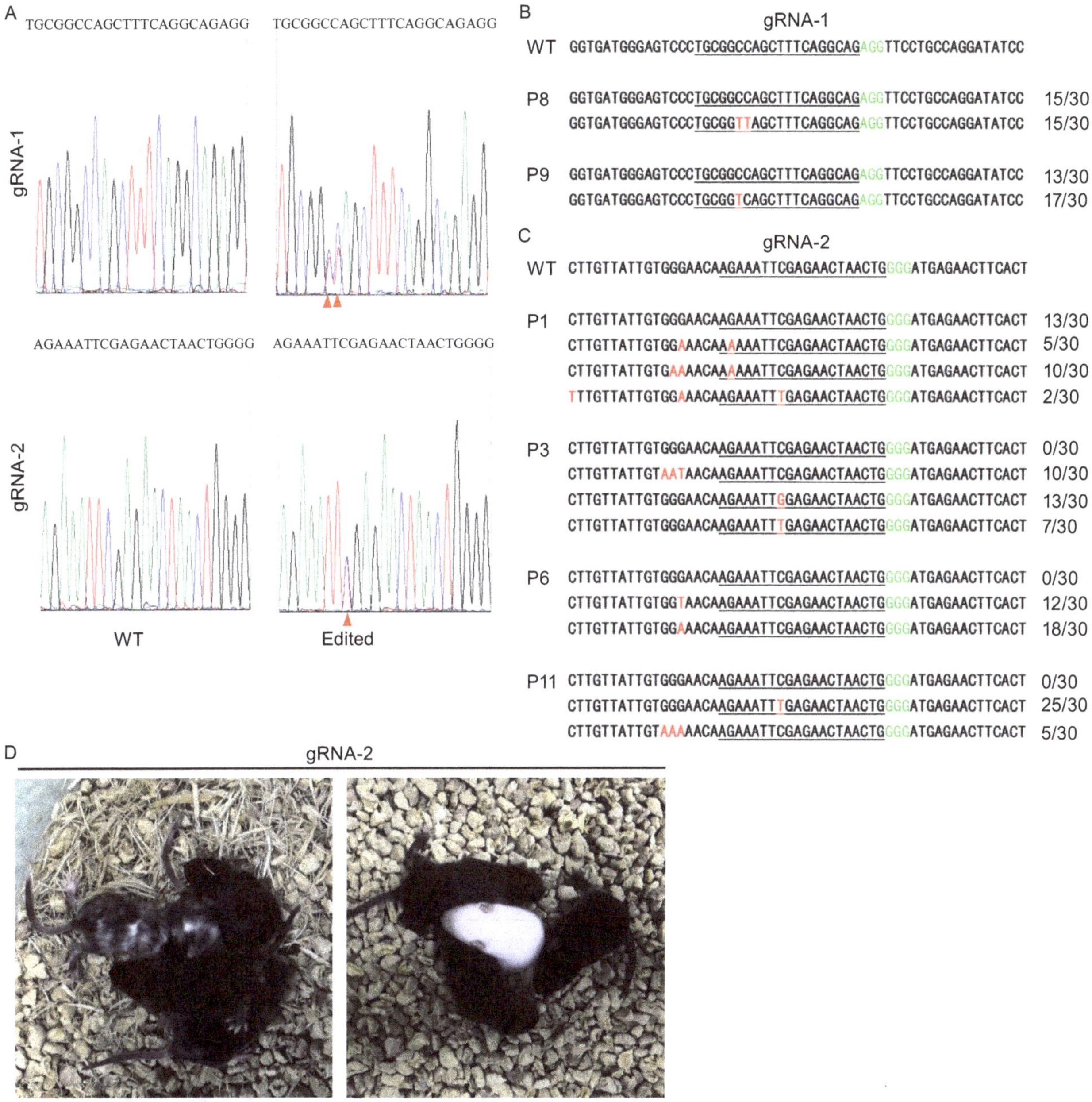

Figure 3. Generation of base-edited mice using HF2-BE2. (A) Representative sequencing chromatographs of the PCR amplicons of target sites from the founder mice are shown here. WT, wild-type embryo. Edited, embryo edited by HF2-BE2, with the successfully edited base indicated by red arrowheads. (B) PCR amplicons of gRNA-1 target sites from the genomic DNA of selected founder mice were subcloned into pGEM-T vectors and sequenced. The number of clones for each sequence pattern is indicated. Underlined, gRNA target regions. Green, PAM sequence. Red, point mutations. (C) PCR amplicons of gRNA-2 target sites from the genomic DNA of selected founder mice were subcloned into pGEM-T vectors and sequenced. (D) Founder pups (10 days old) from the gRNA-2 group.

Of the 11 pups in gRNA-2 group, 4 (P1, P3, P5, P11) were chimeras (coat-color mosaic) and 1 (P6) was albino (Fig. 3D). In this albino pup, the GAA codon (Gln) −2 bp from the gRNA target site was converted respectively to TAA (stop) and AAA (lysine), implicating this glutamine residue as being critical for the activity and function of tyrosinase (Fig. 3C). When the biallelic mutant P11 pup was mated with WT mouse, we found that the mutant allele could be successfully transmitted to the next generation (Fig. S6).

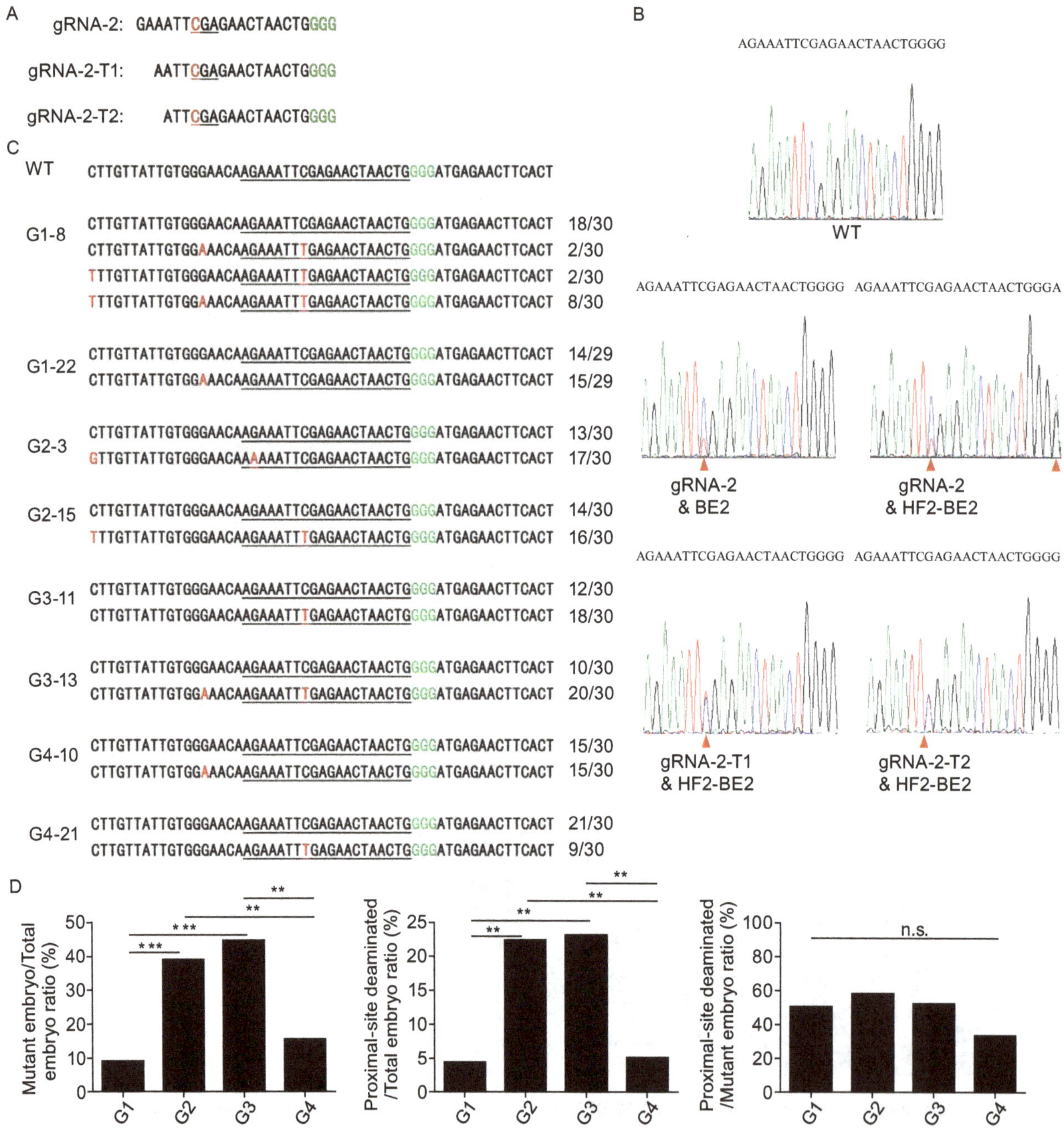

Figure 4. Examination of proximal-site deamination by HF2-BE2. (A) Design of truncated gRNA variants of gRNA-2 (gRNA-2-T1 and gRNA-2-T2). The codon to be modified is underlined, with the nucleotide to be deaminated in red. PAM is in green. (B) Representative sequencing chromatographs of the PCR amplicons of target sites from the embryos are shown here. Bases successfully edited are indicated by red arrowheads. (C) PCR amplicons of target sites from embryos were subcloned into pGEM-T vectors and sequenced. The number of clones for each sequence pattern is indicated. Underlined, gRNA target regions. Green, PAM sequence. Red, point mutations. (D) Statistical analysis of base editing by HF2-BE2 and BE2 in mouse embryos. n.s., not significant, *P* values > 0.05. $^{**}P < 0.01$, $^{***}P < 0.001$, statistical significance was determined using the χ^2 test.

Proximal-site deamination of cytidines near gRNA binding sites by HF2-BE2

It is postulated that rAPOBEC1 catalyzes C-to-T conversions at exposed single-stranded DNA regions displaced by gRNAs (Conticello, 2008; Harris et al., 2002; Komor et al., 2016; Saraconi et al., 2014), truncating gRNAs may therefore reduce proximal-site deamination by unwinding smaller stretches of DNA (Fu et al., 2014a; Fu et al., 2014b). To better assess how manipulating gRNA/dCas9 targeting

Table 2. Summary of base editing by HF2-BE2 and BE2 in mouse embryos

Group	gRNA + base editor	Total embryos	Mutant embryos (%)	Proximal-site deaminated embryos	Proximal-site deaminated/ Mutant embryos ratio (%)	Proximal-site deaminated/ Total embryos ratio (%)
1	gRNA-2 + BE2 mRNA	68	6 (8.8)	3	50.0	4.4
2	gRNA-2 + HF2-BE2 mRNA	67	26 (38.8)	15	57.7	22.4
3	gRNA-2-T1 + HF2-BE2 mRNA	56	25 (44.6)	13	52.0	23.2
4	gRNA-2-T2 + HF2-BE2 mRNA	59	9 (15.3)	3	33.30	5.1

affects the activity of HF2-BE2, we generated two truncated versions of gRNA-2 (gRNA-2-T1 and gRNA-2-T2) with 16 or 17 nucleotides of guide sequence (Fig. 4A). HF2-BE2 mRNA was individually co-injected into mouse embryos with gRNA-2, gRNA-2-T1, and gRNA-2-T2 (groups 2, 3, & 4, G2, 3, & 4). In addition, we also co-injected conventional BE2 mRNA (containing no HF mutations) with gRNA-2 (group 1, G1) as a comparison. The embryos were genotyped 48 h after injection (Fig. 4B and 4C). Of the different combinations, group 2 with HF2-BE2 mRNA and full-length gRNA-2 clearly had higher editing efficiency, as well as the highest number of proximal-site base edits (Fig. 4D and Table 2). However, the ratio of proximal-site deaminated embryos *vs.* mutant embryos was similar between groups 1 and 2, suggesting that lower proximal-site deamination in BE2-edited embryos was most likely a result of lower overall deamination activity of BE2 compared to HF2-BE2. While group 3 showed similar editing efficiency as group 2, group 4 was significantly reduced, suggesting that excessive truncation of gRNA sequences drastically decreased the efficiency of base editing by HF2-BE2 (Fig. 4D). Moreover, the differences in the ratios of proximal-site deaminated/mutant embryo were not statistically significant between the groups (P values > 0.05), implying that truncating gRNAs will decrease the efficiency of on-target as well as proximal-site deamination, consistent with findings in human cells (Kim et al., 2017b). These data demonstrate that base editors may deaminate cytidines proximal to gRNA binding sites, similar to observations in *E. coli* cells, where the zinc-finger domain-guided cytidine deaminase induced off-target deamination at cytidines ~150-bp away from the target site (Yang et al., 2016).

DISCUSSION

In this study, we present data that highlight important similarities and differences between base editors. While this manuscript was under preparation, Kim et al. reported generating fully base-edited mouse embryos using BE3 (Kim et al., 2017a). Similar to base editor 3 (BE3), both BE2 and HF2-BE2 could efficiently modify bases in mouse zygotes. Indeed, we were able to obtain homozygous edited embryos free of mosacism using HF2-BE2 (Fig. 2D). Unlike BE3, BE2 and HF2-BE2 are able to convert C to T on both strands. This difference is likely caused by the ability of BE3 to cleave the modified target strand, rendering it incapable of serving as a DNA repair template. Base editing by deaminases is constrained by the presence of cytidines on a given target. Our data suggest that BE2 and HF2-BE2 may expand the choice of target nucleotides thanks to their abilities to deaminate C on both strands of target DNA. Previous studies using BE3 did not find proximal-site deamination (Kim et al., 2017a; Komor et al., 2016; Li et al., 2016; Lu and Zhu, 2016; Zong et al., 2017). In this study, we found that BE2 and HF2-BE2 could both lead to proximal-site deamination at cytidines close to gRNA target sites, which may occur as a result of spontaneous or Cas9-catalyzed DNA unwinding that exposes single-stranded DNA to rAPOBEC1. To avoid proximal-site deamination, reducing the amount of HF2-BE2 (or BE2) mRNAs and gRNA and/or truncating gRNAs may be useful at some sites (Kim et al., 2017b). Recently, a base editor variant with a narrower deamination window was described, such modifications may bring more specificity without sacrificing efficiency at genomic sites (Kim et al., 2017b). Additionally, we found higher editing efficiency for HF2-BE2 than BE2 in mouse embryos, when guided by full-length gRNA-2. A more comprehensive comparison of HF2-BE2 *vs.* BE2 and other base editors is needed to fully understand the effects of different Cas9 proteins on the efficiency and specificity of base editors.

We were surprised to find low-frequency (2/44 embryos) base insertions and deletions at target sites with HF2-BE2, even though dCas9 has no demonstrable nuclease activity (Hsu et al., 2014; Mali et al., 2013; Qi et al., 2013). Such indels are unlikely the result of any potential residual enzymatic activities of dCas9, because they were far from Cas9 cleavage sites (≥12 bp upstream of the PAM sequence) (Jinek et al., 2012). Taken together with previous findings that fusing UGI to nCas9-PmCDA1 suppressed indel formation (Nishida et al., 2016), we hypothesize that the indels might have been caused by deamination and base excision repair. If the base excision repair pathway is active when deamination occurs simultaneously on both strands, base excision will create abasic sites on both strands and ultimately DNA double-strand breaks (DSB) (Kingma et al., 1995; Ma et al., 2009). It has been shown that even a single abasic site is capable of inducing DNA DSBs (Kidane et al., 2014). Subsequent DSB repair can result in indel formation. Consequently, to elucidate how DNA repair machineries are utilized following base conversions will be crucial to reducing and eliminating unwanted indels in genome editing mediated by base editors. Our study highlights the advantages of gene editing using base editor 2 and sheds light on possible new avenues of research for precise gene editing.

MATERIALS AND METHODS

Plasmids

BE2 and HF2-BE2 expression cassette were synthesized and ligated into pcDNA3.1 (-) vector by IGE BIOTECHNOLOGY LTD. To construct gRNA expression vector, gRNA backbone with U6 promoter were amplified from pX330, and then ligated into the pGEM-T vector (Promega).

Animals

All animal experiments were performed according to protocols approved by the Committee on Animal Care at the School of Life Sciences, Sun Yat-Sen University. Superovulated C57BL/6J mice (6–8 week old females) were mated with C57BL/6 J males. Plugged females were sacrificed by cervical dislocation. Zygotes (0.5 day) were collected using potassium simplex optimized medium (KSOM) containing N′-2-Hydroxyethylpiperazine-N′-2-ethanesulfonic acid and sodium bicarbonate (HKSOM), and cultured in KSOM until genotyping or transplantation. Embryos were *in vitro* cultured for 48 h before genotyping or whole genome amplification. CD1 female mice (6–8 weeks old) that were mated with sterilized CD1 male mice were used as foster mothers.

In vitro transcription

HF2-BE2 mRNA was transcribed using the mMESSAGE mMACHINE T7 ULTRA kit (Life Technologies) following the manufacturer's instruction. gRNA-1 and gRNA-2 (Table S2) were cloned into the pDR274 vector (Addgene) and transcribed using the MEGAshortscript T7 kit (Life Technologies) following the manufacturer's instruction. mRNAs and gRNAs were subsequently purifed using the MEGAclear kit (Life Technologies) and resuspended in RNase-free water.

Intracytoplasmic injection of HF2-BE2 mRNA and gRNA

The mixture of HF2-BE2 mRNA (200 ng/μL) and gRNA (100 ng/μL) was injected into 0.5-day 1-cell zygotes of C57BL/6 J mice. The injected zygotes were transplanted into the oviduct of 0.5-day pseudopregnant mothers ~2 h after injection.

Single embryo PCR amplification and mouse genotyping

Single embryo PCR amplification was performed as described before (Zhang et al., 2016). Briefly, each embryo was transferred into a PCR tube containing 1 μL lysis buffer, and then incubated at 65°C for 3 h followed by 95°C for 10 min. The lysis product was then amplified using primers listed in Table S3. Mouse genotyping was done by PCR and sequencing of tail-snips using the Mouse Genotyping Kit (KAPA Biosystems) and primers listed in Table S2.

Genomic DNA analysis

Target sites were PCR amplified using primers listed in Table S2. The PCR products were then used in T7 endonuclease I (T7EI) cleavage assay as described before (Zhang et al., 2016). Primers for direct sequencing of the PCR products, which reveal the presence of double peaks and/or indels, are listed in Table S3. PCR products with double peaks were then TA cloned into the pGEM-T vector (Promega) for plasmid DNA extraction and Sanger sequencing.

Whole genome sequencing, data processing, and off-target analysis

Whole genome amplification of embryos was performed using the PEPLI-g Midi Kit (Qiagen). Briefly, embryos were transferred into PCR tubes containing reconstituted buffer D2 (7 μL), and then incubated at 65°C for 10 min, before the addition of stop solution (3.5 μL) and MDA master mix (40 μL) and incubation at 30°C for 16 h. Whole genome sequencing (WGS) was done on an Illumina HiSeq 2000 PE150 as paired-end 150 bp reads. The reads were aligned to the mouse reference genome (UCSC, mm10) by means of BWA with default parameters (v0.7.13) (Li and Durbin, 2010). Samtools (v1.3, http://samtools.sourceforge.net) and Picard tools (version 2.2.2, http://picard.sourceforge.net) were used to build indices and remove duplicates. Base score recalibration (BaseRecalibrator) was applied by GATK (The Genome Analysis ToolKit, version 3.5-0) (McKenna et al., 2010) to enhance accuracy in identifying indels and single nucleotide variants (SNVs). GATK HaplotypeCaller was used to call variants for two samples and the variants were then divided into indels and SNVs by SelectVariants. Low-quality variants (indels and SNVs) and those appeared in dbSNP (build 142) were marked by VariantFiltration and discarded by Python.

To avoid false positive calls that overlap with repeat sequences and/or include homopolymers (Bansal and Libiger, 2011), we removed indels and SNVs that overlapped with low-complexity regions as defined by RepeatMasker (UCSC Genome Browser) and filtered out indels and SNVs containing homopolymers (>7 bp) in the low-complexity flanking region (±100 bp). To more definitively assign indels and SNVs, we searched regions flanking potential indel or SNV sites (±100 bp) for possible off-target sites. Bowtie1 (version 1.1.2, http://bowtie-bio.sourceforge.net) was used to align gRNA sequences (20 bp) to the ±100 bp sequences, allowing for ≤5 mismatches or perfect match of the last 10 nt 3′ of the gRNA. Successfully aligned sites with an NGG PAM were deemed as on/off-target sites. No potential off-target site indel or SNV was found in the homozygous embryo (Table S1).

ACKNOWLEDGMENTS

This work was supported by the National Natural Science Foundation of China (Grant Nos. 91640119, 31601196, 81330055, 31371508, and 31671540), the Natural Science Foundation of Guangdong Province (2016A030310206 and 2014A030312011), the Science and Technology Planning Project of Guangdong Province (2015B020228002 and 2015A020212005), the Guangzhou Science and Technology Project (201605030012 and 201707010085), and the Fundamental Research Funds for the Central Universities (161gzd13 and 161gpy31). We would also like to acknowledge the support of CA211653, CPRIT RP160462, the Welch Foundation Q-1673, and the C-BASS Shared Resource at the Dan L. Duncan Cancer Center (DLDCC) of Baylor College of Medicine (P30CA125123).

ABBREVIATIONS

BE, base editor; C, cytidine; DSB, double-strand breaks; HR, homologous recombination; HF2-BE2, high-fidelity version of base editor 2; KSOM, potassium simplex optimized medium; NHEJ, non-homologous end joining; SNVs, single nucleotide variations; T, thymidine; T7EI, T7 endonuclease I; U, uridine; UGI, uracil DNA glycosylase inhibitor; WGS, Whole genome sequencing; WT, wild-type

COMPLIANCE WITH ETHICS GUIDELINES

Puping Liang, Hongwei Sun, Ying Sun, Xiya Zhang, Xiaowei Xie, Jinran Zhang, Zhen Zhang, Yuxi Chen, Chenhui Ding, Yuanyan Xiong, Wenbin Ma, Dan Liu, Junjiu Huang, and Zhou Songyang declare that they have no conflict of interest. All institutional and national guidelines for the care and use of laboratory animals were followed.

All animal experiments were performed according to protocols approved by the Committee on Animal Care at the School of Life Sciences, Sun Yat-Sen University.

REFERENCES

Anders C, Niewoehner O, Duerst A, Jinek M (2014) Structural basis of PAM-dependent target DNA recognition by the Cas9 endonuclease. Nature 513:569–573

Bansal V, Libiger O (2011) A probabilistic method for the detection and genotyping of small indels from population-scale sequence data. Bioinformatics 27:2047–2053

Capecchi MR (2005) Gene targeting in mice: functional analysis of the mammalian genome for the twenty-first century. Nat Rev Genet 6:507–512

Cho SW, Kim S, Kim JM, Kim JS (2013) Targeted genome engineering in human cells with the Cas9 RNA-guided endonuclease. Nat Biotechnol 31:230–232

Cong L, Ran FA, Cox D, Lin S, Barretto R, Habib N, Hsu PD, Wu X, Jiang W, Marraffini LA et al (2013) Multiplex genome engineering using CRISPR/Cas systems. Science 339:819–823

Conticello SG (2008) The AID/APOBEC family of nucleic acid mutators. Genome Biol 9:229

Fu Y, Reyon D, Joung JK (2014a) Targeted genome editing in human cells using CRISPR/Cas nucleases and truncated guide RNAs. Methods Enzymol 546:21–45

Fu Y, Sander JD, Reyon D, Cascio VM, Joung JK (2014b) Improving CRISPR-Cas nuclease specificity using truncated guide RNAs. Nat Biotechnol 32:279–284

Gaj T, Gersbach CA, Barbas CF 3rd (2013) ZFN, TALEN, and CRISPR/Cas-based methods for genome engineering. Trends Biotechnol 31:397–405

Harris RS, Petersen-Mahrt SK, Neuberger MS (2002) RNA editing enzyme APOBEC1 and some of its homologs can act as DNA mutators. Mol Cell 10:1247–1253

Hsu PD, Lander ES, Zhang F (2014) Development and applications of CRISPR-Cas9 for genome engineering. Cell 157:1262–1278

Inui M, Miyado M, Igarashi M, Tamano M, Kubo A, Yamashita S, Asahara H, Fukami M, Takada S (2014) Rapid generation of mouse models with defined point mutations by the CRISPR/Cas9 system. Sci Rep 4:5396

Jiang W, Bikard D, Cox D, Zhang F, Marraffini LA (2013) RNA-guided editing of bacterial genomes using CRISPR-Cas systems. Nat Biotechnol 31:233–239

Jinek M, Chylinski K, Fonfara I, Hauer M, Doudna JA, Charpentier E (2012) A programmable dual-RNA-guided DNA endonuclease in adaptive bacterial immunity. Science 337:816–821

Jinek M, East A, Cheng A, Lin S, Ma E, Doudna J (2013) RNA-programmed genome editing in human cells. Elife 2:e00471

Kidane D, Murphy DL, Sweasy JB (2014) Accumulation of abasic sites induces genomic instability in normal human gastric epithelial cells during Helicobacter pylori infection. Oncogenesis 3:e128

Kim H, Kim JS (2014) A guide to genome engineering with programmable nucleases. Nat Rev Genet 15:321–334

Kim K, Ryu SM, Kim ST, Baek G, Kim D, Lim K, Chung E, Kim S, Kim JS (2017a) Highly efficient RNA-guided base editing in mouse embryos. Nat Biotechnol

Kim YB, Komor AC, Levy JM, Packer MS, Zhao KT, Liu DR (2017b) Increasing the genome-targeting scope and precision of base editing with engineered Cas9-cytidine deaminase fusions. Nat Biotechnol

Kingma PS, Corbett AH, Burcham PC, Marnett LJ, Osheroff N (1995) Abasic sites stimulate double-stranded DNA cleavage mediated by topoisomerase II. DNA lesions as endogenous topoisomerase II poisons. J Biol Chem 270:21441–21444

Kleinstiver BP, Prew MS, Tsai SQ, Topkar VV, Nguyen NT, Zheng Z, Gonzales AP, Li Z, Peterson RT, Yeh JR et al (2015) Engineered CRISPR-Cas9 nucleases with altered PAM specificities. Nature 523:481–485

Kleinstiver BP, Pattanayak V, Prew MS, Tsai SQ, Nguyen NT, Zheng Z, Joung JK (2016) High-fidelity CRISPR-Cas9 nucleases with no detectable genome-wide off-target effects. Nature 529:490–495

Komor AC, Kim YB, Packer MS, Zuris JA, Liu DR (2016) Programmable editing of a target base in genomic DNA without double-stranded DNA cleavage. Nature 533:420–424

Komor AC, Badran AH, Liu DR (2017) CRISPR-based technologies for the manipulation of eukaryotic genomes. Cell 168:20–36

Li H, Durbin R (2010) Fast and accurate long-read alignment with Burrows-Wheeler transform. Bioinformatics 26:589–595

Li J, Sun Y, Du J, Zhao Y, Xia L (2016) Generation of targeted point mutations in rice by a modified CRISPR/Cas9 system. Mol Plant

Lu Y, Zhu JK (2016) Precise editing of a target base in the rice genome using a modified CRISPR/Cas9 system. Mol Plant

Ma W, Panduri V, Sterling JF, Van Houten B, Gordenin DA, Resnick MA (2009) The transition of closely opposed lesions to double-strand breaks during long-patch base excision repair is prevented by the coordinated action of DNA polymerase delta and Rad27/Fen1. Mol Cell Biol 29:1212–1221

Ma Y, Zhang J, Yin W, Zhang Z, Song Y, Chang X (2016) Targeted AID-mediated mutagenesis (TAM) enables efficient genomic diversification in mammalian cells. Nat Methods 13:1029–1035

Mali P, Aach J, Stranges PB, Esvelt KM, Moosburner M, Kosuri S, Yang L, Church GM (2013) CAS9 transcriptional activators for target specificity screening and paired nickases for cooperative genome engineering. Nat Biotechnol 31:833–838

McKenna A, Hanna M, Banks E, Sivachenko A, Cibulskis K, Kernytsky A, Garimella K, Altshuler D, Gabriel S, Daly M et al (2010) The Genome Analysis Toolkit: a MapReduce framework for analyzing next-generation DNA sequencing data. Genome Res 20:1297–1303

Nishida K, Arazoe T, Yachie N, Banno S, Kakimoto M, Tabata M, Mochizuki M, Miyabe A, Araki M, Hara KY, et al. (2016) Targeted nucleotide editing using hybrid prokaryotic and vertebrate adaptive immune systems. Science 353

Nishimasu H, Ran FA, Hsu PD, Konermann S, Shehata SI, Dohmae N, Ishitani R, Zhang F, Nureki O (2014) Crystal structure of Cas9 in complex with guide RNA and target DNA. Cell 156:935–949

Porteus MH (2006) Mammalian gene targeting with designed zinc finger nucleases. Mol Ther 13:438–446

Porteus MH, Carroll D (2005) Gene targeting using zinc finger nucleases. Nat Biotechnol 23:967–973

Qi LS, Larson MH, Gilbert LA, Doudna JA, Weissman JS, Arkin AP, Lim WA (2013) Repurposing CRISPR as an RNA-guided platform for sequence-specific control of gene expression. Cell 152:1173–1183

Saraconi G, Severi F, Sala C, Mattiuz G, Conticello SG (2014) The RNA editing enzyme APOBEC1 induces somatic mutations and a compatible mutational signature is present in esophageal adenocarcinomas. Genome Biol 15:417

Suzuki K, Tsunekawa Y, Hernandez-Benitez R, Wu J, Zhu J, Kim EJ, Hatanaka F, Yamamoto M, Araoka T, Li Z et al (2016) In vivo genome editing via CRISPR/Cas9 mediated homology-independent targeted integration. Nature 540:144–149

Tesson L, Usal C, Menoret S, Leung E, Niles BJ, Remy S, Santiago Y, Vincent AI, Meng X, Zhang L et al (2011) Knockout rats generated by embryo microinjection of TALENs. Nat Biotechnol 29:695–696

Thomas KR, Capecchi MR (1987) Site-directed mutagenesis by gene targeting in mouse embryo-derived stem cells. Cell 51:503–512

Wu Y, Liang D, Wang Y, Bai M, Tang W, Bao S, Yan Z, Li D, Li J (2013) Correction of a genetic disease in mouse via use of CRISPR-Cas9. Cell stem cell 13:659–662

Yang H, Wang H, Shivalila CS, Cheng AW, Shi L, Jaenisch R (2013) One-step generation of mice carrying reporter and conditional alleles by CRISPR/Cas-mediated genome engineering. Cell 154:1370–1379

Yang L, Briggs AW, Chew WL, Mali P, Guell M, Aach J, Goodman DB, Cox D, Kan Y, Lesha E et al (2016) Engineering and optimising deaminase fusions for genome editing. Nat Commun 7:13330

Zhang X, Liang P, Ding C, Zhang Z, Zhou J, Xie X, Huang R, Sun Y, Sun H, Zhang J et al (2016) Efficient production of gene-modified mice using *Staphylococcus aureus* Cas9. Sci Rep 6:32565

Zong Y, Wang Y, Li C, Zhang R, Chen K, Ran Y, Qiu JL, Wang D, Gao C (2017) Precise base editing in rice, wheat and maize with a Cas9- cytidine deaminase fusion. Nat Biotechnol

14

A peep into mitochondrial disorder: multifaceted from mitochondrial DNA mutations to nuclear gene modulation

Chao Chen[1], **Ye Chen**[1,2]✉, **Min-Xin Guan**[1,2]

[1] School of Medicine, Institute of Genetics, Zhejiang University, Hangzhou 310058, China
[2] Collaborative Innovation Center for Diagnosis and Treatment of Infectious Diseases, Zhejiang University, Hangzhou 310058, China
✉ Correspondence: yechency@zju.edu.cn (Y. Chen)

ABSTRACT

Mitochondrial genome is responsible for multiple human diseases in a maternal inherited pattern, yet phenotypes of patients in a same pedigree frequently vary largely. Genes involving in epigenetic modification, RNA processing, and other biological pathways, rather than "threshold effect" and environmental factors, provide more specific explanation to the aberrant phenotype. Thus, the double hit theory, mutations both in mitochondrial DNA and modifying genes aggravating the symptom, throws new light on mitochondrial dysfunction processes. In addition, mitochondrial retrograde signaling pathway that leads to reconfiguration of cell metabolism to adapt defects in mitochondria may as well play an active role. Here we review selected examples of modifier genes and mitochondrial retrograde signaling in mitochondrial disorders, which refine our understanding and will guide the rational design of clinical therapies.

KEYWORDS mitochondrial disorder, mitochondrial DNA mutation, nuclear modifier gene, mitochondrial retrograde signaling

INTRODUCTION

Mitochondria are essential organelles inside cells that are responsible for cellular energy production. Through a variety of pathways, the mitochondria provide fuel (adenosine triphosphate) for cell survival. In addition, mitochondria now are recognized as a fundamental platform in cellular signaling, with crucial roles in a number of metabolic and developmental process, including cell autophagy, cell apoptosis (death), calcium, copper and iron homeostasis, and cell cycle regulation (Chan, 2006, Newmeyer and Ferguson-Miller, 2003, Nunnari and Suomalainen, 2012, Rubinsztein et al., 2011). Chronic, multi-symptom illness arises when sufficient numbers of mitochondrion were damaged. Mitochondrial disorders has been linked to an enormous variety of disease, such as MELAS (mitochondrial encephalomyopathy, lactic acidosis, and stroke-like episodes) syndrome, MERRF (myoclonic epilepsy with ragged red fibers) syndrome, LOHN (Leber's hereditary optic neuropathy), deafness, diabetes, Alzheimer disease, and Parkinson disease (Wallace, 2005). Moreover, progressive mitochondrial dysfunction has also been implicated in the aging process (Ross et al., 2014).

Mitochondrial damage can be inherited via mutations both in maternal DNA (mtDNA) and nuclear DNA, and present at birth or remain latent until triggered later in life. In this review, we will focus specifically on disorders caused by primary mutations of mtDNA and highlight a few major recent and ongoing developments of genetic modification, which may offer insights into the research in progress, as well as suggestions regarding further advances needed.

AN OVERVIEW OF THE MITOCHONDRIAL DNA MUTATIONS

The human mtDNA is a 16,569 base pairs double stranded circular molecule (cytosine-rich light (L) and guanine-rich heavy (H) strands). This remarkably compact genome contains 13 protein-coding genes (core subunits of respiratory chain complexes), 22 tRNA genes, and 2 ribosomal genes (12S rRNA and 16S rRNA) (Fig. 1) (Wallace and Chalkia,

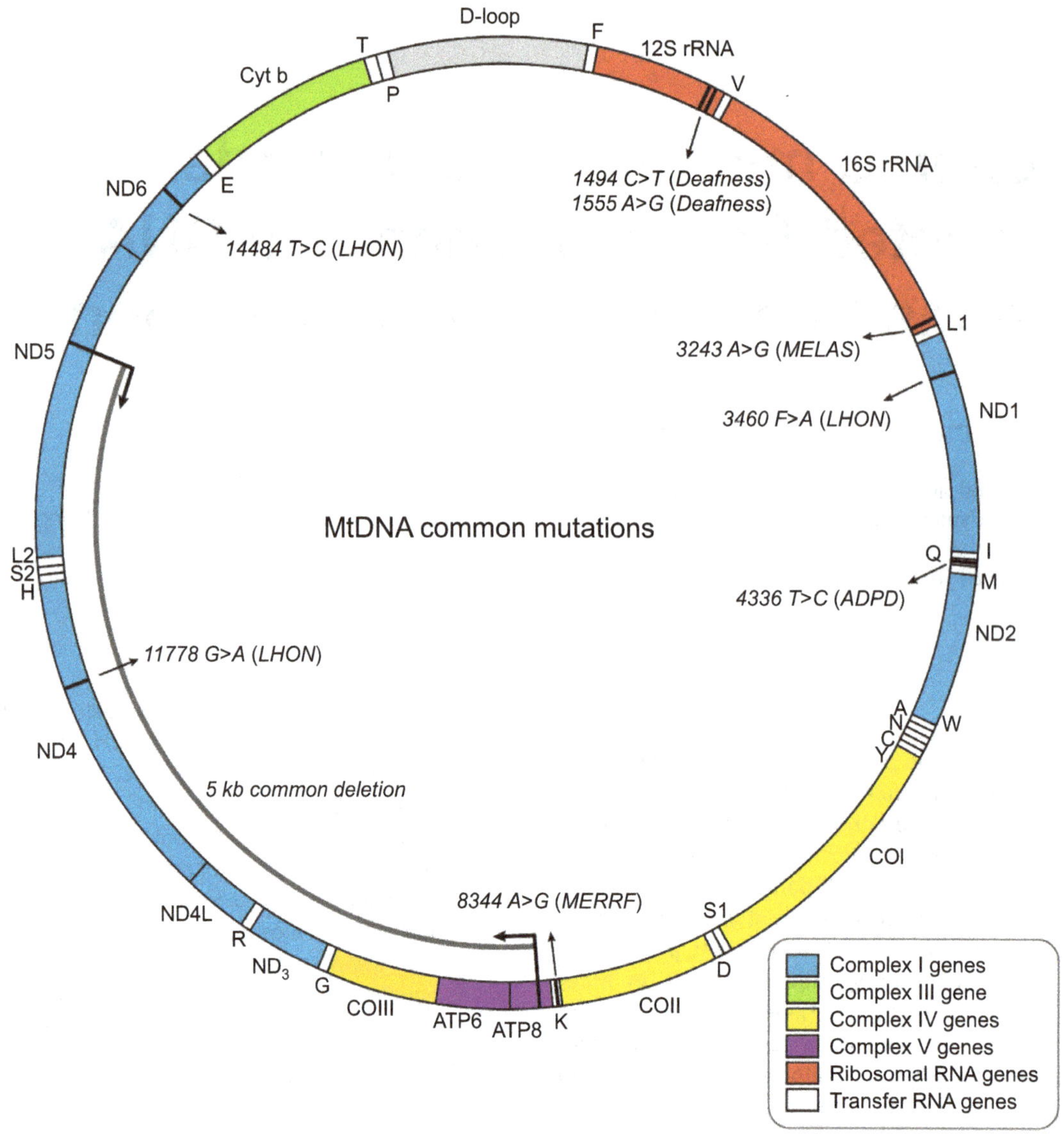

Figure 1. Human mitochondrial genome. Represented is a schematic diagram of the 16.6 kb circular, double-stranded human mitochondrial genome. The D-loop region, or non-coding control region, is vital for the initiation of mtDNA replication and transcription. The two ribosomal RNAs (12S rRNA and 16S rRNA) are shown in *red* and 22 tRNAs are shown in *white* and denoted by *single letter* codes. The subunits of complex I (ND1–ND6 and ND4L) are shown in *blue*; cytochrome b (Cyt b) of complex III is shown in *green*; cytochrome c oxidase (COI–COIII) is shown in *yellow*; and the subunits of the ATP synthase (ATP6 and ATP8) are shown in *purple*. The positions of mutations referred to in the text are marked by *black lines* and *arrows*. LHON, Leber's hereditary optic neuropathy; MELAS, mitochondrial encephalomyopathy, lactic acidosis, and stroke-like episodes; MERRF, myoclonic epilepsy and ragged *red* muscle fibers; ADPD, Alzeimer's disease and Parkinsons's disease.

2013). Due to the oxidative damage and lacking of protective histones, the mitochondrial genome has a very high mutation rate, 10- to 17-fold higher than that observed in nuclear DNA. In normal tissues, usually all the mtDNA molecules are identical, known as homoplasmy; when a mixture of wild type and mutant mtDNA is encountered, results in heteroplasmy. In heteroplasmic cells, the mtDNA genotype can shift during cell replication. Consequently, some lineages drift toward wild type mtDNA and become homoplasmy, while others remain heteroplasmic.

Most mtDNA alterations are neutral polymorphisms, and this type of DNA sequence variation has been categorized into "haplogroups", which have been proved useful in the reconstruction of historic population movements and practical applications such as forensics (Parson and Bandelt, 2007). The first pathogenic mtDNA mutations were identified in 1988 in patients with mitochondrial myopathies (Holt et al., 1988). Since then, mtDNA mutations have been increasingly recognized as an important contributor to an array of diseases. Over 250 pathogenic mutations (point mutations and rearrangements) have been identified and characterized in mtDNA (http://mitomap.org/MITOMAP), which cause a wide variety of disorders with heterogeneity of phenotypes and a variable age of onset.

The mtDNA point mutations are usually heteroplasmic and maternally inherited. These can occur with mtDNA-encoded proteins, tRNAs, or ribosomal RNA (rRNA), and consequently affect the replication, transcription, or RNA processing. However, more than half of reported disease-related mtDNA point mutations are located within the tRNA genes. The most common sites for mt-tRNA mutation are tRNA$^{Leu(UUR)}$ (*MT-TL1*), and tRNA^{Lys} (*MT-TK*). The m.3243A>G mutation in the tRNA$^{Leu(UUR)}$ gene was first identified in patients with MELAS syndrome (Goto et al., 1990), and now it has also been proved as one of the important causes of maternally inherited diabetes and deafness (van den Ouweland et al., 1992). To investigate the molecular pathogenic mechanism of the mitochondrial mutations, cell hybrids (namely cybrids) of mtDNA-deficient cells and enucleated lymphoblastoid cells from patients are generated. Analysis of cybrids harboring homoplasmic m.3243A>G mutation revealed that the level of aminoacylated tRNA$^{Leu(UUR)}$ was reduced 70%–75%, which is mainly due to a shortage of tRNA$^{Leu(UUR)}$, leading to the reduced rate of mitochondrial protein synthesis and respiration defects (Picard et al., 2014). The most common mutation in tRNA^{Lys} is m.8344A>G missense mutation, which is the main cause of MERRF syndrome that accounts for 80% of affect individuals. Similar to m.3243A>G, decreased steady-state levels and aminoacylation of the tRNA^{Lys} were observed (Enriquez et al., 1995). In addition, m.8356T>G and m.8363G>A mutations in the tRNA^{Lys} are associated with MERRF, compatible with hearing loss. Yasukawa et al. reported that both m.3243A>G and m.8344A>G mutations lead to similar uridine modification defects at the anticodon wobble position, respectively (Yasukawa et al., 2000a, Yasukawa et al., 2000b). Biochemical studies of tRNAs isolated from MELAS or MERRF patient cybrid cells revealed that the taurine and 2-thio modifications of uridine stabilize the codon-anticodon pairing. Thus, these structural tRNA modifications are critical for efficient and accurate decoding process and further overall mitochondrial translation (Suzuki and Nagao, 2011).

Primary mutations in mtDNA-encoded proteins, namely subunits of respiratory chain complexes, also have been linked to inherited diseases. m.11778G>A, m.3460G>A, and m.14484T>C, respectively altering NADH dehydrogenase subunit *MT-ND4*, *MT-ND1*, *MT-ND6*, are present in at least 90% of LHON cases. When there is a greater percentage of m.8993T>G or m.8993T>C mutation in *MT-ATP6*, maternally inherited Leigh syndrome (MILS) is observed, which particularly affects the brainstem, diencephalon, and basal ganglia. The m.1555A>G mutation in the12S rRNA (MT-RNR1) was first identified in 1993 in a large Arab-Israeli pedigree (Prezant et al., 1993), which is the first example of point mutation in mitochondrial rRNA mtDNA, and subsequently found in many families of various ethnic backgrounds. The m.1555A>G or m.1494C>T mutation is located in the decoding site of 12S rRNA and is predicted to cause a change in the secondary rRNA structure. This alteration impairs protein synthesis and enhances an interaction with aminoglycoside antibiotics (Zhao et al., 2004).

Among different mtDNA rearrangement mutations, large-scale deletions, varying in size from 1.3 to 8 kb, were the majority patterns. With these large-scale deletions, patients are more likely to suffer from Kearns-Sayre syndrome, chronic progressive external ophthalmoplegia (CPEO), or Pearson syndrome (Taylor and Turnbull, 2005). The most common mtDNA deletion is a 5-kb deletion (m.8470-m.13447), which is present in approximately one third of patients. Despite different origins, most mtDNA deletions occur between the regions of replication that is flanked by short direct repeat sequences (Samuels et al., 2004).

DOUBLE HIT THEORY: THE PENETRANCE OF MITOCHONDRIAL DISORDER MAY BE SUBJECT TO THE EFFECTS OF NUCLEAR MODIFIER GENES

Given the mitochondrion's important role in cell growth and survival, patients with mitochondrial dysfunction appear to have a highly diverse multi-organ symptom. Mitochondrial dysfunction is usually considered as one of many risk factors in multifactorial disease, as it predisposes for disease throughout the entire system. And considering the mitochondrion's role in energy production, high energy dependent tissues such as brain, heart, liver, and muscles, are most susceptible to direct mitochondrial damage. "Threshold effect" is one of the specific features of mitochondrial disease, since majority of mtDNA mutations are found in some but not all mitochondria genomes that refer to "heteroplasmy". Biochemical defects and tissue dysfunction will not be apparent until the mutated mtDNAs reach a minimum critical proportion, and this threshold level varies among different tissues. The incomplete penetrance can be influenced by pharmaceutical or environmental exposures and nutrient or cofactor deficits. Although "threshold effect" and environmental factors can partly explain the various disease phenotypes observed in patients harboring same mtDNA mutation, the exact correlation and mechanism of tissue specification are still lacking.

In addition, family members with same background of mtDNA genome were found to be affected in different ways. For example, LHON is usually due to a homoplasmic mtDNA mutation and all maternal offspring will inherit the mutation; however, whilst 50% of male offspring are affected with

LHON, only 10% of female offspring will develop visual loss (Harding et al., 1995, Riordan-Eva et al., 1995). In a very recent study, Jaime and colleagues reported that inherited mtDNA sequence variation combined with somatic mtDNA mutagenesis has an additive effect in creating phenotypes relevant for pathology and ageing (Ross et al., 2014). To explain the pathogenesis of these cases, the most accepted hypothesis will be the double hit theory: first hit, predisposition for disease as a consequence of primary mtDNA mutation, and second hit, the effects of genetic modifier.

Over the past few years, many studies have been done on genetic modification, and major advances have occurred in both understanding and practice with regard to targeting modifier genes in various diseases such as cancer, arrythmia, and cystic fibrosis (Gusella et al., 2014, Luhmann e al., 2015). These modifier genes often have at least two alleles, one of which exacerbates disease, and one that suppresses disease. It has been proved that mitochondrial genome combined with a poorly co-adapted nucleus will lead to reduced fitness/lifespan in animal models (Ross et al., 2013). Given that mitochondrial disorders generally refer to diseases caused by dysfunctional bioenergetics, the modifier genes are more likely to be associated with OXPHOS systems, including mitochondrial DNA, RNA, and protein dynamics (Table 1). Scientists from various countries are now at different stages in researching potential genetic modifiers responsible for the phenotypic variances. Various nuclear genes have been confirmed to cause mitochondrial diseases (Koopman et al., 2012), yet few studies have focus on their interplay with primary mitochondrial mutations. Once the modifier genes that suppress mitochondrial dysfunction are identified, the door opens to new potential therapeutic targets, since these modifier genes are more amenable to administrate than the primary mutant mtDNAs.

Function of modifier gene in mitochondrial hearing loss was first reported by Bykhovskaya et al. in a non-syndromic hearing loss (NSHL) Arab-Israeli family bearing 12S rRNA (m.1555A>G) mutation (Bykhovskaya et al., 2004). The m.1555A>G mutation, locates in the decoding site of the mitochondrial small subunit (SSU) ribosomal RNA, is the first identified homoplasmic mitochondrial mutation. In addition, the mutation is predicted to cause an alteration in the second structure, which impairs protein synthesis and enlarges sensitivity to aminoglycoside ototoxicity (Prezant et al., 1993). It is also well accepted that m.1555A>G mutation presents as a key cause of antibiotic-induced hearing loss; however, the mutation alone typically does not lead to disease. Among these Arab-Israeli family members without previous exposure to aminoglycosides, the m.1555A>G mutation induced various clinical phenotypes ranging from severe congenital deafness, to moderate progressive hearing loss of later onset, to completely normal hearing. It was characterized by Guan et al. that there's more severe biochemical defects in the lymphoblastoid cells derived from symptomatic individuals than those from asymptomatic ones of the Arab-Israeli family bearing m.1555A>G mutation (Guan et al., 1996). On the other hand, an identical degree of mitochondrial dysfunction was observed when they compared the cybrids cell lines derived from symptomatic and asymptomatic individuals (Guan et al., 2001). These findings strongly indicate that the m.A1555G mutation as a primary cause of hearing loss and nuclear modifier genes play a role in modulating the phenotypic expression (Guan et al., 2006). Naturally, the most promising candidates would be these nuclear genes encode the subunits of respiratory chain complex, proteins involved in mitochondrial protein synthesis, and proteins involved in mtDNA replication and maintenance. *TFB1M* (transcription factor B1), encoding a mitochondrial rRNA methyltransferase, has been putatively identified as a possible nuclear modifier of the m.1555A>G mutation, suggesting a connection between 12S rRNA methylation and hearing loss (Raimundo et al., 2012).

Mutations in mitochondrial tRNAs have been reported to be associated with various mitochondrial disease states (Abbott et al., 2014). With disrupted structures, mt tRNAs mutations would cause defective translation and impaired mt protein synthesis, leading to defects in OXPHOS systems. Post-transcriptional processing, including maturation of primary tRNA, multiple chemical residue modifications, and aminoacylation, are critical to accurate and effective

Table 1. Putative modifier genes reviewed in text

Gene name	Function	Reference
TFB1M	rRNA methylation	Raimundo et al., (2012)
TRMU	tRNA base modification	Guan et al., (2006)
MTO1		Li et al., (2002)
GTPBP3		Li and Guan, (2003)
KARS	tRNA aminoacylation	McMillan et al., (2014)
YARS2		Nakajima et al., (2014)
VARS2		Diodato et al., (2014)
TARS2		Diodato et al., (2014)
LARS2		Perli et al., (2014)

translation. Thus enzymes involved in these processing are highly possible modifier genes. The penetrance is much higher in the presence of nuclear mutations involved in transfer RNA base modification (*MTO1*, *TRMU-MTO2*, and *GTPBP3* genes) (Guan et al., 2006, Li and Guan, 2003, Li et al., 2002); however, additional supporting evidence is still needed to firmly confirm their role as genetic modifier. Establishment of ideal animal models may help discover their functions in mitochondrial diseases and explain their tissue specificity. Recent studies largely expand the phenotypic spectrum associated with different aminoacyl-tRNA synthetases (ARS). McMilan et al. reported congenital visual impairment and progressive microcephaly has been associated with *KARS* mutations (McMillan et al., 2014) and Nakajima et al. reported a homozygous *YARS2* causes severe myopathy, lactic acidosis, and sideroblastic anemia 2 (Nakajima et al., 2014). Another whole-exome sequencing study reveals that mutations in *VASR2* and *TARS2* are the causes of mitochondrial encephalomyopathies (Diodato et al., 2014). Perli et al. further reported that isolated non-catalytic C-terminal of *LASR2* can improve both viability and bioenergetic proficiency of cybrid cells carrying pathogenic mutations in mt-tRNAs (Perli et al., 2014). These findings strongly suggest the group of aminoacyl-tRNA synthetases as active modifying players in mitochondrial disorders, and may lead to further understanding of tissue specific mitochondrial diseases.

In general, our knowledge of modifier genes involved in mitochondrial disorders has increased substantially in the past decade. Several mitochondrial rRNA methyltransferase and mitochondrial tRNA modifications have been identified in human, but the proteins involved in these modifications are far from being all identified. Understanding how the cells modulate biological processes to accommodate the adverse effects of mtDNA dysfunction is important as it may provide vital clues in the search for modifier genes as well as therapeutic targets.

MITOCHONDRIAL RETROGRADE SIGNALING IN MITOCHONDRIAL DISORDERS

Mitochondria play a central role not only in energy production but also in the integration of metabolic pathways as well as signals for apoptosis and autophagy. Mitochondria-to-nucleus retrograde signaling was first discovered in yeast by Parikh et al. (Parikh et al., 1987) and subsequently described in mammalian cells, which also known as mitochondrial stress signaling (Gomes et al., 2013). The mitochondrial metabolism perturbation is due to dysfunctional OXPHOS system or mtDNA mutations leading to loss of mitochondrial membrane potential (Δψ) and abnormal ROS generation. These stress signals mitochondrial dysfunction to cytosol by unbalanced levels of ATP and NADH and the release of Ca^{2+}, which results in active calcium-sensitive proteins and further the activation of downstream transcription factors. Consequently, the adaptive modulating the expression of nuclear genes such as metabolic enzyme genes and stress response genes leads to a compensation for the metabolic reconfiguration, in which various mitochondrial or cellular effects can be achieved (Fig. 2) (Butow and Avadhani, 2004).

Over the past few years, the awareness of the biological processing in mitochondrial dysfunction disease has grown exponentially along with the development of sequencing technology. It has been reported that the retrograde signaling interacts with several other signaling pathways, such as target of rapamycin (TOR) signaling, AMP-dependent protein kinase (AMPK) pathway, and mitochondrial unfolded protein response (Butow and Avadhani, 2004, Pellegrino et al., 2013, Ryan and Hoogenraad, 2007). A wide spectrum of genes were affected under mitochondrial stress, including those involved in Ca^{2+} storage and release (RyR1, RyR2, calreticulin, calsequestrin), in glucose uptake and metabolism (Glut 4, IGF1R, hexokinase, IRS1), in oncogenesis (TGFβ1, p53, and cMyc), and in apoptosis (Bcl-2, Survivin, BAD, Bax, Bid) (Table 2) (Bers, 2008, Biswas et al., 2005, Wallace, 2012, Youle and Strasser, 2008).

The increase of mitochondrial biogenesis is one of the compensatory strategies to mitochondrial dysfunction commonly observed, as exemplified by the significantly increased proliferation of mitochondria in skeletal muscle fibers from patients with LHON (DiMauro and Schon, 2003). In a recent study, Giordano et al. showed a much higher mitochondrial DNA content and increased mitochondrial biogenesis in multiple tissues in unaffected mutation (m.11778G>A) carriers, which differentiates the unaffected carries from LHON affected patients and healthy individuals (Giordano et al., 2014). The gene expression analysis showed a scale of increasing expression of transcription factors (NRF1 and TFAM) and the PPRC1 from controls to affected individuals to carriers. A significant difference was reached for PPRC1 and TFAM, comparing carriers to controls. Considering that PPRC1 are transcriptional co-activator upstream of NRF1 and TFAM, both regarded as key regulators of mitochondrial biogenesis, these results provide a reasonable explanation that efficient mitochondrial biogenesis may account for the incomplete penetrance in LHON.

Although the compensatory mechanism may improve the efficiency of the mitochondrial translation, the function of the OXPHOS system can still remain impaired in cell with mtDNA mutations. The loss of mitochondrial transmembrane potential (Δψm) acts as inducers of mitochondrial retrograde signaling. Using ρ° human fibrosarcoma 143B cells and a MERRF cybrid cell line carrying the mutated mitochondrial $tRNA^{Lys}$ (m.8344A>G), Arnould et al. showed that the respiratory deficiency induced the activation of CaMK IV, which in turn activated CREB by protein phosphorylation (Arnould et al., 2002). Recently, some models suggest that mitochondrial reactive oxygen species (ROS) also acts as an important messengers in mitochondria-nucleus crosstalk. ROS is the natural by-products of oxygen metabolism and is

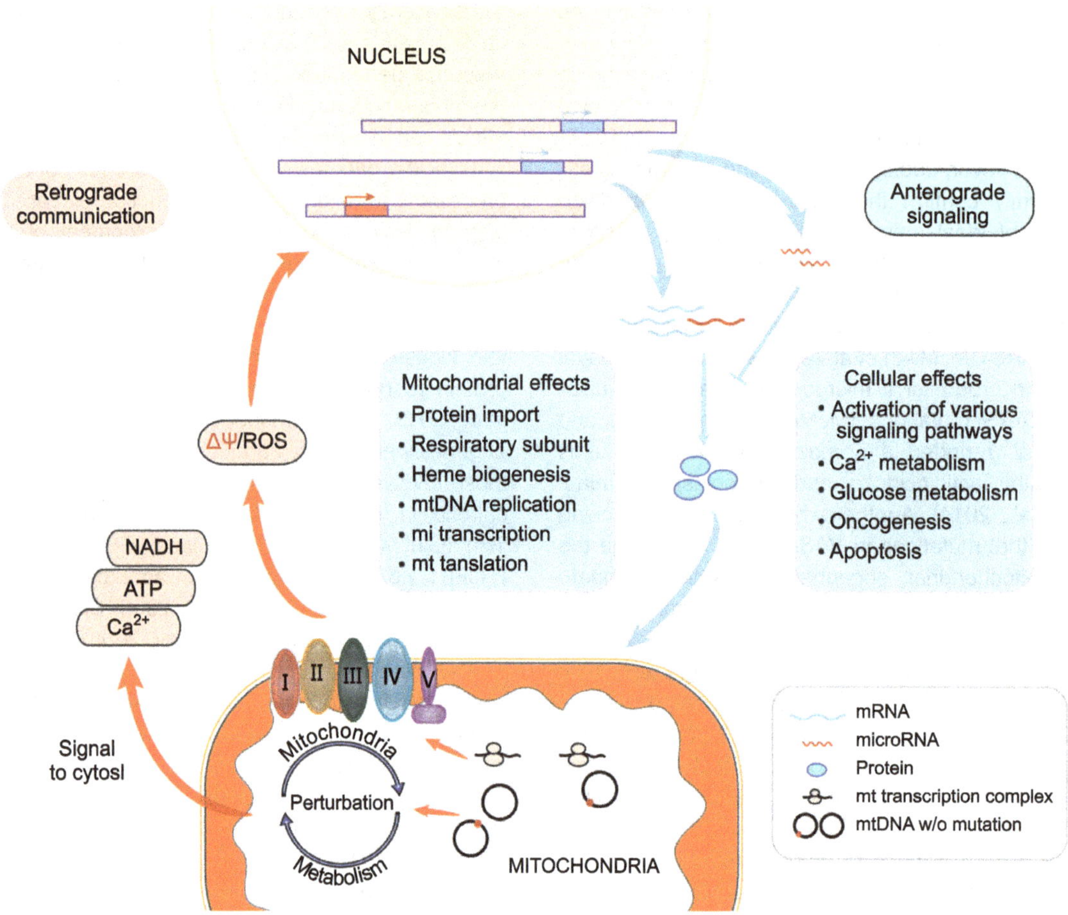

Figure 2. A diagram of the communication between mitochondria and nucleus. Multiple genes are involved in the nuclear-mitochondrial cross talk and respond to oxidative stress manifested due to impaired mitochondrial function.

Table 2. Pathways and processing in retrograde signaling

Name	Genes	Reference
TOR signaling pathway	Tor1, Tor2	Butow and Avadhani, (2004)
AMPK pathway	PGC1α, UCP1, p53	Ryan and Hoogenraad, (2007)
Unfolded protein response	JNK2, AKT	Pellegrino et al., (2013)
Ca^{2+} metabolism	RyR1, RyR2, calreticulin, calsequestrin	Bers, (2008)
Glucose metabolism	Glut 4, IGF1R, hexokinase, IRS1	Biswas et al., (2005)
Oncogenesis	TGFβ1, p53, and cMyc	Wallace, (2012)
Apoptosis	Bcl-2, Survivin, BAD, Bax, Bid	Youle and Strasser, (2008)

important to a number of basic cell and life processes, such as signaling and the defense against pathogens, but ROS levels must be kept in strict balance. Raimundo et al. showed that increased level of mitochondrial ROS in m.1555A>G cybrids activates the proapoptotic nuclear transcription factor E2F1 in an AMPK dependent manner; and in the animal study by using Tg-mtTFB1 transgenic (to model pathogenesis due to increased mitochondrial12S rRNA methylation), progressive hearing loss was observed associated with tissue-specific upregulation of E2F1, as well as the apoptosis of critical cells in inner ear (Raimundo et al., 2012). Importantly, blocking the pathway at any level in cultured cells ablates apoptosis susceptibility, which was equally observed *in vivo* (Raimundo et al., 2012). Gradually researchers are changing their focus from decreased mitochondrial performance to pathogenic signaling elicited by mitochondria dysfunction (Raimundo, 2014). It was recently reported by Meseguer et al. that mt-DNA mutation can directly affect microRNA expression (Meseguer et al., 2015). The authors found that enhanced ROS level in MELAS cells induced a post-transcriptional miRNA mediated response which is responsible for the regulation of mt-tRNA-modifying enzymes. MicroRNA-9/9* expression was significantly induced through a ROS/NFkB signaling pathway in cybrids with m.3243A>G mutation, which negatively regulate GTPBP3, MTO1, and TRMU, leading to mt-tRNA hypo-modification and contributes to the MELAS phenotype.

It has been reported that single mtDNA point mutation can cause different cellular transcriptional responses within cells of same nuclear background. Picard et al. demonstrated that continuous changes in mtDNA heteroplasmy (m.3243A>G) result in discontinuous remodeling of nuclear DNA and mtDNA gene expression profiles due to alterations in both the signal transduction and epigenetic regulatory processes. It was reported that individuals harboring 10%–30% m.3243A>G mutation manifest diabetes and occasionally autism, individuals with 50%–90% mutant mtDNAs manifest encephalomyopathies, and these cases with 90%–100% mutant mtDNAs face perinatal lethality (Picard et al., 2014). This result provides an alternative perspective on the cellular basis of phenotypic heterogeneity in mtDNA diseases.

CONCLUSION

Unlike chromosomal genes, the mtDNA can be present in hundreds to thousands of copies. Relatively subtle changes in the proportion of mutant mtDNA can lead to dramatic effects on a patient's phenotype; however, the mtDNA mutation doesn't do it alone. In spite of environmental factors, the discovery of nuclear modifier genes provides additional information about pathways in which the primary mutation functions, as well as new entry points for understanding the pathological effects of certain disease gene.

That mitochondrial dysfunction can modulate nuclear gene expression has been demonstrated in different species. Existing studies indicate that the retrograde response accumulates overall a cell's lifespan, which compensates for mitochondrial dysfunction as mitochondrial quality control (Owusu-Ansah et al., 2013, Raimundo, 2014). In addition, the mitochondrial retrograde signaling triggers both adaptive and maladaptive cellular responses, which constitutes a complex network of processes (Guha and Avadhani, 2013). However, our current understanding of the mechanistic details is far from complete, a systems approach is needed.

In this review, we have presented the examples of modifier genes for human mitochondrial disorders. While chromosomal location of some modifier effects has been identified, cloning of modifier genes still remains to be difficult. Studies of the mitochondrial retrograde signaling will provide insight into mechanisms of genetic interactions and facilitate the identification of potential modifier genes. Finally, the elucidation of modifier genes associated with the suppression of mitochondrial defects could be useful in designing new therapeutics, improving prediction of risk factors for susceptibility, and eventual prevention of disease manifestation.

ACKNOWLEDGEMENTS

The authors were supported by the National Basic Research Program (973 Program) (No. 2014CB541700), the project from National Natural Science Foundation of China (Grant No. 81470685) and the Natural Science Foundation of Zhejiang province (LR15H080001).

ABBREVIATIONS

AMPK, AMP-dependent protein kinase; ARS, aminoacyl-tRNA synthetases; CPEO, chronic progressive external ophthalmoplegia; LOHN, Leber's hereditary optic neuropathy; MELAS, mitochondrial encephalomyopathy, lactic acidosis, and stroke-like episodes; MERRF, myoclonic epilepsy with ragged red fibers; MILS, maternally inherited Leigh syndrome; mtDNA, maternal DNA; NSHL, non-syndromic hearing loss; rRNA, ribosomal RNA; SSU, small subunit.

COMPLIANCE WITH ETHICS GUIDELINES

Chao Chen, Ye Chen, and Min-Xin Guan declare that they have no conflict of interest. This article does not contain any studies with human or animal subjects performed by the any of the authors.

REFERENCES

Abbott JA, Francklyn CS, Robey-Bond SM (2014) Transfer RNA and human disease. Front Genet 5:158

Arnould T, Vankoningsloo S, Renard P, Houbion A, Ninane N, Demazy C et al (2002) CREB activation induced by mitochondrial dysfunction is a new signaling pathway that impairs cell proliferation. EMBO J 21:53–63

Bers DM (2008) Calcium cycling and signaling in cardiac myocytes. Annu Rev Physiol 70:23–49

Biswas G, Guha M, Avadhani NG (2005) Mitochondria-to-nucleus stress signaling in mammalian cells: nature of nuclear gene targets, transcription regulation, and induced resistance to apoptosis. Gene 354:132–139

Butow RA, Avadhani NG (2004) Mitochondrial signaling: the retrograde response. Mol Cell 14:1–15

Bykhovskaya Y, Mengesha E, Wang D, Yang H, Estivill X, Shohat M et al (2004) Human mitochondrial transcription factor B1 as a modifier gene for hearing loss associated with the mitochondrial A1555G mutation. Mol Genet Metab 82:27–32

Chan DC (2006) Mitochondria: dynamic organelles in disease, aging, and development. Cell 125:1241–1252

DiMauro S, Schon EA (2003) Mitochondrial respiratory-chain diseases. N Engl J Med 348:2656–2668

Diodato D, Melchionda L, Haack TB, Dallabona C, Baruffini E, Donnini C et al (2014) VARS2 and TARS2 mutations in patients with mitochondrial encephalomyopathies. Hum Mutat 35:983–989

Enriquez JA, Chomyn A, Attardi G (1995) MtDNA mutation in MERRF syndrome causes defective aminoacylation of tRNA(Lys) and premature translation termination. Nat Genet 10:47–55

Giordano C, Iommarini L, Giordano L, Maresca A, Pisano A, Valentino ML et al (2014) Efficient mitochondrial biogenesis drives incomplete penetrance in Leber's hereditary optic neuropathy. Brain 137:335–353

Gomes AP, Price NL, Ling AJY, Moslehi JJ, Montgomery MK, Rajman L et al (2013) Declining NAD+ induces a pseudohypoxic state disrupting nuclear-mitochondrial communication during aging. Cell 155:1624–1638

Goto Y, Nonaka I, Horai S (1990) A mutation in the tRNALeu(UUR) gene associated with the MELAS subgroup of mitochondrial encephalomyopathies. Nature 348:651–653

Guan M-X, Fischel-Ghodsian N, Attardi G (1996) Biochemical evidence for nuclear gene involvement in phenotype of non-syndromic deafness associated with mitochondrial 12S rRNA mutation. Hum Mol Genet 5:963–971

Guan MX, Fischel-Ghodsian N, Attardi G (2001) Nuclear background determines biochemical phenotype in the deafness-associated mitochondrial 12S rRNA mutation. Hum Mol Genet 10:573–580

Guan MX, Yan Q, Li X, Bykhovskaya Y, Gallo-Teran J, Hajek P et al (2006) Mutation in TRMU related to transfer RNA modification modulates the phenotypic expression of the deafness-associated mitochondrial 12S ribosomal RNA mutations. Am J Hum Genet 79:291–302

Guha M, Avadhani NG (2013) Mitochondrial retrograde signaling at the crossroads of tumor bioenergetics, genetics and epigenetics. Mitochondrion 13:577–591

Gusella JF, MacDonald ME, Lee J-M (2014) Genetic modifiers of Huntington's disease. Mov Disord 29:1359–1365

Harding AE, Sweeney MG, Govan GG, Riordan-Eva P (1995) Pedigree analysis in Leber hereditary optic neuropathy families with a pathogenic mtDNA mutation. Am J Hum Genet 57:77–86

Holt IJ, Harding AE, Morgan-Hughes JA (1988) Deletions of muscle mitochondrial DNA in patients with mitochondrial myopathies. Nature 331:717–719

Koopman WJ, Willems PH, Smeitink JA (2012) Monogenic mitochondrial disorders. N Engl J Med 366:1132–1141

Li X, Guan M-X (2003) Identification and characterization of mouse GTPBP3 gene encoding a mitochondrial GTP-binding protein involved in tRNA modification. Biochem Biophys Res Commun 312:747–754

Li X, Li R, Lin X, Guan MX (2002) Isolation and characterization of the putative nuclear modifier gene MTO1 involved in the pathogenesis of deafness-associated mitochondrial 12 S rRNA A1555G mutation. J Biol Chem 277:27256–27264

Luhmann UF, Carvalho LS, Holthaus SM, Cowing JA, Greenaway S, Chu CJ et al (2015) The severity of retinal pathology in homozygous Crb1rd8/rd8 mice is dependent on additional genetic factors. Hum Mol Genet 24:128–141

McMillan HJ, Humphreys P, Smith A, Schwartzentruber J, Chakraborty P, Bulman DE, et al. (2014) Congenital visual impairment and progressive microcephaly due to lysyl-transfer ribonucleic acid (rna) synthetase (kars) mutations: the expanding phenotype of aminoacyl-transfer RNA synthetase mutations in human disease. J Child Neurol. doi:10.1177/0883073814553272

Meseguer S, Martinez-Zamora A, Garcia-Arumi E, Andreu AL, Armengod ME (2015) The ROS-sensitive microRNA-9/9* controls the expression of mitochondrial tRNA-modifying enzymes and is involved in the molecular mechanism of MELAS syndrome. Hum Mol Genet 24:167–184

Nakajima J, Eminoglu TF, Vatansever G, Nakashima M, Tsurusaki Y, Saitsu H et al (2014) A novel homozygous YARS2 mutation causes severe myopathy, lactic acidosis, and sideroblastic anemia 2. J Hum Genet 59:229–232

Newmeyer DD, Ferguson-Miller S (2003) Mitochondria: releasing power for life and unleashing the machineries of death. Cell 112:481–490

Nunnari J, Suomalainen A (2012) Mitochondria: in sickness and in health. Cell 148:1145–1159

Owusu-Ansah E, Song W, Perrimon N (2013) Muscle mitohormesis promotes longevity via systemic repression of insulin signaling. Cell 155:699–712

Parikh VS, Morgan MM, Scott R, Clements LS, Butow RA (1987) The mitochondrial genotype can influence nuclear gene expression in yeast. Science 235:576–580

Parson W, Bandelt HJ (2007) Extended guidelines for mtDNA typing of population data in forensic science. Forensic Sci Int Genet 1:13–19

Pellegrino MW, Nargund AM, Haynes CM (2013) Signaling the mitochondrial unfolded protein response. Biochim et Biophys Acta 1833:410–416

Perli E, Giordano C, Pisano A, Montanari A, Campese AF, Reyes A et al (2014) The isolated carboxy-terminal domain of human mitochondrial leucyl-tRNA synthetase rescues the pathological phenotype of mitochondrial tRNA mutations in human cells. EMBO Mol Med 6:169–182

Picard M, Zhang J, Hancock S, Derbeneva O, Golhar R, Golik P et al (2014) Progressive increase in mtDNA 3243A>G heteroplasmy causes abrupt transcriptional reprogramming. Proc Natl Acad Sci USA 111:E4033–E4042

Prezant TR, Agapian JV, Bohlman MC, Bu X, Oztas S, Qiu WQ et al (1993) Mitochondrial ribosomal RNA mutation associated with both antibiotic-induced and non-syndromic deafness. Nat Genet 4:289–294

Raimundo N (2014) Mitochondrial pathology: stress signals from the energy factory. Trends Mol Med 20:282–292

Raimundo N, Song L, Shutt TE, McKay SE, Cotney J, Guan MX et al (2012) Mitochondrial stress engages E2F1 apoptotic signaling to cause deafness. Cell 148:716–726

Riordan-Eva P, Sanders MD, Govan GG, Sweeney MG, Da Costa J, Harding AE (1995) The clinical features of Leber's hereditary optic neuropathy defined by the presence of a pathogenic mitochondrial DNA mutation. Brain 118(Pt 2):319–337

Ross JM, Stewart JB, Hagstrom E, Brene S, Mourier A, Coppotelli G et al (2013) Germline mitochondrial DNA mutations aggravate ageing and can impair brain development. Nature 501:412–415

Ross JM, Coppotelli G, Hoffer BJ, Olson L (2014) Maternally transmitted mitochondrial DNA mutations can reduce lifespan. Sci Rep 4:6569

Rubinsztein DC, Marino G, Kroemer G (2011) Autophagy and aging. Cell 146:682–695

Ryan MT, Hoogenraad NJ (2007) Mitochondrial-nuclear communications. Annu Rev Biochem 76:701–722

Samuels DC, Schon EA, Chinnery PF (2004) Two direct repeats cause most human mtDNA deletions. Trends Genet 20:393–398

Suzuki T, Nagao A (2011) Human mitochondrial tRNAs: biogenesis, function, structural aspects, and diseases. Annu Rev Genet 45:299–329

Taylor RW, Turnbull DM (2005) Mitochondrial DNA mutations in human disease. Nat Rev Genet 6:389–402

van den Ouweland JMW, Lemkes HHPJ, Ruitenbeek W, Sandkuijl LA, de Vijlder MF, Struyvenberg PAA et al (1992) Mutation in mitochondrial tRNALeu(UUR) gene in a large pedigree with maternally transmitted type II diabetes mellitus and deafness. Nat Genet 1:368–371

Wallace DC (2005) A mitochondrial paradigm of metabolic and degenerative diseases, aging, and cancer: a dawn for evolutionary medicine. Annu Rev Genet 39:359–407

Wallace DC (2012) Mitochondria and cancer. Nature reviews. Cancer 12:685–698

Wallace DC, Chalkia D (2013) Mitochondrial DNA genetics and the heteroplasmy conundrum in evolution and disease. Cold Spring Harb Perspect Biol 5:a021220

Yasukawa T, Suzuki T, Ishii N, Ueda T, Ohta S, Watanabe K (2000a) Defect in modification at the anticodon wobble nucleotide of mitochondrial tRNA(Lys) with the MERRF encephalomyopathy pathogenic mutation. FEBS Lett 467:175–178

Yasukawa T, Suzuki T, Ueda T, Ohta S, Watanabe K (2000b) Modification defect at anticodon wobble nucleotide of mitochondrial tRNAs(Leu)(UUR) with pathogenic mutations of mitochondrial myopathy, encephalopathy, lactic acidosis, and stroke-like episodes. J Biol Chem 275:4251–4257

Youle RJ, Strasser A (2008) The BCL-2 protein family: opposing activities that mediate cell death. Nat Rev Mol Cell Biol 9:47–59

Zhao H, Li R, Wang Q, Yan Q, Deng JH, Han D et al (2004) Maternally inherited aminoglycoside-induced and nonsyndromic deafness is associated with the novel C1494T mutation in the mitochondrial 12S rRNA gene in a large Chinese family. Am J Hum Genet 74:139–152

15

MicroRNA-495 induces breast cancer cell migration by targeting JAM-A

Minghui Cao[1], Weiwei Nie[2], Jing Li[1], Yujing Zhang[1], Xin Yan[3], Xiaoxiang Guan[2], Xi Chen[1], Ke Zen[1], Chen-yu Zhang[1✉], Xiaohong Jiang[1✉], Dongxia Hou[1✉]

[1] State Key Laboratory of Pharmaceutical Biotechnology, Jiangsu Engineering Research Center for microRNA Biology and Biotechnology, School of Life Sciences, Nanjing University, Nanjing 210093, China
[2] Department of Medical Oncology, Jinling Hospital, School of Medicine, Southern Medical University, Guangzhou 510282, China
[3] The Comprehensive Cancer Center of Drum Tower Hospital, Medical School of Nanjing University and Clinical Cancer Institute of Nanjing University, Nanjing 210008, China
✉ Correspondence: cyzhang@nju.edu.cn (C. Zhang), xiaohong.jiang@yahoo.com (X. Jiang), dxhou128@nju.edu.cn (D. Hou)

ABSTRACT

MicroRNAs (miRNAs) are small, non-coding RNAs that function as post-transcriptional regulators of gene expression. The deregulated expression of miRNAs is associated with a variety of diseases, including breast cancer. In the present study, we found that miR-495 was markedly up-regulated in clinical breast cancer samples by quantitative real time-PCR (qRT-PCR). Junctional adhesion molecule A (JAM-A) was predicted to be a potential target of miR-495 by bioinformatics analysis and was subsequently verified by luciferase assay and Western blotting. JAM-A was found to be negatively correlated with the migration of breast cancer cells through loss-of-function and gain-of-function assays, and the inhibition of JAM-A by miR-495 promoted the migration of MCF-7 and MDA-MB-231 cells. Furthermore, overexpression of JAM-A could restore miR-495-induced breast cancer cell migration. Taken together, our findings suggest that miR-495 could facilitate breast cancer progression through the repression of JAM-A, making this miRNA a potential therapeutic target.

KEYWORDS miR-495, JAM-A, breast cancer, migration

Minghui Cao and Weiwei Nie have contributed equally to this work.

INTRODUCTION

Breast cancer is the most common malignancy and the leading cause of cancer death among females worldwide, accounting for ~36% of female primary malignant tumors (Jemal et al., 2011). Although advances in diagnosis such as regular mammography and appropriately systemic treatments have improved the prognosis (DeSantis et al., 2014), distant metastases usually occur several years after the primary breast cancer, causing approximately 90% of breast cancer mortality (Bendre et al., 2003). Considering that the precise mechanisms of breast cancer metastasis remain largely unknown, additional investigation is necessary to clarify the progression of this phenomenon.

MicroRNAs (miRNAs) are a class of endogenous, non-coding RNAs with lengths of 18–25 nt (Bartel, 2004). By base-pairing with 3′-untranslated region (3′-UTR) of messenger RNA (mRNA), miRNAs can repress the expression of target genes by inhibiting translation or by destabilizing the mRNA (Bartel, 2009; Fabian et al., 2010). Under normal circumstances, miRNAs participate in a broad range of biological processes, such as cell proliferation and differentiation, development, metabolism, immunity and stress responses (Ambros, 2001). In recent years, the deregulated expression of miRNAs has been widely reported in many diseases, especially in cancer (Lu et al., 2005; Osada and Takahashi, 2007). With an increasing number of target genes of miRNAs being validated by experimental assays and verified in clinical samples, miRNAs have gradually emerged as a new important regulator of tumorigenesis and have been found to be involved in various aspects of cancer progression including tumor metastasis (Lujambio and Lowe,

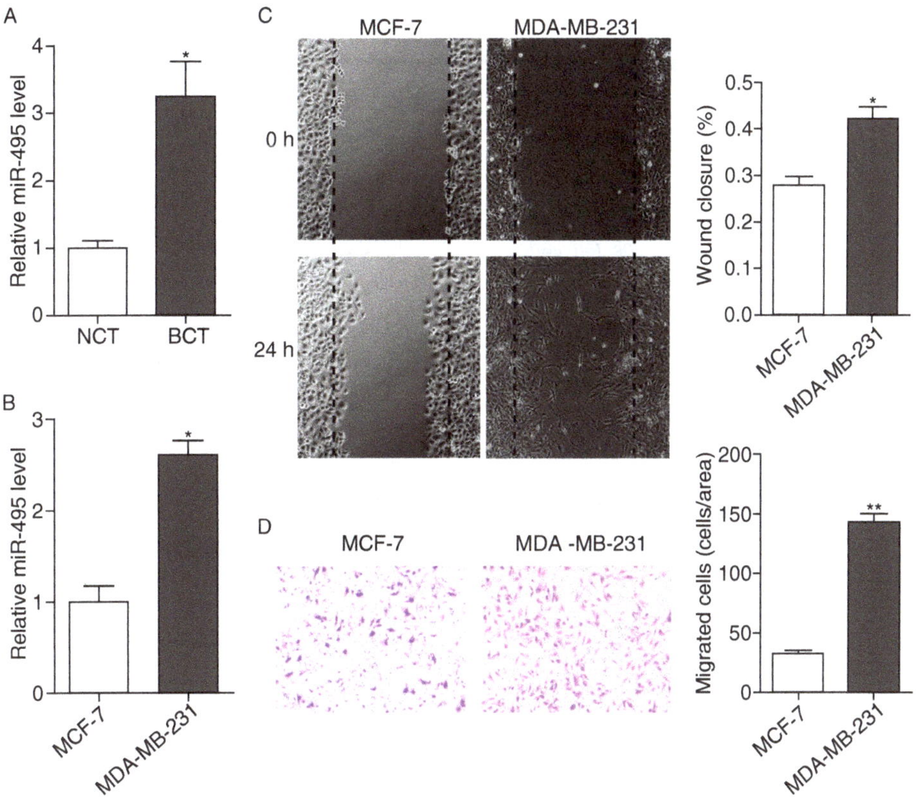

Figure 1. The expression of miR-495 was increased in breast cancer tissues and was positively correlated with the mobility of breast cancer cells. (A) Quantitative real time-PCR analysis of the relative expression of miR-495 in seven pairs of breast cancer tissue (BCT) and non-cancerous tissue (NCT) samples. (B) Quantitative real time-PCR analysis of the relative expression of miR-495 in breast cancer cell lines MCF-7 and MDA-MB-231 cells. (C) Left panel: Representative image of wound healing assay of MCF-7 and MDA-MB-231 cells. Right panel: Quantitative analysis of the wound closure rates. (D) Left panel: Representative image of Transwell assay of MCF-7 and MDA-MB-231 cells. Right panel: Quantitative analysis of the migration rates. $^{*}P < 0.05$; $^{**}P < 0.01$.

2012). For example, miR-10b is specifically up-regulated in metastatic breast cancer cells, resulting in the increased expression of a well-characterized pro-metastatic gene, *RHOC*, by inhibiting the translation of HoxD10 (Ma et al., 2007). Moreover, the expression of miR-206 is selectively reduced during breast cancer metastasis, and this reduction is positively correlated with a low metastasis-free survival in patients (Li et al., 2013; Vimalraj et al., 2013). These findings highlight the need for thorough investigations of miRNAs that are aberrantly expressed during breast cancer progression, especially miRNAs associated with breast cancer metastasis (Vimalraj et al., 2013).

In this study, we demonstrated that miR-495 is significantly up-regulated in primary breast cancer tissue samples when compared with noncancerous tissue samples. By manipulating miR-495 levels in MCF-7 and MDA-MB-231 cells, we proved that miR-495 promotes the mobility of breast cancer cells. JAM-A was predicted to be a target of miR-495, which was verified by luciferase assay and Western blotting; the function of *JAM-A* in breast cancer metastasis was validated by overexpression or knock down of the JAM-A protein. Finally, the rescued expression of JAM-A could reverse the observed effects of miR-495. Our study demonstrates that miR-495 acts as a metastasis promoter by directly targeting JAM-A, suggesting that miR-495 has potential therapeutic value for breast cancer treatment.

RESULTS

MiR-495 is up-regulated in clinical breast cancer specimens and is positively correlated with the mobility of breast cancer cells

First, the level of miR-495 in clinical breast cancer tissue samples was determined using quantitative real time-PCR (qRT-PCR), and we found that the level of miR-495 in breast cancer tissues was markedly higher than in paired adjacent normal breast tissues (Fig. 1A), suggesting that miR-495 is associated with the progression of breast cancer. The level of miR-495 in two different breast cancer cell lines MCF-7 and MDA-MB-231 cells was then detected, and we found that miR-495 was significantly up-regulated in MDA-MB-231

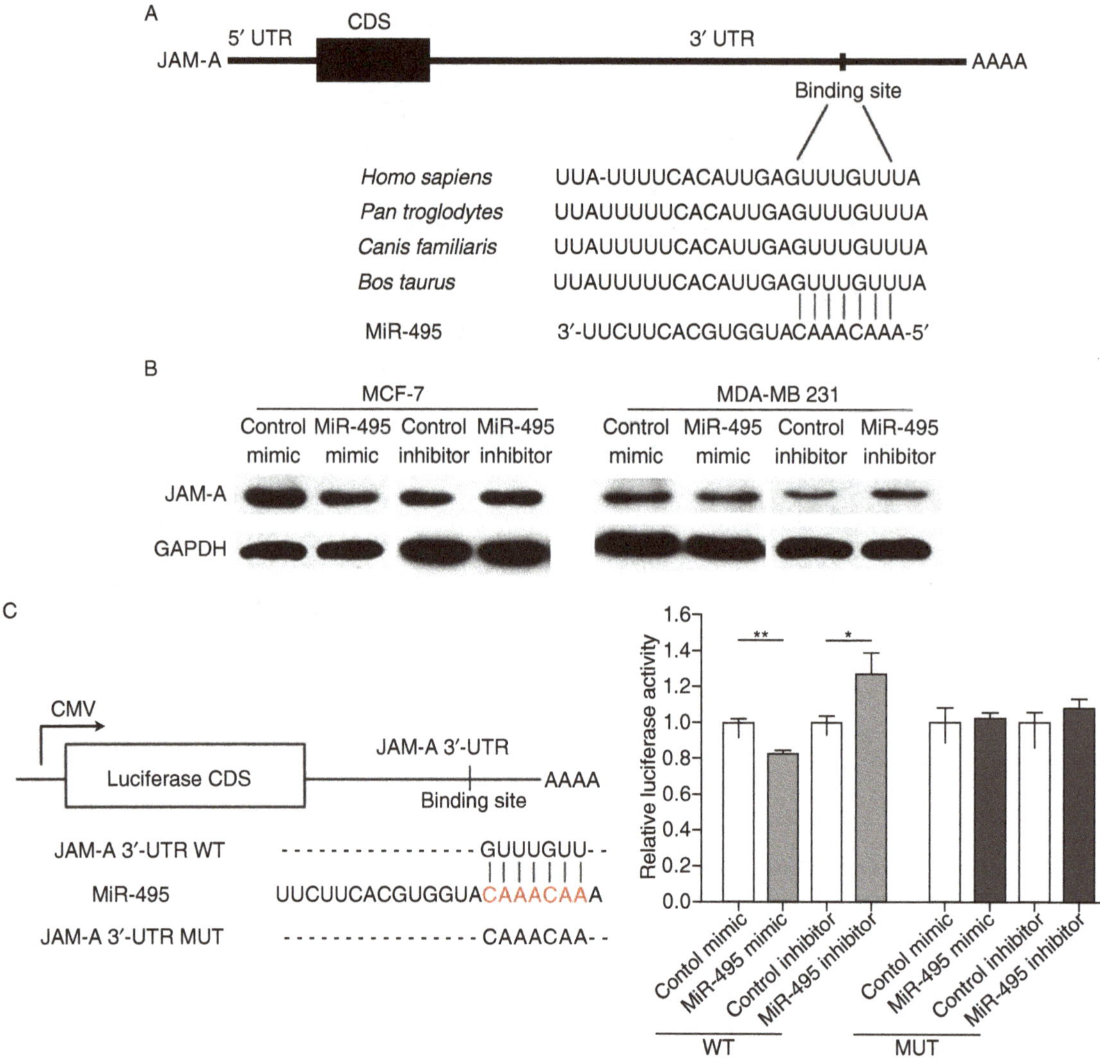

Figure 2. JAM-A is a target gene of miR-495 in breast cancer cells. (A) Schematic illustration of the conserved miR-495 binding sites. The JAM-A 3′-UTR contains one predicted miR-495 binding sites. The seed regions of miR-495 and the seed-recognizing sites in the JAM-A 3′-UTR are indicated in red, and all nucleotides in seed-recognizing sites are completely conserved across several species. (B) Western blotting analysis of JAM-A protein levels in MCF-7 and MDA-MB-231 cells transfected with miR-495 mimic or inhibitor. (C) Direct recognition of the JAM-A 3′-UTR by miR-495. Firefly luciferase reporters containing either wild-type (WT) or mutant (MUT) miR-495 binding sites in the JAM-A 3′-UTR were co-transfected into MDA-MB-231 cells with the scrambled negative control RNA, miR-495 mimic or inhibitor. At 24 h post-transfection, the cells were assayed using a luciferase assay kit. The results were calculated as the ratio of firefly luciferase activity in the miR-495-transfected cells normalized to the control cells. *$P < 0.05$; **$P < 0.01$.

cells (Fig. 1B). MDA-MB-231 cells exhibited a higher mobility in wound healing assays and Transwell assays (Fig. 1C and 1D), indicating that miR-495 was positively correlated with the mobility of breast cancer cells.

JAM-A is a potential target of miR-495 in breast cancer cells

The *in silico* approaches TargetScan (Lewis et al., 2003) and miRanda (John et al., 2004) were used in combination to predict target genes of miR-495, and junctional adhesion molecule A (JAM-A) was identified as a potential one. The putative binding sites for miR-495 in the 3′-UTR of JAM-A mRNA are shown in Fig. 2A. The seed region (the core sequences that encompass the first 2–8 bases of the mature miRNA) of miR-495 perfectly base-pairs with 3′-UTR of JAM-A mRNA. Furthermore, the miR-495 binding sequences in the 3′-UTR of JAM-A mRNA are highly conserved across species.

To assess whether JAM-A could be regulated by miR-495, we investigated the effect of miR-495 on JAM-A

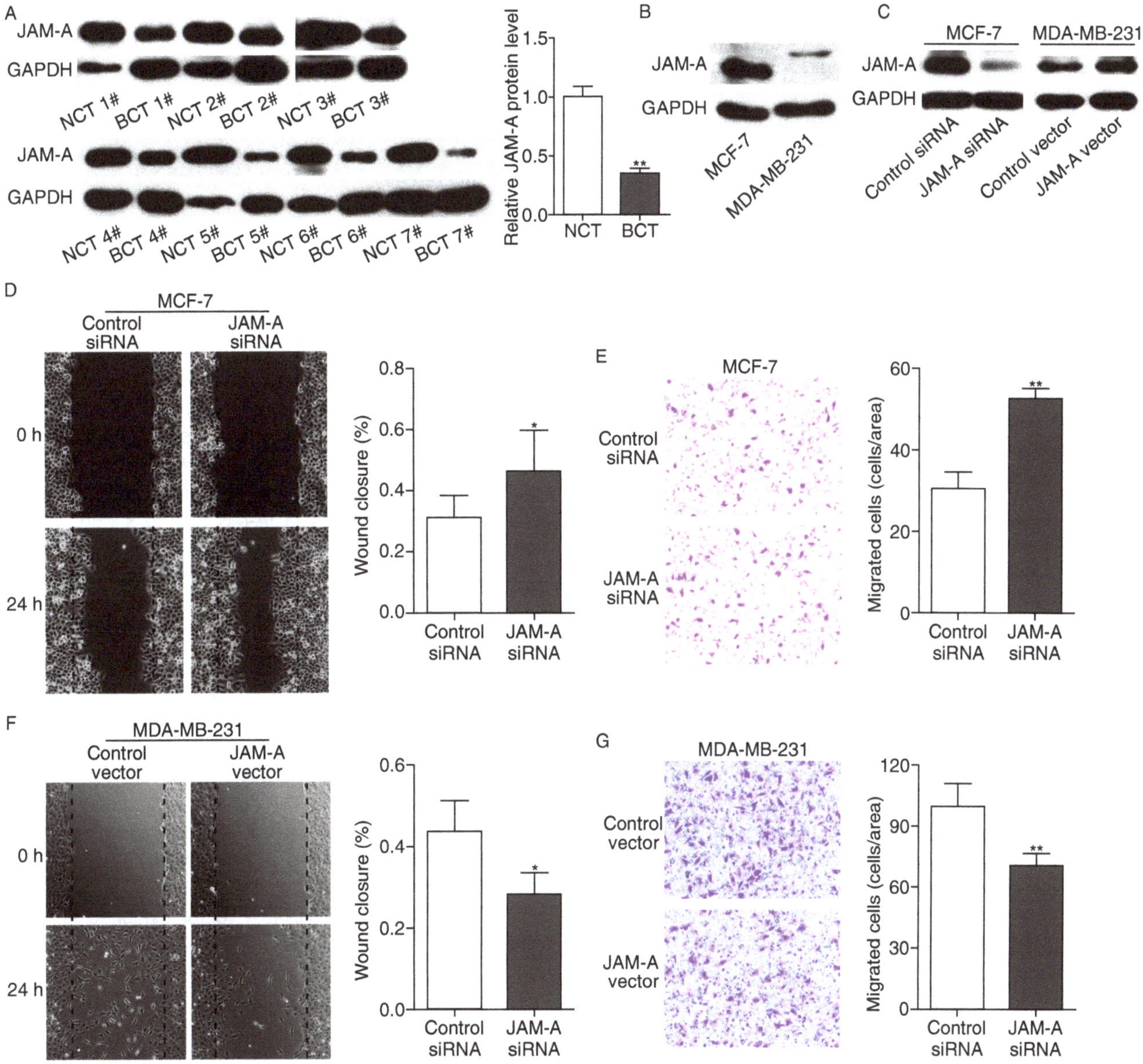

Figure 3. JAM-A expression is decreased in breast cancer tissues and is inversely correlated with the mobility of breast cancer cells. (A) Western blotting analysis and quantification of JAM-A protein levels in seven pairs of breast cancer tissue (BCT) and non-cancerous tissue (NCT) samples. (B) Western blotting analysis of JAM-A protein levels in MCF-7 and MDA-MB-231 cells. (C) Left panel: Western blotting analysis of JAM-A protein levels in MCF-7 cells transfected with control siRNA or JAM-A siRNA. Right panel: Western blotting analysis of JAM-A protein levels in MDA-MB-231 cells transfected with control plasmid or JAM-A overexpression plasmid. (D) Left panel: Representative image of wound healing assay of MCF-7 cells transfected with control siRNA or JAM-A siRNA. Right panel: Quantitative analysis of the wound closure rates. (E) Left panel: Representative image of Transwell assay of MDA-MB-231 cells transfected with control plasmid or JAM-A overexpression plasmid. Right panel: Quantitative analysis of the migration rates. (F) Left panel: Representative image of wound healing assay of MDA-MB-231 cells transfected with control plasmid or JAM-A overexpression plasmid. Right panel: Quantitative analysis of the wound closure rates. (G) Left panel: Representative image of Transwell assay of MDA-MB-231 cells transfected with control plasmid or JAM-A overexpression plasmid. Right panel: Quantitative analysis of the migration rates. $^*P < 0.05$; $^{**}P < 0.01$.

protein level in MCF-7 and MDA-MB-231 cells. As shown in Fig. 2B, the level of JAM-A protein was reduced by the induction of miR-495 mimic but significantly increased by transfection with miR-495 inhibitor in both cell lines.

To ascertain whether miR-495 directly regulates JAM-A expression by binding with JAM-A 3′-UTR, the full-length 3′-UTR of JAM-A was amplified by PCR and then fused downstream of the firefly luciferase gene in a reporter plasmid.

The reporter plasmid was transfected into MDA-MB-231 cells along with a transfection control plasmid (β-gal) and miR-495 mimic or inhibitor. As expected, overexpression of miR-495 resulted in approximately a 20% reduction in luciferase reporter activity, whereas inhibition of miR-495 resulted in a 1.3-fold increase in reporter activity compared with the cells transfected with control inhibitor (Fig. 2C). Furthermore, we introduced point mutations into the corresponding complementary sites in the JAM-A 3′-UTR to eliminate the predicted miR-495 binding sites. This mutated luciferase reporter was unaffected by either the overexpression or knockdown of miR-495 (Fig. 2C). In conclusion, the results demonstrate that miR-495 inhibits JAM-A expression by binding to the 3′-UTR of JAM-A.

JAM-A expression is decreased in breast cancer tissues and is inversely correlated with the mobility of breast cancer cells

MiRNAs are generally thought to have an expression pattern that is opposite to that of their targets (Olsen and Ambros, 1999). As miR-495 expression was increased in breast cancer tissue samples, we next investigated whether JAM-A protein level was decreased. After detecting the protein level of JAM-A in the same seven pairs of breast cancer and corresponding noncancerous tissue samples, we found that JAM-A protein level was dramatically lower in the breast cancer samples (Fig. 3A). Moreover, we determined the level of JAM-A protein in MCF-7 and MDA-MB-231 cells, and higher level of JAM-A protein was detected in MCF-7 cells which showed a lower level of miR-495 (Fig. 3B). These findings further suggest that the level of JAM-A protein is negatively correlated with the miR-495 level and that JAM-A expression is regulated by miR-495.

As an adhesion molecule participating in comprising tight junctions, JAM-A was reported to be associated with metastasis of breast cancer cells; however, there are conflicting reports. For example, Naik et al. (Naik et al., 2008) and Wang et al. (Wang and Lui, 2012) reported that attenuation of JAM-A contributes to breast cancer cell invasion, whereas McSherry et al. revealed that JAM-A drives breast cancer cell migration (McSherry et al., 2011). To further elucidate the function of JAM-A in regulating the mobility of breast cancer cells, loss-of-function assay was performed by transfecting an siRNA against JAM-A into MCF-7 cells. The left panel of Fig. 3C shows that the protein level of JAM-A was knocked-down in MCF-7 cells with JAM-A siRNA. Then, wound healing and Transwell assays were used to investigate MCF-7 cell migration. Within 24 h, MCF-7 cells with a lower level of JAM-A protein occupied approximately 46% of the wound, and MCF-7 cells with control siRNA covered approximately 31% of the wound, indicating that knockdown of JAM-A promoted the mobility of MCF-7 cells (Fig. 3D). Similarly, Transwell assays showed that more MCF-7 cells migrated through the porous membrane when JAM-A expression was inhibited by siRNA (Fig. 3E).

Gain-of-function assay was also conducted by transfecting JAM-A cDNA plasmid into MDA-MBA-231 cells; the increase in JAM-A protein is shown in the right panel of Fig. 3C. As expected, overexpression of JAM-A undermined the migration capability of MDA-MB-231 cells, as demonstrated by wound healing and Transwell assays (Fig. 3F and 3G). Altogether, JAM-A was found to be down-regulated in breast cancer and to negatively regulate the mobility of breast cancer cells.

MiR-495 induces breast cancer cell migration by targeting JAM-A

After establishing that JAM-A is involved in the migration of breast cancer cells, the biological significance of miR-495 in breast cancer was then investigated. As shown by wound healing assays in Fig. 4A, more MCF-7 cells migrated into the scratch on the cell monolayer when transfected with miR-495 mimic; consistently, knockdown of miR-495 by miRNA inhibitor showed the opposite effects. The same biological function of miR-495 was found in MDA-MB-231 cells (Fig. 4C). In addition, Transwell assays revealed a significant increase in cells migrating through the membrane when MCF-7 cells and MDA-MB-231 cells were transfected with miR-495 mimic (Fig. 4B and 4D). As anticipated, the mobility of MCF-7 and MDA-MB-231 cells was clearly inhibited by transfection with miR-495 inhibitor, shown by Fig. 4B and 4D, respectively.

To further confirm that the effects of miR-495 are mediated by repression of JAM-A in breast cancer cells, knockdown of JAM-A in MCF-7 cells by miR-495 mimic was restored by transfecting JAM-A cDNA vector, the expression of which is not regulated by miR-495 due to its lack of the 3′-UTR (Fig. 5A). Compared with the cells transfected with miR-495 mimic and control vector, MCF-7 cells transfected with miR-495 mimic and JAM-A vector exhibited impaired migration ability (Fig. 5B and 5C). Moreover, JAM-A siRNA was transfected into MDA-MB-231 cells to counteract the increase of JAM-A protein caused by miR-495 inhibitor (Fig. 5D). Fig. 5E and 5F show that the inhibition of MDA-MB-231 cell migration by miR-495 inhibitor can be reversed by JAM-A siRNA. Taken together, these results demonstrated that miR-495 promotes breast cancer cell migration by inhibiting JAM-A.

DISCUSSION

MiRNAs are a class of small, non-coding RNAs that regulate the expression of specific mRNAs by either translational inhibition or mRNA degradation (Bartel, 2004; Fabian et al., 2010). More than 50% of miRNAs are reported to be located in cancer-associated genomic break points (Calin et al., 2004), and these miRNAs can function as tumor suppressor genes or oncogenes, depending on the target genes regulated by the miRNA (Croce, 2009). MiR-495 was reported to act as an oncogene in multiple models of cancer. For example, up-regulation of miR-495 contributes to lower MAT1 expression and enhanced tumorigenesis in hepatocellular

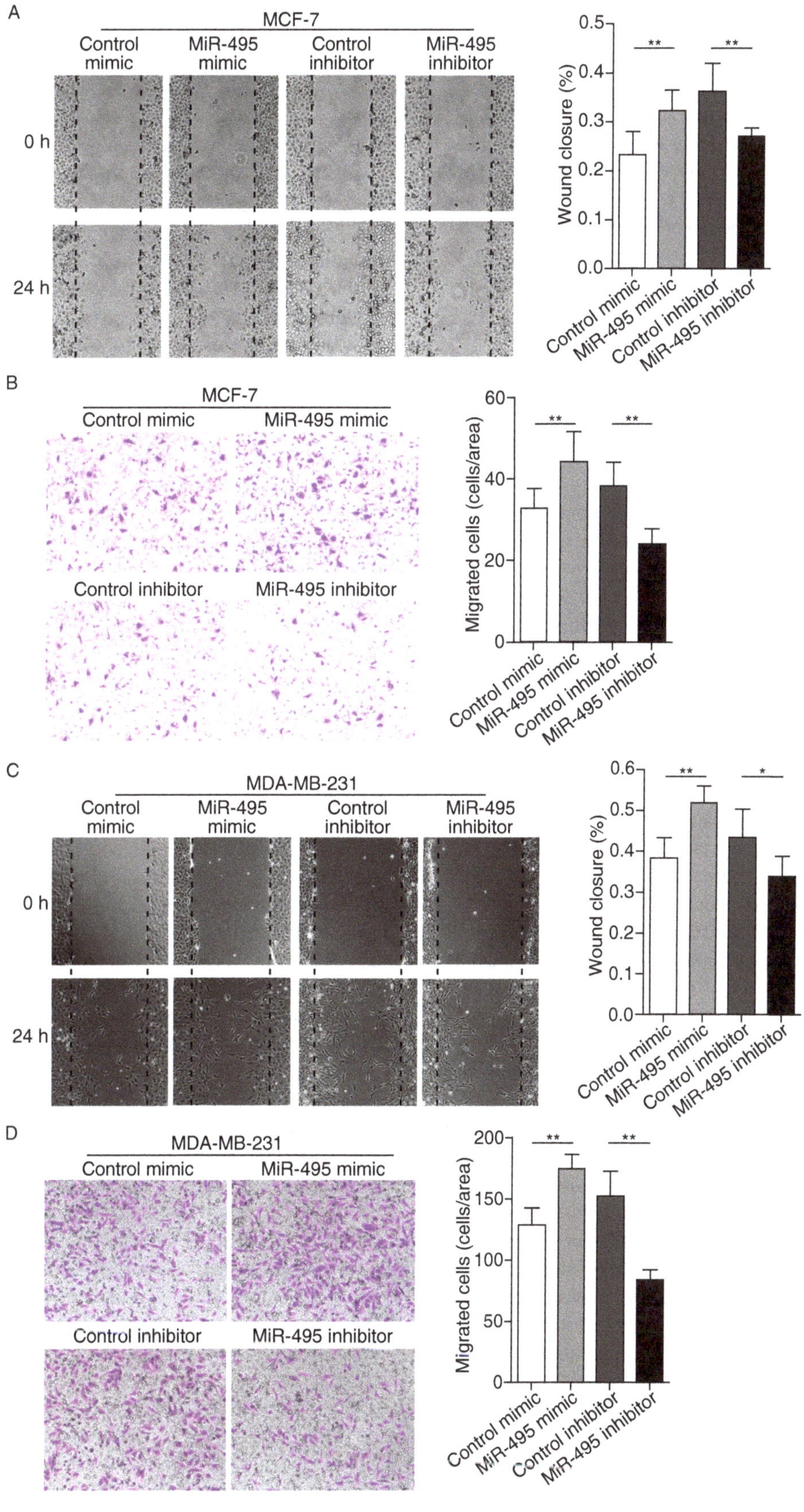
A
MCF-7
Control mimic
MiR-495 mimic
Control inhibitor
MiR-495 inhibitor
0 h
24 h
Wound closure (%)
0.0
0.1
0.2
0.3
0.4
0.5
**
**
Control mimic
MiR-495 mimic
Control inhibitor
MiR-495 inhibitor
B
MCF-7
Control mimic
MiR-495 mimic
Control inhibitor
MiR-495 inhibitor
Migrated cells (cells/area)
0
20
40
60
**
**
C
MDA-MB-231
Control mimic
MiR-495 mimic
Control inhibitor
MiR-495 inhibitor
0 h
24 h
Wound closure (%)
0.0
0.1
0.2
0.3
0.4
0.5
0.6
**
*
D
MDA-MB-231
Control mimic
MiR-495 mimic
Control inhibitor
MiR-495 inhibitor
Migrated cells (cells/area)
0
50
100
150
200
**
**

Figure 4. Effects of miR-495 on breast cancer cell migration. (A) Left panel: Representative image of wound healing assay of MCF-7 cells transfected with control mimic, miR-495 mimic, control inhibitor or miR-495 inhibitor. Right panel: Quantitative analysis of the wound closure rates. (B) Left panel: Representative image of Transwell assay of MCF-7 cells transfected with control mimic, miR-495 mimic, control inhibitor or miR-495 inhibitor. Right panel: Quantitative analysis of the migration rates. (C) Left panel: Representative image of wound healing assay of MDA-MB-231 cells transfected with control mimic, miR-495 mimic, control inhibitor or miR-495 inhibitor. Right panel: Quantitative analysis of the wound closure rates. (D) Left panel: Representative image of Transwell assay of MDA-MB-231 cells transfected with control mimic, miR-495 mimic, control inhibitor or miR-495 inhibitor. Right panel: Quantitative analysis of the migration rates. $*P < 0.05$; $**P < 0.01$.

carcinoma (HCC) and may represent a potential target for HCC therapy (Yang et al., 2013). In addition, up-regulation of miR-495 by E12/E47 in breast cancer stem cells promotes oncogenesis and hypoxia resistance via the down-regulation of E-cadherin and REDD1 (Hwang-Verslues et al., 2011). However, miR-495 was also shown to suppress tumorigenesis in different types of cancer. For example, miR-495 functions as a tumor suppressor by targeting essential leukemia-related genes in MLL-rearranged leukemia (Jiang et al., 2012) and by down-regulating cyclin-dependent kinase 6 in glioblastoma multiforme cells (Chen et al., 2013). Moreover, miR-495 inhibits the migration of gastric cancer cells and lung cancer cells by targeting PRL-3 and MTA 3, respectively (Chu et al., 2014; Li et al., 2012). Although the function of miR-495 is still controversial in different types of cancer, in this study, we found that the expression of miR-495 was up-regulated in clinical breast cancer samples, indicating that miR-495 may be associated with the progression of breast cancer.

Breast cancer cell line MCF-7 was classified into luminal A subtype, which exhibited a characteristic epithelial cobblestone-like morphology with high expression of cell-cell adhesion molecules such as E-cadherin, and has been identified as a representative breast cancer cell line with low-invasive ability. By contrast, MDA-MB-231 was classified into claudin-low subtype, and this cell line displayed a high invasive potential with the characteristics of epithelial-mesenchymal transition (EMT), such as expressing a high level of vimentin (Holliday and Speirs, 2011; Neve et al., 2006). In the present study, the highly tumorigenic breast cancer cell line MDA-MB-231 was found to present a high level of miR-495 when being compared with the less tumorigenic cell MCF-7. These results suggest that miR-495 is positively associated with the malignant phenotypes of breast cancer, including the migration of breast cancer cells, and may serve as a marker of potential progression of the cancer.

According to the bioinformatics analysis, junction adhesion molecule A (JAM-A, JAM-1, F11 receptor, CD321) was predicted as a potential target gene of miR-495. Consistently, the level of JAM-A protein was decreased in human breast cancer tissues, which is coincident with the up-regulation of miR-495. Furthermore, MCF-7 cells with a lower level of miR-495 expressed a higher level of JAM-A protein than MDA-MB-231 cells. Moreover, Western blotting analysis showed that miR-495 inhibits JAM-A translation in breast cancer cells. We also found that luciferase activity of the reporter plasmid containing JAM-A 3′-UTR can be remarkably decreased by miR-495 mimic, while knockdown of miR-495 in MDA-MB-231 cells resulted in an increase of luciferase activity. These results proved that miR-495 can fine-tune the expression of JAM-A through binding to 3′-UTR of JAM-A.

Originally characterized by the monoclonal antibody F11 on platelets, JAM-A is a type I transmembrane glycoprotein involved in numerous biological and pathological processes (Mandell and Parkos, 2005). JAM-A is a cell-adhesion protein predominantly expressed at the tight junctions of epithelial cells, including those of the mammary epithelium (Feigin and Muthuswamy, 2009). Under normal circumstances, JAM-A participates in the formation of tight junctions (Schneeberger and Lynch, 2004). The disruption of focal adhesion is associated with EMT, which leads to the promotion of cell migration (Lee et al., 2006). However, the role of JAM-A in the migration of tumor cells remains a controversial issue. Koshiba et al. reported that JAM-A expression was reduced in high-grade or advanced endometrial carcinoma and may constitute a poor prognostic factor (Koshiba et al., 2009). In addition, Fong et al. found that low JAM-A expression is associated with metastasis and poor survival in pancreatic cancer (Fong et al., 2012). Nevertheless, another report published recently showed that high JAM-A expression in non-small cell lung cancer (NSCLC) tissues is positively correlated with NSCLC progression (Zhang et al., 2013). These data suggest that JAM-A expression is regulated in a tissue-dependent manner. Additionally, conflicting data about the function of JAM-A in regulating cancer cell migration have been reported with regard to breast cancer. Naik et al. initially reported that up-regulation of JAM-A expression decreased migration and invasion in breast cancer cells, whereas JAM-A knockdown enhanced invasiveness (Naik et al., 2008). It was therefore hypothesized that the loss of JAM-A may correlate with poor clinical prognosis. Similar results were reported by Wang et al., who demonstrated that transforming growth factor-β1 (TGF-β1) induced breast cancer cell invasion by down-regulating JAM-A expression (Wang and Lui, 2012). However, McSherry and colleagues reported that the expression of JAM-A is positively correlated with poor prognosis in invasive breast cancer patients, indicating that JAM-A promotes cell motility in breast cancer by activating Rap 1 GTPase (McSherry et al., 2011). In our study, loss-of-function assay using MCF-7 cells transfected with JAM-A siRNA showed that knockdown of JAM-A promotes the migration of breast cancer cells, whereas gain-of-function assay by overexpressing JAM-A in MDA-MB-231 cells showed that

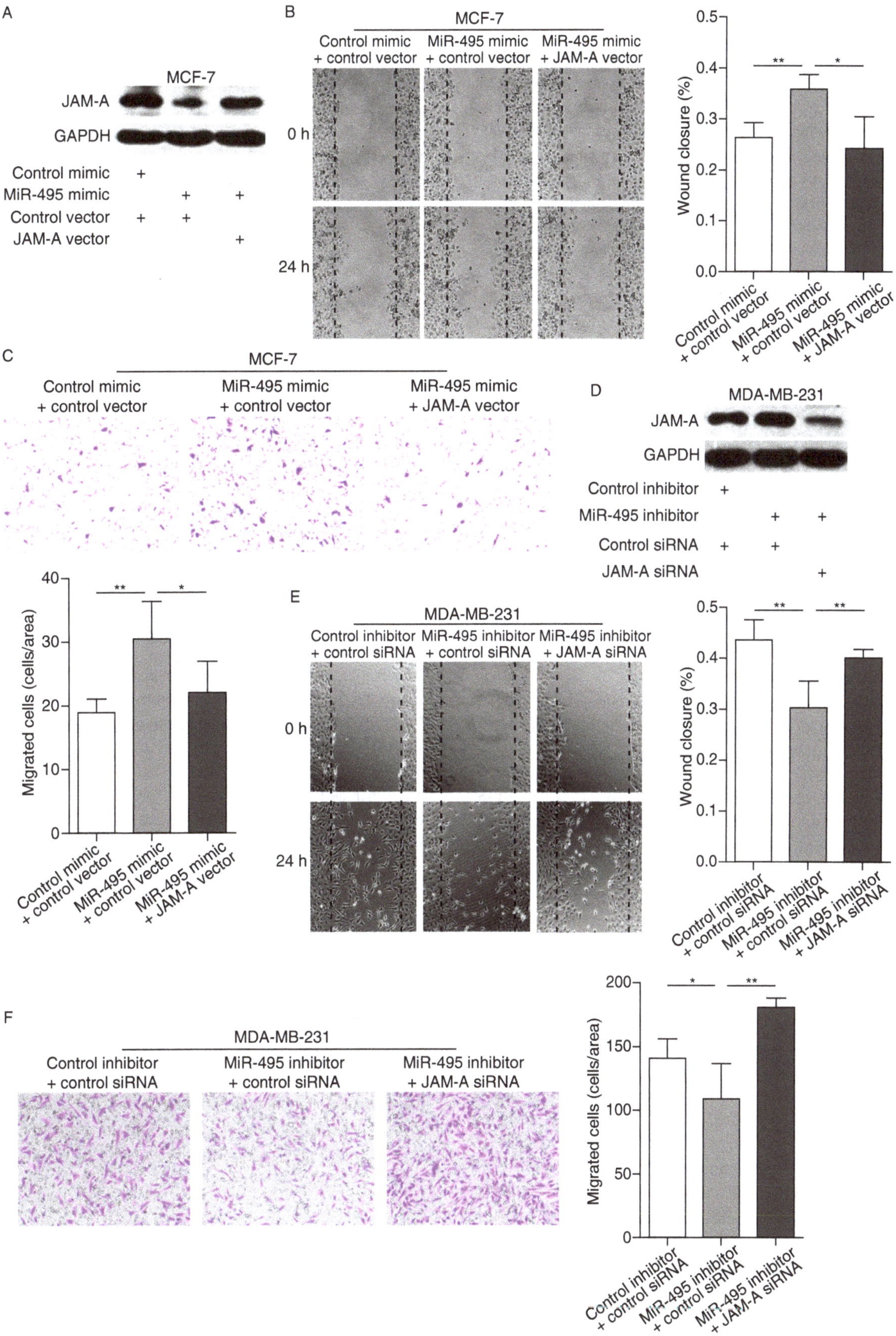
A
MCF-7
JAM-A
GAPDH
Control mimic
MiR-495 mimic
Control vector
JAM-A vector
B
MCF-7
Control mimic + control vector
MiR-495 mimic + control vector
MiR-495 mimic + JAM-A vector
0 h
24 h
Wound closure (%)
C
MCF-7
Control mimic + control vector
MiR-495 mimic + control vector
MiR-495 mimic + JAM-A vector
Migrated cells (cells/area)
D
MDA-MB-231
JAM-A
GAPDH
Control inhibitor
MiR-495 inhibitor
Control siRNA
JAM-A siRNA
E
MDA-MB-231
Control inhibitor + control siRNA
MiR-495 inhibitor + control siRNA
MiR-495 inhibitor + JAM-A siRNA
0 h
24 h
Wound closure (%)
F
MDA-MB-231
Control inhibitor + control siRNA
MiR-495 inhibitor + control siRNA
MiR-495 inhibitor + JAM-A siRNA
Migrated cells (cells/area)

◀ **Figure 5. MiR-495 induces breast cancer cell migration by targeting JAM-A.** (A) Western blotting analysis of the protein levels of JAM-A in MCF-7 cells transfected with control mimic plus control plasmid, miR-495 mimic plus control plasmid or miR-495 mimic plus JAM-A overexpression plasmid. (B) Left panel: Representative image of wound healing assay of MCF-7 cells transfected with control mimic plus control plasmid, miR-495 mimic plus control plasmid or miR-495 mimic plus JAM-A overexpression plasmid. Right panel: Quantitative analysis of the wound closure rates. (C) Top panel: Representative image of Transwell assay of MCF-7 cells transfected with control mimic plus control plasmid, miR-495 mimic plus control plasmid or miR-495 mimic plus JAM-A overexpression plasmid. Bottom panel: Quantitative analysis of the migration rates. (D) Western blotting analysis of the protein levels of JAM-A in MDA-MB-231 cells transfected with control inhibitor plus control siRNA, miR-495 inhibitor plus control siRNA or miR-495 inhibitor plus JAM-A siRNA. (E) Left panel: Representative image of wound healing assay of MDA-MB-231 cells transfected with control inhibitor plus control siRNA, miR-495 inhibitor plus control siRNA or miR-495 inhibitor plus JAM-A siRNA. Right panel: Quantitative analysis of the wound closure rates. (F) Left panel: Representative image of Transwell assay of MDA-MB-231 cells transfected with control inhibitor plus control siRNA, miR-495 inhibitor plus control siRNA or miR-495 inhibitor plus JAM-A siRNA. Right panel: Quantitative analysis of the migration rates. $^{*}P < 0.05$; $^{**}P < 0.01$.

JAM-A overexpression attenuates migration. These results suggest the role of JAM-A as a negative regulator of migration in breast cancer cells.

Regarding the mechanisms of the down-regulation of JAM-A protein in breast cancer, previous report showed that TGF-β1 inhibits JAM-A gene transcription via the activation of Smads (Wang and Lui, 2012). Considering that the 3′-UTR of human JAM-A is as long as 3.6 kb, we hypothesized that JAM-A expression may be down-regulated by miRNAs, such as miR-495, in breast cancer; indeed, a negative correlation between the expression of miR-495 and JAM-A was established in the present study. The overexpression of miR-495 promoted the mobility of breast cancer cells by down-regulating JAM-A. Additionally, the restoration of JAM-A protein by expressing JAM-A cDNA vector, which lacked 3′-UTR and thus not regulated by miRNAs, reversed the migration promotion exerted by miR-495, further confirming that miR-495 regulates the migration of breast cancer cells by targeting JAM-A. Moreover, comparison of the mobility of MDA-MB-231 cells transfected with miR-495 inhibitor plus control siRNA or miR-495 inhibitor plus JAM-A siRNA revealed that miR-495 inhibitor down-regulated the migration of breast cancer cells by regulating JAM-A expression.

Taken together, our data reveal a new role for miR-495 as an oncogenic miRNA in breast carcinogenesis. Identification of the miR-495-targeting JAM-A pathway provides a potential new therapeutic target in the treatment of breast cancer.

MATERIALS AND METHODS

Ethics statement

Breast cancer tissue and paired adjacent noncancerous tissue samples were collected from the Affiliated Drum Tower Hospital of Nanjing University Medical School and the Affiliated Jinling Hospital of Southern Medical University (Nanjing, China). Written informed consent was obtained from all patients. The Ethics Committee of Nanjing University approved all aspects of this study. The tissue fragments were immediately frozen in liquid nitrogen at the time of surgery and stored at −80°C. The clinical features of the patients are listed in Table S1.

RNA isolation and quantitative real time-PCR (qRT-PCR)

Total RNA was extracted from cultured cells or tissue samples with TRIzol Reagent (Invitrogen) according to the manufacturer's instructions. Assays for miR-495 quantification were conducted by using gene-specific TaqMan miRNA Assay Probes (Applied Biosystems, Foster City, CA). After the real-time PCR reaction, the cycle threshold (C_T) data were determined using fixed threshold settings; the mean C_T was determined from triplicate PCRs. A comparative C_T method was used to compare each condition to the controls. The U6 snRNA was used as an internal control, and the relative amount of miR-495 normalized to U6 was calculated using the equation $2^{-\Delta\Delta CT}$, where $\Delta\Delta C_T = (C_{T\ miR\text{-}495} - C_{T\ U6})_{target} - (C_{T\ miR\text{-}495} - C_{T\ U6})_{control}$.

Overexpression or knockdown of JAM-A

A full-length *JAM-A* cDNA expression plasmid (EX-U0777-M02) lacking the 3′-UTR was purchased from GeneCopoeia (Rockville, MD, USA), and the empty plasmid served as the negative control. The siRNA sequence targeting human *JAM-A* mRNA was designed and synthesized by RiboBio (Guangzhou, China), with a scrambled siRNA included as negative control. The sequences of the *JAM-A* siRNA were as follows: 5′-GGAUAGUGAUGCCUACGAAdTdT-3′ (sense); 5′-dTdTCCUAUCACUACGGAUGCUU-3′ (antisense).

The *JAM-A* overexpression plasmid or siRNA was transfected into MCF-7 or MDA-MB-231 cells using Lipofectamine 2000 (Invitrogen) according to the manufacturer's instructions. The concentration for transfection of *JAM-A* cDNA plasmid or siRNA was 0.05 μg/mL or 50 nmol/L, respectively.

Overexpression or knockdown of miR-495

The overexpression or knockdown of miR-495 was accomplished by transfecting cells with a synthetic miR-495 mimic or inhibitor purchased from Genephama (Shanghai, China). The transfection concentrations were 50 nmol/L for the miR-495 mimic and 200 nmol/L for the miR-495 inhibitor, which were also adopted when the control mimic or inhibitor was transfected.

Western blotting

Western blotting was carried out as described previously (Cao et al., 2014). An anti-JAM-A antibody was obtained from Epitomics (#1840-1, Burlingame, CA, USA). An antibody against GAPDH, as the control, was obtained from Santa Cruz Biotechnology (Santa Cruz, CA, USA).

Luciferase reporter assay

The entire 3′-UTR of human JAM-A was amplified by PCR using human genomic DNA as the template, and the PCR products were inserted into the p-MIR-reporter plasmid (Ambion, Austin, TX, USA). The insertion was confirmed by sequencing. To test the binding specificity, the sequences that interact with the miR-495 seed sequence were mutated (from GUUUGUU to CAAACAA), and the mutant JAM-A 3′-UTR was inserted into an equivalent luciferase reporter plasmid. For the luciferase reporter assay, MDA-MB-231 cells were seeded in 24-well plates, and each well was transfected with 0.2 μg of luciferase reporter plasmid, 0.4 μg of β-galactosidase (β-gal) expression plasmid and 40 pmol of miR-495 mimic, 120 pmol of miR-495 inhibitor or scrambled negative control RNAs using Lipofectamine 2000 (Invitrogen). The β-gal plasmid was used as the transfection control. At 24 h after transfection, the cells were harvested and analyzed for luciferase activity using luciferase assay kits (Promega, Madison, WI, USA).

Wound healing assays

Cell migration was assessed in classical wound healing assays with some modifications (Rodriguez et al., 2005). Briefly, cells were seeded in 6-well plates and transfected when they were attached. After transfection, the cells were allowed to culture to confluence. Then, the cell layer was gently wounded using a plastic pipette tip (P200) and rinsed with PBS before the culture medium was replaced. The bottom of the wells was marked to indicate where the initial images of the wounded area were captured. At 24 h of incubation, images (10×) of the same areas were recorded using a photomicroscope (BX51 Olympus, Japan), and closure of the wounds was processed using Image-Pro Plus 6.0.

Transwell assays

The migration ability of cells was also tested in a Transwell Boyden Chamber (6.5 mm, Costar, USA). The polycarbonate membranes (8-μm pore size) on the bottom of the upper compartment of the Transwells were coated with 1% human fibronectin (R&D Systems 1918-FN, USA). At 24 h after transfection, the cells were harvested, counted and suspended in FBS-free DMEM medium. Then, 3×10^4 cells in 200 μL DMEM medium were added to each upper chamber; 0.6 mL of DMEM with 10% FBS was added to the lower compartment, and the Transwell-containing plates were incubated at 37°C and 5% CO_2 for 8 h. After incubation, the cells that had entered the lower surface of the membrane were fixed with 4% paraformaldehyde for 20 min at room temperature, washed 3 times with distilled water and stained with 0.1% crystal violet in 0.1 mol/L borate and 2% ethanol for 15 min at room temperature. The non-migrant cells remaining on the upper surface of the filter membrane were scraped off gently with a cotton swab. The lower surfaces (with migrant cells) were captured using a photomicroscope (5 fields per chamber) (BX51 Olympus, Japan), and the cells were counted blindly.

Statistical analysis

All images of Western blotting, wound healing assays and Transwell assays are representative of at least three independent experiments. Quantitative RT-PCR and the luciferase reporter assay were performed in triplicate, and each experiment was repeated at least three times. The results are presented as the mean ± SD. The differences between groups were calculated using Student's *t*-test, and $P < 0.05$ was defined as statistically significant.

ACKNOWLEDGEMENTS

This work was supported by grants from the National Basic Research Program of China (973 Program) (No. 2014CB542300), the National Natural Science Foundation of China (Grant Nos. 31301060 and 81250044), Specialized Research Fund for the Doctoral Program of Higher Education (No. 20130091120033) and the Research Special Fund for Public Welfare Industry of Health (No. 201302018).

ABBREVIATIONS

EMT, epithelial-mesenchymal transition; HCC, hepatocellular carcinoma; JAM-A, junctional adhesion molecule A; miRNAs, microRNAs; qRT-PCR, quantitative real time-PCR; TGF-β1, transforming growth factor-β1.

COMPLIANCE WITH ETHICS GUIDELINES

Minghui Cao, Weiwei Nie, Jing Li, Yujing Zhang, Xin Yan, Xiaoxiang Guan, Xi Chen, Ke Zen, Chen-Yu Zhang, Xiaohong Jiang and Dongxia Hou declare that they have no conflict of interest.

All procedures followed were in accordance with the ethical standards of the responsible committee on human experimentation (institutional and national) and with the Helsinki Declaration of 1975, as revised in 2000. Informed consent was obtained from all patients for being included in the study.

REFERENCES

Ambros V (2001) microRNAs: tiny regulators with great potential. Cell 107:823–826

Bartel DP (2004) MicroRNAs: genomics, biogenesis, mechanism, and function. Cell 116:281–297

Bartel DP (2009) MicroRNAs: target recognition and regulatory functions. Cell 136:215–233

Bendre M, Gaddy D, Nicholas RW, Suva LJ (2003). Breast cancer metastasis to bone: it is not all about PTHrP. Clin Orthop Relat Res S39–S45

Calin GA, Sevignani C, Dumitru CD, Hyslop T, Noch E, Yendamuri S, Shimizu M, Rattan S, Bullrich F, Negrini M et al (2004) Human microRNA genes are frequently located at fragile sites and genomic regions involved in cancers. Proc Natl Acad Sci USA 101:2999–3004

Cao M, Hou D, Liang H, Gong F, Wang Y, Yan X, Jiang X, Wang C, Zhang J, Zen K et al (2014) miR-150 promotes the proliferation and migration of lung cancer cells by targeting SRC kinase signalling inhibitor 1. Eur J Cancer 50:1013–1024

Chen SM, Chen HC, Chen SJ, Huang CY, Chen PY, Wu TW, Feng LY, Tsai HC, Lui TN, Hsueh C et al (2013) MicroRNA-495 inhibits proliferation of glioblastoma multiforme cells by downregulating cyclin-dependent kinase 6. World J Surg Oncol 11:87

Chu H, Chen X, Wang H, Du Y, Wang Y, Zang W, Li P, Li J, Chang J, Zhao G et al (2014) MiR-495 regulates proliferation and migration in NSCLC by targeting MTA3. Tumour biol: J Int Soc Oncodev Biol Med 35:3487–3494

Croce CM (2009) Causes and consequences of microRNA dysregulation in cancer. Nat rev Genet 10:704–714

DeSantis C, Ma J, Bryan L, Jemal A (2014) Breast cancer statistics, 2013. Cancer J Clin 64:52–62

Fabian MR, Sonenberg N, Filipowicz W (2010) Regulation of mRNA translation and stability by microRNAs. Annu Rev Biochem 79:351–379

Feigin ME, Muthuswamy SK (2009) Polarity proteins regulate mammalian cell-cell junctions and cancer pathogenesis. Curr Opin Cell biol 21:694–700

Fong D, Spizzo G, Mitterer M, Seeber A, Steurer M, Gastl G, Brosch I, Moser P (2012) Low expression of junctional adhesion molecule A is associated with metastasis and poor survival in pancreatic cancer. Ann Surg Oncol 19:4330–4336

Holliday DL, Speirs V (2011) Choosing the right cell line for breast cancer research. Breast Cancer Res 13:215

Hwang-Verslues WW, Chang PH, Wei PC, Yang CY, Huang CK, Kuo WH, Shew JY, Chang KJ, Lee EY, Lee WH (2011) miR-495 is upregulated by E12/E47 in breast cancer stem cells, and promotes oncogenesis and hypoxia resistance via downregulation of E-cadherin and REDD1. Oncogene 30:2463–2474

Jemal A, Bray F, Center MM, Ferlay J, Ward E, Forman D (2011) Global cancer statistics. Cancer J Clin 61:69–90

Jiang X, Huang H, Li Z, He C, Li Y, Chen P, Gurbuxani S, Arnovitz S, Hong GM, Price C et al (2012) MiR-495 is a tumor-suppressor microRNA down-regulated in MLL-rearranged leukemia. Proc Natl Acad Sci U S A 109:19397–19402

John B, Enright AJ, Aravin A, Tuschl T, Sander C, Marks DS (2004) Human MicroRNA targets. PLoS Biol 2:e363

Koshiba H, Hosokawa K, Kubo A, Tokumitsu N, Watanabe A, Honjo H (2009) Junctional adhesion molecule A [corrected] expression in human endometrial carcinoma. Int J Gynecol Cancer 19: 208–213

Lee JM, Dedhar S, Kalluri R, Thompson EW (2006) The epithelial-mesenchymal transition: new insights in signaling, development, and disease. J Cell Biol 172:973–981

Lewis BP, Shih IH, Jones-Rhoades MW, Bartel DP, Burge CB (2003) Prediction of mammalian microRNA targets. Cell 115: 787–798

Li Z, Cao Y, Jie Z, Liu Y, Li Y, Li J, Zhu G, Liu Z, Tu Y, Peng G et al (2012) miR-495 and miR-551a inhibit the migration and invasion of human gastric cancer cells by directly interacting with PRL-3. Cancer Lett 323:41–47

Li Y, Hong F, Yu Z (2013) Decreased expression of microRNA-206 in breast cancer and its association with disease characteristics and patient survival. J Int Med Res 41:596–602

Lu J, Getz G, Miska EA, Alvarez-Saavedra E, Lamb J, Peck D, Sweet-Cordero A, Ebert BL, Mak RH, Ferrando AA et al (2005) MicroRNA expression profiles classify human cancers. Nature 435:834–838

Lujambio A, Lowe SW (2012) The microcosmos of cancer. Nature 482:347–355

Ma L, Teruya-Feldstein J, Weinberg RA (2007) Tumour invasion and metastasis initiated by microRNA-10b in breast cancer. Nature 449:682–688

Mandell KJ, Parkos CA (2005) The JAM family of proteins. Adv Drug Deliv Rev 57:857–867

McSherry EA, Brennan K, Hudson L, Hill AD, Hopkins AM (2011) Breast cancer cell migration is regulated through junctional adhesion molecule-A-mediated activation of Rap1 GTPase. Breast Cancer Res 13:R31

Naik MU, Naik TU, Suckow AT, Duncan MK, Naik UP (2008) Attenuation of junctional adhesion molecule-A is a contributing factor for breast cancer cell invasion. Cancer Res 68:2194–2203

Neve RM, Chin K, Fridlyand J, Yeh J, Baehner FL, Fevr T, Clark L, Bayani N, Coppe JP, Tong F et al (2006) A collection of breast cancer cell lines for the study of functionally distinct cancer subtypes. Cancer Cell 10:515–527

Olsen PH, Ambros V (1999) The lin-4 regulatory RNA controls developmental timing in Caenorhabditis elegans by blocking LIN-14 protein synthesis after the initiation of translation. Dev Biol 216:671–680

Osada H, Takahashi T (2007) MicroRNAs in biological processes and carcinogenesis. Carcinogenesis 28:2–12

Rodriguez LG, Wu X, Guan JL (2005) Wound-healing assay. Methods Mol Biol 294:23–29

Schneeberger EE, Lynch RD (2004) The tight junction: a multifunctional complex. Am J Physiol Cell physiol 286:C1213–C1228

Vimalraj S, Miranda PJ, Ramyakrishna B, Selvamurugan N (2013) Regulation of breast cancer and bone metastasis by microRNAs. Dis Markers 35:369–387

Wang Y, Lui WY (2012) Transforming growth factor-beta1 attenuates junctional adhesion molecule-A and contributes to breast cancer cell invasion. Eur J Cancer 48:3475–3487

Yang H, Cho ME, Li TW, Peng H, Ko KS, Mato JM, Lu SC (2013) MicroRNAs regulate methionine adenosyltransferase 1A expression in hepatocellular carcinoma. J Clin Investig 123:285–298

Zhang M, Luo W, Huang B, Liu Z, Sun L, Zhang Q, Qiu X, Xu K, Wang E (2013) Overexpression of JAM-A in non-small cell lung cancer correlates with tumor progression. PloS One 8:e79173

Epigenetic reprogramming, gene expression and *in vitro* development of porcine SCNT embryos are significantly improved by a histone deacetylase inhibitor—*m*-carboxycinnamic acid bishydroxamide (CBHA)

Yuran Song[1,2], Tang Hai[1], Ying Wang[1], Runfa Guo[1], Wei Li[1], Liu Wang[1], Qi Zhou[1]✉

[1] State Key Laboratory of Reproductive Biology, Institute of Zoology, Chinese Academy of Sciences, Beijing 100101, China
[2] Graduate University of the Chinese Academy of Sciences, Beijing 100049, China
✉ Correspondence: zhouqi@ioz.ac.cn (Q. Zhou)
Received October 15, 2013 Accepted November 12, 2013

ABSTRACT

Insufficient epigenetic reprogramming of donor nuclei is believed to be one of the most important causes of low development efficiency of mammalian somatic cell nuclear transfer (SCNT). Previous studies have shown that both the *in vitro* and *in vivo* development of mouse SCNT embryos could be increased significantly by treatment with various histone deacetylase inhibitors (HDACi), including *Trichostatin A*, *Scriptaid*, and *m*-carboxycinnamic acid bishydroxamide (CBHA), in which only the effect of CBHA has not yet been tested in other species. In this paper we examine the effect of CBHA treatment on the development of porcine SCNT embryos. We have discovered the optimum dosage and time for CBHA treatment: incubating SCNT embryos with 2 μmol/L CBHA for 24 h after activation could increase the blastocyst rate from 12.7% to 26.5%. Immunofluorescence results showed that the level of acetylation at histone 3 lysine 9 (AcH3K9), acetylation at histone 3 lysine 18 (AcH3K18), and acetylation at histone 4 lysine 16 (AcH4K16) was raised after CBHA treatment. Meanwhile, CBHA treatment improved the expression of development relating genes such as *pou5f1*, *cdx2*, and the imprinted genes like *igf2*. Despite these promising *in vitro* results and histone reprogramming, the full term development was not significantly increased after treatment. In conclusion, CBHA improves the *in vitro* development of pig SCNT embryos, increases the global histone acetylation and corrects the expression of some developmentally important genes at early stages. As in mouse SCNT, we have shown that nuclear epigenetic reprogramming in pig early SCNT embryos can be modified by CBHA treatment.

KEYWORDS swine, nuclear transfer, epigenetic reprogramming, histone deacetylase inhibitor

INTRODUCTION

Somatic cell nuclear transfer (SCNT) has been a successful technology to derive cloned animals in many species (Wilmut et al., 1997; Kato et al., 1998; Wakayama et al., 1998; Baguisi et al., 1999; Onishi et al., 2000; Chesne et al., 2002; Shin et al., 2002; Galli et al., 2003; Woods et al., 2003; Zhou et al., 2003; Lee et al., 2005; Li et al, 2006). It is a valuable tool in fundamental research, biomedical and reproduction fields (Verma et al., 2011; Bendixen et al., 2013). As pigs share much similar physiological characteristics with human beings, it is a potential animal model for human disease and xenotransplantation (Gock et al., 2011; Ekser et al., 2012; Luo et al., 2012). SCNT is a practical way to produce targeted genetic modification in pigs.

However, like in other animals, SCNT in pigs is inefficient, due to the low cloning efficiency, fetus abnormality, and placenta deficiency (Polejaeva et al., 2000; Walker et al.,

Table 1. Pre-implantation development of pig SCNT embryos under treatment with different CBHA concentration

CBHA Treatment (μmol/h)	No. of reconstructed embryos	No. of 2-cell (% reconstructed)	No. of blastocysts (% 2-cell)	No. of blastocysts (% reconstructed)
0	212	132 (62.3)	27 (20.5)	27 (12.7 ± 5.7)[a]
0.02/24	58	40 (69.0)	8 (20.0)	8 (13.8 ± 1.4)[a]
0.2/24	123	75 (61.0)	25 (33.3)	25 (20.3 ± 9.0)[b]
2/24	211	117 (55.4)	56 (47.9)	56 (26.5 ± 12.3)[c]
10/24	90	60 (66.7)	17 (28.3)	17 (18.9 ± 17.9)[b]
20/24	213	110 (51.6)	35 (31.8)	35 (16.4 ± 10.7)[b]
50/24	90	35 (38.9)	12 (34.3)	12 (13.3 ± 3.8)[a]
200/24	65	1 (1.5)	0 (–)	0 (–)
400/24	65	0 (–)	0 (–)	0 (–)

Significant χ^2 comparisons: a versus c, $P < 0.05$; a versus b, $P > 0.05$; b versus c, $P > 0.05$.

2002). It is widely believed that the crucial cause of poor developmental capacity in nuclear transfer embryos is the aberrant nuclear reprogramming. There is a close relationship between embryo reprogramming and epigenetic modification. DNA methylation and histone acetylation are the main modification patterns in epigenetic reprogramming. Compared with normal fertilized zygotes, in SCNT embryos nuclear, reprogramming has to occur within a short time span, so a more "open" chromatin configuration may be beneficial to SCNT embryos (William et al., 2001; Whitworth et al., 2010; Ogura et al., 2013). Histone deacetylase inhibitor (HDACi), has been proved to increase the acetylation level in somatic cells or in nuclear transfer embryos, as it makes the chromatin more flexible and allowing the combination of transcriptional factors (Kishigami et al., 2006; Lager et al., 2008; Dai et al., 2010).

In pigs, some histone deacetylase inhibitors, such as Trichostatin (TSA), *Scriptaid*, and Valproic acid (VPA), have been proved to increase the *in vitro* development of SCNT embryos (Li et al., 2008; Miyoshi et al., 2010; Zhao et al., 2010; Kim et al., 2011). Our previous study has shown that *m*-carboxycinnamic acid bishydroxamide (CBHA), a member of type II HDACi, could improve the *in vitro* and full term development of mouse SCNT embryos and improve the cell line establishment efficiency of embryonic stem cells (ES) derived from SCNT embryos (SCNT-ESCs) (Dai et al., 2010). However, CBHA has not yet been studied in domestic animals. In this study we use this HDACi to treat porcine nuclear transfer embryos to investigate whether it affects the developmental efficiency in pigs.

We have examined the optimum treatment program of CBHA on pig nuclear transfer embryos and observed the development rates in various treated and non-treated groups. To further explore how it improves the development of porcine nuclear transfer embryos, we detected the global acetylation levels of Histone 3 at Lysine 9 (AcH3K9), Histone 3 at Lysine 18 (AcH3K18), and Histone 4 at Lysine 16 (AcH4K16) using immunofluorescence in the CBHA treated embryos and control embryos. Then we analyzed the effect of CBHA treatment on the transcription level of some development-related genes, histone deacetylase genes, and imprinted genes using quantitative PCR. Finally, we detected the full term developmental efficiency of CBHA treated and non-treated embryos.

RESULTS

In vitro development of porcine SCNT embryos are improved by CBHA treatment

To test whether CBHA could improve the early development of porcine SCNT embryos, we treated the SCNT embryos with different concentrations of CBHA and calculated the developmental rates from 2-cell to blastocyst stage (Table 1). We found that proper CBHA treatment had no effect on the cleavage rates of porcine SCNT embryos, but could significantly increase the *in vitro* development rate to blastocyst stage. The optimum treatment concentration of CBHA was 2 μmol/L, while 200 μmol/L or higher concentrations were detrimental to embryos. After treatment with 2 μmol/L CBHA for 24 h, the blastocyst ratio of CBHA-SCNT embryos was two folds (26.5%) of the embryos in control group (12.7%).

Next, to optimize the protocol for CBHA treatment, we examined the effect of different CBHA incubation times on the blastocyst ratios of SCNT embryos. The SCNT embryos were treated with 2 μmol/L or 10 μmol/L CBHA for 16 h, 20 h, and 24 h, respectively. There was no significant difference among the three groups, but the treatment of 2 μmol/L, 24 h gave the best blastocyst ratio (Table 2).

Blastocyst quality was improved after CBHA treatment

To unravel the improvement effects of CBHA treatment on SCNT embryo development, we examined the blastocyst cell number in two groups (Fig. 1). Though the cell numbers in the treated group were higher than the non-treated group, there was no significant difference between the two groups

Table 2. Effect of time duration of CBHA treatment on the pre-implantation development of SCNT embryos

Concentration of CBHA (μmol /L)	Time duration (h)	No. of reconstructed embryos	No. of 2-cell (% reconstructed)	No. of blastocysts (% 2-cell)	No. of blastocysts (% reconstructed)
2	16	77	40 (52.0)	11 (27.5)	11 (14.3 ± 12.5)
	20	78	44 (56.4)	17 (38.6)	17 (21.8 ± 3.8)
	24	77	34 (44.2)	16 (47.1)	16 (20.8 ± 7.2)
20	16	73	34 (46.6)	13 (38.2)	13 (17.8 ± 13.2)
	20	80	38 (47.5)	13 (34.2)	13 (16.3 ± 13.9)
	24	78	36 (45.2)	10 (27.8)	10 (12.8 ± 4.0)
0	–	80	33 (41.3)	14 (42.4)	14 (17.5 ± 6.7)

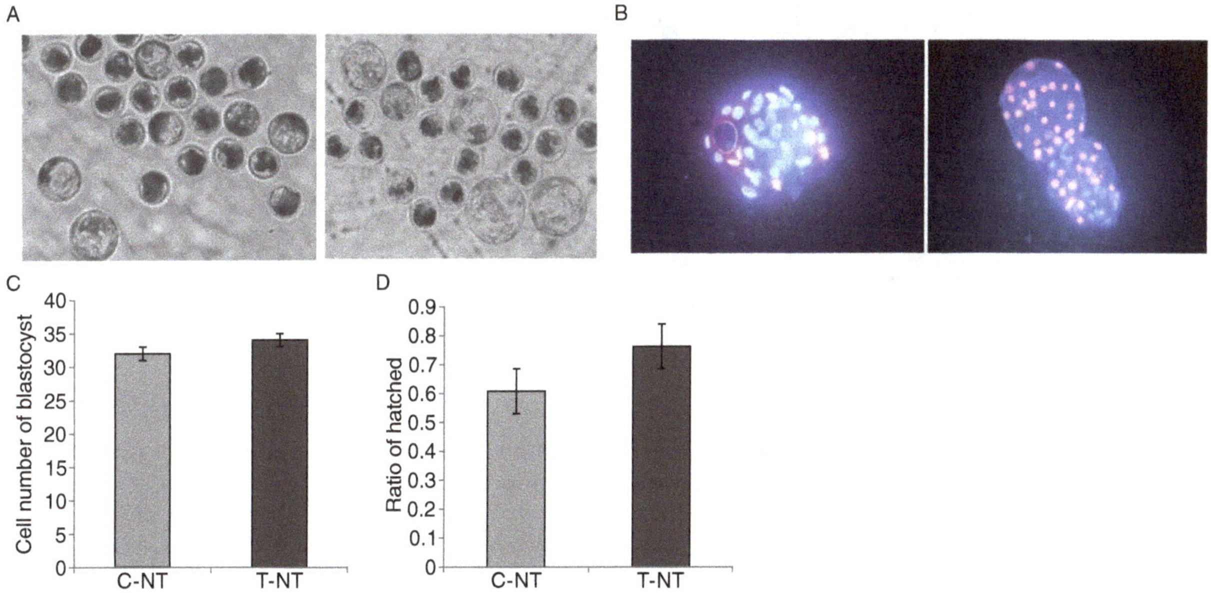

Figure 1. Effect of CBHA on hatching rate and cell number of SCNT blastocysts. (A) Representative photographs of pig blastocysts. Day 7 blastocysts developed from 0 μmol/L CBHA treated SCNT embryos (C-NT group) and 2 μmol/L CBHA treated SCNTembryos (T-NT group). Original magnification was 100×. (B) Fluorescent photographs of pig blastocysts stained with Hoechst. Original magnification was 200×. (C) Statistics of the number of cells in blastocyst from 0 μmol/L CBHA treated SCNT embryos (C-NT group) and 2 μmol/L treated embryos (T-NT group). (D) Ratio of hatched blastocysts from 0 μmol/L CBHA treated SCNT embryos (C-NT group) and 2 μmol/L treated embryos (T-NT group).

(34 vs. 32). Cell numbers from both groups were lower than their *in vivo* and *in vitro* fertilized counterparts.

Embryo implantation and further development are largely related to the blastocyst-hatching rate and the communication between embryo and the uterus. Hence we examined the blastocyst-hatching rate in the two groups. The hatched blastocysts were observed under a stereomicroscope at 168 h after activation (Fig. 1). Although the hatching rate in the CBHA treated group (76.25%) was much higher than those in the control group (67%), there was no significant difference between these two groups. Nevertheless, we found that embryos treated with CBHA could reach the blastocyst stage much earlier than those in the control groups, indicating that at the developmental phase they are more like the *in vivo* embryos, who reach the blastocyst stage on about day 5.

CBHA treatment increases global histone acetylation levels of pig SCNT embryos

As shown in embryos, the global histone acetylation level is closely related to the reprogramming state of chromatin. Meanwhile, we showed that an open, decondensed chromatin state was relevant to the pluripotency of iPSCs

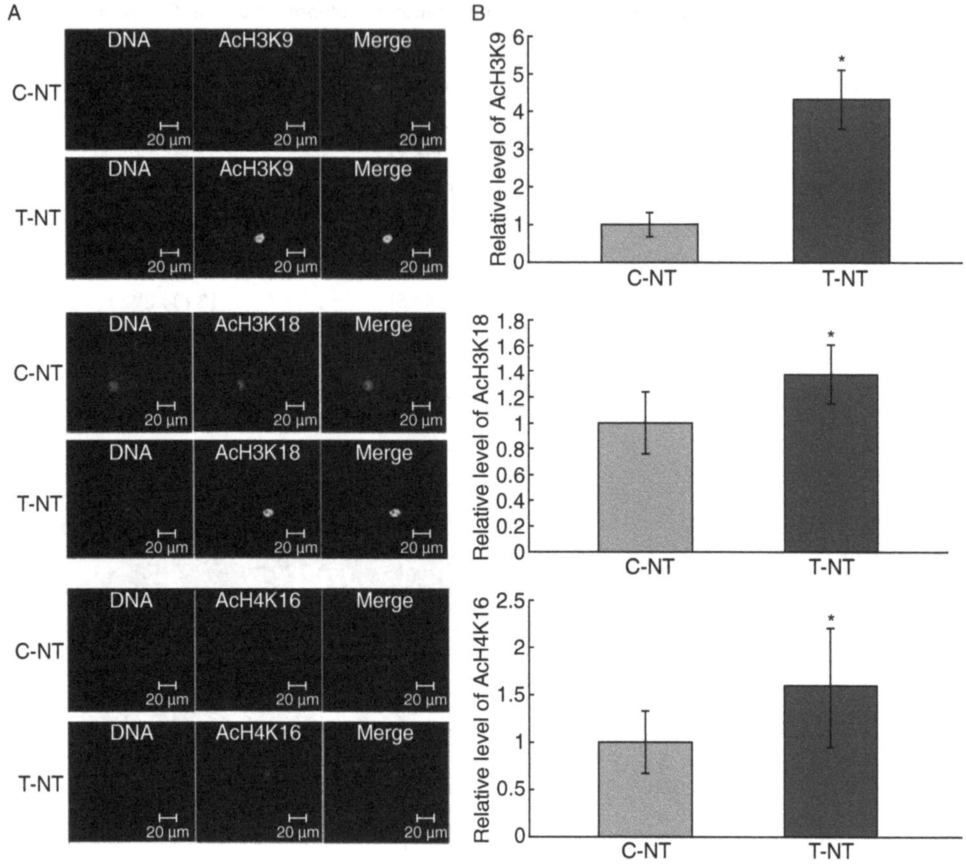

Figure 2. The global AcH3K9, AcH3K18, and AcH4K16 levels of 1-cell stage embryos. (A) Staining of AcH3K9, AcH3K18, and AcH4K16 (green) in non-treated control SCNT embryos (C-NT), and 2 µmol/L CBHA treated embryos (T-NT) at 1-cell stage. Each sample was counterstained with DAPI to visualize DNA (red). Original magnification was 200×. (B) Quantification of AcH3K9 and AcH3K18 signal intensities in C-NT (gray bars) and T-NT (black bars) embryos. Labeling intensity was expressed relative to that of the C-NT embryos (set as 100%). $^*P < 0.05$. The experiments were replicated 3 times. n = 8–10 per group.

(induced pluripotent stem cells) and ESCs (embryonic stem cells). We further examined the embryo histone acetylation levels by assaying three epigenetic markers: AcH3K9, AcH3K18, and AcH4K16 at 1-cell, 2-cell, and blastocyst stages. We found that CBHA treatment increased the level of AcH3K9, AcH3K18, and AcH4K16 at 1-cell (Fig. 2) and 2-cell stages (Fig. 3). However, apart from AcH3K18, at the blastocyst stage there were no differences between the control and experiment groups (Fig. 4).

Relative transcription level of development related genes, imprinted genes, and histone deacetylase genes

In mammalian early embryonic development, *pou5f1* (*oct4*), an indicator for inner cell mass, is widely accepted to be a key factor in embryonic quality and pluripotency maintenance. Meanwhile, *cdx2*, a pivotal trophectoderm (TE) marker, is important for placenta formation. The expression level of imprinted genes and histone deacetylase genes is also vital for nuclear reprogramming. So, the transcription levels of the five genes (*pou5f1*, *cdx2*, *igf2*, *igf2r*, and *hdac2)* whose functions are mentioned above were detected by quantitive PCR (Fig. 5) at the blastocyst stage. The expression of two key developmental genes, *pou5f1* and *cdx2,* was significantly increased by CBHA treatment. However, the expression of these two genes in both groups was significantly lower than the IVF (*in vitro*, fertilized) group. The expression of the imprinted gene *IGF2* was increased after the CBHA treatment, which was similar to the IVF ones and significantly higher than the control group. However, the expression of another imprinted gene, *igf2r*, and the histone deacetylase gene, *hdac2*, was not affected by CBHA treatment, and both of them were significantly lower than the IVF ones.

Full term development of porcine SCNT embryos following CBHA treatment

To determine whether CBHA could improve the full term development of porcine SCNT embryos, 2-cell stage embryos with and without CBHA treatment were transferred to the oviducts of surrogates on the day of or 1 day after the onset of estrus. In the first batch of experiments, we

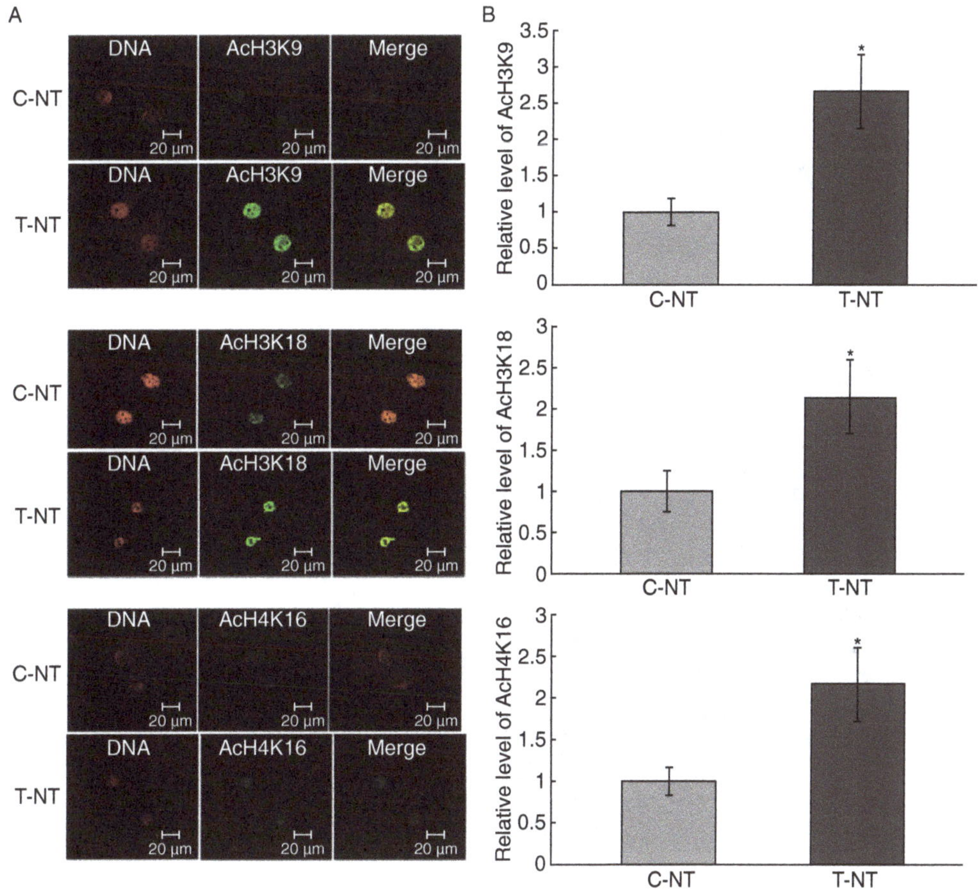

Figure 3. The global AcH3K9, AcH3K18, and AcH4K16 levels of 2-cell stage embryos. (A) Staining of AcH3K9, AcH3K18, and AcH4K16 (green) in non-treated control SCNT embryos (C-NT), and 2 μmol/L CBHA treated embryos (T-NT) at 2-cell stage. Each sample was counterstained with DAPI to visualize DNA (red). Original magnification was 200×. (B) Quantification of AcH3K9 and AcH3K18 signal intensities in C-NT (gray bars) and T-NT (black bars) embryos. Labeling intensity was expressed relative to that of the C-NT embryos (set as 100%). $^*P < 0.05$, $^†P < 0.01$. The experiments were replicated 3 times. n = 8–10 per group.

transferred more than 200 embryos to each surrogate. We found that there was no significant difference between the two groups (Table 3), which is inconsistent with the beneficial effect of CBHA in the *in vitro* development. To further exclude the noise caused by the excessive number of embryos transferred, we designed a second batch of experiments, which transferred only 100 embryos to each recipient (Table 4). We obtained a total of 15 live, 6 dead, and 1 mummy from the CBHA treated group and 17 live, and 7 dead from the control group. The full term development ratio of CBHA-treated embryos was 4.2%, which is not significantly different to the control group (4.4%), suggesting CBHA treatment does not improve the full term development of porcine SCNT embryos.

DISCUSSION

Nowadays, somatic cells can be reprogrammed in many ways, the most common being somatic cell nuclear transfer (SCNT) and induced pluripotent stem cells (iPSCs) (Wilmut et al., 1997; Kato et al., 1998; Wakayama et al., 1998; Baguisi et al., 1999; Onishi et al., 2000; Chesne et al., 2002; Shin et al., 2002; Han et al., 2003; Galli et al., 2003; Woods et al., 2003; Zhou et al., 2003; Lee et al., 2005; Li et al, 2006; Takahashi et al., 2006; Yu et al., 2007). After being transfected with pluripotent genes or small molecules, a fraction of somatic cells could be reprogrammed to a pluripotent state like embryonic stem cells (ESCs). However, it has been reported that there are differences in the epigenetic modification and expression of genes between iPSCs and ESCs and the expression of exogenous pluripotent factors is likely to cause tumors (Okita et al., 2007; Bock et al., 2011; Menendez et al., 2011). Although several iPS cell lines have been reported, the silencing of exogenous genes and germline transmission remain unsolved. Thus nuclear transfer is still an indispensable method for somatic cell reprogramming. Meanwhile for pigs, besides therapeutic cloning, SCNT is a useful tool in transgenic research and especially in xenotransplantation and disease models. However, the

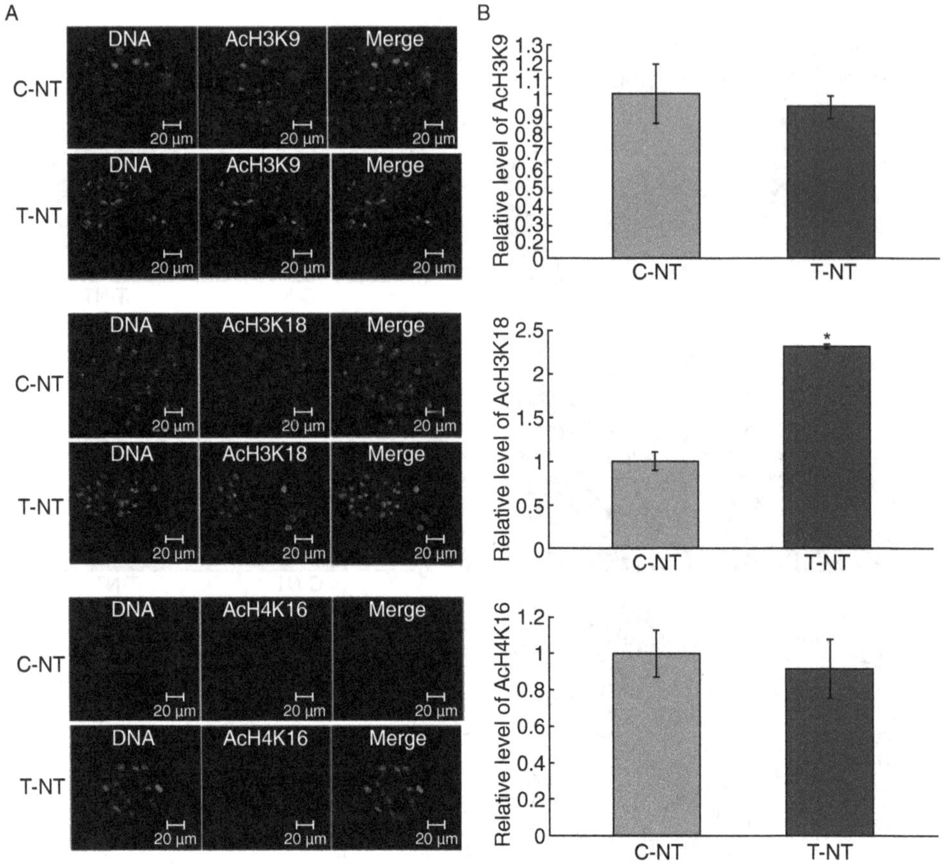

Figure 4. The global AcH3K9, AcH3K18, and AcH4K16 levels of blastocyst stage embryos. (A) Staining of AcH3K9, AcH3K18, and AcH4K16 (green) in non-treated control SCNT embryos (C-NT), and 2 μmol/L CBHA treated embryos (T-NT) at blastocyst stage. Each sample was counterstained with DAPI to visualize DNA (red). Original magnification was 200×. (B) Quantification of AcH3K9 and AcH3K18 signal intensities in C-NT (gray bars) and T-NT (black bars) embryos. Labeling intensity was expressed relative to that of the C-NT embryos (set as 100%). $^*P < 0.05$. The experiments were replicated 3 times. n = 8–10 per group.

efficiency of SCNT is still low and the mechanism of nuclear reprogramming is poorly understood.

To improve the efficiency of reprogramming, numerous factors or molecules have been used to alter the epigenetic modification in somatic cells or reconstructed embryos. HDACi is a widely used epigenetic modifier in cancer treatment and somatic cell nuclear reprogramming such as SCNT and iPSCs derivation. However, not all HDACi are beneficial to SCNT embryos, with some of them being unprofitable or detrimental. Generally, type II HDACi is used in epigenetic reprogramming (Monteriro et al., 2011).

Here we used CBHA, a hybrid polar compound, to improve the development of pig SCNT embryos. We found that the efficiency of SCNT and the quality of SCNT embryos are improved after CBHA treatment. Compared with non-treated embryos, CBHA treated SCNT embryos showed a higher global histone acetylation level. Expression of the development-related genes and the imprinted genes at blastocyst stage was also improved with the CBHA treatment.

In mouse SCNT, the optimum CBHA treatment protocol is 20 μmol/L incubation for 10 h post activation (Dai et al., 2010). However, in pig SCNT, we found that the best protocol was 2–10 μmol/L treated for 20–24 h after activation. This difference may be due to the differing developmental time spans between mice and pigs.

Histone acetylation is a significant epigenetic modification in embryo development and nuclear reprogramming. In previous studies, it has been shown that global hyperacetylation associated with a more permissive chromatin state which is an indicator of more fully reprogramming state in nuclear transfer and iPSCs (Li et al., 2002; Rybouchkin et al., 2006). In iPS cells, compared to partially reprogrammed ones, the fully reprogrammed iPSCs exhibit a more open and decondensed chromatin state and are more identical to ES cells (Mattout et al., 2011). We tested the histone acetylation level of reconstructed embryos after the CBHA treatment to examine the effect of this HDACi. Compared with the control SCNT embryos, the CBHA SCNT embryos showed a higher level of global histone acetylation. AcH3K9,

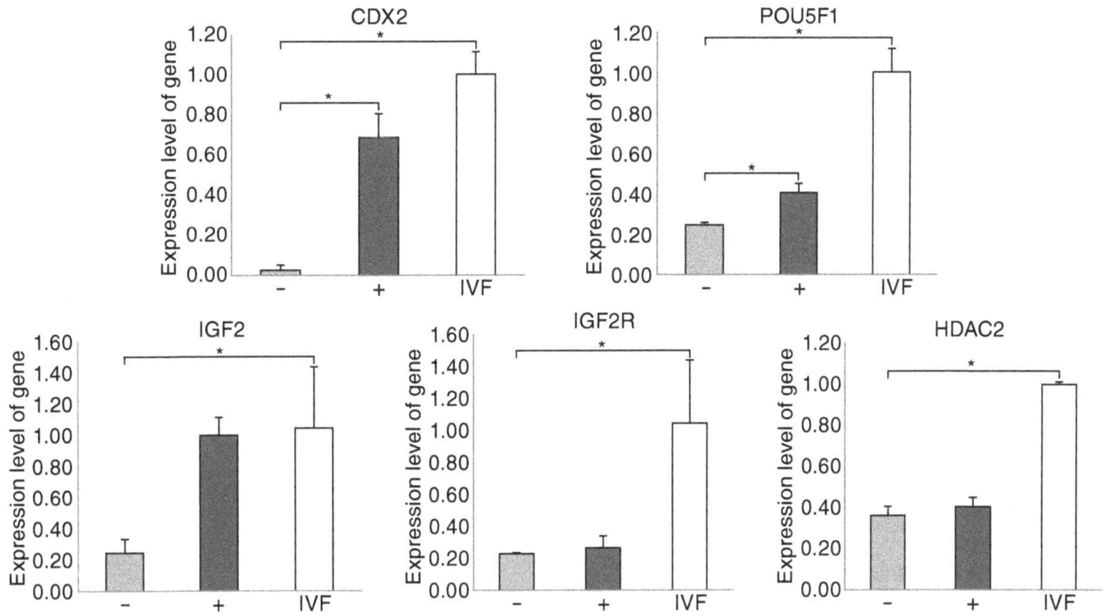

Figure 5. Relative expression at the transcriptional level of histone deacetylase genes, development-related, and imprinted genes. Blastocysts, IVF (open bars), non-treated control SCNT embryos (C-NT) (gray bars), and 2 μmol/L CBHA treated embryos (T-NT) (black bars) were collected on day 7. *$P < 0.05$; $n = 15$.

Table 3. Full-term development of CBHA treated SCNT embryos

Cell type	No. of transferred embryos	CBHA treatment[a]	Pregnancy check[b]	Piglets born	Cloning efficiency (%)[c]
PEFCs	231	+	–	0	0.94
PEFCs	187	+	+	2 dead + 1 mummy	
PEFCs	276	+	+	1 live + 1 dead + 2 mummy	
PEFCs	263	+	+	1 live + 1 dead	
PEFCs	262	–	–	0	
PEFCs	280	–	+	2 live + 5 dead	1.12
PEFCs	259	–	+	1 live + 3 dead	
PEFCs	269	–	+	2 live + 1 dead	
PEFCs	272	–	+	1 dead	

[a] +, Treated; –, untreated.

[b] +, pregnant; –, not pregnant.

[c] No. of piglets/No. of embryos transferred.

AcH3K18, and AcH4K16 were all increased at the 1-cell and 2-cell stages, but there were no significant differences at the blastocyst stage except in AcH3K18. We speculated that CBHA depresses the expression of certain histone deacetylases, thus limiting the modification of deacetylation and increasing the global acetylation level. The hyperacetylation of histone may affect the chromatin remodeling and provide a more permissible and accessible configuration for transcription factors or chaperones combining (Vignon et al., 2002; Lee et al., 2004; Deshmukh et al., 2012; Mason et al., 2012). Embryonic gene activation (EGA) in pigs happened at the 4-cell stage (Maalouf et al., 2009), and the increase of global histone acetylation was mainly discovered at the 1-cell and 2-cell stages, which suggested that a more "open" chromatin structure might be beneficial for EGA.

To further explore the effect of CBHA on SCNT embryo development, we tested the expression of five important genes at the transcription level in the blastocysts of SCNT embryos, CBHA-SCNT embryos, and IVF embryos. This included two development-related genes: *cdx2* and *pou5f1*,

Table 4. Full-term development of CBHA treated SCNT embryos with minimum transfer volume

Cell type	No. of transferred embryos	CBHA treatment[a]	Pregnancy check[b]	Piglets born	Cloning efficiency (%)[c]
PEFCs	100	+	+	6 live + 3 dead + 1 mummy	4.2
PEFCs	100	+	+	6 live + 3 dead	
PEFCs	100	+	+	3 live	
PEFCs	100	+	–	0	
PEFCs	100	+	–	0	
PEFCs	100	–	+	5 live + 1 dead	4.4
PEFCs	100	–	+	5 live + 1 dead	
PEFCs	100	–	+	2 live	
PEFCs	100	–	+	5 live + 3 dead	
PEFCs	100	–	–	0	

[a] +, Treated; –, untreated.

[b] +, pregnant; –, not pregnant.

[c] No. of piglets/No. of embryos transferred.

two imprinted genes: *igf2* and *igf2r*, and histone deacetylase 2 (*hdac2*). In IVF embryos, the expression of *cdx2* and *pou5f1* are significantly higher than the control SCNT and CBHA treated SCNT embryos. However, CBHA treatment improved the expression of the two genes to a large extent compared with the control SCNT embryos. *Cdx2* plays a key role in embryo placental development, and in mouse early embryo development, interaction between *oct3/4* and *cdx2* determines the level of trophectoderm differentiation (Tomanck et al., 1989; Chawengsaksophac et al., 2004; Strumpf et al., 2005). The extremely low expression of *cdx2* in the control SCNT embryos may explain the abnormal placental development in some cloned embryos. While *pou5f1* (*oct4*) is a well-known gene in embryo development, it interacts with regulatory enzymes and epigenetic histone markers in mice. As a central regulator of pluripotency and cell differentiation, it positively modulates the transcription of other recognized stem cell regulators, and it plays a vital role in inner cell mass (ICM) maintenance and ES cell self-renewal (Niwa et al., 2005; Loh et al., 2006). A previous study has shown that in porcine SCNT embryos, the expression of *oct4* and *oct4*-related genes are aberrant (Pesce et al., 2001). It has been reported that in mice, CBHA could improve the derivation efficiency of ESCs from SCNT embryos (Dai et al., 2010). In this process, improved expression of *pou5f1* may act. In our experiment, although still lower than the IVF embryos, the expression of *pou5f1* in CBHA-SCNT embryos was significantly increased compared to the control SCNT embryos. For the importance of embryo development, the two best studied imprinted genes, *igf2* and *igf2r,* have been discussed here. It has been reported that these two genes are related to EGA and involved in fetal growth regulation and are essential for normal development (Lee et al., 2006). Aberrant expression patterns of these imprinted genes may be responsible for the abnormal development found in fetuses and offspring originating from SCNT embryos (Latham et al., 1994; Yang et al., 2005). After CBHA treatment, we observed that the expression of *igf2* increased to the same level as the IVF embryos, yet the expression of *igf2r* was not affected. It can be concluded that CBHA can improve the transcription activity of these genes, which should be related to the more "open" chromatin configuration. CBHA "corrects" the expression of the pivotal genes. It is known that histone deacetylase (*hdacs*) and histone acetylase (*hats*) mediate histone acetylation-deacetylation (Peserico et al., 2011). Previous studies on IVF and *in vivo* porcine embryos have reported that *hdac2* exhibited a low level of expression at 4-cell stage, which steadily increased to its maximum level at the blastocyst stage, however this trend was reversed in SCNT embryos (Kumar et al., 2007; Peserico et al., 2011). Indeed, our results showed that the highest expression of *hdac2* occurred in the IVF blastocyts. However, we also observed that the expression of *hdac2* was hardly affected by the CBHA treatment. This may be because the *hdac2* was not the target protein of CBHA and it is not the only deacetylase to regulate histone deacetylation.

Nevertheless, unlike the improved development *in vitro*, full term development efficiency was hardly affected by the CBHA treatment. The cloning efficiency between CBHA-SCNT and the control SCNT groups is not significantly different. One reason may be that the sample volume was not sufficient to show the difference between the two groups. Another may be because we transferred the embryos at the 2-cell stage. While the *in vivo* and *in vitro* conditions for pre-implanted embryo development are different, the *in vivo* environment may correct the aberrant development from the 2-cell stage onwards to some extent. Another possibility is that the donor cells we transferred are in a good state, therefore the embryos are well developed *in vivo* even without the CBHA treatment. Thus in our experiment, the piglet birth rate is much higher than other reports (0–3.7%) (Zhao et al., 2010), regardless of the CBHA-SCNT (4.2%) or the control SCNT groups (4.4%).

To sum up, our present study indicates that, as a HDACi, CBHA can improve the *in vitro* development of pig nuclear transfer embryos. The presence of CBHA can increase the global acetylation level in AcH3K9, AcH3K18, and AcH4K16. It influences the modification of chromatin structure and improves the expression of important genes in development and imprinting, rendering a more similar expression profile to IVF ones. It enhances the nuclear reprogramming and *in vitro* development of pig SCNT embryos, which may provide better qualified embryos for ES cell derivation in the case of low quality donors. Further studies are needed to discuss the effect of CBHA on chromatin configuration modifications.

MATERIALS AND METHODS

Chemicals

All chemicals were purchased from Sigma Chemical Co. (St. Louis, MO), unless otherwise stated.

Preparation of somatic cells for SCNT

Porcine fetal fibroblast cells were established from a 35-day old fetus. Cells at passage 2–5 were grown to 90% confluency in Dulbecco's Modified Eagle Medium (DMEM; Gibco BRL, Life Technologies, Grand Island, USA) and supplemented with 10% fetal bovine serum (Gibco BRL, Paisley, UK). Fibroblast cells were dissociated with 0.25% trypsin-EDTA (Invitrogen, Carlsbad, CA, USA) at 37°C for about 3–5 min, then transferred to a centrifuge tube with Hepes-buffered Tissue Culture Medium 199 (TCM-199; BioWhittake, Walkersville, MD, USA) and supplemented with 2% cattle serum (CS; Danish Veterinary Institute, DTU, Frederiksberg, Denmark). The suspension was stored in ice until use.

In vitro maturation of oocytes

Ovaries were collected from prepubertal gilts at a local slaughterhouse and transported to the laboratory at 37°C. Follicles between 3 mm and 6 mm in diameter were aspirated with an 18-gauge needle attached to a 10-mL syringe. The cumulus-oocyte complexes (COCs) in the follicular fluid were allowed to settle by gravity. The COCs were rinsed in Hepes-buffered Tyrode's medium containing 0.01% PVA three times. Only the COCs with multiple layers of intact cumulus cells and uniform ooplasm were selected for *in vitro* maturation (IVM). After washing three times in IVM medium, a group of 50 COCs was placed into each well of twenty-four-well cell culture plates containing 500 μL of IVM medium. The COCs were matured for 42 h at 38.5°C and 5% CO_2 in air with 100% humidity. Matured COCs were then vortexed in 0.1% hyaluronidase in Hepes-buffered Tyrode's medium containing 0.01% PVA for 3 min to remove the cumulus cells. Only the matured oocytes having an extruded first polar body (PB) with uniform cytoplasm were used for the SCNT or IVF embryos.

Somatic cell nuclear transfer

Matured oocytes used for SCNT were placed in manipulation medium (MAN) supplemented with 7.5 μg/mL Cytochalasin B, and overlaid with warm mineral oil. Oocytes were enucleated by aspirating the polar body (PB), metaphase II chromosomes, and a small amount of surrounding cytoplasm using a beveled glass pipette with an inner diameter of 17–20 μm. A single intact donor cell was injected into the perivitelline space and placed adjacent to the recipient cytoplasm. Karyoplast-cytoplast complexes (KCCs) were placed into embryo culture medium until fusion and activation occurred. The fusion and activation of the KCCs were accomplished with two direct current pulses (1 s interval) of 1.2 kV/cm for 30 microseconds provided by Eppendorf Multiporator in fusion medium (0.3 mol/L mannitol, 1.0 mmol/L $CaCl_2$, 0.1 mmol/L $MgCl_2$, and 0.5 mmol/L Hepes [pH adjusted to 7.0–7.4]). Oocytes were then incubated for 20 min in PZM3 and evaluated for fusion under a stereomicroscope. Only the fused embryos were placed into four-well cell culture plates containing 500 μL of PZM3 at 38.5°C and 5% CO_2 with 100% humidified air.

Post-activation treatment and embryo culture

Stock solutions of CBHA were dissolved in dimethyl sulfoxide (DMSO) at 100 mmol/L and stored at −80°C. Following electrical activation, the SCNT embryos were treated with various concentrations (from 0.02 μmol/L to 400 μmol/L) of CBHA for different times (from 16 h to 24 h) in PZM3. After treatment, embryos were washed three times and then transferred into a four-well cell culture plate containing 500 μL PZM3 medium. They were cultured at 38°C and 5% CO_2 with 100% humidity either overnight or for 6 days. Cleavage and blastocyst formation were evaluated on days 2 and 7, respectively, with the day of SCNT designated day 0.

Embryo transfer

Day 2 SCNT embryos were transferred to the oviducts of surrogates on the day of, or 1 day after, the onset of estrus. Pregnancy was diagnosed on day 25, and the surrogates were checked regularly at 2-week intervals by ultrasound examination. All of the cloned piglets were delivered by either natural birth or cesarean section on day 117 of gestation and hand raised. All animals were treated according to preapproved institutional animal care and use protocols.

Blastocyst cell number counting

Expanded day 6 blastocysts derived from SCNT were selected for cell number comparison. After being fixed in 4% paraformaldehyde in PBS for 15 min at room temperature, embryos were mounted on slides in mounting medium containing 4,6-diamidino-2-phenylindole. At least 10 oocytes or embryos were processed for each separate sample, and the experiments were replicated three times. Slides were analyzed under an epifluorescent microscope (LEICA DM2500) equipped with a digital camera.

Detection of histone acetylation in SCNT embryos

Embryos treated without (control) or with the optimal CBHA concentration were collected at different stages of development: 1-cell stage (4 h after fusion); 2-cell stage (28–40 h after fusion); and blastocyst (140–150 h after fusion); and fixed in 4% paraformaldehyde in phosphate buffered saline (PBS) for 1 h at 4–8°C. Subsequently, embryos were washed in PBS containing 0.1% Tween 20 (TPBS) for 30 min and permeabilized by 1% Triton X-100 in PBS overnight at 4–8°C. After washing, unspecific antibody binding was blocked by incubation in 2% bovine serum albumin (BSA) in PBS for 1 h, followed by 1 h incubation with primary antibody at RT (anti-acH3K9, 1:500; anti-acH4K16, 1:500; anti-acH3K18, 1:500). After washing, the embryos were incubated for 1 h with Alexa Flour 594-conjugated secondary antibody (1:100, Jackson, USA). Immunofluorescent labeling of AcH3K9, AcH3K18, and AcH4K16 in TSA treated and control embryos were evaluated under Zeiss LSM 510 Laser Scanning Microscope System with the same exposure times and adjustments of the microscope. Fluorescence was measured by analyzing the embryo pictures with Image J software. Average pixel intensity value was measured in 5 random nucleoplasm regions (excluding the nucleolar regions) subtracting the average value measured in the same process within the cytoplasm. Negative control was performed by omitting the primary antibody and led in all cases to the lack of labeling.

Quantitative PCR

At day 15, seven blastocysts were used per sample. The total RNA of the embryos was isolated using the Phenol-chloroform extraction protocol. The mRNA levels were quantified using SYBR Green Jump Start Taq Ready Mix on a real-time PCR detection system at the following thermal cycling conditions: 94°C for 2 min, followed by 40 PCR cycles of 94°C for 15 s, 60°C for 30 s, and 72°C for 30 s. Transcripts were quantified in 3 replicates for each sample and calculated relative to the transcription of the housekeeping gene, β-actin, in every sample. The $2^{-\Delta\Delta CT}$ was used to quantify the relative mRNA levels. The primers used in the quantitative PCR are listed in the Table S1.

Statistical analysis

Data were analyzed using SAS (9.1) with one-way ANOVA. A probability of $P < 0.05$ was considered to be statistically significant.

ACKNOWLEDGEMENTS

This study was supported by grants from the National Natural Science Foundation of China (Grant Nos.90919060 to Q.Z. and 31201112 to T.H.), and the "Strategic Priority Research Program" of the Chinese Academy of Sciences (Grant No. XDA01020101 to Q.Z.).

We thank Yuanxuan Zhao and Zechen Li for nuclear transfer and *in vitro* fertilization assistance; Xiangnan Miao for embryo transfer; and Shiwen Li for confocal microscopic studies.

ABBREVIATIONS

CBHA, *m*-carboxycinnamic acid bishydroxamide; COCs, cumulus-oocyte complexes; EGA, embryonic gene activation; ESCs, embryonic stem cells; HDACi, histone deacetylase inhibitors; iPSC, induced pluripotent stem cells; KCCs, karyoplast-cytoplast complexes; PBS, phosphate buffered saline; SCNT, somatic cell nuclear transfer.

COMPLIANCE WITH ETHICS GUIDELINES

Yuran Song, Tang Hai, Ying Wang, Runfa Guo, Wei Li, Liu Wang, Qi Zhou declare that they have no conflict of interest.

All experiments comply with all of the relevant institutional and national animal welfare laws.

REFERENCES

Baguisi A, Behboodi E, Melican DT, Pollock JS, Destrempes MM et al (1999) Production of goats by somatic cell nuclear transfer. Nat Biotechnol 17:456–461

Bendixen E, Danielsen M, Larsen K, Bendixen C et al (2013) Advances in porcine genomics and proteomics—a toolbox for developing the pig as a model organism for molecular biomedical research. Brief Funct Genomics 9(3):208–219

Bock C, Kiskinis E, Verstappen G et al (2011) Reference maps of human ES and iPS cell variation enable high-throughput characterization of pluripotent cell lines. Cell 144:439–452

Chawengsaksophac K, Graaff W, Beck F et al (2004) Cdx2 is essential for axial elongation in mouse development. PNAS 101:7641–7645

Chesne P, Adenot PG, Viglietta C, Baratte M, Boulanger L et al (2002) Cloned rabbits produced by nuclear transfer from adult somatic cells. Nat Biotechnol 20:366–369

Dai XP, Hao J, Hou XJ, Hai T, Zhou Q et al (2010) Somatic nucleus reprogramming is significantly improved by m-carboxycinnamic acid bishydroxamide, a histone deacetylase inhibitor. J Biol Chem 285:31002–31010

Deshmukh RS, Ostrup O, Hyttel P et al (2012) Early aberrations in chromatin dynamics in embryos produced under in vitro conditions. Cell Reprogram 14:225–234

Ekser B, Ezzelarab M, Hara H, Windt D, Wijkstrom M, Bottino R, Trucco M, Cooper DKC et al (2012) Clinical xenotransplantation: the next medical revolution? Lancet 379:672–683

Galli C, Lagutina I, Crotti G, Colleoni S, Turini P et al (2003) A cloned horse born to its dam twin (vol 424, p 635, 2003). Nature 425:680

Gock H, Nottle M, Lew AM, d'Apice AJ, Cowan P et al (2011) Genetic modification of pigs for solid organ xenotransplantation. Transplant Rev 25:9–20

Han YM, Kang YK, Koo DB, Lee KK et al (2003) Nuclear reprogramming of cloned embryos produced in vitro. Theriogenology 59:33–34

Kato Y, Tani T, Sotomaru Y, Kurokawa K, Kato J et al (1998) Eight calves cloned from somatic cells of a single adult. Science 282:2095–2098

Kim YJ, Ahn KS, Kim M, Shim H et al (2011) Comparison of potency between histone deacetylase inhibitors trichostatin A and valproic acid on enhancing in vitro development f porcine somatic cell nuclear transfer embryos. In Vitro Cell Dev Biol Anim 47:283–289

Kishigami S, Mizutani E, Ohta H et al (2006) Significant improvement of mouse cloning technique by treatment with trichostain A after somatic nuclear transfer. Biochem Biophys Res Commun 340:183–189

Kumar BM, Jin HF, Kim JG, Ock SA, Hong Y, Balasubramanian S et al (2007) Differential gene expression patterns in porcine nuclear transfer embryos reconstructed with fetal fibroblasts and mesenchymal stem cells. Dev Dyn 236:435–446

Lager AE, Ragina NP, Cibelli JB et al (2008) Trichostatin A improves histone acetylation in bovine somatic cell nuclear transfer early embryos. Cloning Stem Cells 10:371–379

Latham KE, Doherty AS, Scott CD, Schultz RM (1994) Igf2r and Igf2 gene expression in androgenetic, gynogenetic, and parthenogenetic preimplantation mouse embryos: absence of regulation by genomic imprinting. Genes Dev 8:290–299

Lee JH, Hart SRL, Skalnik DG (2004) Histone deacetylase activity is required for embryonic stem cell differentiation. Genesis 38:32–38

Lee BC, Kim MK, Jang G, Oh HJ, Yuda F et al (2005) Dogs cloned from adult somatic cells. Nature 436:641

Lee E, Lee SH, Lee BC et al (2006) Analysis of nuclear reprogramming in cloned miniature pig embryos by expression of Oct-4 and Oct-4 related genes. Biochem Biophys Res Commun 348:1419–1428

Li E et al (2002) Chromatin modification and epigenetic reprogramming in mammalian development. Nature 3:662–673

Li ZY, Sun XS, Chen J, Liu XM, Wisely SM et al (2006) Cloned ferrets produced by somatic cell nuclear transfer. Dev Biol 293:439–448

Li J, Svarcova O, Villemoes K, Vajta G et al (2008) High in vitro development after somatic cell nuclear transfer and trichostatin A treatment of reconstructed porcine embryos. Theriogenology 70:800–808

Loh YH, Wu Q, Ng HH et al (2006) The Oct4 and Nanog transcription network regulated pluripotency in mouse embryonic stem cells. Nat Genet 38:431–440

Luo YL, Lin L, Bolund L, Jesen TG, Sorensen CB et al (2012) Genetically modified pigs for biomedical research. J Inherit Metab Dis 35:695–713

Maalouf W, Liu ZC, Zink D et al (2009) Trichostatin A treatment of cloned mouse embryos improves constitutive heterochromatin remodeling as well as developmental potential to term. BMC Dev Biol 9:11

Mason K, Liu ZC, Beaujean N et al (2012) Chromatin and epigenetic modifications during early mammalian development. Anim Reprod Sci 134:45–55

Mattout A, Biran A, Meshorer E et al (2011) Global epigenetic changes during somatic cell reprogramming to iPS cells. J Mol Biol 3:341–350

Menendez S, Camus S, Herreria A et al (2011) Increased dosage of tumor suppressors limits the tumorigenicity of cells without affecting their pluripotency. Aging Cell 11:41–50

Miyoshi K, Mori H, Mizobe Y, Sato M et al (2010) Benefical effects of reversine on in vitro development of miniature pig somatic cell nuclear transfer embryos. J Reprod Dev 56:291–296

Monteriro FM, Oliveira CS, Oliveira LZ, Garcia JM et al (2011) Chromatin modifying agents in the vitro production of bovine embryos. SAGE-Hindawi access to research veterinary medicine international

Niwa H, Toyooka Y, Rossant J et al (2005) Interaction between Oct3/4 and Cdx2 determines trophectodern differentiation. Cell 123:917–929

Ogura A, Inoue K, Wakayama T et al (2013) Recent advancements in cloning by somatic cell nuclear transfer. Philos Trans R Soc 368:20110329

Okita K, Ichisaka T, Yamanaka S et al (2007) Generation of germline-competent induced pluripotent stem cells. Nature 448:313–318

Onishi A, Iwamoto M, Akita T, Mikawa S, Takeda K et al (2000) Pig cloning by microinjection of fetal fibroblast nuclei. Science 289:1188–1190

Pesce M, Scholer HR et al (2001) Oct-4: gatekeeper in the beginnings of mammalian development. Stem Cells 19:271–278

Peserico A, Simone C et al (2011) Physical and functional HAT/HDAC interplay regulates protein acetylation balance. J Biomed Biotechnol 2011:371832

Polejaeva IA, Chen SH, Vaught TD, Page RL, Mullins J, Ball S, Dal Y, Boone J et al (2000) Cloned pigs produced by nuclear transfer from adult somatic cells. Nature 407:86–90

Rybouchkin A, Kato Y, Tsunoda Y et al (2006) Role of histone acetylation in reprogramming of somatic nuclei following nuclear transfer. Biol Reprod 74:1083–1089

Shin T, Kraemer D, Pryor J, Liu L, Rugila J et al (2002) A cat cloned by nuclear transplantation. Nature 415:859

Strumpf D, Mao CA, Yamanaka Y, Rossant J et al (2005) Cdx2 is required for correct cell fate specification and differentiation of trophectoderm in the mouse blastocyst. Development 132:2093–2102

Takahashi K, Yamanaka S et al (2006) Induction of pluripotent stem cells from mouse embryonic and adult fibroblast cultures by defined factors. Cell 126:663–676

Tomanck M, Kopecny V, Kanka J et al (1989) Genome reactivation in developing early embryos: an ultrastructural and autoradiaphic analysis. Anat Embryol 180:309–316

Verma N, Rettenmeier AW, Spanke SS et al (2011) Recent advances in the use of *Sus scrofa* (pig) as a model system for proteomic studies. Proteomics 11:776–793

Vignon X, Zhou Q, Renard JP et al (2002) Chromatin as a regulative architecture od the early developmental functions of mammalian embryos after fertilization or nuclear transfer. Cloning Stem Cells 4:363–377

Wakayama T, Perry ACF, Zuccotti M, Johnson KR, Yanagimachi R (1998) Full-term development of mice from enucleated oocytes injected with cumulus cell nuclei. Nature 394:369–374

Walker SC, Shin T, Zaunbrecher GM, Romano JE, Johnson GA, Bazer FW, Piedrahita JA et al (2002) A highly efficient method for porcine cloning by nuclear transfer using in vitro-matured oocytes. Cloning Stem Cells 4:105–112

Whitworth K, Prather RS et al (2010) Somatic cell nuclear transfer efficiency: how can it be improved through nuclear remodeling and reprogramming? Mol Reprod Dev 77:1001–1015

William M, Rideout III, Eggan K, Jaenisch R et al (2001) Nuclear cloning and epigenetic reprogramming of the genome. Science 293:1093–1098

Wilmut I, Schnieke AE, McWhir J, Kind AJ, Campbell KHS (1997) Viable offspring derived from fetal and adult mammalian cells (vol 385, p 810, 1997). Nature 386:200

Woods GL, White KL, Vanderwall DK, Li GP, Aston KI et al (2003) A mule cloned from fetal cells by nuclear transfer. Science 301:1063

Yang L, Palmer PC, Tian XC et al (2005) Expression of imprinted genes is aberrant in deceased newborn cloned calves and relatively normal in surviving adult clones. Mol Reprod Dev 71:431–438

Yu JY, Vodyanik MA, Smuga-Otto K et al (2007) Induced pluripotent stem cell lines derived from human somatic cells. Science 318:1916–1920

Zhao JG, Hao YH, Prather RS et al (2010) Histone deacetylase inhibitors improve in vitro and in vivo developmental competence of somatic cell nuclear transfer porcine embryos. Cell Reprogram 12:75–83

Zhou Q, Renard JP, Le Friec G, Brochard V, Beaujean N et al (2003) Generation of fertile cloned rats by regulating oocyte activation. Science 302:1179

Permissions

All chapters in this book were first published in P&C, by Springer; hereby published with permission under the Creative Commons Attribution License or equivalent. Every chapter published in this book has been scrutinized by our experts. Their significance has been extensively debated. The topics covered herein carry significant findings which will fuel the growth of the discipline. They may even be implemented as practical applications or may be referred to as a beginning point for another development.

The contributors of this book come from diverse backgrounds, making this book a truly international effort. This book will bring forth new frontiers with its revolutionizing research information and detailed analysis of the nascent developments around the world.

We would like to thank all the contributing authors for lending their expertise to make the book truly unique. They have played a crucial role in the development of this book. Without their invaluable contributions this book wouldn't have been possible. They have made vital efforts to compile up to date information on the varied aspects of this subject to make this book a valuable addition to the collection of many professionals and students.

This book was conceptualized with the vision of imparting up-to-date information and advanced data in this field. To ensure the same, a matchless editorial board was set up. Every individual on the board went through rigorous rounds of assessment to prove their worth. After which they invested a large part of their time researching and compiling the most relevant data for our readers.

The editorial board has been involved in producing this book since its inception. They have spent rigorous hours researching and exploring the diverse topics which have resulted in the successful publishing of this book. They have passed on their knowledge of decades through this book. To expedite this challenging task, the publisher supported the team at every step. A small team of assistant editors was also appointed to further simplify the editing procedure and attain best results for the readers.

Apart from the editorial board, the designing team has also invested a significant amount of their time in understanding the subject and creating the most relevant covers. They scrutinized every image to scout for the most suitable representation of the subject and create an appropriate cover for the book.

The publishing team has been an ardent support to the editorial, designing and production team. Their endless efforts to recruit the best for this project, has resulted in the accomplishment of this book. They are a veteran in the field of academics and their pool of knowledge is as vast as their experience in printing. Their expertise and guidance has proved useful at every step. Their uncompromising quality standards have made this book an exceptional effort. Their encouragement from time to time has been an inspiration for everyone.

The publisher and the editorial board hope that this book will prove to be a valuable piece of knowledge for researchers, students, practitioners and scholars across the globe.

List of Contributors

Gang Li
Department of Biochemistry and Molecular Biology, School of Basic Medical Sciences, Peking University Health Science Center, Beijing 100191, China

Zhongmin Liu
Department of Biochemistry and Molecular Biology, School of Basic Medical Sciences, Peking University Health Science Center, Beijing 100191, China
Tsinghua-Peking Joint Center for Life Sciences, School of Life Sciences, Tsinghua University, Beijing 100084, China

Jia Wang and Hong-Wei Wang
Tsinghua-Peking Joint Center for Life Sciences, School of Life Sciences, Tsinghua University, Beijing 100084, China

Lili Liu
Department of Pharmacology and Chemical Biology, University of Pittsburgh School of Medicine, Pittsburgh, PA 15213, USA
Department of Biochemistry and Molecular Genetics, University of Colorado School of Medicine, Aurora, CO 80045, USA

Mingxia Huang
Department of Biochemistry and Molecular Genetics, University of Colorado School of Medicine, Aurora, CO 80045, USA

Chengying Ma, Kaige Yan, Ningning Li, Yixiao Zhang, Yi Yuan, Zhifei Li, Jianlin Lei and Ning Gao
School of Life Sciences, Tsinghua University, Beijing 100084, China

Dan Tan and Meng-Qiu Dong
National Institute of Biological Sciences, Beijing 102206, China
Graduate Program in Chinese Academy of Medical Sciences and Peking Union Medical College, Beijing 100730, China

Xiaojie Ma, Linghao Kong and Saiyong Zhu
Life Sciences Institute, Zhejiang University, Hangzhou 310058, China

Logan A. Walsh
Human Oncology and Pathogenesis Program, Memorial Sloan-Kettering Cancer Center, New York, NY 10065, USA

David M. Roy
Human Oncology and Pathogenesis Program, Memorial Sloan-Kettering Cancer Center, New York, NY 10065, USA
Weill Cornell/Rockefeller/Sloan-Kettering Tri-Institutional MD-PhD Program, New York, NY 10065, USA

Timothy A. Chan
Human Oncology and Pathogenesis Program, Memorial Sloan-Kettering Cancer Center, New York, NY 10065, USA
Department of Radiation Oncology, Memorial Sloan-Kettering Cancer Center, New York, NY 10065, USA

Jiaming Su, Yi Sui, Jian Ding, Fuqiang Li, Shuang Shen, Yang Yang, Zeming Lu and Fei Wang
School of Life Sciences, Jilin University, 2699 Qianjin Street, Changchun 130012, China

Jingji Jin and Yong Cai
School of Life Sciences, Jilin University, 2699 Qianjin Street, Changchun 130012, China
National Engineering Laboratory for AIDS Vaccine, Jilin University, 2699 Qianjin Street, Changchun 130012, China
Key Laboratory for Molecular Enzymology and Engineering, The Ministry of Education, Jilin University, 2699 Qianjin Street, Changchun 130012, China

Lingling Cao
School of Life Sciences, Jilin University, 2699 Qianjin Street, Changchun 130012, China
The first Hospital and Institute of Epigenetic Medicine, Jilin University, Changchun 130061, China

Xiaoxia Liu
Department of Gynecologic Oncology, The First Clinical Hospital of Jilin University, 71 Xinmin Street, Changchun 130021, China

Xiuling Xu
National Laboratory of Biomacromolecules, CAS Center for Excellence in Biomacromolecules, Institute of Biophysics, Chinese Academy of Sciences, Beijing 100101, China

Zhenyu Ju
Key Laboratory of Regenerative Medicine of Ministry of Education, Institute of Aging and Regenerative Medicine, Jinan University, Guangzhou 510632, China

Lixia Wang
National Laboratory of Biomacromolecules, CAS Center for Excellence in Biomacromolecules, Institute of Biophysics, Chinese Academy of Sciences, Beijing 100101, China
State Key Laboratory of Stem Cell and Reproductive Biology, Institute of Zoology, Chinese Academy of Sciences, Beijing 100101, China
University of Chinese Academy of Sciences, Beijing 100049, China

Lina Fu, Jiping Yang and Si Wang
National Laboratory of Biomacromolecules, CAS Center for Excellence in Biomacromolecules, Institute of Biophysics, Chinese Academy of Sciences, Beijing 100101, China
University of Chinese Academy of Sciences, Beijing 100049, China

Guang-Hui Liu
National Laboratory of Biomacromolecules, CAS Center for Excellence in Biomacromolecules, Institute of Biophysics, Chinese Academy of Sciences, Beijing 100101, China
University of Chinese Academy of Sciences, Beijing 100049, China
Key Laboratory of Regenerative Medicine of Ministry of Education, Institute of Aging and Regenerative Medicine, Jinan University, Guangzhou 510632, China
Beijing Institute for Brain Disorders, Capital Medical University, Beijing 100069, China

Jing Qu
State Key Laboratory of Stem Cell and Reproductive Biology, Institute of Zoology, Chinese Academy of Sciences, Beijing 100101, China
University of Chinese Academy of Sciences, Beijing 100049, China

Fei Yi
Department of Molecular and Cellular Physiology, Stanford University School of Medicine, Stanford, CA 94305, USA

Zhaoxia Wang and Yun Yuan
Department of Neurology, Peking University First Hospital, Beijing 100034, China

Juan Carlos Izpisua Belmonte
Gene Expression Laboratory, Salk Institute for Biological Studies, 10010 North Torrey Pines Road, La Jolla, CA 92037, USA

Keiichiro Suzuki
Gene Expression Laboratory, Salk Institute for Biological Studies, 10010 North Torrey Pines Road, La Jolla, CA 92037, USA
Universidad Católica San Antonio de Murcia (UCAM), Campus de los Jerónimos, N 135 Guadalupe, 30107 Murcia, Spain

Yang Yu and Jie Qiao
Department of Gynecology and Obstetrics, Peking University Third Hospital, Beijing 100191, China

Liang Sun and Ze Yang
Beijing Hospital of the Ministry of Health, Beijing 100730, China

Ping Wang
National Laboratory of Biomacromolecules, CAS Center for Excellence in Biomacromolecules, Institute of Biophysics, Chinese Academy of Sciences, Beijing 100101, China
University of Chinese Academy of Sciences, Beijing 100049, China

Jingyi Li and Weiqi Zhang
National Laboratory of Biomacromolecules, CAS Center for Excellence in Biomacromolecules, Institute of Biophysics, Chinese Academy of Sciences, Beijing 100101, China
University of Chinese Academy of Sciences, Beijing 100049, China
National Clinical Research Center for Geriatric Disorders, Xuanwu Hospital of Capital Medical University, Beijing 100053, China

Guang-Hui Liu
National Laboratory of Biomacromolecules, CAS Center for Excellence in Biomacromolecules, Institute of Biophysics, Chinese Academy of Sciences, Beijing 100101, China
University of Chinese Academy of Sciences, Beijing 100049, China
National Clinical Research Center for Geriatric Disorders, Xuanwu Hospital of Capital Medical University, Beijing 100053, China
Institute of Stem cell and Regeneration, Chinese Academy of Sciences, Beijing 100101, China
Key Laboratory of Regenerative Medicine of Ministry of Education, Institute of Aging and Regenerative Medicine, Jinan University, Guangzhou 510632, China

Zunpeng Liu and Xiaoqian Zhang
State Key Laboratory of Stem Cell and Reproductive Biology, Institute of Zoology, Chinese Academy of Sciences, Beijing 100101, China
University of Chinese Academy of Sciences, Beijing 100049, China

Jing Qu
State Key Laboratory of Stem Cell and Reproductive Biology, Institute of Zoology, Chinese Academy of Sciences, Beijing 100101, China
University of Chinese Academy of Sciences, Beijing 100049, China
Institute of Stem cell and Regeneration, Chinese Academy of Sciences, Beijing 100101, China

Moshi Song
State Key Laboratory of Membrane Biology, Institute of Zoology, Chinese Academy of Sciences, Beijing 100101, China
University of Chinese Academy of Sciences, Beijing 100049, China
Institute of Stem cell and Regeneration, Chinese Academy of Sciences, Beijing 100101, China

Piu Chan
National Clinical Research Center for Geriatric Disorders, Xuanwu Hospital of Capital Medical University, Beijing 100053, China

Liang Sun and Jian Li
The MOH Key Laboratory of Geriatrics, Beijing Hospital, National Center of Gerontology, Beijing 100730, China

Ruo-Ran Wang, Ran Pan and Wenjing Zhang
Department of Pathology and Pathophysiology, Key Laboratory of Disease Proteomics of Zhejiang Province, School of Medicine, Chronic Disease Research Institute of School of Public Health, Zhejiang University, Hangzhou 310058, China

Zhuo-Xian Meng
Department of Pathology and Pathophysiology, Key Laboratory of Disease Proteomics of Zhejiang Province, School of Medicine, Chronic Disease Research Institute of School of Public Health, Zhejiang University, Hangzhou 310058, China
Life Sciences Institute and Department of Cell & Developmental Biology, University of Michigan Medical Center, Ann Arbor, MI 48109, USA

Junfen Fu
Children's Hospital, Zhejiang University School of Medicine, Hangzhou 310003, China

Jiandie D. Lin
Life Sciences Institute and Department of Cell & Developmental Biology, University of Michigan Medical Center, Ann Arbor, MI 48109, USA

Yi Yang, Xiangjin Kang, Geng An, Yali Liu, Qian Yu, Zhenglai Ma, Yaoyong Chen, Xiaofang Sun and Yong Fan
Key Laboratory for Major Obstetric Diseases of Guangdong Province, Key Laboratory of Reproduction and Genetics of Guangdong Higher Education Institutes, The Third Affiliated Hospital of Guangzhou Medical University, Guangzhou 510150, China

Qingfeng Chen
Key Laboratory for Major Obstetric Diseases of Guangdong Province, Key Laboratory of Reproduction and Genetics of Guangdong Higher Education Institutes, The Third Affiliated Hospital of Guangzhou Medical University, Guangzhou 510150, China
Disease Modeling and Therapeutics Laboratory, A*STAR Institute of Molecular and Cell Biology, 61 Biopolis Drive Proteos, Singapore 138673, Singapore

Boon Seng Soh
Key Laboratory for Major Obstetric Diseases of Guangdong Province, Key Laboratory of Reproduction and Genetics of Guangdong Higher Education Institutes, The Third Affiliated Hospital of Guangzhou Medical University, Guangzhou 510150, China
Disease Modeling and Therapeutics Laboratory, A*STAR Institute of Molecular and Cell Biology, 61 Biopolis Drive Proteos, Singapore 138673, Singapore
Department of Biological Sciences, National University of Singapore, Singapore 117543, Singapore

Han Wu, Yanhui Liang, Ting Lan, Jiangyun Peng, Quanjun Zhang and Xiaoping Li
Key Laboratory of Regenerative Biology of the Chinese Academy of Sciences and Guangdong Provincial Key Laboratory of Stem Cells and Regenerative Medicine, South China Institute for Stem Cell Biology and Regenerative Medicine, Guangzhou
Institutes of Biomedicine and Health, Chinese Academy of Sciences, Guangzhou 510530, China

Tianjie Li, Ying Lian, Ping Liu, Rong Li, Xiumei Zhen, Ping Liu and Yang Yu
Center of Reproductive Medicine, Department of Obstetrics and Gynecology, Peking University Third Hospital, Beijing 100191, China

Tao Tan
Yunnan Key Laboratory of Primate Biomedical Research, Institute of Primate Translational Medicine, Kunming University of Science and Technology, Kunming 650500, China

Jianying Guo, Rujin Huang, Jia Ming, Min Ye, Kehkooi Kee and Jie Na
Department of Basic Medical Sciences, School of Medicine, Center for Stem Cell Biology, Tsinghua University, Beijing 100084, China

Dacheng Ma and Zhen Xie
MOE Key Laboratory of Bioinformatics and Bioinformatics Division, Center for Synthetic and System Biology, TNLIST/Department of Automation, Tsinghua University, Beijing 100084, China

Hosuk Lee, Sungwook Han and Daeyoup Lee
Department of Biological Sciences, Korea Advanced Institute of Science and Technology, Daejeon 305-701, Korea

Chang Seob Kwon
Department of Chemistry and Biology, Korea Science Academy of KAIST, Busan 614-822, Korea

Hongwei Sun, Ying Sun, Xiya Zhang, Xiaowei Xie, Jinran Zhang, Yuxi Chen, Yuanyan Xiong and Wenbin Ma
Key Laboratory of Gene Engineering of the Ministry of Education, Guangzhou Key Laboratory of Healthy Aging Research and State Key Laboratory of Biocontrol, SYSU-BCM Joint Research Center, School of Life Sciences, Sun Yat-sen University, Guangzhou 510275, China
State Key Laboratory of Ophthalmology, Zhongshan Ophthalmic Center, Sun Yat-sen University, Guangzhou 510275, China
Key Laboratory of Reproductive Medicine of Guangdong Province, School of Life Sciences and the First Affiliated H ospital, S un Y at-sen University, Guangzhou 510275, China

Junjiu Huang
Key Laboratory of Gene Engineering of the Ministry of Education, Guangzhou Key Laboratory of Healthy Aging Research and State Key Laboratory of Biocontrol, SYSU-BCM Joint Research Center, School of Life Sciences, Sun Yat-sen University, Guangzhou 510275, China
State Key Laboratory of Ophthalmology, Zhongshan Ophthalmic Center, Sun Yat-sen University, Guangzhou 510275, China

Puping Liang
Key Laboratory of Gene Engineering of the Ministry of Education, Guangzhou Key Laboratory of Healthy Aging Research and State Key Laboratory of Biocontrol, SYSU-BCM Joint Research Center, School of Life Sciences, Sun Yat-sen University, Guangzhou 510275, China
State Key Laboratory of Ophthalmology, Zhongshan Ophthalmic Center, Sun Yat-sen University, Guangzhou 510275, China
Key Laboratory of Reproductive Medicine of Guangdong Province, School of Life Sciences and the First Affiliated Hospital, Sun Yat-sen University, Guangzhou 510275, China

Zhen Zhang
Key Laboratory of Gene Engineering of the Ministry of Education, Guangzhou Key Laboratory of Healthy Aging Research and State Key Laboratory of Biocontrol, SYSU-BCM Joint Research Center, School of Life Sciences, Sun Yat-sen University, Guangzhou 510275, China
Guangzhou Magigen Biotechnology Co.Ltd, Guangzhou 510320, China

Zhou Songyang
Key Laboratory of Gene Engineering of the Ministry of Education, Guangzhou Key Laboratory of Healthy Aging Research and State Key Laboratory of Biocontrol, SYSU-BCM Joint Research Center, School of Life Sciences, Sun Yat-sen University, Guangzhou 510275, China
State Key Laboratory of Ophthalmology, Zhongshan Ophthalmic Center, Sun Yat-sen University, Guangzhou 510275, China
Key Laboratory of Reproductive Medicine of Guangdong Province, School of Life Sciences and the First Affiliated H ospital, S un Y at-sen University, Guangzhou 510275, China
Verna and Marrs Mclean Department of Biochemistry and Molecular Biology, Baylor College of Medicine, One Baylor Plaza, Houston, TX 77030, USA

Chenhui Ding
Key Laboratory of Reproductive Medicine of Guangdong Province, School of Life Sciences and the First Affiliated Hospital, Sun Yat-sen University, Guangzhou 510275, China

Dan Liu
Verna and Marrs Mclean Department of Biochemistry and Molecular Biology, Baylor College of Medicine, One Baylor Plaza, Houston, TX 77030, USA

Chao Chen
School of Medicine, Institute of Genetics, Zhejiang University, Hangzhou 310058, China

Ye Chen and Min-Xin Guan
School of Medicine, Institute of Genetics, Zhejiang University, Hangzhou 310058, China
Collaborative Innovation Center for Diagnosis and Treatment of Infectious Diseases, Zhejiang University, Hangzhou 310058, China

Minghui Cao, Jing Li, Yujing Zhang, Xi Chen, Ke Zen, Chen-yu Zhang, Xiaohong Jiang and Dongxia Hou
State Key Laboratory of Pharmaceutical Biotechnology, Jiangsu Engineering Research Center for microRNA Biology and Biotechnology, School of Life Sciences, Nanjing University, Nanjing 210093, China

Weiwei Nie and Xiaoxiang Guan
Department of Medical Oncology, Jinling Hospital, School of Medicine, Southern Medical University, Guangzhou 510282, China

Xin Yan
The Comprehensive Cancer Center of Drum Tower Hospital, Medical School of Nanjing University and Clinical Cancer Institute of Nanjing University, Nanjing 210008, China

Yuran Song
State Key Laboratory of Reproductive Biology, Institute of Zoology, Chinese Academy of Sciences, Beijing 100101, China
Graduate University of the Chinese Academy of Sciences, Beijing 100049, China

Tang Hai, Ying Wang, Runfa Guo, Wei Li, Liu Wang and Qi Zhou
State Key Laboratory of Reproductive Biology, Institute of Zoology, Chinese Academy of Sciences, Beijing 100101, China

Index

P

Q

R

S

T

U

V

W

CPSIA information can be obtained
at www.ICGtesting.com
Printed in the USA
LVHW060833060222
710136LV00022B/99

9 781639 893744